Stem Cell Biology and Regenerative Medicine

Volume 70

Series Editor

Kursad Turksen, Ottawa Hospital Research Institute, Ottawa, ON, Canada

Our understanding of stem cells has grown rapidly over the last decade. While the apparently tremendous therapeutic potential of stem cells has not yet been realized, their routine use in regeneration and restoration of tissue and organ function is greatly anticipated. To this end, many investigators continue to push the boundaries in areas such as the reprogramming, the stem cell niche, nanotechnology, biomimetics and 3D bioprinting, to name just a few. The objective of the volumes in the Stem Cell Biology and Regenerative Medicine series is to capture and consolidate these developments in a timely way. Each volume is thought-provoking in identifying problems, offering solutions, and providing ideas to excite further innovation in the stem cell and regenerative medicine fields.

More information about this series at https://link.springer.com/bookseries/7896

Irma Virant-Klun
Editor

Stem Cells in Reproductive Tissues and Organs

From Fertility to Cancer

Humana Press

Editor
Irma Virant-Klun
Clinical Research Center
University Medical Center Ljubljana
Ljubljana, Slovenia

ISSN 2196-8985 ISSN 2196-8993 (electronic)
Stem Cell Biology and Regenerative Medicine
ISBN 978-3-030-90110-3 ISBN 978-3-030-90111-0 (eBook)
https://doi.org/10.1007/978-3-030-90111-0

This Humana imprint is published by the registered company Springer Nature Switzerland AG
The registered company address is: Gewerbestrasse 11, 6330 Cham, Switzerland

Preface

The field of stem cells is extremely interesting. Research is advancing very rapidly and is already leading to successful cell therapies for the treatment of various degenerative diseases in regenerative medicine. Stem cells and the possibilities of their use in reproductive medicine are less well known, but the chapters in this book show unimaginable possibilities of their use for better understanding and treatment of infertility and more individualized and advanced cancer treatment in humans. The book was written during the difficult Corona time and it proves that life, human thoughts, and science go on, although it is not easy. I sincerely thank all the authors who contributed chapters for this book, as the presentation of this work to others is important and shows the importance of this kind of research in reproductive medicine and I wish them to persevere on this interesting path that is already leading to the first cell therapies in the field of reproductive medicine. I also hope that the book will be interesting for readers and also pave some new ideas and research paths in this field.

Ljubljana, Slovenia | Irma Virant-Klun

Contents

Contributors

Mehdi Amirian Medical Faculty, Institute for Anatomy and Cell Biology, University of Heidelberg, Heidelberg, Germany

Hossein Azizi Faculty of Biotechnology, Amol University of Special Modern Technologies, Amol, Iran

Surinder K. Batra Department of Biochemistry and Molecular Biology, University of Nebraska Medical Center, Omaha, NE, USA

Jure Bedenk Department of Obstetrics and Gynecology, University Medical Center Ljubljana, Ljubljana, Slovenia

Deepa Bhartiya Stem Cell Biology Department, ICMR-National Institute for Research in Reproductive Health, Mumbai, India

Benedetta Bussolati Department of Molecular Biotechnology and Health Sciences, University of Torino, Torino, Italy

Stefano Canosa Obstetrics and Gynecology 1U, Physiopathology of Reproduction and IVF Unit, Department of Surgical Sciences, S. Anna Hospital, University of Torino, Torino, Italy

Andrea Roberto Carosso Obstetrics and Gynecology 1U, Physiopathology of Reproduction and IVF Unit, Department of Surgical Sciences, S. Anna Hospital, University of Torino, Torino, Italy

Irene Cervelló Fundación Instituto Valenciano de Infertilidad (FIVI), La Fe Health Research Institute, La Fe University Hospital, Valencia, Spain

Sabine Conrad Tübingen, Germany

Cristina Eguizabal Cell Therapy, Stem Cells and Tissues Group, Biocruces Bizkaia Health Research Institute, Barakaldo, Spain;
Research Unit, Basque Center for Blood Transfusion and Human Tissues, Osakidetza, Galdakao, Spain

Branka Golubić-Ćepulić Department for Transfusion Medicine and Transplantation Biology, University of Zagreb, University Hospital "Zagreb", Zagreb, Croatia

Maryam Hatami Medical Faculty, Institute for Anatomy and Cell Biology, University of Heidelberg, Heidelberg, Germany

Davor Ježek Department for Transfusion Medicine and Transplantation Biology, University of Zagreb, University Hospital "Zagreb", Zagreb, Croatia;
Department of Histology and Embryology, School of Medicine, University of Zagreb, Zagreb, Croatia

Sham S. Kakar Department of Physiology and Brown Cancer Center, University of Louisville, Louisville, KY, USA

Ankita Kaushik Stem Cell Biology Department, ICMR-National Institute for Research in Reproductive Health, Mumbai, India

Kazuhiro Kawamura Advanced Reproduction Research Center, Department of Obstetrics and Gynecology, International University of Health and Welfare Graduate School of Medicine, Narita Chiba, Japan

Mateja Erdani Kreft Institute of Cell Biology, Faculty of Medicine, University of Ljubljana, Ljubljana, Slovenia

Magda Kucia Stem Cell Institute at James Graham Brown Cancer Center, University of Louisville, Louisville, KY, USA;
Department of Regenerative Medicine, Center for Preclinical Research and Technology, Medical University of Warsaw, Warszawa, Poland

Antonio Simone Laganà Department of Obstetrics and Gynecology, "Filippo Del Ponte" Hospital, University of Insubria, Varese, Italy

Myriam Martin-Inaraja Cell Therapy, Stem Cells and Tissues Group, Biocruces Bizkaia Health Research Institute, Barakaldo, Spain

Lucía de Miguel Gómez Fundación Instituto Valenciano de Infertilidad (FIVI), La Fe Health Research Institute, La Fe University Hospital, Valencia, Spain;
University of Valencia, Valencia, Spain

Antoine Naem Faculty of Medicine, Damascus University, Damascus, Syria

Seema C. Parte Department of Biochemistry and Molecular Biology, University of Nebraska Medical Center, Omaha, NE, USA

Antonio Pellicer University of Valencia, Valencia, Spain;
IVIRMA Rome, Rome, Italy

Martina Perše Medical Faculty, Medical Experimental Center, University of Ljubljana, Ljubljana, Slovenia

Moorthy P. Ponnusamy Department of Biochemistry and Molecular Biology, University of Nebraska Medical Center, Omaha, NE, USA

Taja Železnik Ramuta Institute of Cell Biology, Faculty of Medicine, University of Ljubljana, Ljubljana, Slovenia

Mariusz Z. Ratajczak Stem Cell Institute at James Graham Brown Cancer Center, University of Louisville, Louisville, KY, USA;
Department of Regenerative Medicine, Center for Preclinical Research and Technology, Medical University of Warsaw, Warszawa, Poland

Janina Ratajczak Stem Cell Institute at James Graham Brown Cancer Center, University of Louisville, Louisville, KY, USA;
Department of Regenerative Medicine, Center for Preclinical Research and Technology, Medical University of Warsaw, Warszawa, Poland

Alberto Revelli Obstetrics and Gynecology 1U, Physiopathology of Reproduction and IVF Unit, Department of Surgical Sciences, S. Anna Hospital, University of Torino, Torino, Italy

Marta Sestero Obstetrics and Gynecology 1U, Physiopathology of Reproduction and IVF Unit, Department of Surgical Sciences, S. Anna Hospital, University of Torino, Torino, Italy

Diksha Sharma Stem Cell Biology Department, ICMR-National Institute for Research in Reproductive Health, Mumbai, India

Pushpa Singh Stem Cell Biology Department, ICMR-National Institute for Research in Reproductive Health, Mumbai, India

Thomas Skutella Medical Faculty, Institute for Anatomy and Cell Biology, University of Heidelberg, Heidelberg, Germany

Nataša Kenda Šuster Department of Obstetrics and Gynecology, University Medical Center Ljubljana, Ljubljana, Slovenia

Larisa Tratnjek Institute of Cell Biology, Faculty of Medicine, University of Ljubljana, Ljubljana, Slovenia

Željka Večerić-Haler Department of Nephrology, University Medical Center Ljubljana, Ljubljana, Slovenia

Marija Vilaj Department for Transfusion Medicine and Transplantation Biology, University of Zagreb, University Hospital "Zagreb", Zagreb, Croatia

Irma Virant-Klun Clinical Research Center, University Medical Center Ljubljana, Ljubljana, Slovenia

Tuyen Kim Cat Vo Advanced Reproduction Research Center, Department of Obstetrics and Gynecology, International University of Health and Welfare Graduate School of Medicine, Narita Chiba, Japan

Chapter 1
Challenges of Stem Cell Therapies for the Treatment of Infertility in Reproductive Medicine

Myriam Martin-Inaraja and Cristina Eguizabal

Abstract

Introduction: Stem cells, especially mesenchymal stem cells (MSCs), have demonstrated great potential and availability for treating infertility in animal and human studies among pre-clinical and clinical trials, respectively. Allogenic placental and cord blood derived-stem cells, autologous bone marrow and hematopoietic derived-stem cells, and adipose-derived MSCs are especially useful because they are not only easily obtained, but also avoid graft rejection after transplantation.
Methods: The use of stem cells for the treatment of infertility in reproductive medicine has been reviewed.
Results: Recently, several clinical trials have been performed with promising results to treat infertility related disorders in male and female patients. However, there are several issues related to the adult stem cell therapies to treat infertility that have to be investigated: (1) Most studies have been done on small animals, and there is a serious lack of valuable research in large animal models that more closely mimic the ovarian or testicular pathophysiology of human infertility. Furthermore, a randomized controlled trial should be conducted to confirm the therapeutic effect of MSCs in fertility medicine. (2) The mechanism of MSCs in treating dysfunction of reproductive organs is still unknown. Possibilities include promoting angiogenesis, differentiating into functional cells, and a paracrine mechanism. However, beneficial paracrine factors remain unknown and multiple mechanisms may be synergistic. (3) While MSC therapy is promising, the limited survival and engraftment of these cells together with biomaterials remains still unsolved.
Conclusions: Nevertheless, additional work is needed to optimize this approach. Many promising experimental stem cell-based approaches are under development to

---

M. Martin-Inaraja · C. Eguizabal (✉)
Cell Therapy, Stem Cells and Tissues Group, Biocruces Bizkaia Health Research Institute, Barakaldo, Spain
e-mail: CRISTINA.EGUIZABALARGAIZ@osakidetza.eus

C. Eguizabal
Research Unit, Basque Center for Blood Transfusion and Human Tissues, Osakidetza, Galdakao, Spain

I. Virant-Klun (ed.), *Stem Cells in Reproductive Tissues and Organs*, Stem Cell Biology and Regenerative Medicine 70, https://doi.org/10.1007/978-3-030-90111-0_1

restore male fertility in prepuberal boys, regeneration of endometrium, and ovaries in primary ovarian insufficiency.

Keywords Cell therapy · Infertility · Reproductive medicine · Stem cells

Abbreviations

AR	Acrosome reaction
ART	Assisted reproductive technology
BM	Bone marrow
CD	Cluster of differentiation
CFTR	Transmembrane conductance regulator
CUA	Congenital uterine abnormalities
ESC	Embryonic stem cell
FGF	Fibroblast growth factor
FSH	Follicle stimulating hormone
GnRH	Gonadotropin-releasing hormone
HLA	Human leukocyte antigen
HSC	Hematopoietic stem cell
ICSI	Intracytoplasmic sperm injection
iPSC	Induced pluripotent stem cell
IUA	Intrauterine adhesion
IUI	Intrauterine insemination
IVF	*In vitro* Fertilization
KAL	Kallmann syndrome
LH	Luteinizing hormone
MNC	Mononuclear cells
MSC	Mesenchymal stem cell
Oct4	Octamer-binding transcription factor 4
OSC	Oogonial stem cell
PCOS	Polycystic ovary syndrome
PID	Pelvic inflammatory disease
POF	Premature ovarian failure
POI	Primary ovarian insufficiency
PROK	Prokineticin
PSC	Pluripotent stem cell
R	Receptor
SSC	Spermatogonial stem cell
UCB	Umbilical cord blood
WHO	World Health Organization

Introduction

By definition of World Health Organization, Infertility is a disease of the reproductive system defined by the failure to achieve a clinical pregnancy after 12 months. Primary infertility is infertility in a couple who have never had a child. Secondary infertility is failure to conceive following a previous pregnancy.

In males, 25% of infertility cases are caused by male factors, principally due to disorders related with the sperm, including mature puberty, congenital diseases, and testicular structural problems such as genital injury or damage driving to sperm dysfunction or any psychological and environmental factors. In females, infertility cases are caused by female factors including endometritis, ovulation dysfunction, abnormal fallopian tube or uterus, pelvic adhesions, and primary ovarian insufficiency [1]. Current infertility treatments differ from pharmacotherapy to assisted reproductive technology (ART), determined by the patient and origin aspects. Treatment for male infertility comprises behavior changes, surgery, drugs, supplements, and sperm remodeling. The prime treatment for female infertility is hormone therapy, supplements, surgery, or an ART procedure, comprising in vitro fertilization-embryo transfer (IVF), intrauterine insemination (IUI), and intracytoplasmic sperm injection (ICSI).

Recently, stem cells are increasingly being studied in the field of infertility [2]. The origin and potency of stem cells can be different and the stem cells can be divided and differentiated into several cell types during entire life with the aim to develop repair and regenerate tissues and organs. Researches in experimental models have demonstrated that treating infertility with stem cell-based therapy is promoting recognition [3]. Several female infertility studies using stem cells from diverse sources were newly initiated. Pre-clinical studies on sexual infertility-related diseases have proposed novel ways to study for the treatment of infertility [4]. Studies utilizing experimental models have reported the capacity of stem cell therapy for treating infertility and proved these results [5, 6]. In this book chapter, we will outline present knowledge concerning the use of stem cells and the current clinical trials in reproductive medicine for treating infertility related diseases.

Overview of Stem Cells: Cell Types and Therapeutic Applications in Reproductive Medicine

Stem cells are cells that are not differentiated yet with the capacity to broadly expand and differentiate into distinct specific cells [6, 7]. There are five principal stem cell types that comprised from less differentiated/more potency stem cells to more differentiated/less potency stem cells such as: embryonic stem cells (ESCs), induced pluripotent stem cells (iPSCs), mesenchymal stem cells (MSCs), hematopoietic stem cells (HSCs), and spermatogonial or oogonial stem cells (SSCs and OSC, respectively). See all stem cell types in Fig. 1.1 and Table 1.1.

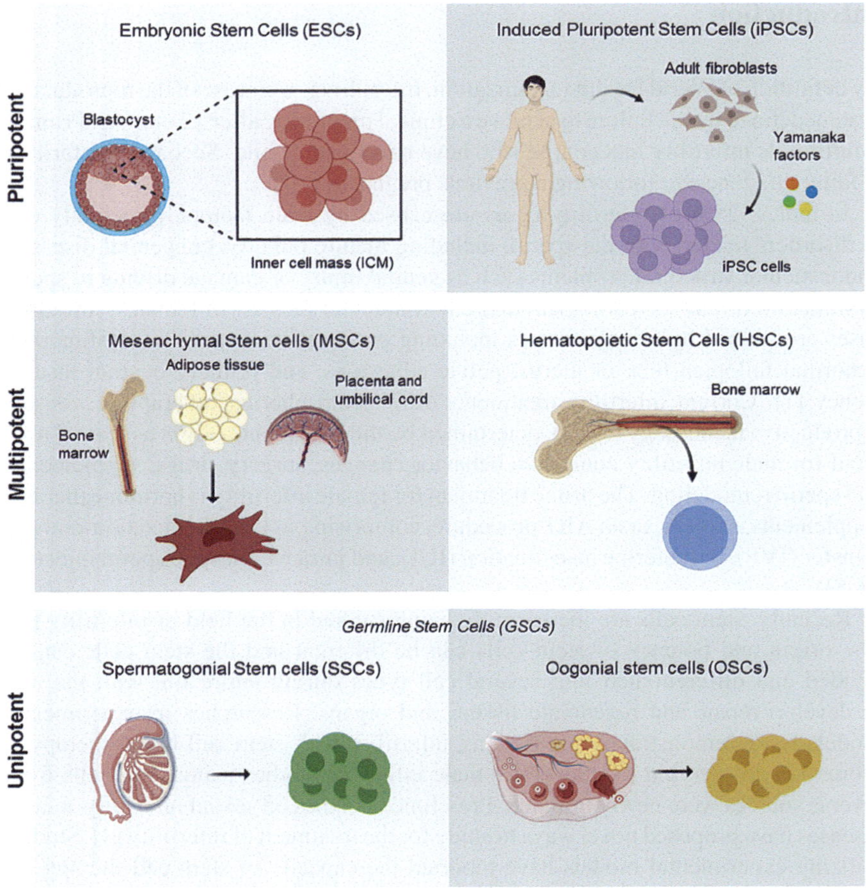

Fig. 1.1 Types of stem cells with possible uses in reproductive medicine

Embryonic Stem Cells

Embryonic stem cells (ESCs) are pluripotent cells derived from the inner cell mass of the blastocyst [8]. ESCs' principal features contain self-renewal capacity, extended proliferation in the undifferentiated state, and are able to differentiate into three germ layers (endoderm, mesoderm, and ectoderm) and germ cells. Prof. Shinya Yamanaka achieved induced pluripotent stem cell lines (iPS) by reprograming somatic cells with four transcription factors Oct4, Sox2, Klf4, and c-Myc [9, 10]. These cells have typical features with ESCs. In spite of sharing identical characteristics, the purpose of ESCs has ethical involvement whereas iPSCs do not involve in this trouble. Currently, there exist further than 33 clinical trials that are applying treatment for several diseases by utilizing differentiated cells from pluripotent stem cells (PSCs)

Table 1.1 Characteristics of stem cell types for infertility related disease treatments

	ESCs	iPSCs	MSCs	HSCs	SSCs-OSCs
Cell source and origin	Derived from the inner cell mass of the blastocyst	Derived from human somatic cells	Derived from bone marrow, placenta, cord blood, adipose tissue	Derived from bone marrow	Derived from testicular and ovarian tissue
Potency and differentiation capacity	Pluripotent—differentiate into cells of all three germ layers and germ cells	Pluripotent—differentiate into cells of all three germ layers and germ cells	Multipotent—differentiate into cells of mesoderm	Multipotent—differentiate into cells of mesoderm	Unipotent—differentiate into male and female gametes
Clinical applications	In few clinical trials (not in infertility)	In few clinical trials (not in infertility)	Widely used	Widely used	Not yet
Immunonicity	Yes	Yes/No	No	No (autologous)	No
Ethical concerns	Yes	No	No	No	No

[7], but none of them are yet to treat infertility related issues. The in vitro creation of "artificial" gametes from PSC (ESCs and iPSCs) is still a distant prospect in humans. The achievement of PGCs from PSCs is the first differentiating point toward post-meiotic oocytes and spermatozoa together with a creation of a testicular or ovarian niche and correct epigenetic and germline genes expression with a correct meiosis progression in order to get functional gametes in vitro.

Nevertheless, advance differentiation towards mature gametes has demonstrated challenging in vitro, although many efforts have been made in differentiation protocols further investigations are required to obtain functional male and female gametes from PSCs for forthcoming applications.

Mesenchymal Stem Cells

Mesenchymal stem cells (MSCs) are originally derived from mesoderm layer retaining restricted self-renewal and differentiation capacity. MSCs are adult multipotent stem cells that can be isolated from various adult tissues, such as adipose tissue, bone marrow, placenta, peripheral, and umbilical cord blood [6, 11, 12]. Their principal benefit is the efficient ability of isolation from adult tissues bypassing ethical issues. Even though MSCs are widely used to treat different diseases, their long-term maintenance in culture is hard to be managed [6]. To address the use of these multipotent stem cells, the Mesenchymal and Tissue Stem Cell Committee of the International Society for Cellular Therapy has suggested basic principles to designate human MSCs. Firstly, MSCs must have the ability to be adherent into plastic surface during regular culture conditions. Secondly, MSCs must express CD73, CD90, and CD105, and lose expression of CD34, CD45, CD11b or CD14, CD19 or CD79a, and HLA-DR surface antigens. Thirdly, MSCs must in vitro differentiate into adipocytes, osteoblasts, and chondroblasts [13]. Pre-clinical studies and clinical trials are presently utilizing MSCs for broad disease applications along with promising application to infertility related diseases such as endometrial disorders, ovarian dysfunction, and erectile dysfunction [11, 12]. Principally, therapeutic approaches with MSCs for treating some diseases related with female and male infertility are provoking considerable excitement. More notably, these approaches may support an attractive experimental model for decoding the fundamental mechanism of the use of MSCs for treating female and male infertility issues. This supports the hypothetical ground for future investigations and clinical MSC cell-therapy.

Hematopoietic Stem Cells

Another mesoderm-derived multipotent cells are hematopoietic stem cells (HSCs). HSCs are the adult stem cells that can be differentiated into any blood cells. This developmental process is named hematopoiesis. This progression takes place in the bone marrow, located within the majority of bones. The HSCs from umbilical cord blood (UCB) and bone marrow (BM) are expressing both CD34 and CD133 surface markers. These multipotent stem cells also provide such regenerative characteristics, specifically CD133$^+$ cells from BM, which are recognized as the remarkably immature HSCs. Recently, clinical trials by using CD133$^+$ cells to treat endometrial disorders demonstrate cell proliferation and neoangiogenesis in the endometrium [14].

Spermatogonial and Oogonial Stem Cells

Spermatogonial and oogonial stem cells are germline stem cells located in the testis and ovary tissue, respectively [6]. SSCs and OSCs are unipotent stem cells that can give rise to fully differentiated gametes (sperm and oocytes). Even though these stem cells would be an excellent proposal to treat infertility related diseases, the amount of SSCs and OSCs in both reproductive organs is very low which is not easy to maintain and grow them in vitro in comparison with other stem cell types (iPSCs or ESCs, MSCs) [1]. The spermatogenesis is a strictly organized differentiating program that occurs within the seminiferous tubules in the testis. This process initiates with spermatogonial stem cells (SSCs) that either self-renew to maintain the stem cell pool perfectly or differentiate to make sure the regular generation of haploid sperm cells among life. In humans, SSC therapy is not administered yet in clinics. However, testicular biopsies cryopreservation is at the present proposed at several fertility clinics worldwide. Translational investigations on specific stem cell therapy features are still mandatory before offering SSC-based therapy to the clinic [2].

Oogonial stem cells (OSCs) are suggested to be an inherent adult stem cell population that could be located in ovaries. As well as SSCs, these cells would be unipotent stem cells able to self-renew, proliferate and regenerate the group of oocytes in the ovary. Nevertheless, the theoretical existence of an OSC pool still continues to be questionable. Despite commercial companies are strongly studying the feasibility of the use of presumed OSCs to administer in infertility issues, there is presently no general agreement on their presence, origin, and function [2]. Additional investigations are required to argue the questions linked with the OSCs existence.

Among the several existing stem cell types such as patient-specific iPSCs, ESCs, MSCs, HSCs, SSCs or OSCs, may be used as stem cell therapy for infertility related disease treatments (Fig. 1.1 and Table 1.1) [12, 15]. In the next section, we will comprise the stem cells uses and the current clinical trials for treating female and male infertility related disorders.

Infertility Related Disorders in Female Reproductive Organs

A broad international research to establish the gender classification and etiologies of infertility was performed by the World Health Organization (WHO). Within the 37% of infertile couples, female infertility was induced by female factors, such as ovulatory dysfunction (25%), endometriosis (15%), pelvic adhesions (12%), tubal blockage (11%), other uterine/tubal anomalies (11%), and hyperprolactinemia (7%) [16]. See all female infertility disorders in Fig. 1.2.

Ovulatory Disorders

Previously mentioned ovulatory disorders are 25% of the female infertility causes. Anovulation or Oligo-ovulation are two ovulatory disorders with a lack of an oocyte release monthly, consequently there is no chance for fertilization and gestation. To encourage with therapy and classification, the WHO separated ovulatory disorders into four types: *Normogonadotropic normoestrogenic anovulation* (e.g. polycystic ovary syndrome (PCOS)); *Hypogonadotropic hypogonadal anovulation* (e.g. hypothalamic amenorrhea); *Hypergonadotropic hypoestrogenic anovulation* (e.g. premature ovarian failure (POF) and premature ovarian insufficiency (POI) and Turner syndrome); and *Hyperprolactinemic anovulation* (e.g. pituitary adenoma).

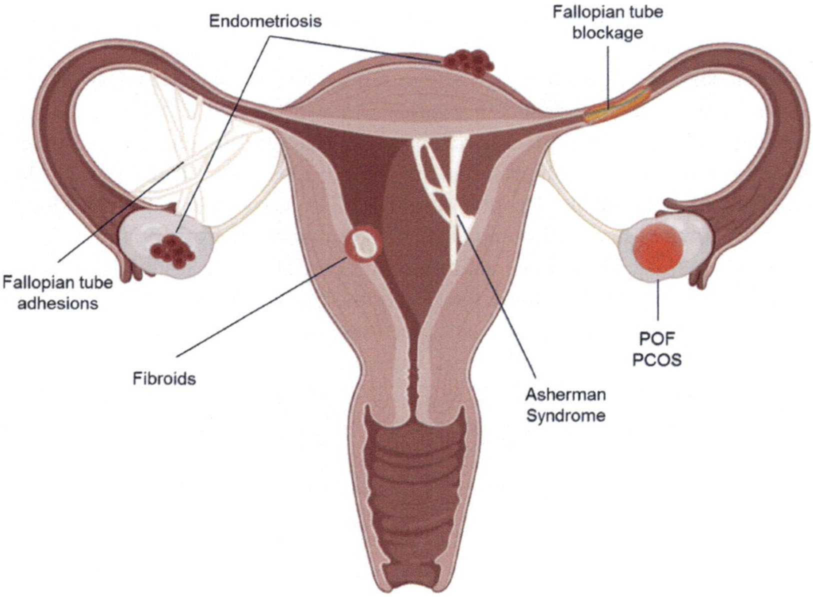

Fig. 1.2 Female infertility causes diagram

Endometriosis

Endometriosis is a disorder that occurs when endometrial tissue is outside the uterine cavity, most frequently raised in the pelvis but can be expanded throughout the whole abdomen and 10–15% of women can suffer this disorder in reproductive age. Approximately, 40–50% of women with endometriosis will suffer infertility. Endometriosis disorder is classified into four stages, from minimal (stage I) to severe (stage IV).

Tubal/Pelvic Adhesions

Tubal and pelvic adhesions, together with tubal and uterine anomalies caused huge female infertility related disorders. The primary causes of pelvic/tubal adhesions are the infectious growth within the abdomen; being the most regular infectious process is pelvic inflammatory disease (PID) affecting infertility. Normally, the PID is caused by the microorganism named Chlamydia trachomatis. A tubal anomaly, named hydrosalpinges, caused by acute and chronic inflammation can damage the unity of the fallopian tubes.

Uterine Factors

Uterine factors of infertility are related with either reduced endometrial receptivity or space-occupying lesions. For example, congenital uterine abnormalities (CUA) and uterine leiomyomas (fibroids), although exceptional, are also related with infertility. Also Asherman's syndrome is an acquired condition that refers to having scar tissue in the uterus or in the cervix (the opening to the uterus). This scar tissue induces the walls of these organs attach together and diminishes the uterus size. Asherman's syndrome is also identified as intrauterine synechiae (means adhesions) or uterine synechiae. Asherman's syndrome is also named intrauterine adhesions (IUA).

Present Stem-Cell Based Therapies for Female Infertility

Research in stem cells has shown potential therapies to treat several diseases [2]. The first clinical trial with cells from hESCs was launched in 2010 for patients with spinal cord injury and the first major improvement was reported on patients with retinal pigment epithelium treated with human embryonic stem cell-derived [17] and few years later was launched the first clinical trial in Japan using iPS to treat macular degeneration. This shows that the technology can be applied in multiple clinical trials worldwide but not yet in infertility related diseases. However, there are many stem-cell therapies ongoing for the treatment of female infertility shown in Table 1.2 that we will discuss in this section.

To understand the development of human germ cells, well-characterized hES cell lines could be used as therapeutic tools for female infertility [18]. Also, for future female infertility purposes, it could be possible to generate artificial gametes from pluripotent stem cells [2] but actually are not yet ready [19, 20]. These studies have been induced into primordial germ cell-like cells (hPGCLCs) from human iPS cells and differentiate more into oogonia; nevertheless, further in vitro differentiation to mature oocytes has not been obtained yet. More investigations are required to achieve functional female gametes from human PSCs for prospective applications.

Clinical trials using MSCs and HSCs provide the potential application of these cells that may have to treat female infertility. Several clinical trials using autologous MSCs from bone marrow and adipose tissue and allogenic cell therapy using umbilical cord MSCs for patients with primary ovarian failure (POF) have shown promising results in rescuing of overall ovarian function, evidenced by increasing the ovary volume, menstrual and endometrium changes, elevated estradiol concentrations, improved follicular development, and increased number of antral follicles [21–23]. Also, it was reported that the bone MSC implantation procedure was very well tolerated with any reported adverse events [24].

Recently, a study with patients with Intrauterine adhesion (IUA) or endometrial dysplasia has shown promising results regarding the efficacy and safety of transplanting clinical-grade UC-MSCs added onto a degradable collagen scaffold into the uterine cavity following adhesiolysis surgery for recurrent IUA patients. By clustering stem cells, retaining their viability, and gaining the length of contact with the damaged place of the endometrium, the ability for endometrial proliferation and differentiation was raised greatly by using a scaffold [25].

The intraovarian injection of MSC increases estrogen production and reduces menopausal symptoms in women with POF [24]. MSCs can migrate in the ovary and stimulate with several factors to its recovery [12]. In addition, ovarian specific stem cells, OSCs have been proposed as a future clinical treatment [26]. For premature ovarian insufficiency (POI), it is important to characterize the OSCs present in the ovary in order to use them as therapeutic tools [27].

Regarding the clinical trial results of stem cell-based therapies for Asherman Syndrome treatment, the main outcome is the beneficial effects of the autologous $CD133^+$ cell therapy which increased the mature vessel density and the endometrial thickness with the spontaneous pregnancy of some patients after the stem cell treatment [14].

Finally, the uterine niche is described as a triangular anechoic complex at the place of the scar in the myometrium at the place of a prior cesarean area. The principal clinical indications are postmenstrual spotting and intrauterine infection, which may critically influence the day-to-day life of non-pregnant women. Recently, a trial has demonstrated greater efficacy and safety for the promising UC-MSCs as a therapeutic alternative for scar reconstruction [28].

The outcome of these clinical trials is very promising to improve female fertility treatments and to use regularly in all these patients in the future.

Table 1.2 Clinical trials with stem cell-therapies to treat female reproductive dysfunctions

Disease	Stem cell therapy	Title of clinical trial	Clinical trial reference	Status	Country	Publications related to the clinical trial results
Primary ovarian failure (POF)	Autologous bone marrow transplantation	Autologous bone marrow transplantation for premature ovarian insufficiency (BMT-POI)	NCT02779374	Unknown	Egypt	
	Autologous bone marrow-derived mesenchymal stem cells (BMMSCs)	Autologous bone marrow-derived stem cell transplantation in patients with premature ovarian failure (POF)	NCT03069209	Active, not recruiting	Jordan	
	Intraovarian injection of adipose-derived stromal cells (ADSCs)	Autologous adipose-derived mesenchymal stromal cells transplantation in women with premature ovarian failure (POF)	NCT02603744	Unknown	Iran	
	Bone marrow-derived stem cells	Rejuvenation of premature ovarian failure with stem cells (ROSE-1)	NCT02696889	Active not recruiting	USA	[24]
	Umbilical cord-derived mesenchymal stem cells (UC-MSCs)	Transplantation of HUC-MSCs with injectable collagen scaffold for POF	NCT02644447	Completed	China	[21]
	Autologous bone marrow-derived mesenchymal stem cells (BMMSCs)	Autologous mesenchymal stem cells transplantation in women with premature ovarian failure	NCT02062931	Unknown	Egypt	[22]
	Bone Marrow Transplantation	Bone marrow transplantation to promote follicle recruitment in poor ovarian reserve	NCT02240342	Unknown	Spain	[23, 29]

(continued)

Table 1.2 (continued)

Disease	Stem cell therapy	Title of clinical trial	Clinical trial reference	Status	Country	Publications related to the clinical trial results
Intrauterine adhesion (IUA), endometrial dysplasia	Umbilical cord-derived mesenchymal stem cells (UC-MSCs)	Treatment of infertility by collagen scaffold loaded with umbilical cord derived mesenchymal stem cells	NCT02313415	Completed	China	[25]
	Autologous bone marrow stem cells	Treatment of infertility by collagen scaffold loaded with autologous bone marrow stem cells	NCT02204358	Unknown	China	
Asherman syndrome	Bone marrow-derived mesenchymal stem cells (BMMSCs) and hormonal replacement therapy	Autologous bone marrow-derived mesenchymal stem cells for atrophic endometrium in patients with repeated IVF failures	NCT03166189	Completed	Russia	
	Autologous bone marrow mononuclear cells (ABMNC)	Treatment of severe Asherman syndrome by collagen scaffold loaded with autologous bone marrow mononuclear cells	NCT02680366	Unknown	China	
	Bone marrow-derived $CD133^{+}$ stem cells	Bone marrow stem cell treatment for Asherman syndrome and endometrial atrophy (BMSCT)	NCT02144987	Completed	Spain	[14, 30]
Infertility of uterine origin	Bone marrow-derived mononuclear cells (ABMNC)	Clinical study of in situ regeneration of endometrium	NCT04233892	Recruiting	China	

(continued)

Table 1.2 (continued)

Disease	Stem cell therapy	Title of clinical trial	Clinical trial reference	Status	Country	Publications related to the clinical trial results
Thin endometrium or endometrial scarring	Umbilical cord-derived mesenchymal stem cells (UC-MSCs)	The efficacy and safety of collagen scaffold loaded with umbilical cord-derived mesenchymal stem cells in infertile women with thin endometrium or endometrial scarring	NCT03592849	Enrolling by invitation	China	
Uterine scar	Umbilical cord-derived mesenchymal stem cells (UC-MSCs)	Safety and efficacy of umbilical cord mesenchymal stem cell local intramuscular injection for treatment of uterine scars	NCT02968459	Recruiting	China	[28]

Infertility Related Disorders in Male Reproductive Organs

Regarding the several causes of male infertility, they are classified according to their etiology such as: 2% to 5% are endocrine disorders (e.g. hypogonadism), 5% are sperm transport disorders (e.g. vasectomy), 65% to 80% are primary testicular defects (e.g. aberrant sperm parameters), and 10% to 20% are idiopathic (e.g. normal sperm and semen parameters) [31]. A resume of particular etiologies is indexed below and in Fig. 1.3:

Endocrinological Cause: Prader Willi Syndrome, Laurence-Moon-Beidl syndrome, Kallmann syndrome, iron overload syndrome, familial cerebellar Ataxia, intracranial radiation, prolactinoma, testosterone supplementation, or hyperthyroidism.

Idiopathic Cause: 10–20% of idiopathic male infertility is found with normal semen parameters but the male continues to be infertile.

Genetic Factors: Specific mutations in genes such as: CFTR, Kal-1, Kal-2, FSH, LH, FGFS, GnRH1/GNRHR PROK2/PROK2R, AR and Klinefelter syndrome, Kallman Syndrome, Young syndrome, Sertoli cell-only syndrome, primary ciliary dyskinesia, chromosomal anomalies, Y chromosome microdeletion, and gr/gr deletions.

Congenital Urogenital Anomalies: Dysfunctional, obstructed, or absent epididymis, vas deferens congenital anomalies, ejaculatory duct disorders (e.g. cysts), primitive testicular dysfunction, cryptorchidism, and atrophic testes.

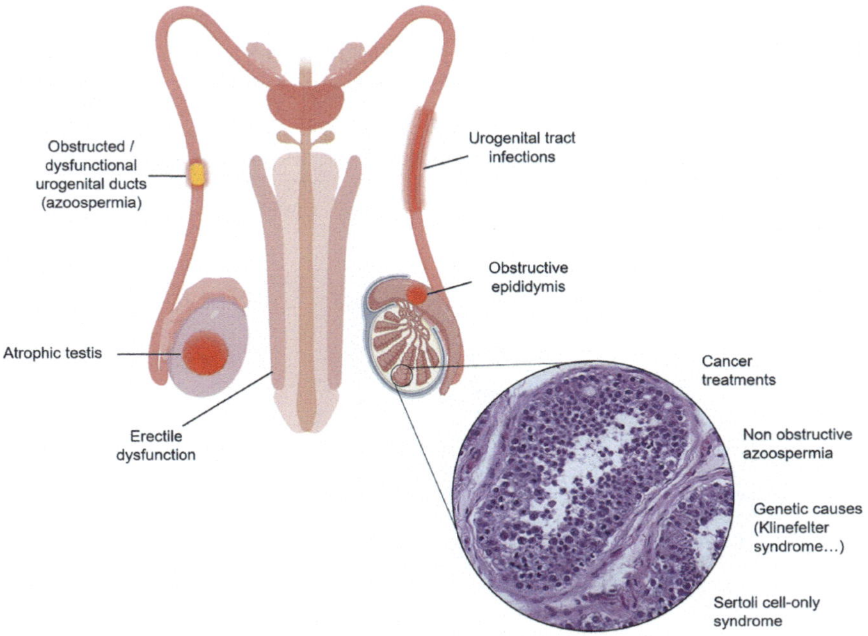

Fig. 1.3 Male infertility causes diagram

Acquired Urogenital Anomalies: Bilateral obstruction or vas deferens ligation, varicocele, epididymitis, bilateral orchiectomy, retrograde ejaculation.

Immunological Cause: Hemosiderosis, lymphocytic hypophysitis, hemochromatosis, histiocytosis, sarcoidosis, fungal infections, tuberculosis.

Urogenital Tract Infections: Syphilis, chlamydia, gonococci, periodic urogenital infections, prostatitis, viral mumps orchitis, and frequent prostatovesiculitis.

Sexual Dysfunction: Anejaculation, premature ejaculation, and erectile dysfunction.

Malignancies: Cancer in general (testicular tumors or adrenal tumors), chemotherapy, and radiation treatment.

Medications or Drugs or Environmental Toxins: In general, drugs that can cause inhibition of sexual hormones (cannabinoids, opioids, psychotropic drugs), GnRH analogs and antagonists utilized in prostatic carcinoma, androgenic steroids supplementation, glucocorticoid therapy, alkylating agents, cimetidine, ketoconazole, antiandrogens, smoking, excess alcohol.

Present Stem-Cell Based Therapies for Male Infertility

Worldwide, between 8 and 12% of couples are affected by infertility. One-third of infertility cases are due to male factor [1]. Research in stem cells has shown potential

Table 1.3 Clinical trials with stem cell-therapies to treat male reproductive dysfunctions

Disease	Stem cell	Title	Clinical trial reference	Status	Country	Publication
Non-obstructive Azoospermia Azoospermia Oligospermia	Autologous bone marrow-derived $CD34^+$, $CD133^+$ Hematopoietic Stem cells (HSCs), and Mesenchymal Stem Cells (MSCs)	Intra-testicular transplantation of autologous stem cells for treatment of non-obstructive azoospermia male infertility	NCT02641769	Recruiting	Jordan	
	Bone Marrow-derived Mesenchymal Stem Cells	Intra-testicular artery injection of bone marrow stem cell in management of azoospermia	NCT02008799	Unknown	Egypt	
	Bone marrow-derived mesenchymal stem cells (BMMSCs)	Testicular injection of autologous bone marrow-derived stem cells	NCT02041910	Unknown	Egypt	
	Adipose-derived adult stromal vascular cell	Autologous adipose-derived adult stromal vascular cell administration for male patients with infertility	NCT03762967	Enrolling by invitation	Russia	
	Bone marrow-derived mesenchymal stem cells (BMMSCs)	MSCs for treatment of Azoospermic patients	NCT02025270	Unknown	Egypt	
Klinefelter Syndrome Azoospermia	Bone marrow derived mesenchymal stem cells (BMMSCs)	Sperm production in klinefelter syndrome patients after mesenchymal stem cell injection	NCT02414295	Completed	Egypt	

(continued)

Table 1.3 (continued)

Disease	Stem cell	Title	Clinical trial reference	Status	Country	Publication
Young male. Cancer. Autoimmune disorders	Spermatogonial stem cells (SSCs)	Spermatogonial stem cell (SSC) transplant and testicular tissue grafting	NCT04452305	Recruiting	USA	
	Testicular tissue cryopreservation	Testicular tissue cryopreservation for fertility preservation	NCT02972801	Recruiting	USA	
Erectile dysfunction Peyronie's disease	Autologous adipose MSCs	Management of Peyronie's disease with adipose tissue stem cell	NCT02414308	Unknown	Egypt	
	Placental matrix-derived mesenchymal stem cells (PMD-MSCs)	Evaluate the safety and feasibility of injecting PMD-MSC into the penis to treat the symptoms of PD (PMD-MSC-PD-01)	NCT02395029	Completed	USA	[46]
	Autologous adipose-derived stromal vascular fraction (SVF)	Safety and clinical outcomes study: SVF deployment for orthopedic, neurologic, urologic, and cardio-pulmonary conditions	NCT01953523	Completed	USA	[40, 41]
	Placental matrix-derived mesenchymal stem cells (PMD-MSCs)	Evaluate the safety and feasibility of injecting PMD-MSC into the penis to treat the symptoms of mild to moderate ED (PMD-MSC-ED-01)	NCT02398370	Completed	USA	[45]

(continued)

Table 1.3 (continued)

Disease	Stem cell	Title	Clinical trial reference	Status	Country	Publication
	Bone marrow mononucleated cells (BMMNC)	Intracavernous bone marrow stem-cell injection for post prostatectomy erectile dysfunction (INSTIN)	NCT01089387	Completed	France	[39, 40, 47]
	Adipose-derived stem cells	Can fat derived stem cells (SVF) be used in the treatment of erectile dysfunction after prostatectomy	NCT02240823	Unknown	Denmark	[40, 42]
	Bone marrow-derived mesenchymal stem cells (BMMSCs)	Bone marrow mesenchymal stem cells in erectile dysfunction (ED)	NCT02945462	Completed	Jordan	[38]
	Adipose-derived mesenchymal stem cells	Administration of adipose-derived mesenchymal stem cells and platelet lysate in erectile dysfunction: a single center pilot study	N/A		Greece	[43]
	Umbilical cord-derived mesenchymal stem cells (UC MSCs)	Treatment of diabetic impotence with umbilical cord blood stem cell intracavernosal transplant: preliminary report of 7 cases	N/A		Korea	[44]

therapies to treat male infertility [2]. There are many clinical trials using stem cells ongoing for male infertility shown in Table 1.3 that we will discuss in this section.

Embryonic stem cells (ESC) and induced pluripotent stem cells (iPSC) are the most studied origin of stem cells. hES cell lines may be used to treat patients with azoospermia and oligospermia [18] and also hiPS is used to treat patients with Klinefelter syndrome [32] and azoospermia with Y chromosome azoospermia factor (AZF) [33]. Up to date, several protocols have been described the potential in vitro generation of male haploid-like cells generated from hiPS cells [2, 34, 35], however, the in vitro generated cells are not functionally yet ready. More studies are required to obtain functional male gametes from human PSCs for future applications.

Several studies have characterized MSCs as cell therapy candidates for male infertility [36]. Clinical trials using intra-testicular transplantation of autologous hematopoietic and mesenchymal stem cells from bone marrow and adipose tissue to treat non-obstructive azoospermia, azoospermia, oligospermia, and Klinefelter Syndrome patients are shown in Table 1.3. Spermatogonial stem cell-based approaches for male infertility are an option in patients where it is not possible to store sperm. Then, characterization and proliferation of the SSCs are being investigated properly [36]. Cancer is a principal death cause in children and adolescents. Nevertheless, as a consequence of important enhancements in therapies, cancer death percentages have decreased fairly in childhood and adolescence. The outcomes indicate an expectation of an 80% survival of children and adolescents diagnosed with cancer in Europe and America. Unluckily, the cancer treatments applied, such as chemotherapy and radiotherapy, due to their risk, dose, and gonadotoxic reaction, can injure the stem cell population of SSCs in the patient´s testis, provoking long-term infertility issues, in the majority of patients, resulting in permanent infertility problems. Besides these malignant diseases, there are also genetic syndromes as well, such as Klinefelter Syndrome (KS) that can contribute to the early damage of germ stem cells in prepuberal boys' testis.

Diverse approaches have been established to protect the young patients´ fertility; firstly, sperm cryopreservation, as a first-choice fertility cryopreservation procedure, in the adolescent patients, was accomplished regularly. Nonetheless, for prepuberal males and some adolescents, this therapeutic choice is not feasible. For both type of patients, the testicular tissue cryopreservation is the unique experimental possible choice that remains to safeguard their fertility [37].

As mentioned above, once the testicular tissue has been preserved and the patients will expect fertility issues in the next years, then following the current approaches described therefore should be suitable to recover their fertility.

Nowadays, three strategies to restore fertility are considered by scientific community: (1) in vitro SSCs expansion and later testicular autotransplant, (2) autotransplant of entire testicular biopsy, and (3) in vitro spermatogenesis. All strategies are shown in Fig. 1.4. However, there are still problems among the several experimental fertility recovery approaches already described, that must be clarified as a result of the optimal combination of both basic investigation linked with the clinic work in the near future. In addition, fertility preservation programs in adolescents and children are progressively noticed, in our experience, since 2016 we have initiated a pioneer

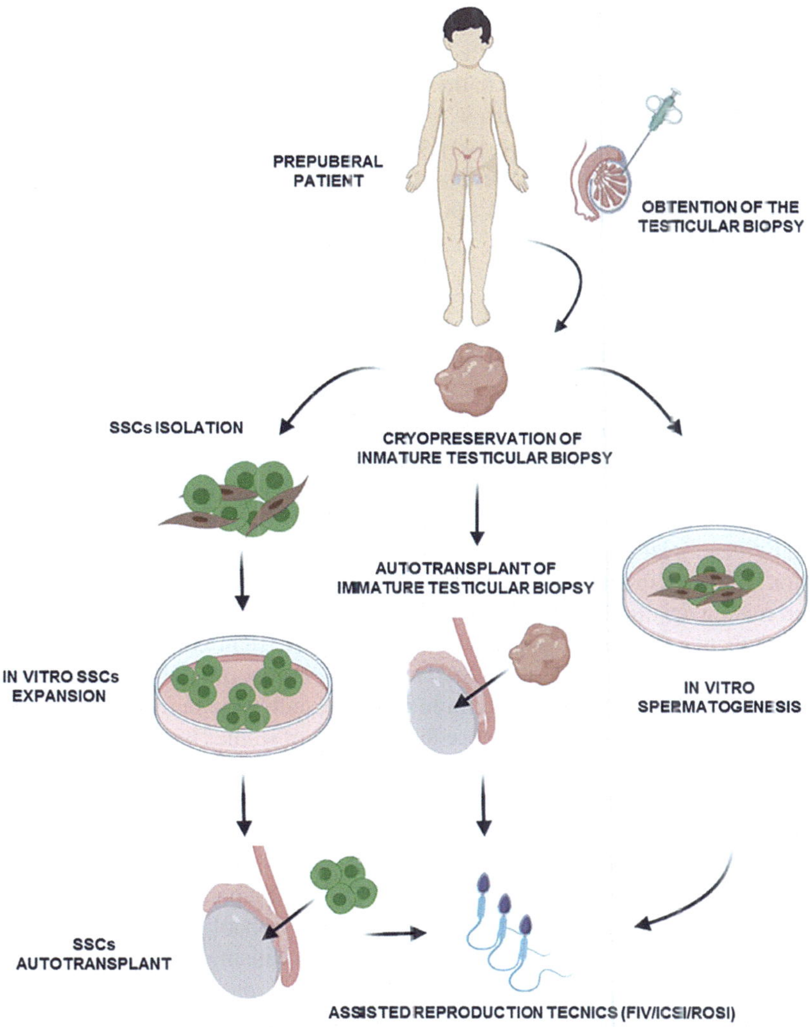

Fig. 1.4 Male prepuberal fertility preservation approaches

Fertility preservation program in Osakidetza (Public Basque Health Service) located at Basque Center for Blood Transfusion and Human Tissues (Spain) concentrated on prepubertal boys with cancer and Klinefelter syndrome patients, together with other pathologies, but unfortunately, there are still hospitals or centers that do not realize about the existence of these Fertility preservation programs, and for that reason it is crucial to correctly advertise and generate guidelines or recommendations for the future patient benefits. Finally, patients with sexual dysfunction (erectile dysfunction) were under stem cell therapy in nine clinical trials published. The results showed significant improvements in penile hemodynamics and progresses in erectile function scores with no major adverse effects showing efficacy in this therapy. The stem cells administered at several doses in these clinical trials were derived from multiple cell sources, autologous bone marrow [38–41], adipose tissue [40–43], MSCs and allogenic MSCs from cord blood [44], and placenta [45]. However, the specific mechanisms underlying stem cell efficacy in this sexual dysfunction treatment has not been evaluated in these clinical studies. Further, studies are needed to elucidate the mechanism beyond this benefit.

Conclusions and Future Outlook

Stem cells, especially MSCs, have indicated considerable potential and availability for treating infertility in animal and human studies among pre-clinical and clinical trials, respectively. Allogenic placental and cord blood derived-stem cells, autologous bone marrow and hematopoietic derived-stem cells, and adipose-derived MSCs are especially beneficial because they are not only easily obtained, but also avoid graft rejection after transplantation. Recently, several clinical trials have been obtained with promising results to treat infertility related disorders in male and female patients. However, there are several issues related to the adult stem cell therapies to treat infertility that have to be investigated: (1) there is an important absence of treasure research in large animal models that more firmly resemble the human ovary or testis with infertility related diseases because the majority of studies have been performed in small animals. Moreover, more randomized controlled clinical trials should be conducted to verify the beneficial and therapeutic effect of MSCs in infertility related diseases. (2) The mechanism of MSCs behind the treatment of treating reproductive organ disorders is still undiscovered. Several explanations can occur such as paracrine mechanism, inducing angiogenesis, or differentiation into functional cells. However, beneficial paracrine factors persist unexplained and probably several mechanisms can be synergistic. (3) Although MSC therapy is promising, the defined survival and engraftment of these cells together with biomaterials remains still unsolved. However, supplementary work is required to optimize this approach.

In addition, many encouraging preliminary stem cell-based procedures are under progress to recover male fertility in prepubertal young men. Currently, relevant achievements have been accomplished in this research area; SSC transplantation has been translated to human cadaver testis, preliminary data has been reached for testicular tissue grafting in non-human primates, and substantial advances have been led in demonstrating in vitro spermatogenesis. Nevertheless, until outstanding clinical trials proving efficacy and safety of fertility recovery have been managed, testicular tissue cryopreservation for fertility care should continue experimental approach.

Finally, even though in vitro germ cell generation from hPSCs is frequently suggested as a tentative future alternative for infertility therapy, it has to consider that the current in vitro protocols do not support enough differentiation of entirely matured sperm or oocytes, and even if achievable, confirming the functionality of in vitro generated sperm by fertilizing oocytes and contributing offspring will present many ethical controversies, which requires to be discussed very cautiously. However, iPSCs modeling approaches to investigate the elemental features of germ cell specification can be contemplated as important points to obtain more learning on differentiation methods and germ cell development. Lastly, the generation of artificial gametes from human iPS cells is still a "distant prospect".

References

1. Wang J, Liu C, Fujino M, Tong G, Zhang Q, Li X-K, Yan H (2019) Stem cells as a resource for treatment of infertility-related diseases. Curr Mol Med 19:539–546. https://doi.org/10.2174/1566524019666190709172636
2. Vassena R, Eguizabal C, Heindryckx B, Sermon K, Simon C, van Pelt AMM, Veiga A, Zambelli F, ESHRE special interest group Stem Cells (2015) Stem cells in reproductive medicine: ready for the patient? Hum Reprod 30:2014–2021. https://doi.org/10.1093/humrep/dev181
3. Somigliana E, Paffoni A, Busnelli A, Filippi F, Pagliardini L, Vigano F, Vercellini P (2016) Age-related infertility and unexplained infertility: an intricate clinical dilemma. Hum Reprod 31:1390–1396. https://doi.org/10.1093/humrep/dew066
4. Vanni VS, Viganò P, Papaleo E, Mangili G, Candiani M, Giorgione V (2017) Advances in improving fertility in women through stem cell-based clinical platforms. Expert Opin Biol Ther 17:585–593. https://doi.org/10.1080/14712598.2017.1305352
5. Brunauer R, Alavez S, Kennedy BK (2017) Stem cell models: a guide to understand and mitigate aging? Gerontology 63:84–90. https://doi.org/10.1159/000449501
6. Volarevic V, Bojic S, Nurkovic J, Volarevic A, Ljujic B, Arsenijevic N, Lako M, Stojkovic M (2014) Stem cells as new agents for the treatment of infertility: current and future perspectives and challenges. BioMed Res Int 2014:507234. https://doi.org/10.1155/2014/507234
7. Eguizabal C, Aran B, Chuva de Sousa Lopes SM, Geens M, Heindryckx B, Panula S, Popovic M, Vassena R, Veiga A (2019) Two decades of embryonic stem cells: a historical overview. Hum Reprod 2019:hoy024. https://doi.org/10.1093/hropen/hoy024
8. Thomson JA, Itskovitz-Eldor J, Shapiro SS, Waknitz MA, Swiergiel JJ, Marshall VS, Jones JM (1998) Embryonic stem cell lines derived from human blastocysts. Science 282:1145–1147. https://doi.org/10.1126/science.282.5391.1145
9. Takahashi K, Yamanaka S (2006) Induction of pluripotent stem cells from mouse embryonic and adult fibroblast cultures by defined factors. Cell 126:663–676. https://doi.org/10.1016/j.cell.2006.07.024
10. Takahashi K, Tanabe K, Ohnuki M, Narita M, Ichisaka T, Tomoda K, Yamanaka S (2007) Induction of pluripotent stem cells from adult human fibroblasts by defined factors. Cell 131:861–872. https://doi.org/10.1016/j.cell.2007.11.019
11. He M, von Schwarz ER (2020) Stem-cell therapy for erectile dysfunction: a review of clinical outcomes. Int J Impot Res 1–7. https://doi.org/10.1038/s41443-020-0279-8
12. Zhao Y, Chen S, Su P, Huang F, Shi Y, Shi Q, Lin S (2019) Using mesenchymal stem cells to treat female infertility: an update on female reproductive diseases. Stem Cells Int 2019:9071720. https://doi.org/10.1155/2019/9071720
13. Dominici M, Le Blanc K, Mueller I, Slaper-Cortenbach I, Marini F, Krause D, Deans R, Keating A, Prockop D, Horwitz E (2006) Minimal criteria for defining multipotent mesenchymal stromal cells. the international society for cellular therapy position statement. Cytotherapy 8:315–317. https://doi.org/10.1080/14653240600855905
14. Santamaria X, Cabanillas S, Cervelló I, Arbona C, Raga F, Ferro J, Palmero J, Remohí J, Pellicer A, Simón C (2016) Autologous cell therapy with $CD133^+$ bone marrow-derived stem cells for refractory Asherman's syndrome and endometrial atrophy: a pilot cohort study. Hum Reprod 31:1087–1096. https://doi.org/10.1093/humrep/dew042

15. Ilic D, Telfer EE, Ogilvie C, Kolundzic N, Khalaf Y (2019) What can stem cell technology offer to IVF patients? BJOG Int J Obstet Gynaecol 126:824–827. https://doi.org/10.1111/1471-0528.15638
16. Walker MH, Tobler KJ (2020) Female infertility. In: StatPearls. StatPearls Publishing, Treasure Island (FL)
17. Schwartz SD, Hubschman J-P, Heilwell G, Franco-Cardenas V, Pan CK, Ostrick RM, Mickunas E, Gay R, Klimanskaya I, Lanza R (2012) Embryonic stem cell trials for macular degeneration: a preliminary report. Lancet 379:713–720. https://doi.org/10.1016/S0140-6736(12)60028-2
18. Pickering SJ, Braude PR, Patel M, Burns CJ, Trussler J, Bolton V, Minger S (2003) Preimplantation genetic diagnosis as a novel source of embryos for stem cell research. Reprod Biomed Online 7:353–364. https://doi.org/10.1016/s1472-6483(10)61877-9
19. Yamashiro C, Sasaki K, Yabuta Y, Kojima Y, Nakamura T, Okamoto I, Yokobayashi S, Murase Y, Ishikura Y, Shirane K, Sasaki H, Yamamoto T, Saitou M (2018) Generation of human oogonia from induced pluripotent stem cells in vitro. Science 362:356–360. https://doi.org/10.1126/science.aat1674
20. Yamashiro C, Sasaki K, Yokobayashi S, Kojima Y, Saitou M (2020) Generation of human oogonia from induced pluripotent stem cells in culture. Nat Protoc 15:1560–1583. https://doi.org/10.1038/s41596-020-0297-5
21. Xie Y, Liu W, Liu S, Wang L, Mu D, Cui Y, Cui Y, Wang B (2020) The quality evaluation system establishment of mesenchymal stromal cells for cell-based therapy products. Stem Cell Res Ther 11:176. https://doi.org/10.1186/s13287-020-01696-6
22. Edessy M, Hosni H, Wafa Y, Bakry S, Shady Y, Kamel M (2014) Stem cells transplantation in premature ovarian failure. World J Med Sci 10:12–16. https://doi.org/10.5829/idosi.wjms.2014.10.1.1137
23. Herraiz S, Buigues A, Díaz-García C, Romeu M, Martínez S, Gómez-Seguí I, Simón C, Hsueh AJ, Pellicer A (2018) Fertility rescue and ovarian follicle growth promotion by bone marrow stem cell infusion. Fertil Steril 109:908-918.e2. https://doi.org/10.1016/j.fertnstert.2018.01.004
24. Igboeli P, El Andaloussi A, Sheikh U, Takala H, ElSharoud A, McHugh A, Gavrilova-Jordan L, Levy S, Al-Hendy A (2020) Intraovarian injection of autologous human mesenchymal stem cells increases estrogen production and reduces menopausal symptoms in women with premature ovarian failure: two case reports and a review of the literature. J Med Case Reports 14:108. https://doi.org/10.1186/s13256-020-02426-5
25. Cao Y, Sun H, Zhu H, Zhu X, Tang X, Yan G, Wang J, Bai D, Wang J, Wang L, Zhou Q, Wang H, Dai C, Ding L, Xu B, Zhou Y, Hao J, Dai J, Hu Y (2018) Allogeneic cell therapy using umbilical cord MSCs on collagen scaffolds for patients with recurrent uterine adhesion: a phase I clinical trial. Stem Cell Res Ther 9:192. https://doi.org/10.1186/s13287-018-0904-3
26. Radhakrishnan G (2017) Stem cells—The new agents in infertility treatment: The light at the end of the tunnel? Fertil Sci Res 4:70. https://doi.org/10.4103/fsr.fsr_16_18
27. van Kasteren YM, Schoemaker J (1999) Premature ovarian failure: a systematic review on therapeutic interventions to restore ovarian function and achieve pregnancy. Hum Reprod Update 5:483–492. https://doi.org/10.1093/humupd/5.5.483
28. Fan D, Wu S, Ye S, Wang W, Guo X, Liu Z (2017) Umbilical cord mesenchyme stem cell local intramuscular injection for treatment of uterine niche: Protocol for a prospective, randomized, double-blinded, placebo-controlled clinical trial. Medicine 96:e8480. https://doi.org/10.1097/MD.0000000000008480
29. Herraiz S, Romeu M, Buigues A, Martínez S, Díaz-García C, Gómez-Seguí I, Martínez J, Pellicer N, Pellicer A (2018) Autologous stem cell ovarian transplantation to increase reproductive potential in patients who are poor responders. Fertil Steril 110:496-505.e1. https://doi.org/10.1016/j.fertnstert.2018.04.025
30. Cervelló I, Gil-Sanchis C, Santamaría X, Cabanillas S, Díaz A, Faus A, Pellicer A, Simón C (2015) Human CD133(+) bone marrow-derived stem cells promote endometrial proliferation in a murine model of Asherman syndrome. Fertil Steril 104(1552–1560):e1-3. https://doi.org/10.1016/j.fertnstert.2015.08.032

31. Leslie SW, Siref LE, Khan MA (2020) Male infertility. In: StatPearls. StatPearls Publishing, Treasure Island (FL)
32. Shimizu T, Shiohara M, Tai T, Nagao K, Nakajima K, Kobayashi H (2015) Derivation of integration-free iPSCs from a Klinefelter syndrome patient. Reprod Med Biol 15:35–43. https://doi.org/10.1007/s12522-015-0213-9
33. Ramathal C, Durruthy-Durruthy J, Sukhwani M, Arakaki JE, Turek PJ, Orwig KE, Reijo Pera RA (2014) Fate of iPSCs derived from Azoospermic and fertile men following xenotransplantation to murine seminiferous tubules. Cell Rep 7:1284–1297. https://doi.org/10.1016/j.celrep.2014.03.067
34. Eguizabal C, Montserrat N, Vassena R, Barragan M, Garreta E, Garcia-Quevedo L, Vidal F, Giorgetti A, Veiga A, Izpisua Belmonte JC (2011) Complete meiosis from human induced pluripotent stem cells. Stem Cells 29:1186–1195. https://doi.org/10.1002/stem.672
35. Kurek M, Albalushi H, Hovatta O, Stukenborg J-B (2020) Human pluripotent stem cells in reproductive science-a comparison of protocols used to generate and define male germ cells from pluripotent stem cells. Int J Mol Sci 21:1028. https://doi.org/10.3390/ijms21031028
36. Gauthier-Fisher A, Kauffman A, Librach CL (2020) Potential use of stem cells for fertility preservation. Andrology 8:862–878. https://doi.org/10.1111/andr.12713
37. Goossens E, Jahnukainen K, Mitchell R, van Pelt A, Pennings G, Rives N, Poels J, Wyns C, Lane S, Rodriguez-Wallberg K, Rives A, Valli-Pulaski H, Steimer S, Kliesch S, Braye A, Andres M, Medrano J, Ramos L, Kristensen S, Andersen C, Bjarnason R, Orwig K, Neuhaus N, Stukenborg J (2020) Fertility preservation in boys: recent developments and new insights. Hum Reprod 2020:hoaa016. https://doi.org/10.1093/hropen/hoaa016
38. Al Demour S, Jafar H, Adwan S, AlSharif A, Alhawari H, Alrabadi A, Zayed A, Jaradat A, Awidi A (2018) Safety and potential therapeutic effect of two intracavernous autologous bone marrow derived mesenchymal stem cells injections in diabetic patients with erectile dysfunction: an open label phase I clinical trial. Urol Int 101:358–365. https://doi.org/10.1159/000492120
39. Yiou R, Hamidou L, Birebent B, Bitari D, Le Corvoisier P, Contremoulins I, Rodriguez A-M, Augustin D, Roudot-Thoraval F, de la Taille A, Rouard H (2017) Intracavernous injections of bone marrow mononucleated cells for postradical prostatectomy erectile dysfunction: final results of the INSTIN clinical trial. Eur Urol Focus 3:643–645. https://doi.org/10.1016/j.euf.2017.06.009
40. Khera M, Albersen M, Mulhall JP (2015) Mesenchymal stem cell therapy for the treatment of erectile dysfunction. J Sex Med 12:1105–1106. https://doi.org/10.1111/jsm.12871
41. Lander EB, Berman MH, See JR (2016) Stromal vascular fraction combined with shock wave for the treatment of Peyronie's disease. Plast Reconstr Surg Glob Open 4:e631. https://doi.org/10.1097/GOX.0000000000000622
42. Haahr MK, Jensen CH, Toyserkani NM, Andersen DC, Damkier P, Sørensen JA, Lund L, Sheikh SP (2016) Safety and potential effect of a single intracavernous injection of autologous adipose-derived regenerative cells in patients with erectile dysfunction following radical prostatectomy: an open-label phase I clinical trial. EBioMedicine 5:204–210. https://doi.org/10.1016/j.ebiom.2016.01.024
43. Protogerou V, Michalopoulos E, Mallis P, Gontika I, Dimou Z, Liakouras C, Stavropoulos-Giokas C, Kostakopoulos N, Chrisofos M, Deliveliotis C (2019) Administration of adipose derived mesenchymal stem cells and platelet lysate in erectile dysfunction: a single center pilot study. Bioengineering 6:21. https://doi.org/10.3390/bioengineering6010021
44. Bahk JY, Jung JH, Han H, Min SK, Lee YS (2010) Treatment of diabetic impotence with umbilical cord blood stem cell intracavernosal transplant: preliminary report of 7 cases. Exp Clin Transplant Off J Middle East Soc Organ Transplant 8:150–160
45. Levy JA, Marchand M, Iorio L, Cassini W, Zahalsky MP (2016) Determining the feasibility of managing erectile dysfunction in humans with placental-derived stem cells. J Am Osteopath Assoc 116:e1-5. https://doi.org/10.7556/jaoa.2016.007

46. Levy JA, Marchand M, Iorio L, Zribi G, Zahalsky MP (2015) Effects of stem cell treatment in human patients with peyronie disease. J Am Osteopath Assoc 115:e8-13. https://doi.org/10.7556/jaoa.2015.124
47. Yiou R, Hamidou L, Birebent B, Bitari D, Lecorvoisier P, Contremoulins I, Khodari M, Rodriguez A-M, Augustin D, Roudot-Thoraval F, de la Taille A, Rouard H (2016) Safety of intracavernous bone marrow-mononuclear cells for postradical prostatectomy erectile dysfunction: an open dose-escalation pilot study. Eur Urol 69:988–991. https://doi.org/10.1016/j.euro.2015.09.026

Chapter 2
Regeneration of Human Ovaries Through Mesenchymal Stem Cell Transplantation is Becoming a Reality

Irma Virant-Klun

Abstract

Introduction: There are various human health conditions associated with reduced ovarian function and consequently infertility. One such condition is premature ovarian insufficiency (premature ovarian failure), which can be a natural condition caused by various factors (genetic cause, autoimmunity, viruses, environmental influences, etc.) or caused by oncotherapy (chemotherapy and radiotherapy).
Methods: Possibilities for human ovarian regeneration by mesenchymal stem cell transplantation were investigated.
Results: Numerous studies on animal models show that it is possible to successfully regenerate the non-functional ovaries by transplanting human mesenchymal stem cells from various sources. Not only that, the first research has also been done in humans and the results are quite promising.
Conclusions: Based on the existing results, we estimate that the regeneration of non-functional ovaries by stem cell transplantation is worth further research, as it represents a realistic possibility and the first applications in humans have already been performed.

Keywords Human · Infertility · Mesenchymal stem cells · Ovary · Premature ovarian insufficiency · Regeneration · Transplantation

Abbreviations

AD-MSC	Amnion-derived mesenchymal stem cell
AEC	Amniotic epithelial cell
AFC	Antral follicle count
AFC	Amniotic fluid cell
AMH	Anti-Müllerian hormone

I. Virant-Klun (✉)
Clinical Research Center, University Medical Center Ljubljana, Ljubljana, Slovenia
e-mail: irma.virant@kclj.si

I. Virant-Klun (ed.), *Stem Cells in Reproductive Tissues and Organs*, Stem Cell Biology and Regenerative Medicine 70, https://doi.org/10.1007/978-3-030-90111-0_2

AMPK	5′ Adenosine monophosphate-activated protein kinase
AMSC	Amniotic mesenchymal stem cell
BAX	BCL2 Associated X, Apoptosis Regulator
BCL2	BCL2 apoptosis regulator
BMSC	Bone marrow mesenchymal stem cell
BMD-MSC	Bone marrow-derived mesenchymal stem cell
BrdUrd	Bromodeoxyuridine
CD	Cluster of differentiation
CDDP	Cisplatin
CFA	Freund's adjuvant
CK	Cytokeratin
CP-MSC	Chorionic plate-derived mesenchymal stem cell
CTX	Cyclophosphamide
DDX4	DEAD-box helicase 4 (VASA)
Dil	1,1′-Dioctadecyl-3,3,3′,3′-tetramethylindocarbocyanine perchlorate
E2	Estradiol
EGF	Epidermal growth factor
eMSC	Endometrial mesenchymal stem cell
EnSC	Endometrial mesenchymal stem cell from menstrual blood
ES-MSC	Human embryonic stem cell-derived mesenchymal stem cell
Exos	Exosome
FGF	Fibroblast growth factor
FIA	Freund's incomplete adjuvant
fMSC	Fetal liver mesenchymal stem cell
FORKO	Follitropin receptor knockout (mouse)
FOXO1	Forkhead box O1
FOXO3	Forkhead box O3
FSH	Follicle-stimulating hormone
FSHR	FSH receptor
GADD45B	Growth arrest and DNA damage inducible beta
GC	Granulosa cell
GFP	Green fluorescent protein
GSC	Germline stem cell
h	Human
HCB-MSC	Human umbilical cord blood mesenchymal stem cell
hESC	Human embryonic stem cell
HGF	Hepatocyte growth factor
HIV	Human immunodeficiency virus
HOVEC	Human primary ovarian endothelial cells
HPLC	High-performance liquid chromatography
HuAFC	Human amniotic fluid stem cell
hUCMSC	Human umbilical cord mesenchymal stem cell
hUCV-MSC	Human umbilical cord vein mesenchymal stem cell
HuMenSC	Human menstrual blood stem cell
hUMSC	Human umbilical cord mesenchymal stem cell

IGF-1	Insulin-like growth factor 1
IL-2	Interleukin 2
IFN-γ	Interferon gamma
IRE1α	Serine/threonine-protein kinase/endoribonuclease inositol-requiring enzyme 1 α
IU	International unit
IVF	In vitro fertilization
JNK1	Mitogen-activated protein kinase 8 (MAPK8)
LIPUS	Low-intensity pulsed ultrasound
MII	Metaphase II
MenSC	Menstrual blood-derived stromal cell
mRNA	Messenger ribonucleic acid
microRNA	Micro-ribonucleic acid (miRNA)
MSC	Mesenchymal stem cell
MT1	Metallothionein-1
mTOR	Mammalian/mechanistic target of rapamycin
Mvh	Mouse Vasa homologue
NGF	Nerve growth factor
NOA	Natural ovarian aging
OSSC	Ovarian stroma stem cell
PARP	Poly(ADP-Ribose) polymerase
PCNA	Proliferating cell nuclear antigen
PD-MSC	Placenta-derived mesenchymal stem cell
PMSC	Placenta mesenchymal stem cell
PI3K	Phosphoinositide 3-kinase
P	Progesterone
PKH26	Red fluorescent cell linker kit for general cell membrane
POI	Premature ovarian insufficiency
POF	Premature ovarian failure
RFP	Red fluorescent protein
SCID	Severe combined immune deficiency
siRNA	Small interfering RNA
SMAD3	SMAD family member 3
Tc17	IL-17-expressing/secreting $CD8^+$ T cells
TGF-β	Transforming growth factor beta
Th1	T helper type 1 cell
Th2	T helper type 2 cell
Th17	T helper 17 cell
Tie2	TEK receptor tyrosine kinase (Tek)
Treg	Regulatory T cell
TrkA	Tropomyosin receptor kinase A
TUNEL	Terminal deoxynucleotidyl transferase dUTP nick-end labeling
TVUS	Transvaginal ultrasound
uNK	Uterine natural killer cells
VASA	DEAD-box helicase 4 (DDX4)

VEGF	Vascular endothelial growth factor
VEGFR2	Vascular endothelial growth factor receptor 2
VSEL	Very small embryonic-like stem cell
ZP2	Zona pellucida glycoprotein 2
ZP3	Zona pellucida glycoprotein 3

Introduction

One of the biggest problems of modern times is infertility, which also includes female infertility. The causes of female infertility can be different and curable in different ways. Despite intensive developments in the treatment of infertility, including in vitro fertilization (IVF), some forms of infertility, especially ovarian infertility, remain poorly curable or incurable. The purpose of this chapter is to review the literature and current findings to determine if ovarian infertility could be cured in a new way—by cell therapy based on mesenchymal stem cell (MSC) transplantation for ovarian regeneration.

Ovarian Infertility

Ovarian regeneration by stem cell transplantation could be very important in patients facing various forms of ovarian infertility, especially premature ovarian insufficiency as well as other forms of ovarian infertility.

Premature Ovarian Insufficiency

The most severe form of ovarian infertility is premature ovarian insufficiency (POI), a condition in which a woman experiences amenorrhea or absence of regular menstrual cycles before the age of 40, high concentrations of gonadotropins, especially follicle-stimulating hormone (FSH), and declined estradiol (E2) concentrations in the serum [1]. The global prevalence of POI was estimated to be 1 to 3.7% [2]. Most of the women with POI do not have mature ovarian follicles and oocytes suitable for fertilization, embryonic development, and pregnancy, so they are infertile. POI can be caused by a variety of causes such as being born with non-optimal birth characteristics (small for gestational age, preterm, or with low birth weight) [3], genetic abnormalities (e.g., Fragile X premutation) [4], pelvic surgery [5], autoimmunity [6, 7], infection with various viruses (e.g., HIV) and vaccination [8–12], and exposure to various chemical agents [5, 13] including endocrine disruptors [14] and air pollutants [15], but in most cases it remains unexplained.

POI can also be caused by the treatment of cancer with chemotherapy and radiotherapy, which is successful, but can cause poorer ovarian function or POI in patients. At a median of 5.0 years from initial breast cancer diagnosis, 49% patients after adjuvant chemotherapy with anthracyclines and taxanes and 11% after therapy with tamoxifen had become post- and peri-menopausal [16]. Decreased ovarian follicle reserve occurs in more than one-third of patients after breast cancer treatment resulting in permanent infertility [17]. In long-term female survivors of pediatric hematologic malignancies, 26.7% experienced POI and face infertility after cancer treatment [18]. The situation is similar with other cancers; cancer therapy is the cause of POI in 25% of women with this diagnosis [19]. It is estimated that approximately one-third of all cancer patients develop POI and face infertility after completion of cancer treatment and cure.

One of the possible therapies for better health and quality of life in patients with POI is hormone (estrogen and progesterone) replacement therapy which ensures the development of secondary sex characteristics, acquisition of peak bone mass, and promotion of uterine growth and maturation [20]. Given the extent to which oocyte depletion or dysfunction is variable, there is the possibility of spontaneous ovulation after hormone replacement therapy and subsequent pregnancy, however, the chance of pregnancy is very low. In women with residual ovarian follicles, infertility can be treated with some new experimental methods, such as non-chemical or chemical activation of follicles in vitro [21, 22], while in women without residual ovarian follicles, infertility can only be treated by fertilizing donated oocytes and transferring the resulting embryos into the uterus in the IVF program. This is not always possible, as donated oocytes are not acceptable for many women. On the other hand, there is a shortage of donated oocytes that would be needed by women with POI.

Diminished Ovarian Reserve

Other causes of reduced ovarian reserve with decreased serum anti-Müllerian hormone (AMH) levels are also known, such as surgical treatment of ovarian endometriomas [23, 24]. In these women, fertility can be greatly reduced. This is thought to be due to hyperactivation of dormant primordial follicles. In addition, many patients with reduced ovarian reserve, which is not explained, are treated in the IVF programs. Their ovaries respond poorly to hormonal stimulation to obtain oocytes, which is consequently associated with a small number and poor quality of oocytes and low chance of pregnancy in the IVF program [25–27]. Last but not the least, age reduces the ovarian reserve, quality of oocytes, and the possibility of pregnancy in elderly patients who, for various personal reasons, decide to have a child in their late 40 s or even later [28, 29].

Regeneration of Ovaries by Transplantation of Human Stem Cells in Animal Models: Potential Sources, Transplantation, and Improvements

Due to the limited possibilities for the treatment of POI, stem cell therapies are proving to be very interesting and promising. Various types of human mesenchymal stem cells (hMSCs) have been tested on different animal models in terms of regeneration of non-functional ovaries. There is also a lot of research on in vitro oogenesis or in vitro development of oocytes from various pluripotent stem cells, including germinal stem cells, but ovarian regeneration through stem cell transplantation appears to be more realistic and clinically acceptable.

Mesenchymal Stem Cells

Numerous studies have been found in the literature in which hMSCs from various sources have been used to regenerate inactive ovaries due to POI in various animal (mammalian) models. Animal models, hMSC origins, transplantation, and stem cell transfer effects are shown in Table 2.1 [30–68]. In all of these studies, POI was simulated in a variety of ways, primarily by treating animals with chemotherapy—combination of cyclophosphamide (CTX) and busulfan [30, 35, 50, 56, 63, 66], cisplatin (CDDP) [31, 33, 34, 67], cyclophosphamide [32, 37, 39, 41, 43, 49, 53–55, 57–59, 64, 65], paclitaxel [42], or busulfan alone [61, 62]. Rarely, ovarian failure was simulated by natural ovarian aging [52]; immunization with the ovarian antigen—zona pellucida glycoprotein 3 (ZP3) peptide emulsified in complete Freund's adjuvant (CFA) [36, 38, 45–48]; ovarian removal (ovariectomy) [44]; hydrogen peroxide treatment [51]; or FSHR (−/−) mouse model (FORK) with follitropin receptor knockout [68] with elevated FSH, decreased estrogen levels, and with sterility because of the absence of folliculogenesis, thin uteri, and small non-functional ovaries. The most common animal models for investigating ovarian regeneration with MSCs were the mouse (i.e., ICR, BALB/c, C57BL/6, SCID-NOD) and the rat (i.e., Wistar albino, Sprague–Dawley), aged 6 to 12 weeks.

hMSCs from various sources, such as human embryonic stem cell (hESC)-derived MSCs [30, 31], ovarian stromal cells [32], umbilical cord [33–43], placenta [44–48], fetus (i.e., fetal liver) [49], amnion/amniotic fluid [50–57], chorionic plate [58], menstrual blood (endometrium) [59–64], and bone marrow [65–68], are used in ovarian regeneration research in animal models simulating POI. In such studies, the most commonly used stem cells are from human cord blood (Fig. 2.1). All these sources of MSCs are of interest for potential clinical practice, especially those that would allow the transplantation of autologous stem cells (i.e., bone marrow, menstrual blood/endometrium). In animal models (mice, rats), hMSCs were mostly administered by injection into the tail vein [30–39, 41–44, 47–50, 52–55, 58–63, 66, 67] and less frequently intraperitoneally [32, 46, 51] or into the ovaries [50, 56,

Table 2.1 Regeneration of ovaries by human mesenchymal stem cell (MSC) transplantation in animal models with simulated premature ovarian insufficiency

Reference	Animal	Type of human stem cells	Simulation of premature ovarian insufficiency	Stem cell application
Bahrehbar et al. 2010 [30]	Mouse	Human embryonic stem cell-derived mesenchymal stem cells(ES-MSCs)	– Cyclophosphamide and busulfan	Tail-vein injection
Yoon et al. 2020 [31]	Mouse	Human embryonic stem cell-derived mesenchymal stem cells(ES-MSCs)	– Cisplatin	Tail-vein injection
Besikcioglu et al. 2019 [32]	Rat (Wistar Albino)	Human ovarian stroma stem cells (OSSCs) and bone marrow mesenchymal stem cells (BMMSCs) from fetal tissue (day 16)	– Cyclophosphamide	Intraperitoneal injection
Lu et al. 2020 [33]	Rat (Wistar rats, 6 weeks old)	Human umbilical cord-derived mesenchymal stem cells (hUMSCs)	– Cisplatin	Tail-vein injection
Cui et al. 2020 [34]	Rat (Wistar rats, 7 weeks old)	Human umbilical cord mesenchymal stem cells (hUMSCs)	– Cisplatin	Tail-vein injection
Zhao et al. 2020 [35]	Mouse (ICR, 6 weeks old)	Human umbilical cord MSCs (hUCMSCs)	– Cyclophosphamide and busulfan	Tail-vein injection
Wang et al. 2020 [36]	Rat (specific-pathogen-free-grade Sprague–Dawley rats, aged 8 weeks)	Human umbilical cord MSCs (hUCMSCs)	– Autoimmune factor, induced by subcutaneous injection of ovarian antigen three times, once every 10 days	Tail-vein injection

(continued)

Table 2.1 (continued)

Reference	Animal	Type of human stem cells	Simulation of premature ovarian insufficiency	Stem cell application
Shen et al. 2020 [37]	Mouse (BALB/c mice, 7–8 weeks old)	Human umbilical cord mesenchymal stem cells (hUCMSCs)	– Cyclophosphamide	Tail-vein injection
Lu et al. 2019 [38]	Mouse	Human umbilical cord blood-derived stem cells (hUMSCs)	– Autoimmune, induced by zona pellucida 3 (ZP3) peptide emulsified in Freund's incomplete adjuvant (FIA)	Tail-vein injection
Zheng et al. 2019 [39]	Rat (specific pathogen-free Sprague–Dawley rats, 12 weeks old)	Human umbilical cord mesenchymal stem cells (UCMSCs)	– Cyclophosphamide	Tail-vein injection
Yang et al. 2019 [40]	Mouse (C57BL/6 mice)	Human umbilical cord-derived mesenchymal stem cells on a collagen scaffold (collagen/UC-MSCs)	– Cyclophosphamide	Collagen scaffold loaded with human umbilical cord-derived mesenchymal stem cells (collagen/UC-MSCs) and transplanted (injected) into the core of ovaries
Jalalie et al. 2019 [41]	Mouse	CM-DiI-labeled human umbilical cord vein MSCs (hUCV-MSCs)	- Cyclophosphamide	-Lateral tail-vein injection
Elfayomy et al. 2016 [42]	Rat	Human umbilical cord blood mesenchymal stem cells (HCB-MSCs)	– Paclitaxel	Tail-vein injection
Song et al. 2016 [43]	Rat (specific pathogen-free-grade Wistar rats)	Human umbilical cord mesenchymal stem cells (UCMSCs)	– Cyclophosphamide	Tail-vein injection

(continued)

Table 2.1 (continued)

Reference	Animal	Type of human stem cells	Simulation of premature ovarian insufficiency	Stem cell application
Seok et al. 2020 [44]	Rat	Human placenta-derived MSCs (PD-MSCs)	– Ovariectomy	Tail-vein injection
Li et al. 2019 [45]	Mouse (BALB/c mice, 6–8 weeks old)	Human placenta-derived mesenchymal stem cells (hPMSCs)	– Autoimmune, induced by ZP3 peptide	Subcutaneous injection
Yin et al. 2018 [46]	Mouse (BALB/c mice, 6 weeks old for in-vivo study and 3 weeks old for in-vitro study)	Human placenta-derived mesenchymal stem cell s (hPMSCs)	– Autoimmune, induced by ZP3 peptide emulsified in complete Freund's adjuvant (CFA)	Intraperitoneal injection
Zhang et al. 2018 [47]	Mouse (Balb/c mice, 6–8 weeks old)	Human placenta mesenchymal stem cells (hPMSCs)	– Autoimmune, induced by ZP3 peptide which was synthesized by an automatic peptide synthesizer; peptide purity was determined by HPLC analysis; the amino acid composition was checked by amino acid analysis	Tail-vein injection
Yin et al. 2018 [48]	Mouse	Human placenta-derived mesenchymal stem cells (hPMSCs)	– Autoimmune, induced by ZP3 peptide treatment	Tail-vein injection
Huang et al. 2019 [49]	Mouse (ICR mice, 7 to 8 weeks old)	Human fetal liver MSCs (fMSCs) (first trimester)	– Cyclophosphamide	Tail-vein injection

(continued)

Table 2.1 (continued)

Reference	Animal	Type of human stem cells	Simulation of premature ovarian insufficiency	Stem cell application
Feng et al. 2020 [50]	Rat	Human amnion-derived mesenchymal stem cells (hAD-MSCs)	– Cyclophosphamide and busulfan	Tail-vein or ovary injection
Liu et al. 2019 [51]	Mouse	Human amniotic mesenchymal stem cells (hAMSCs)	– Hydrogen peroxide	Intraperitoneal injection
Ding et al. 2018 [52]	Mouse (C57BL/6 old mice of SPF class, 12–14 months old and young mice, 3–5 months old)	Human amniotic mesenchymal stem cells (hAMSCs)	– Natural aging	– Injection of hAMSCs into mice – Co-culture of human granulosa cells from patients with NOA and control (tubal occlusion) with hAMSCs
Ding et al. 2017 [53]	Mouse	Human amniotic mesenchymal stem cells (hAMSCs) and human amniotic epithelial cells (hAECs)	– Cyclophosphamide	Tail-vein injection of Dil-labeled cells (Screened with a live imaging system (Xenogen IVIS 100 in vivo Imaging System, PerkinElmer, USA) to characterize, quantify, and visualize Dil-labeled cells in organism)
Ling et al. 2019 [54]	Rat	Human amnion-derived mesenchymal stem cells (hAD-MSCs)	– Cyclophosphamide	Tail-vein injection (PKH26-labeled hAD-MSCs)
Ling et al. 2017 [55]	Rat (Sprague–Dawley rats, 10–12 weeks old)	Human amnion-derived mesenchymal stem cells (hAD-MSCs)	– Cyclophosphamide	Tail-vein injection

(continued)

Table 2.1 (continued)

Reference	Animal	Type of human stem cells	Simulation of premature ovarian insufficiency	Stem cell application
Lai et al. 2013 [56]	Mouse	Human amniotic fluid cells (hAFCs)	– Cyclophosphamide and busulfan	Injection into both ovaries
Liu et al. 2012 [57]	Mouse	$CD44^+/CD105^+$ human amniotic fluid cells (HuAFCs)	– Cyclophosphamide	Injection into the ovaries (red fluorescent protein (RFP)-transduced $CD44^+/CD105^+$ HuAFCs)
Li et al. 2018 [58]	Mouse (specific pathogen-free C57/BL6 mice, 8 weeks old)	Human chorionic plate-derived mesenchymal stem cells (CP-MSCs)	– Cyclophosphamide	Tail-vein injection
Feng et al. 2019 [59]	Mouse	Human menstrual blood-derived stromal cells (MenSCs)	– Cyclophosphamide	Tail-vein injection
Reig et al. 2019 [60]	Mouse (C57BL6 mice)	Endometrial mesenchymal stem cells (eMSCs) from menstrual blood	– Cyclophosphamide and busulfan	Retro-orbital injection
Noory et al. 2019 [61]	Rat (Wistar albino rats, 6–8 weeks old)	Human menstrual blood stem cells (HuMenSCs)	– Busulfan	Tail-vein injection
Manshadi et al. 2019 [62]	Rat (Wistar rats)	Human endometrial menstrual blood stem cells (HuMenSCs)	– Busulfan	Tail-vein injection
Lai et al. 2015 [63]	Mouse	Human endometrial mesenchymal stem cells from menstrual blood (EnSCs)	– Cyclophosphamide and busulfan	Tail-vein injection (GFP-labeled EnSCs measured by live imaging and immunofluorescent methods)
Liu et al. 2014 [64]	Mouse	Human endometrial stem cells (HuMenSCs)	– Cyclophosphamide	

(continued)

Table 2.1 (continued)

Reference	Animal	Type of human stem cells	Simulation of premature ovarian insufficiency	Stem cell application
Mohamed et al. 2018 [65]	Mouse (C57BL6 mice, 4 to 6 weeks old)	Human bone marrow mesenchymal stem cells (BMSCs)	– Cyclophosphamide	Intraovarian injection (laparotomy)
Herraiz et al. 2018 [66]	Mouse (immuno-deficient NOD/SCIDmice)	Human bone marrow-derived stem cells (hBMDSCs)	– Cyclophosphamide and busulfan	Tail-vein injection
Liu et al. 2014 [67]	Rat	Human bone marrow-derived mesenchymal stem cells (BMSCs)	– Cisplatin	Tail-vein injection
Ghadami et al. 2012 [68]	Mouse	Mouse bone marrow stem cells (BMSCs)	– The phenotype of FSHR (−/−) mouse, FORKO (follitropin receptor knockout); a suitable model to study ovarian failure in humans. Female FORKO mice have elevated FSH, decreased estrogen levels, are sterile because of the absence of folliculogenesis, and display thin uteri and small non-functional ovaries	Tail-vein injection

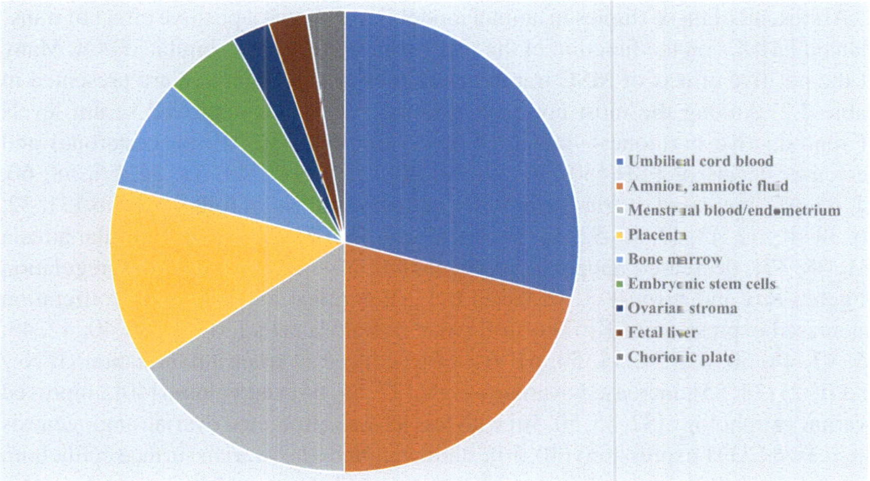

Fig. 2.1 Frequency of human mesenchymal stem cell (MSC) origins in ovarian regeneration studies in animal models of premature ovarian insufficiency (POI) summarized in this section. The most commonly used were human umbilical cord MSCs followed by amnion/amniotic fluid, menstrual blood (endometrium), placenta, bone marrow MSCs, and rarely MSCs from human embryonic stem cells, ovarian stroma, fetal liver, and chorionic plate

57, 65]. Rarely, human stem cells have been transplanted into animals on collagen scaffolds [40]. In some studies, hMSCs have been labeled with different markers such as 1,1′-dioctadecyl-3,3,3′,3′-tetramethylindocarbocyanine perchlorate ('DiI') [53], PKH26 [54], red fluorescent protein (RFP) [57], and green fluorescent protein (GFP) [60], and monitored in the body after transplantation by a live imaging, which proves to be an excellent approach to know more about cells after transplantation. GFP-labeled cells were undetectable 48 h after stem cell transplantation, but were later detected and localized to the ovarian stroma [63]. Stem cells were found to remain alive for a longer time after transplantation. Transplanted stem cells could survive within mouse ovaries for at least 14 days in vivo [64]. BrdUrd incorporation assay and immunofluorescent staining demonstrated that $CD44^{+}$/$CD105^{+}$ human amniotic fluid stem cells (HuAFCs) underwent normal cycles of cell proliferation and self-renewal in the ovarian tissues [57]. Distribution of the transplanted MSCs in different regions of the ovarian tissue was not equal; it was greater in the medulla than the cortex and germinal epithelium [41]. In some studies, human bone marrow stem cells (BMSCs) evenly repopulated the growing follicles [65], while in other studies no bone marrow stem cells were found in the ovarian follicles and corpus lutea after transplantation [67]. Similarly, amnion-derived MSCs were only located in the interstitium of ovaries, rather than in follicles after transplantation [54]. Positive effects in terms of increased number of antral follicles were observed approximately 1 month after MSC transplantation.

Almost all of these studies in animal models have shown a positive effect of transplanted hMSCs on the function of the ovaries of animals with simulated POI. Many of the positive effects of MSC transplantation on ovarian function are presented in Table 2.2. Among the most common positive effects are improved serum levels of reproductive hormones—increased levels of AMH, E2, P (progesterone) and decreased levels of FSH [30, 36, 37, 39, 40, 42, 43, 47, 49, 51, 52, 56, 58, 60, 64, 65, 67], increased ovarian primordial, primary, and antral follicle counts [31, 32, 36, 38–40, 42, 43, 47, 49–52, 56, 59, 60, 62, 64, 65, 67], decreased follicular atresia [31, 38, 47], decreased apoptosis (upregulation of gene *Bcl-2* and downregulation of genes *Bax* and *caspase-3,* increased Bcl-2/Bax ratio) and increased proliferation (increased expression of Ki67) of follicular granulosa cells [30, 36, 39, 40, 42, 43, 45, 47, 48, 50, 52, 54, 55, 59, 61, 64, 66], reduced ovarian inflammation (IFN-γ and IL-2) [38, 55], increased ovarian weight [37, 55, 64] and volume [40], improved ovarian morphology [37, 45, 50, 56] with less lesions, promoted ovarian angiogenesis (increased CD31 expression) [40, 50], stabilization of the ovarian surface epithelium morphostasis [42], and reduced oxidative stress [44], after hMSC transplantation compared to control. Positive effects in terms of increased number of antral follicles were observed approximately 30 days after MSC transplantation [67]. Not only ovarian function, endometrial receptivity was also improved after MSC transplantation [38]. As a result, the normal estrus cycle [35–37, 51, 59], ovulation [31, 58, 62], and fertility were re-established in animals, leading to reduced oocyte zona pellucida remnants [31], higher number of mature (MII) oocytes [65, 66], recovered blastocyst formation rates [31] from ovulated oocytes, more pregnancies [65, 66], and births of healthy pups after mating [30, 31, 51, 60, 65, 66]. The normal offspring mice were also fertile [51].

None of the oocytes or pups incorporated green fluorescent protein (GFP), suggesting that there was no contribution of transplanted stem cells to the oocyte pool [60]. Immunostaining with anti-human antigen-specific antibodies demonstrated that grafted human amniotic fluid cells (hAFCs) survived and differentiated into granulosa cells which directed the oocyte maturation in the mouse model [56]. Stem cell transplantation reduced depletion of the germline stem cell (GSCs) pool induced by chemotherapy [63]. It could be generally accepted that transplanted MSCs can differentiate into granulosa cells and thus support the development and maturation of existing oocytes or their development from germinal stem cells. The improvement in ovarian function after MSC transplantation is attributed to various molecular mechanisms such as inhibition of autophagy of theca-interstitial cells via the AMPK/mTOR signaling pathway [33]; activation of the PI3K pathway [34] and ECM-dependent FAK/AKT signaling pathways [59]; improved ovarian metabolome [34]; paracrine activity by releasing different growth factors such as nerve growth factor (NGF), nerve growth factor receptor (TrkA), epidermal growth factor (EGF), hepatocyte growth factor (HGF), fibroblast growth factor 2 (FGF2), insulin-like growth factor 1 (IGF-1), and vascular endothelial growth factor (VEGF) secreted by MSCs [39, 52–54]; regulation of Treg cells and production of associated cytokines [48]; recovered TGF-β and IFN-γ levels in serum [42, 48] and telomerase activity; expression level of

Table 2.2 Positive effects of human mesenchymal stem cell (hMSC) transplantation on ovarian function in animal models

Reference	Animal	Type of human mesenchymal stem cells	Positive effects of mesenchymal stem cell transplantion on reproductive/ovarian function
Bahrehbar et al. 2010 [30]	Mouse	Human embryonic stem cell-derived mesenchymal stem cells (ES-MSCs)	– Restoration of hormonal secretion and reproductive function – Reduced apoptosis of follicular cells – New offspring
Yoon et al. 2020 [31]	Mouse	Human embryonic stem cell-derived mesenchymal stem cells (ES-MSCs)	– Improved mean primary and primordial follicle counts – Reduced oocyte zona pellucida remnants and apoptotic signs in ovarian follicles – Recovered ovulation – Recovered blastocyst formation rates from ovulated oocytes – Recovered birth rates
Besikcioglu et al. 2019 [32]	Rat	Human ovarian stroma stem cells (OSSCs) and bone marrow mesenchymal stem cells (BMMSCs) from fetal tissue (day 16)	– OSSCs have more protective effect on follicle maturation and primordial follicle counts than BMMSCs
Lu et al. 2020 [33]	Rat	Human umbilical cord-derived mesenchymal stem cells (hUMSCs)	– Protection of ovarian function by regulating autophagy signaling pathway AMPK/mTOR – Restoration of ovarian function by inhibiting autophagy of theca-interstitial cells via the AMPK/mTOR signaling pathway
Cui et al. 2020 [34]	Rat	Human umbilical cord mesenchymal stem cells (hUMSCs)	– TGF-β_1/Smad3 signaling pathway involved in the inhibition of ovarian tissue fibrosis, which contributed to the restoration of ovarian function following transplantation

(continued)

Table 2.2 (continued)

Reference	Animal	Type of human mesenchymal stem cells	Positive effects of mesenchymal stem cell transplantion on reproductive/ovarian function
Zhao et al. 2020 [35]	Mouse	Human umbilical cord MSCs (hUCMSCs)	– Restoration of ovarian function through activating the PI3K pathway and improved ovarian metabolome – Significant improvements in body weight, sex hormone levels, and estrous cycle – Improved reproductive capacity, offspring
Wang et al. 2020 [36]	Rat	Human umbilical cord MSCs (hUCMSCs)	– Estrous cycle returned to normal – Improved follicular development – Increased serum concentrations of 17-estradiol (E2), progesterone (P), and anti-Müllerian hormone (AMH) – Reduced apoptosis and promoted proliferation of granulosa cells; upregulated expression of genes *Bcl-2*, *AMH*, and *FSHR* and downregulated expression of *caspase-3*
Shen et al. 2020 [37]	Mouse	Human umbilical cord mesenchymal stem cells (hUCMSCs)	– Resumption of normal estrous cycle – Heavier ovarian weight – Increased E2 and decreased FSH – Improved ovarian lesions
Lu et al. 2019 [38]	Mouse	Human umbilical cord blood-derived stem cells (hUMSCs)	– Increased serum levels of E2, P, and IL-4 – Decreased serum levels of FSH, IFN-γ, and IL-2 – Increased total number of healthy follicles and decreased number of atretic follicles – Recovered endometrial receptivity regulated by the balance of Th1/Th2 cytokines and expression of uNK cells in the endometrium

(continued)

Table 2.2 (continued)

Reference	Animal	Type of human mesenchymal stem cells	Positive effects of mesenchymal stem cell transplantion on reproductive/ovarian function
Zheng et al. 2019 [39]	Rat	Human umbilical cord mesenchymal stem cells (UCMSCs)	– Recovered hormone secretion – Recovered folliculogenesis – Increased levels of nerve growth factor (NGF) and nerve growth factor receptor (TrkA) – Decreased levels of FSHR and caspase-3 – Improved pregnancy rate in natural conditions
Yang et al. 2019 [40]	Mouse	Human umbilical cord-derived mesenchymal stem cells on a collagen scaffold (collagen/UC-MSCs)	– Improved ovarian function – Increased E2 and AMH levels – Increased ovarian volume – Increased number of antral follicles – Promoted granulosa cell proliferation (increased Ki67 expression) – Promoted ovarian angiogenesis (increased CD31 expression)
Jalalie et al. 2019 [41]	Mouse	CM-DiI-labeled human umbilical cord vein MSCs (hUCV-MSCs)	– Distribution of the transplanted MSCs in different regions of the ovarian tissue is not equal; it is greater in the medulla than the cortex and germinal epithelium
Elfayomy et al. 2016 [42]	Rat	Human umbilical cord blood mesenchymal stem cells (HCB-MSCs)	– Reduction in FSH and elevation in E2 levels – Stabilization of the surface epithelium morphostasis – Increased antral follicle counts – Increased expressions of CK 8/18, TGF ß, and PCNA – Decreased expression of CASP-3

(continued)

Table 2.2 (continued)

Reference	Animal	Type of human mesenchymal stem cells	Positive effects of mesenchymal stem cell transplantion on reproductive/ovarian function
Song et al. 2016 [43]	Rat	Human umbilical cord mesenchymal stem cells (UCMSCs)	– Recovery of disturbed hormone secretion – Recovery of folliculogenesis – Reduced ovarian cell apoptosis – Transplanted cells resided in ovarian tissues and survive for a long time without obvious proliferation
Seok et al. 2020 [44]	Rat	Human placenta-derived MSCs (PD-MSCs)	– Antioxidant activity – Reduced oxidative stress
Li et al. 2019 [45]	Mouse	Human placenta-derived mesenchymal stem cells (hPMSCs)	– Improved structure and function of the ovaries – Suppressed granulosa cell apoptosis induced by endoplasmatic reticulum stress IRE1α signaling pathway
Yin et al. 2018 [46]	Mouse	Human placenta-derived mesenchymal stem cells (hPMSCs)	– PI3K/Akt signaling pathway is involved in the recovery of ovarian function by changing the ratios of Th17/Tc17 and Th17/Treg cells
Zhang et al. 2018 [47]	Mouse	Human placenta mesenchymal stem cells (hPMSCs)	– Improved serum levels of AMH and FSHR in ovaries – Promoted follicular development – Inhibited excessive follicular atresia – Inhibited granulosa cell apoptosis – Improved ovarian reserve capacity

(continued)

Table 2.2 (continued)

Reference	Animal	Type of human mesenchymal stem cells	Positive effects of mesenchymal stem cell transplantion on reproductive/ovarian function
Yin et al. 2018 [48]	Mouse	Human placenta-derived mesenchymal stem cells (hPMSCs)	– Improved ovarian function – Decreased apoptosis of granulosa cells – Increased population of Treg cells which was inhibited by pZP3 treatment – Decreased TGF-β and increased IFN-γ in serum caused by pZP3 treatment have been reversed – Recovery of ovarian function mediated via the regulation of Treg cells and production of associated cytokines following transplantation
Huang et al. 2019 [49]	Mouse	Human fetal liver MSCs (fMSCs) (first trimester)	– Prevention of follicle loss Recovered serum E2, AMH, and FSH levels
Feng et al. 2020 [50]	Rat	Human amnion-derived mesenchymal stem cells (hAD-MSCs)	– Both hAD-MSC (tail vein) and hAD-MSC (ovary in situ) transplantations can repair ovarian injury and improve ovarian function – Improved AMH levels and follicle numbers – Paracrine proteome of hAD-MSCs in the ovarian microenvironment can protect against chemotherapy-induced damage by reducing apoptosis and promoting angiogenesis, cell proliferation, and gene expression

(continued)

Table 2.2 (continued)

Reference	Animal	Type of human mesenchymal stem cells	Positive effects of mesenchymal stem cell transplantion on reproductive/ovarian function
Liu et al. 2019 [51]	Mouse	Human amniotic mesenchymal stem cells (hAMSCs)	– Restoration of the estrous cycle – Increased estrogen levels and decreased FSH levels – Increased ovarian index – Increased fertility rate – Increased population of follicles at different stages – The newborn mice had no deformity and showed normal growth and development – The normal offspring mice were also fertile
Ding et al. 2018 [52]	Mouse	Human amniotic mesenchymal stem cells (hAMSCs)	– Improved mouse ovarian function – Improved follicle numbers – Rescued levels of E2 (95%) and AMH (92%) to the normal levels – Promoted proliferation marker expression of ovarian granulosa cells – Secretion level of epidermal growth factor (EGF) and hepatocyte growth factor (HGF) from hAMSCs (in a supernatant after centrifugation) was higher than other growth factors – Growth factor combination (HGF with EGF) improved the proliferation rate and inhibited the apoptosis rate more powerfully after a co-culture with hGCs – Total follicle numbers and hormone levels were elevated to a normal level after the growth factor combination was injected into the ovaries

(continued)

Table 2.2 (continued)

Reference	Animal	Type of human mesenchymal stem cells	Positive effects of mesenchymal stem cell transplantion on reproductive/ovarian function
Ding et al. 2017 [53]	Mouse	Human amniotic mesenchymal stem cells (hAMSCs) and human amniotic epithelial cells (hAECs)	– hAMSCs are a more effective cell type to improve ovarian function than hAECs attributable to cellular biological characteristics such as telomerase activity, expression level of pluripotent markers, cytokine and collagen secretion that are superior to hAECs, except for immunological rejection
Ling et al. 2019 [54]	Rat	Human amnion-derived mesenchymal stem cells (hAD-MSCs)	– PKH26-labeled hAD-MSCs mainly homed to ovaries after transplantation; they were only located in the interstitium of ovaries, rather than in follicles – Reduced ovarian injury and improved ovarian function at least partly through a paracrine mechanism; the presence of a paracrine mechanism accounting for hAD-MSC-mediated recovery of ovarian function attributed to the growth factors secreted by hAD-MSCs – hAD-MSCs secreted FGF2, IGF-1, HGF, and VEGF, and those growth factors were detected in the hAD-MSC-CM – hAD-MSC-CM injection improved the local microenvironment of ovaries, leading to a decrease in Bax expression and an increase in Bcl-2 and endogenous VEGF expression in ovarian cells, which inhibited chemotherapy-induced granulosa cell apoptosis, promoted angiogenesis and regulated follicular development, thus partly reducing ovarian injury and improving ovarian function

(continued)

Table 2.2 (continued)

Reference	Animal	Type of human mesenchymal stem cells	Positive effects of mesenchymal stem cell transplantion on reproductive/ovarian function
Ling et al. 2017 [55]	Rat	Human amnion-derived mesenchymal stem cells (hAD-MSCs)	– Increased body and reproductive organ weights – Improved ovarian function – Reduced reproductive organ injuries – Increased Bcl-2/Bax ratio – Reduced granulosa cell apoptosis and ovarian inflammation
Lai et al. 2013 [56]	Mouse	Human amniotic fluid cells (hAFCs)	– Restored ovarian morphology – Restored ovaries displayed many follicle-enclosed oocytes at all stages of development – Identification of GFP-labeled cells – Immunostaining with anti–human antigen-specific antibodies demonstrated that grafted hAFCs survived and differentiated into granulosa cells which directed the oocyte maturation – High anti-Müllerian hormone expression after transplantation
Liu et al. 2012 [57]	Mouse	CD44 +/CD105 + human amniotic fluid cells (HuAFCs)	– Stem cells detected by fluorescence microscopy up to three weeks after injection – BrdUrd incorporation assay and immunofluorescent staining demonstrated that $CD44^{+}/CD105^{+}$ HuAFCs underwent normal cycles of cell proliferation and self-renewal in the ovarian tissues
Li et al. 2018 [58]	Mouse	Human chorionic plate-derived mesenchymal stem cells (CP-MSCs)	– Restoration of serum hormone levels and ovarian function – Superovulation of mice

(continued)

Table 2.2 (continued)

Reference	Animal	Type of human mesenchymal stem cells	Positive effects of mesenchymal stem cell transplantion on reproductive/ovarian function
Feng et al. 2019 [59]	Mouse	Human menstrual blood-derived stromal cells (MenSCs)	– Restoration of ovarian function by regulating normal follicle development and estrous cycle – Reduced apoptosis in ovaries to maintain homeostasis of microenvironment and modulating serum sex hormones to a relatively normal status – Activation of ovarian transcriptional expression in ECM-dependent FAK/AKT signaling pathway and thus restored ovarian function to a certain extent
Reig et al. 2019 [60]	Mouse	Endometrial mesenchymal stem cells (eMSCs) from menstrual blood	– Significant increase in oocyte production and serum anti-Müllerian hormone concentrations after 6 weeks, as well as a 19% higher body mass in UC (uterine cell)-treated mice Increased number of pups in mice treated with UCs – None of the oocytes or pups incorporated GFP, suggesting that there was no contribution of these stem cells to the oocyte pool
Noory et al. 2019 [61]	Rat	Human menstrual blood stem cells (HuMenSCs)	– Differentiation into granulosa cells – Decreased TUNEL-positive (apoptotic) cells – Decreased expression level of *Bax* gene
Manshadi et al. 2019 [62]	Rat	Human endometrial menstrual blood stem cells (HuMenSCs)	– Positive effect on follicle formation – Positive effect on ovulation

(continued)

Table 2.2 (continued)

Reference	Animal	Type of human mesenchymal stem cells	Positive effects of mesenchymal stem cell transplantion on reproductive/ovarian function
Lai et al. 2015 [63]	Mouse	Human endometrial mesenchymal stem cells from menstrual blood (EnSCs)	– GFP-labeled cells were undetectable 48 h after cell transplantation, but were later detected and localized to the ovarian stroma – 5' bromodeoxyuridine (BrdU) and mouse VASA homologue (MVH) protein double-positive cells were immunohistochemically detected in mouse ovaries – Stem cell transplantation reduced depletion of the germline stem cell (GSCs) pool induced by chemotherapy
Liu eta al. 2014 [64]	Mouse	Human endometrial stem cells (HuMenSCs)	– Transplanted stem cells could survive within mouse ovaries for at least 14 days in vivo – Higher expression of ovarian markers [AMH, inhibin α/β, and follicle-stimulating hormone receptor (FSHR), and the proliferative marker Ki67 expression – Increased ovarian weight – Increased plasma E2 level – Increased number of normal follicles – Gene (mRNA) expression patterns in the ovarian cells following stimulation of the host ovarian niche became increasingly similar to those observed in human ovarian tissue compared with the pretransplantation expression

(continued)

Table 2.2 (continued)

Reference	Animal	Type of human mesenchymal stem cells	Positive effects of mesenchymal stem cell transplantion on reproductive/ovarian function
Mohamed et al. 2018 [65]	Mouse	Human bone marrow mesenchymal stem cells (BMSCs)	– Higher mean follicle count – Lower FSH levels and higher AMH serum levels – Higher expression of AMH, FSH receptor, inhibin A, and inhibin B in growing follicles – Human BMSCs evenly repopulated the growing follicles – Breeding data showed significant increases in the pregnancies numbers
Herraiz et al. 2018 [66]	Mouse	Human bone marrow-derived stem cells (hBMDSCs)	– Fertility rescue and spontaneous pregnancies in mice with mimicked ovarian insufficiency – Higher numbers of preovulatory follicles, metaphase II oocytes, 2-cell embryos, and healthy pups – Promoted ovarian vascularization and cell proliferation – Reduced apoptosis of follicular cells
Liu et al. 2014 [67]	Rat	Human bone marrow-derived mesenchymal stem cells (BMSCs)	– The number of BMSCs in the ovarian hilum and medulla was greater than in the cortex, but no BMSCs were found in the follicles and corpus lutea – Antral follicle count and E2 levels increased after 30 days – Intravenously delivered BMSCs can home to the ovaries and restore its structure and function
Ghadami et al. 2012 [68]	Mouse	Mouse bone marrow stem cells (BMSCs)	– Increase in total body and reproductive organ weight – Increase in both the maturation and the total number of the follicles – FSH dropped to 40–50% and estrogen increased 4–5.5 times in the serum of treated animals – FSHR mRNA was detected in the ovaries of treated animals

pluripotent markers and cytokine, cytokeratin (CK18/8), collagen, and PCNA secretion [42, 53]. It has been found that TGF-β_1/Smad3 signaling pathway is involved in the inhibition of ovarian tissue fibrosis, which contributed to the restoration of ovarian function following transplantation [34]. PI3K/Akt signaling pathway is involved in the recovery of ovarian function by changing the ratios of Th17/Tc17 and Th17/Treg cells [46]. Suppressed granulosa cell apoptosis was induced by endoplasmatic reticulum stress IRE1α signaling pathway [45]. And not at the end, recovered endometrial receptivity was regulated by the balance of Th1/Th2 cytokines and expression of uNK cells in the endometrium [38].

Extracellular Vesicles of Human Mesenchymal Stem Cells

Recent studies also demonstrate the positive effect of extracellular vesicles of hMSCs on the ovaries. The protective effects of human umbilical cord MSC-derived (HUCMSC) extracellular vesicles on cisplatin-damaged rat (3-week female Sprague–Dawley rats) granulosa cells has been demonstrated [69]. The proportion of surviving cells was higher and the proportion of apoptotic cells lower in cisplatin-damaged granulosa cells treated with extracellular vesicles of HUCMSCs than in untreated controls. HUCMSCs-derived extracellular vesicles could become incorporated to cisplatin-damaged granulosa cells thus playing important roles in promoting resistance to cisplatin-induced granulosa cell apoptosis and restoring synthesis and secretion of steroid hormone in granulosa cells. This might provide a theoretical and experimental basis for use of MSC-derived extracellular vesicles instead of MSCs as a safer and cell-free therapeutic strategy for the patients with POI induced by chemotherapeutic agents. Similarly, it has been evidenced that expression levels of total AKT, p-AKT, and angiogenic cytokines (including VEGF, IGF, and angiogenin) in the ovaries of POI mice were markedly upregulated after HUCMSC-derived microvesicle transplantation, suggesting that transplantation of HUCMSC-derived microvesicles might recover ovarian function by inducing angiogenesis via the PI3K/AKT signaling pathway [70].

Exosomes Derived from Mesenchymal Stem Cells

Exosomes are a class of membranous vesicles with diameters of 30–200 nm that are constitutively released by eukaryotic cells and mediate local cell-to-cell communication by transferring microRNAs and proteins. One study [71] investigated whether human adipose mesenchymal stem cell-derived exosomes (hADSC-Exos) retain the ability to restore ovarian function and how hADSC-exosomes are involved in this process. Exosomes derived from human adipose stem cells were found to improve ovarian function after transplantation into mouse ovaries with simulated premature insufficiency by acting on the SMAD signaling pathway and reduced expression of genes associated with apoptosis [71]. Similarly, exosomes derived from human umbilical cord mesenchymal stem cells (huMSCs) after centrifugation (supernatant) were effectively incorporated by cisplatin-pretreated rat granulosa cells and increased the number of living cells [72] in vitro. The expression of Bcl-2 and caspase-3 was upregulated, whilst the expression of Bax, caspase-3, and PARP was downregulated to protect granulosa cells. These results suggest that hMSCs exosomes can be used to prevent and treat chemotherapy-induced ovarian granulosa cell apoptosis but further research is needed.

Secretome Derived from Mesenchymal Stem Cells

Human primary ovarian endothelial cells (HOVECs) were treated with secretome of human bone marrow mesenchymal stem cells (BM-MSCs) to examine them for the expression of angiogenesis markers (e.g., Endoglin, Tie-2, and VEGF) using FACS and formation of blood vessels by 3D Matrigel tubulogenesis assay [73]. It was found that expression of proliferation marker Ki67 was significantly increased in HOVEC cells that were treated with MSC secretome. MSCs secretome treatment of HOVECs also increased the expression of several angiogenic markers VEGFR2, Tie2/Tek, VE-Cadherin, Endoglin, and VEGF in cells in comparison to control. Moreover, the MSC-derived secretome significantly increased the number of branching points in cells, when tubulogenesis assay was performed. It was proposed that MSC-derived secretome contains some bioactive factors that can increase the ovarian angiogenesis. Better characterization of these factors can enable novel therapeutic possibilities for women with POI and other causes of ovarian infertility.

Growth Factors from Mesenchymal Stem Cells

Human amniotic mesenchymal stem cells (hAMSCs) increased the proliferation rate and reduced apoptosis of granulosa cells from patients with natural ovarian aging (NOA; aged >40 years, an antral follicle count <5 or AMH <1.1 ng/ml and follicle-stimulating hormone (FSH) $\geq$10 mIU/ml), when co-cultured in vitro [52]. After centrifugation of hAMSCs and analysis of the supernatant, it was found that they secrete more HGF and EGF than other growth factors. In the next step, the granulosa cells were treated with these two growth factors and it was found that they also increase proliferation and decrease the apoptosis of the granulosa cells in vitro [52]. Moreover, growth factors HGF and EGF increased follicle count and improved hormonal status (increased E2 and AMH and decreased FSH) after injection into mouse ovaries with simulated POI [52]. It was concluded that hAMSCs improve ovarian function in natural aging through secreting HGF and EGF.

Factors Influencing the Success of Mesenchymal Stem Cell Transplantation

Transplantation of low-intensity pulsed ultrasound (LIPUS) pretreated hAD-MSCs was more effective in improving inflammation, microenvironment, and granulosa cell (GC) apoptosis in rat POI ovaries after chemotherapy than transplantation of stem cells without this pre-treatment [55].

Regeneration of Ovaries by Transplantation of Mesenchymal Stem Cells in Humans

Good results in animal models, however, encouraged the researchers to perform the first MSC transplantations into the ovaries of patients with POI (Table 2.3).

In Vitro Studies: Effects of Mesenchymal Stem Cells on Granulosa Cells

The positive effect of MSCs on the ovaries has been demonstrated both in vitro and in vivo. In one study [49], granulosa cells were isolated from the follicular fluid of women with POI and were handled in the following ways: the first group were control granulosa cells without any treatment, the second group were granulosa cells transfected with MT1 siRNA (hGCs-MT1KD), and the third group were granulosa c-MT1KD cells treated with human fetal mesenchymal stem cells (fMSCs). It was found that fMSCs significantly decreased oxidative damage, increased oxidative protection, improved antiapoptotic effects, and inhibited apoptotic genes in granulosa cells. With melatonin receptor 1 (MT1) knockdown or antagonist treatment of granulosa cells, the expression of proteins JNK1, PCNA, and AMPK and the percentage of granulosa cell proliferation were impaired thus indicating that MT1 may acts as a regulator for treating POI disease. It was concluded that fetal mesenchymal stem cells (fMSCs) might play a crucial role in stimulating the activity of granulosa cells of patients with POI by targeting MT1 [49]. In the study of Yan et al. [74], the effect of MSCs isolated from human menstrual blood (MB-MSCs) on epirubicin-induced damage to human ovarian granulosa cells and its potential mechanisms have been explored. Epirubicin inhibited the secretion of E2, P, AMH, inhibin A, and inhibin B and impaired the proliferation of granulosa cells. Human MB-MSCs repaired epirubicin-induced damage to granulosa cells which might be related to the inhibition of Gadd45b protein expression in granulosa cells. Also, hAMSCs revealed powerful ability to improve the proliferation rate of ovarian granulosa cells (hGCs) from patients with POI when co-cultured. This effect was bigger than after co-culture with human amniotic epithelial cells [53]. Moreover, co-culture of peripheral blood

Table 2.3 Regeneration of ovaries by stem cell transplantation in humans (patients with premature ovarian insufficiency)

Reference	Patients	Type of stem cells	Intervention	Positive effects
Edessy et al. 2016 [76] The first baby of idiopathic premature ovarian insufficiency treatment by autologous bone marrow stem cells	Premature ovarian insufficiency ($n = 10$) (Post-menarche females less than 40 years old, with FSH more than or equal to 20 IU/L, normal karyotyping; ovarian and breast cancer patients were excluded)	Human bone marrow stem cells from iliac crest	Injection into the ovaries through laparoscopy	– Two (20%) patients resumed menstruation after 3 months – One of them (10%) got pregnant after 11 months and delivered a healthy full-term baby – The two menstruating patients showed focal secretory changes of endometrium after being atrophic
Gabr et al. 2016 [77]	Premature ovarian insufficiency ($n = 30$) (aged 18–40 years, high gonadotropin levels, normal karyotypes Exclusion Criteria: patients with secondary ovarian failure (e.g., hypothalamic causes), autoimmune diseases, major medical problems such as malignancy, hepatitis, etc., and abnormal karyotyping (e.g., Turner syndrome))	Human bone marrow was aspirated from the posterior iliac spine and MSCs were isolated under good manufacturing practice (GMP) conditions	Stem cells were injected in one ovary: 3–5 million of stem cells in the ovarian tissue through laparoscope and 3–5 million of stem cells in the ovarian artery through catheter	– Twenty-six out of the 30 patients (86.7%) showed decreased FSH levels and increased estrogen and AMH levels after 4 weeks of injection; this change was maintained throughout the 48-week follow-up period – Eighteen (60%) patients showed ovulation with follicle sizes ranging from 12–20 mm – One patient had spontaneous pregnancy, while three patients were subjected to IVF cycles

(continued)

Table 2.3 (continued)

Reference	Patients	Type of stem cells	Intervention	Positive effects
Herraiz et al. 2018 [78, 79]	Poor responders to hormonal stimulation of the ovaries with very poor prognosis ($n = 17$)	Autologous bone marrow MSCs (hBMMSCs)	Transplantation into the ovaries (infusion)	– Improvement in antral follicle counts (AFCs) 2 weeks after treatment – Ovarian function improved in 81.3% of patients – More oocytes after ovarian stimulation – Five pregnancies: two after in vitro fertilization and three by natural conception
Ding et al. 2018 [80]	Premature ovarian failure ($n = 14$) (women aged between 18 and 39 years, amenorrhea of >1 year duration, with FSH ≥40 mIU mL on two occasions 4–6 weeks apart, and with a normal male partner sperm analysis on three different occasions)	Human umbilical cord MSCs on a collagen scaffold (collagen/UC-MSCs)	Transplantation into the ovaries (Injection into the ovary of patients under transvaginal ultrasonographic (TVUS)-guidance	– Activation of primordial follicles in vitro via phosphorylation of FOXO3a and FOXO1 – Rescued overall ovarian function – Elevated estradiol concentrations – Improved follicular development and increased number of antral follicles – Mature oocytes in three patients – Successful clinical pregnancy in two patients (one on-going pregnancy)

(continued)

Table 2.3 (continued)

Reference	Patients	Type of stem cells	Intervention	Positive effects
Yan et al. 2020 [81]	Premature ovarian insufficiency ($n = 61$)	Human umbilical cord-derived MSCs (hUCMSCs)	Vaginal ultrasound-guided orthotopic injection of stem cells into the ovaries	– Improved follicular development – Oocytes in 15 patients, in vitro fertilization and embryo transfer – Four deliveries of healthy babies – No adverse events
Igboeli et al., 2020 [82]	Premature ovarian insufficiency = premature ovarian failure ($n = 2$)	Autologous bone marrow MSCs (hBMMSCs)	Injection of stem cells into the right ovary	– Increased volume of the ovaries – Increased serum levels of estrogen – Episodes of menses – Improvement of menopausal symptoms
Pellicer De Castellvi et al. 2020 [83]	Premature ovarian insufficiency ($n = 6$)	Autologous bone marrow MSCs (hBMMSCs)	Injection of stem cells into the ovaries	– Increased antral follicle counts in four patients – Ongoing pregnancy in one patient

mononuclear cells (PBMCs) from patients with POI with human adipose-derived mesenchymal stem cells (hADSCs) combined with estrogen showed that this treatment played an immunomodulatory role by promoting regulatory T cells (Tregs) differentiation and proliferation, thereby potentially improving impaired ovarian function in these patients [75].

Clinical Studies: Transplantation of Mesenchymal Stem Cells into the Ovaries of Infertile Women

The principle of ovarian regeneration in patients was intraovarian application of MSCs, mostly laparoscopic injection (Fig. 2.2). As can be seen in Table 2.3, the first child after autologous bone marrow MSC transplantation into the ovaries of a woman with POI was born in y. 2016 [76]. In a study of Edessy et al., the autologous bone marrow MSCs (from iliac crest) were laparoscopically injected into the ovaries of ten women with POI. After that, two women menstruated normally and one of them got pregnant and gave birth to a full-term, healthy baby. This was followed by a study conducted by Gabr et al., where autologous MSCs from bone marrow were injected into the ovaries of 30 patients with POI [77]. After transplantation, 87% of patients had decreased levels of FSH and increased levels of AMH in their serum, and 60% patients showed the restoration of ovulation; one of these patients got pregnant and on-going pregnancy was reported.

To better understand the effect of hMSCs, female immunodeficient SCID/NOD mice were ovariectomized and xenografted with human ovarian cortex from poor-responder patients which was pre-treated with bone marrow-derived mesenchymal

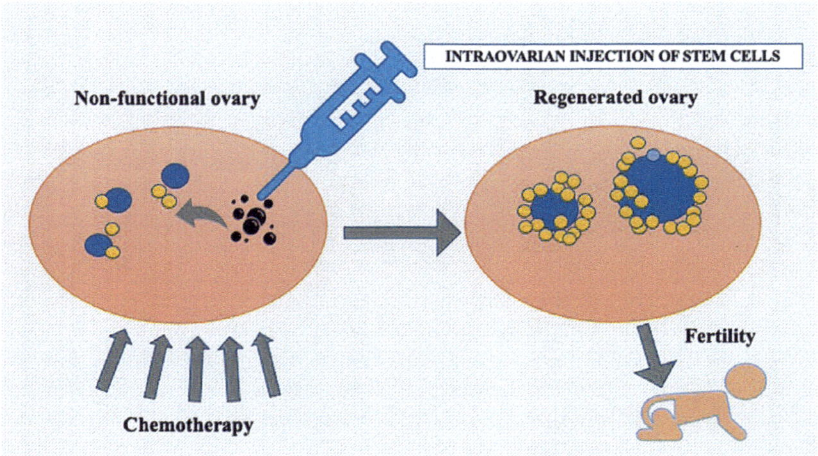

Fig. 2.2 The principle of intraovarian application of autologous hMSCs (e.g., from bone marrow) to the ovaries of patients in clinical studies

stem cells (BMDSCs). In xenografted human ovarian tissues from poor responders, infusion of BMDSCs and their $CD133^+$ fraction into mice led to their engraftment close to follicles, resulting in promotion of follicular growth, increases in E2 secretion, and enhanced local vascularization in human ovarian tissue [66]. In the next step, ovarian transplantation of autologous bone marrow MSCs has been performed in 17 patients included in the IVF program who were poor responders with very poor prognosis in terms of oocyte retrieval and pregnancy [78]. Transplantation of bone marrow MSCs resulted in a significant improvement in antral follicle counts (AFCs) in these patients 2 weeks after treatment. With an increase in AFCs of three or more follicles and/or consecutive increases in AMH levels as success criteria, ovarian function was improved in 81.3% of patients. These positive effects of stem cell transplantation may be associated with the presence of fibroblast growth factor-2 (FGF2) and thrombospondin. During controlled ovarian stimulation, stem cell transplantation increased the number of stimulable antral follicles and oocytes, but the euploidy rate in the IVF-derived embryos was low (16.1%). After stem cell transplantation, five pregnancies were achieved: two after embryo transfer after IVF and three by natural conception. It was concluded that this experimental procedure is effective and safe, and can be used in clinical practice. In the next step, the same research group proposed ovarian bone marrow stem cell transplantation also in patients with POI [79].

In a study of Ding et al. [80], human umbilical cord-derived MSCs on a collagen scaffold were transplanted under transvaginal ultrasonographic-guidance into the ovaries of 14 patients with POI. In these patients, ovarian function was improved after transplantation and elevated estradiol concentrations were observed in addition to an increased number of antral follicles. In three patients, mature oocytes were developed and two pregnancies were achieved (one on-going pregnancy).

Recently, 61 patients diagnosed with POI according to European Society for Human Reproduction and Embryology-ESHRE (oligo/amenorrhea for at least 4 months, an elevated FSH level >25 IU/l on two occasions more than 4 weeks apart) were treated by umbilical cord-derived MSCs (UCMSCs) in a study of Yan et al. [81]. UCMSCs were isolated from pieces (app. 1 mm in size) of newborn umbilical cords (four healthy and full-term human placental samples) and cultured according to good manufacturing practice (GMP) standards in α-MEM culture medium supplemented with 5% KOSR, 5 ng/ml bFGF, and 1 X NEAA; cell cultures were kept in a CO_2-incubator at 37 °C and 5% CO_2 until 80% confluence and then passaged. UCMSCs highly expressed typical MSC markers such as CD73, CD90, CD105; showed low expression of endothelial and hematopoietic markers (CD34, CD45, CD14, CD19); and did not express the MHC Class II molecule HLA-DR; therefore, they were suitable for transplantation according to criteria proposed by the International Society for Cellular Therapies (ISCT) [81]. Stem cells at passage 5 were transplanted into the patients' ovaries by orthotopic injection under the guidance of vaginal ultrasound. Each ovary was injected at three points (35 μl of stem cells per point). All patients received the standard hormone replacement regimen of estradiol during stem cell treatment and up to three stem cell injections: 61 patients one injection, 50 patients two injections, and 30 patients three injections. MSC injections contained 0.5×10^7

stem cells in 100 μl of saline with 5% of AB plasma for unilateral ovary or 1×10^7 stem cells for bilateral ovaries. All patients were followed up to 6 months after stem cell transplantation. The side effects, vital signs, and changes in clinical and collected hematological and imaging parameters were monitored during the follow-ups. After injection of stem cells, all patients showed normal clinical behavior without serious side effects or complications related to the treatment. Transplantation of UCMSCs rescued the ovarian function of patients with POI, as indicated by increased follicular development and improved oocytes collection. Moreover, patients who experienced shorter amenorrhea durations (<1 year) obtained mature follicles more easily after stem cell therapy, and patients with better ovarian conditions including pre-operative antral follicles were more likely to derive the better outcomes by UCMSC injection. Four successful clinical deliveries were achieved in these patients after UCMSC transplantation, and all of these babies are normally developed and healthy [81]. In another study, Igboeli and colleagues reported two patients with POI who resumed ovarian estrogen production and menses 7 months after injection of autologous bone marrow-derived MSCs into their ovaries [82]. The bone marrow-derived MSCs (BM-DMSCs) were retrieved from the iliac crest of each patient. Bone marrow-derived stem cells were concentrated and laparoscopically injected into the right ovary of each patient. Intraovarian stem cell injection resulted in an approximately 50% increase in the volume of treated ovaries compared to the contralateral (left) ovaries that represented the control, which was maintained until the end of the 1-year study. Serum E2 levels were increased by 150% in comparison to pre-injection levels. Both patients with POI experienced a menses episode and reported significant improvement in menopausal symptoms, which persisted until the end of the study. Both patients tolerated intraovarian stem cell injection very well, which proved to be safe and without adverse effects [82]. At the virtual ESHRE Annual Meeting 2020, the first results of autologous bone marrow MSC transplantation into one ovary were presented by the same research group in six patients with POI, with four patients having increased antral follicle counts and one patient with on-going pregnancy after IVF [83].

Mesenchymal stem cells from umbilical cord, bone marrow, and other sources may be a relevant therapeutic alternative for ovary regeneration and follicular development in patients with impaired ovaries such as poor ovarian responders or women diagnosed with POI. Nevertheless, further research is needed to properly evaluate underlying mechanisms, identify best cell sources, and develop less invasive infusion techniques.

Based on previous studies, it may also make sense to consider hormonal support after MSC transplantation. The estrogen-progestin therapy (cyclical sequential combination of estradiol plus dydrogesterone) in women after MSC transplantation for cancer treatment resulted in a dramatic improvement in vasomotor, urogenital, and psychological symptoms related to estrogen deficiency in some patients [84].

Cryopreserved Ovarian Cortex Transplantation in Cancer Patients

Transplantation of cryopreserved ovarian tissue is an important technique to restore endocrine function and fertility, especially for cancer patients. The results of some studies [85–87] have shown that co-transplantation of MSCs may improve the results of this technique, especially in terms of angiogenesis which is usually critical and results in the follicle loss.

The study of Xia et al. [85] has shown that human bone marrow MSCs could enhance the expression levels of VEGF, FGF2, and especially the level of angiogenin, significantly stimulate neovascularization, and increase blood perfusion of the grafts in the cryopreserved ovarian tissue transplantation. Further studies revealed that MSCs could notably reduce the apoptotic rates of primordial follicles and decrease follicle loss in the grafted ovarian tissues. It was found that angiogenin plays a critical role in regulating angiogenesis and follicle survival in xenografted human ovarian tissues and that bone marrow MSC-derived angiogenin promotes primordial follicle survival and angiogenesis in human ovarian tissues after xenografting into female ovariectomized mice with severe combined immune deficiency (SCID) [86]. Recently, it has been published that human adipose tissue-derived stem cells protect the primordial follicle pool from both direct follicle death and abnormal activation after human ovarian tissue transplantation into nude mice by maintaining their quiescence through modulation of the PI3K/Akt pathway [87].

Transplantation of Autologous Mesenchymal Stem Cells and Very Small Embryonic-Like Stem Cells (VSELs) from Human Adult Ovaries?

In this chapter, we focused mainly on ovarian regeneration through transplantation of hMSCs from different sources such as bone marrow, umbilical cord blood, placenta, amniotic membrane and fluid, chorionic plate, as this proves to be realistic and clinically useful. However, there are also some other types of stem cells that would be of interest for ovarian regeneration through transplantation.

Ovarian Mesenchymal Stem Cells

Stem cells with MSC characteristics have already been isolated from adult human ovarian tissue [88]. There are subpopulations of human follicular—granulosa and theca—cells that show the properties of MSCs, including the ability to differentiate into other cell types [89, 90]. Human theca stem cells were even in vitro differentiated into oocyte-like cells [91].

It would be possible to isolate, proliferate, and transplant these cells back into the ovaries, which would enable the transplantation of autologous cells. Stem cells with the properties of MSCs were also isolated from follicular fluid retrieved by ultrasound-guided aspiration of oocytes from ovarian follicles to obtain oocytes for IVF [92–95]. In the IVF program, a large amount of follicular fluid is obtained, which is discarded after oocyte removal, but could be suitable for the isolation of MSCs. Various studies have shown that follicular granulosa and theca cells contain subpopulations of MSCs and are also present in aspirated follicular fluid in the IVF program. MSCs from follicular fluid could be used for ovarian transplantation and regeneration in patients with POI. Perhaps MSCs from the follicular fluid of young patients could be used for ovarian transplantation and regeneration in older women characterized by ovarian aging.

Very Small Embryonic-Like Stem Cells

More studies have shown that there are also very small embryonic-like stem cells (VSELs) in human adult ovaries, which are supposed to persist from the embryonic period of life. These are small round cells, up to 5 μm in diameter, that express a specific pattern of expression of pluripotency and germinal markers and can develop into oocyte-like cells under laboratory conditions [96–98]. These oocyte-like cells expressed germinal markers [96–99] and also responded to the presence of human sperm by releasing a structure similar to the zona pellucida expressing ZP2 gene [100]. VSELs have also been found in the ovarian tissue of patients with premature ovarian insufficiency and have also developed into oocyte-like cells under laboratory conditions [98, 99]. Therefore, it would be possible to isolate [101, 102] these cells from the ovaries of patients with POI to avoid immune blockades in the ovaries, to proliferate them, to grow them into oocyte-like cells, and to transplant them back into the ovaries in addition to proper hormonal stimulation of the ovaries. VSELs have already been proposed as developmental precursors of oogonia and oocytes in above-mentioned publications, so in this way we would actually simulate the natural events. This possibility seems very interesting, but it would require special care and concern that patients do not develop malignant tumors. Namely, it has been proven that VSELs, like ESCs, are very likely to be involved in the development of malignant tumors—ovarian cancer due to their pluripotent nature [101, 103, 104]. These small stem cells (Fig. 2.3) are of great interest for further ovarian regeneration research, but currently MSCs are proving to be more realistic and suitable for clinical use in patients with POI.

Safety of Mesenchymal Stem Cell Transplantation

There are no publications in both animal and human studies showing negative side effects of MSC transplantation. However, caution is needed, as some research suggests that MSCs, especially adipose-derived MSCs, may also promote ovarian cancer and chemoresistance [105–108]. On the other hand, some research has shown that, for example, human endometrial MSCs have shown antitumor activity in epithelial ovarian cancer [109]. Therefore, in clinical trials of MSC transplantation for ovarian regeneration, patient follow-up is essential. Treatment of the ovaries with extracellular vesicles, exosomes, secretions and growth factors from MSCs to improve ovarian function, but with a lower risk of cancer, is also very promising and realistic for human reproductive medicine.

Conclusion

Numerous studies in animal models have shown that transplantation of hMSCs from various sources has many positive effects on the ovaries, manifested by improved folliculogenesis, reduced apoptosis of granulosa cells, development of mature oocytes and offspring after mating of animals with simulated POI. The first human clinical trials are also underway, where healthy children were born to patients with POI or poor ovarian response after autologous MSC transplantation. Cellular therapies for ovarian regeneration, taking into account safety issues, represent one of the biggest challenges in the field of infertility treatment and reproductive medicine.

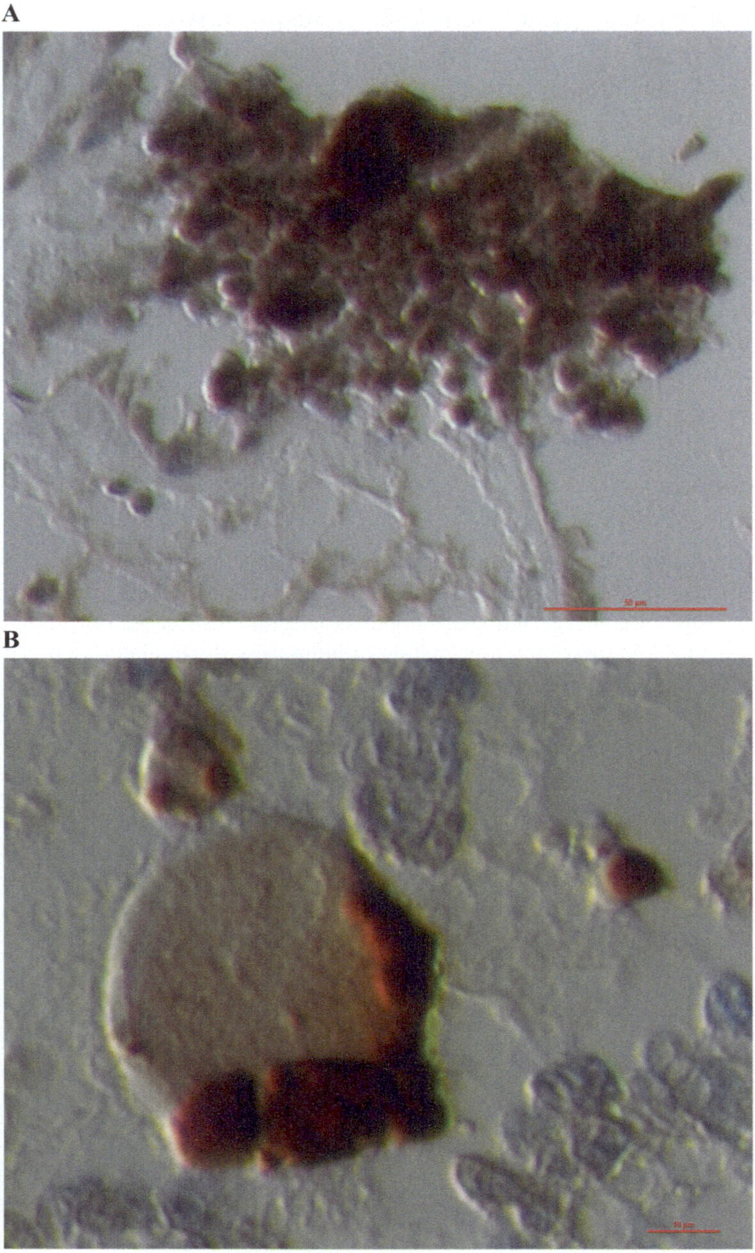

Fig. 2.3 Very small embryonic-like stem cells (VSELs), positively stained for NANOG (brown), forming a cluster **a** and oocyte **b** in sections of human adult ovarian tissue (in situ)

References

1. De Vos M, Devroey P, Fauser BC (2010) Primary ovarian insufficiency. Lancet 376:911–921. https://doi.org/10.1016/S0140-6736(10)60355-8
2. Golezar S, Ramezani Tehrani F, Khazaei S, Ebadi A, Keshavarz Z (2019) The global prevalence of primary ovarian insufficiency and early menopause: a meta-analysis. Climacteric 22:403–411. https://doi.org/10.1080/13697137.2019.1574738
3. Sydsjö G, Bladh M, Rindeborn K, Hammar M, Rodriguez-Martinez H, Nedstrand E (2020) Being born preterm or with low weight implies a risk of infertility and premature loss of ovarian function; a national register study. Ups J Med Sci 125:235–239. https://doi.org/10.1080/03009734.2020.1770380
4. Akande RO, Ibrahim Y (2020) Genetics of primary ovarian insufficiency. Clin Obstet Gynecol 63:687–705. https://doi.org/10.1097/GRF.0000000000000575
5. Wang H, Chen H, Qin Y, Shi Z, Zhao X, Xu J, Ma B, Chen ZJ (2015) Risks associated with premature ovarian failure in Han Chinese women. Reprod Biomed Online 30:401–407. https://doi.org/10.1016/j.rbmo.2014.12.013
6. Oftedal BE, Wolff ASB (2020) New era of therapy for endocrine autoimmune disorders. Scand J Immunol 92:e12961. https://doi.org/10.1111/sji.12961
7. Beitl K, Rosta K, Poetsch N, Seifried M, Mayrhofer D, Soliman B, Marculescu R, Ott J (2021) Autoimmunological serum parameters and bone mass density in premature ovarian insufficiency: a retrospective cohort study. Arch Gynecol Obstet 303:1109–1115. https://doi.org/10.1007/s00404-020-05860-4
8. Dutta D, Sharma LK, Sharma N, Gadpayle AK, Anand A, Gaurav K, Gupta A, Poondla Y, Kulshreshtha B (2017) Occurrence, patterns & predictors of hypogonadism in patients with HIV infection in India. Indian J Med Res 145:804–814. https://doi.org/10.4103/ijmr.IJMR_1926_15
9. Gruber N, Shoenfeld Y (2015) A link between human papilloma virus vaccination and primary ovarian insufficiency: current analysis. Curr Opin Obstet Gynecol 27:265–270. https://doi.org/10.1097/GCO.0000000000000183
10. Colafrancesco S, Perricone C, Tomljenovic L, Shoenfeld Y (2013) Human papilloma virus vaccine and primary ovarian failure: another facet of the autoimmune/inflammatory syndrome induced by adjuvants. Am J Reprod Immunol 70:309–316. https://doi.org/10.1111/aji.12151
11. Ohl J, Partisani M, Demangeat C, Binder-Foucard F, Nisand I, Lang JM (2010) Alterations of ovarian reserve tests in Human Immunodeficiency Virus (HIV)-infected women]. Gynecol Obstet Fertil 38:313–317. https://doi.org/10.1016/j.gyobfe.2009.07.019
12. Cejtin HE, Kalinowski A, Bacchetti P, Taylor RN, Watts DH, Kim S, Massad LS, Preston-Martin S, Anastos K, Moxley M, Minkoff HL (2006) Effects of human immunodeficiency virus on protracted amenorrhea and ovarian dysfunction. Obstet Gynecol 108:1423–1431. https://doi.org/10.1097/01.AOG.0000245442.29969.5c
13. Singh G, Misra R, Aggarwal A (2016) Ovarian insufficiency is major short-term toxicity in systemic lupus erythematosus patients treated with cyclophosphamide. J Assoc Physicians India 64:28–31
14. Monteiro CS, Xavier EBS, Caetano JPJ, Marinho RM (2020) A critical analysis of the impact of endocrine disruptors as a possible etiology of primary ovarian insufficiency. JBRA Assist Reprod 24:324–331. https://doi.org/10.5935/1518-0557.20200005
15. Pan W, Ye X, Yin S, Ma X, Li C, Zhou J, Liu W, Liu J (2019) Selected persistent organic pollutants associated with the risk of primary ovarian insufficiency in women. Environ Int 129:51–58. https://doi.org/10.1016/j.envint.2019.05.023
16. Yeo W, Pang E, Liem GS, Suen JJS, Ng RYW, Yip CCH, Li L, Yip CHW, Mo FKF (2020) Menopausal symptoms in relationship to breast cancer-specific quality of life after adjuvant cytotoxic treatment in young breast cancer survivors. Health Qual Life Outcomes 18:24. https://doi.org/10.1186/s12955-020-1283-x

17. Silvestris E, Dellino M, Cafforio P, Paradiso AV, Cormio G, D'Oronzo S (2020) Breast cancer: an update on treatment-related infertility. J Cancer Res Clin Oncol 146:647–657. https://doi.org/10.1007/s00432-020-03136-7
18. Felicetti F, Castiglione A, Biasin E, Fortunati N, Dionisi-Vici M, Matarazzo P, Ciccone G, Fagioli F, Brignardello E (2020) Effects of treatments on gonadal function in long-term survivors of pediatric hematologic malignancies: a cohort study. Pediatr Blood Cancer 2020:e28709. https://doi.org/10.1002/pbc.28709
19. Yeganeh L, Boyle JA, Gibson-Helm M, Teede H, Vincent AJ (2020) Women's perspectives of early menopause: development of a word cloud. Climacteric 23:417–420. https://doi.org/10.1080/13697137.2020.1730318
20. Collins G, Patel B, Thakore S, Liu J (2017) Primary ovarian insufficiency: current concepts. South Med J 110:147–153. https://doi.org/10.14423/SMJ.0000000000000611
21. Kawamura K, Ishizuka B, Hsueh AJW (2020) Drug-free in-vitro activation of follicles for infertility treatment in poor ovarian response patients with decreased ovarian reserve. Reprod Biomed Online 40:245–253. https://doi.org/10.1016/j.rbmo.2019.09.007
22. Kawamura K, Cheng Y, Suzuki N, Deguchi M, Sato Y, Takae S, Ho CH, Kawamura N, Tamura M, Hashimoto S, Sugishita Y, Morimoto Y, Hosoi Y, Yoshioka N, Ishizuka B, Hsueh AJ (2013) Hippo signaling disruption and Akt stimulation of ovarian follicles for infertility treatment. Proc Natl Acad Sci U S A. 110:17474–17479. https://doi.org/10.1073/pnas.1312830110
23. Matsuzaki S, Pankhurst MW (2020) Hyperactivation of dormant primordial follicles in ovarian endometrioma patients. Reproduction 160:R145–R153. https://doi.org/10.1530/REP-20-0265
24. Zhao F, Lan Y, Chen T, Xin Z, Liang Y, Li Y, Wang S, Zhang J, Yang X (2020) Live birth rate comparison of three controlled ovarian stimulation protocols for in vitro fertilization-embryo transfer in patients with diminished ovarian reserve after endometrioma cystectomy: a retrospective study. J Ovarian Res 13:23. https://doi.org/10.1186/s13048-020-00622-x
25. Hanson BM, Tao X, Zhan Y, Jenkins TG, Morin SJ, Scott RT, Seli EU (2020) Young women with poor ovarian response exhibit epigenetic age acceleration based on evaluation of white blood cells using a DNA methylation-derived age prediction model. Hum Reprod 35:2579–2588. https://doi.org/10.1093/humrep/deaa206
26. Stimpfel M, Vrtačnik-Bokal E, Pozlep B, Kmecl J, Virant-Klun I (2016) Gonadotrophin-releasing hormone agonist protocol of controlled ovarian hyperstimulation as an efficient treatment in Bologna-defined poor ovarian responders. Syst Biol Reprod Med. 62:290–296. https://doi.org/10.3109/19396368.2016.1170229
27. Ferraretti AP, Gianaroli L, Magli MC, Bafaro G, Colacurci N (2000) Female poor responders. Mol Cell Endocrinol 161:59–66. https://doi.org/10.1016/s0303-7207(99)00225-7
28. Sun YF, Zhang J, Xu YM, Luo ZY, Sun Y, Hao GM, Gao BL (2020) Effects of age on pregnancy outcomes in patients with simple tubal factor infertility receiving frozen-thawed embryo transfer. Sci Rep 10:18121. https://doi.org/10.1038/s41598-020-75124-3
29. Zhou SJ, Zhao MJ, Li C, Su X (2020) The comparison of evaluative effectiveness between antral follicle count/age ratio and ovarian response prediction index for the ovarian reserve and response functions in infertile women. Medicine (Baltimore) 99:e21979. https://doi.org/10.1097/MD.0000000000021979
30. Bahrehbar K, Rezazadeh Valojerdi M, Esfandiari F, Fathi R, Hassani SN, Baharvand H (2020) Human embryonic stem cell-derived mesenchymal stem cells improved premature ovarian failure. World J Stem Cells 12:857–878. https://doi.org/10.4252/wjsc.v12.i8.857
31. Yoon SY, Yoon JA, Park M, Shin EY, Jung S, Lee JE, Eum JH, Song H, Lee DR, Lee WS, Lyu SW (2020) Recovery of ovarian function by human embryonic stem cell-derived mesenchymal stem cells in cisplatin-induced premature ovarian failure in mice. Stem Cell Res Ther 11:255. https://doi.org/10.1016/j.tjog.2018.11.010
32. Besikcioglu HE, Sarıbas GS, Ozogul C, Tiryaki M, Kilic S, Pınarlı FA, Gulbahar O (2019) Determination of the effects of bone marrow derived mesenchymal stem cells and ovarian stromal stem cells on follicular maturation in cyclophosphamide induced ovarian failure in rats. Taiwan J Obstet Gynecol 58:53–59. https://doi.org/10.1016/j.tjog.2018.11.010

33. Lu X, Bao H, Cui L, Zhu W, Zhang L, Xu Z, Man X, Chu Y, Fu Q, Zhang H (2020) hUMSC transplantation restores ovarian function in POI rats by inhibiting autophagy of theca-interstitial cells via the AMPK/mTOR signaling pathway. Stem Cell Res Ther 11:268. https://doi.org/10.1186/s13287-020-01784-7
34. Cui L, Bao H, Liu Z, Man X, Liu H, Hou Y, Luo Q, Wang S, Fu Q, Zhang H (2020) hUMSCs regulate the differentiation of ovarian stromal cells via TGF-beta(1)/Smad3 signaling pathway to inhibit ovarian fibrosis to repair ovarian function in POI rats. Stem Cell Res Ther 11:386. https://doi.org/10.1186/s13287-020-01904-3
35. Zhao Y, Ma J, Yi P, Wu J, Zhao F, Tu W, Liu W, Li T, Deng Y, Hao J, Wang H, Yan L (2020) Human umbilical cord mesenchymal stem cells restore the ovarian metabolome and rescue premature ovarian insufficiency in mice. Stem Cell Res Ther 11:466. https://doi.org/10.1186/s13287-020-01972-5
36. Wang Z, Wei Q, Wang H, Han L, Dai H, Qian X, Yu H, Yin M, Shi F, Qi N (2020) Mesenchymal stem cell therapy using human umbilical cord in a rat model of autoimmune-induced premature ovarian failure. Stem Cells Int 2020:3249495. https://doi.org/10.1155/2020/3249495
37. Shen J, Cao D, Sun JL (2020) Ability of human umbilical cord mesenchymal stem cells to repair chemotherapy-induced premature ovarian failure. World J Stem Cells 12:277–287. https://doi.org/10.4252/wjsc.v12.i4.277
38. Lu X, Cui J, Cui L, Luo Q, Cao Q, Yuan W, Zhang H (2019) The effects of human umbilical cord-derived mesenchymal stem cell transplantation on endometrial receptivity are associated with Th1/Th2 balance change and uNK cell expression of uterine in autoimmune premature ovarian failure mice. Stem Cell Res Ther 10:214. https://doi.org/10.1186/s13287-019-1313-y
39. Zheng Q, Fu X, Jiang J, Zhang N, Zou L, Wang W, Ding M, Chen H (2019) Umbilical cord mesenchymal stem cell transplantation prevents chemotherapy-induced ovarian failure via the NGF/TrkA pathway in rats. Biomed Res Int 2019:6539294. https://doi.org/10.1155/2019/6539294
40. Yang Y, Lei L, Wang S, Sheng X, Yan G, Xu L, Liu J, Liu M, Zhen X, Ding L, Sun H (2019) Transplantation of umbilical cord-derived mesenchymal stem cells on a collagen scaffold improves ovarian function in a premature ovarian failure model of mice. In Vitro Cell Dev Biol Anim 55:302–311. https://doi.org/10.1007/s11626-019-00337-4
41. Jalalie L, Rezaie MJ, Jalili A, Rezaee MA, Vahabzadeh Z, Rahmani MR, Karimipoor M, Hakhamaneshi MS (2019) Distribution of the CM-Dil-labeled human umbilical cord vein mesenchymal stem cells migrated to the cyclophosphamide-injured ovaries in C57BL/6 mice. Iran Biomed J 23:200–208. https://doi.org/10.29252/.23.3.200
42. Elfayomy AK, Almasry SM, El-Tarhouny SA, Eldomiaty MA (2016) Human umbilical cord blood-mesenchymal stem cells transplantation renovates the ovarian surface epithelium in a rat model of premature ovarian failure: Possible direct and indirect effects. Tissue Cell 48:370–382. https://doi.org/10.1016/j.tice.2016.05.001
43. Song D, Zhong Y, Qian C, Zou Q, Ou J, Shi Y, Gao L, Wang G, Liu Z, Li H, Ding H, Wu H, Wang F, Wang J, Li H (2016) Human umbilical cord mesenchymal stem cells therapy in cyclophosphamide-induced premature ovarian failure rat model. Biomed Res Int 2016:2517514. https://doi.org/10.1155/2016/2517514
44. Seok J, Park H, Choi JH, Lim JY, Kim KG, Kim GJ (2020) Placenta-derived mesenchymal stem cells restore the ovary function in an ovariectomized rat model via an antioxidant effect. Antioxidants (Basel) 9:591. https://doi.org/10.3390/antiox9070591
45. Li H, Zhao W, Wang L, Luo Q, Yin N, Lu X, Hou Y, Cui J, Zhang H (2019) Human placenta-derived mesenchymal stem cells inhibit apoptosis of granulosa cells induced by IRE1alpha pathway in autoimmune POF mice. Cell Biol Int 43:899–909. https://doi.org/10.1002/cbin.11165
46. Yin N, Wang Y, Lu X, Liu R, Zhang L, Zhao W, Yuan W, Luo Q, Wu H, Luan X, Zhang H (2018) hPMSC transplantation restoring ovarian function in premature ovarian failure mice is associated with change of Th17/Tc17 and Th17/Treg cell ratios through the PI3K/Akt signal pathway. Stem Cell Res Ther 9:37. https://doi.org/10.1186/s13287-018-0772-x

47. Zhang H, Luo Q, Lu X, Yin N, Zhou D, Zhang L, Zhao W, Wang D, Du P, Hou Y, Zhang Y, Yuan W (2018) Effects of hPMSCs on granulosa cell apoptosis and AMH expression and their role in the restoration of ovary function in premature ovarian failure mice. Stem Cell Res Ther 9:20. https://doi.org/10.1186/s13287-017-0745-5
48. Yin N, Zhao W, Luo Q, Yuan W, Luan X, Zhang H (2018) Restoring ovarian function with human placenta-derived mesenchymal stem cells in autoimmune-induced premature ovarian failure mice mediated by treg cells and associated cytokines. Reprod Sci 25:1073–1082. https://doi.org/10.1177/1933719117732156
49. Huang B, Qian C, Ding C, Meng Q, Zou Q, Li H (2019) Fetal liver mesenchymal stem cells restore ovarian function in premature ovarian insufficiency by targeting MT1. Stem Cell Res Ther 10:362. https://doi.org/10.1186/s13287-019-1490-8
50. Feng X, Ling L, Zhang W, Liu X, Wang Y, Luo Y, Xiong Z (2020) Effects of human amnion-derived mesenchymal stem cell (hAD-MSC) transplantation in situ on primary ovarian insufficiency in SD rats. Reprod Sci 27:1502–1512. https://doi.org/10.1007/s43032-020-00147-0
51. Liu R, Zhang X, Fan Z, Wang Y, Yao G, Wan X, Liu Z, Yang B, Yu L (2019) Human amniotic mesenchymal stem cells improve the follicular microenvironment to recover ovarian function in premature ovarian failure mice. Stem Cell Res Ther 10:299. https://doi.org/10.1186/s13287-019-1315-9
52. Ding C, Zou Q, Wang F, Wu H, Chen R, Lv J, Ling M, Sun J, Wang W, Li H, Huang B (2018) Human amniotic mesenchymal stem cells improve ovarian function in natural aging through secreting hepatocyte growth factor and epidermal growth factor. Stem Cell Res Ther 9:55. https://doi.org/10.1186/s13287-018-0781-9
53. Ding C, Li H, Wang Y, Wang F, Wu H, Chen R, Lv J, Wang W, Huang B (2017) Different therapeutic effects of cells derived from human amniotic membrane on premature ovarian aging depend on distinct cellular biological characteristics. Stem Cell Res Ther 8:173. https://doi.org/10.1186/s13287-017-0613-3
54. Ling L, Feng X, Wei T, Wang Y, Wang Y, Wang Z, Tang D, Luo Y, Xiong Z (2019) Human amnion-derived mesenchymal stem cell (hAD-MSC) transplantation improves ovarian function in rats with premature ovarian insufficiency (POI) at least partly through a paracrine mechanism. Stem Cell Res Ther 10:46. https://doi.org/10.1186/s13287-019-1136-x
55. Ling L, Feng X, Wei T, Wang Y, Wang Y, Zhang W, He L, Wang Z, Zeng Q, Xiong Z (2017) Effects of low-intensity pulsed ultrasound (LIPUS)-pretreated human amnion-derived mesenchymal stem cell (hAD-MSC) transplantation on primary ovarian insufficiency in rats. Stem Cell Res Ther 8:283. https://doi.org/10.1186/s13287-017-0739-3
56. Lai D, Wang F, Chen Y, Wang L, Wang Y, Cheng W (2013) Human amniotic fluid stem cells have a potential to recover ovarian function in mice with chemotherapy-induced sterility. BMC Dev Biol 13:34. https://doi.org/10.1186/1471-213X-13-34
57. Liu T, Huang Y, Guo L, Cheng W, Zou G (2012) CD44+/CD105+ human amniotic fluid mesenchymal stem cells survive and proliferate in the ovary long-term in a mouse model of chemotherapy-induced premature ovarian failure. Int J Med Sci 9:592–602. https://doi.org/10.7150/ijms.4841
58. Li J, Yu Q, Huang H, Deng W, Cao X, Adu-Frimpong M, Yu J, Xu X (2018) Human chorionic plate-derived mesenchymal stem cells transplantation restores ovarian function in a chemotherapy-induced mouse model of premature ovarian failure. Stem Cell Res Ther 9:81. https://doi.org/10.1186/s13287-018-0819-z
59. Feng P, Li P, Tan J (2019) Human menstrual blood-derived stromal cells promote recovery of premature ovarian insufficiency via regulating the ECM-dependent FAK/AKT signaling. Stem Cell Rev Rep 15:241–255. https://doi.org/10.1007/s12015-018-9867-0
60. Reig A, Mamillapalli R, Coolidge A, Johnson J, Taylor HS (2019) Uterine cells improved ovarian function in a murine model of ovarian insufficiency. Reprod Sci 26:1633–1639. https://doi.org/10.1177/1933719119875818
61. Noory P, Navid S, Zanganeh BM, Talebi A, Borhani-Haghighi M, Gholami K, Manshadi MD, Abbasi M (2019) Human menstrual blood stem cell-derived granulosa cells participate

in ovarian follicle formation in a rat model of premature ovarian failure in vivo. Cell Reprogram 21:249–259. https://doi.org/10.1089/cell.2019.002

62. Manshadi MD, Navid S, Hoshino Y, Daneshi E, Noory P, Abbasi M (2019) The effects of human menstrual blood stem cells-derived granulosa cells on ovarian follicle formation in a rat model of premature ovarian failure. Microsc Res Tech 82:635–642. https://doi.org/10.1002/jemt.23120
63. Lai D, Wang F, Yao X, Zhang Q, Wu X, Xiang C (2015) Human endometrial mesenchymal stem cells restore ovarian function through improving the renewal of germline stem cells in a mouse model of premature ovarian failure. J Transl Med 13:155. https://doi.org/10.1186/s12967-015-0516-y
64. Liu T, Huang Y, Zhang J, Qin W, Chi H, Chen J, Yu Z, Chen C (2014) Transplantation of human menstrual blood stem cells to treat premature ovarian failure in mouse model. Stem Cells Dev 23:1548–1557. https://doi.org/10.1089/scd.2013.0371
65. Mohamed SA, Shalaby SM, Abdelaziz M, Brakta S, Hill WD, Ismail N, Al-Hendy A (2018) Human mesenchymal stem cells partially reverse infertility in chemotherapy-induced ovarian failure. Reprod Sci 25:51–63. https://doi.org/10.1177/1933719117699705
66. Herraiz S, Buigues A, Díaz-García C, Romeu M, Martínez S, Gómez-Seguí I, Simón C, Hsueh AJ, Pellicer A (2018) Fertility rescue and ovarian follicle growth promotion by bone marrow stem cell infusion. Fertil Steril 109:908–918. https://doi.org/10.1016/j.fertnstert.2018.01.004
67. Liu J, Zhang H, Zhang Y, Li N, Wen Y, Cao F, Ai H, Xue X (2014) Homing and restorative effects of bone marrow-derived mesenchymal stem cells on cisplatin injured ovaries in rats. Mol Cells 37:865–872. https://doi.org/10.14348/molcells.2014.0145
68. Ghadami M, El-Demerdash E, Zhang D, Salama SA, Binhazim AA, Archibong AE, Chen X, Ballard BR, Sairam MR, Al-Hendy A (2012) Bone marrow transplantation restores follicular maturation and steroid hormones production in a mouse model for primary ovarian failure. PLoS ONE 7:e32462. https://doi.org/10.1371/journal.pone.0032462
69. Zhang J, Yin H, Jiang H, Du X, Yang Z (2020) The protective effects of human umbilical cord mesenchymal stem cell-derived extracellular vesicles on cisplatin-damaged granulosa cells. Taiwan J Obstet Gynecol 59:527–533. https://doi.org/10.1016/j.tjog.2020.05.010
70. Yang Z, Du X, Wang C, Zhang J, Liu C, Li Y, Jiang H (2019) Therapeutic effects of human umbilical cord mesenchymal stem cell-derived microvesicles on premature ovarian insufficiency in mice. Stem Cell Res Ther 10:250. https://doi.org/10.1186/s13287-019-1327-5
71. Huang B, Lu J, Ding C, Zou Q, Wang W, Li H (2018) Exosomes derived from human adipose mesenchymal stem cells improve ovary function of premature ovarian insufficiency by targeting SMAD. Stem Cell Res Ther 9:216. https://doi.org/10.1186/s13287-018-0953-7
72. Sun L, Li D, Song K, Wei J, Yao S, Li Z, Su X, Ju X, Chao L, Deng X, Kong B, Li L (2017) Exosomes derived from human umbilical cord mesenchymal stem cells protect against cisplatin-induced ovarian granulosa cell stress and apoptosis in vitro. Sci Rep 7:2552. https://doi.org/10.1038/s41598-017-02786-x
73. Park HS, Ashour D, Elsharoud A, Chugh RM, Ismail N, El Andaloussi A, Al-Hendy A (2019) Towards cell free therapy of premature ovarian insufficiency: human bone marrow mesenchymal stem cells secretome enhances angiogenesis in human ovarian microvascular endothelial cells. HSOA J Stem Cells Res Dev Ther 5:019. https://doi.org/10.24966/srdt-2060/100019
74. Yan Z, Guo F, Yuan Q, Shao Y, Zhang Y, Wang H, Hao S, Du X (2019) Endometrial mesenchymal stem cells isolated from menstrual blood repaired epirubicin-induced damage to human ovarian granulosa cells by inhibiting the expression of Gadd45b in cell cycle pathway. Stem Cell Res Ther 10:4. https://doi.org/10.1186/s13287-018-1101-0
75. Song K, Cai H, Zhang D, Huang R, Sun D, He Y (2018) Effects of human adipose-derived mesenchymal stem cells combined with estrogen on regulatory T cells in patients with premature ovarian insufficiency. Int Immunopharmacol 55:257–262
76. Edessy M, Hosni HN, Shady Y, Waf Y, Bakr S, Kamel M (2016) Autologous stem cells therapy, the first baby of idiopathic premature ovarian failure. Acta Medica Int 3:19–23

77. Gabr H, Elkheir WA, El-Gazzar A (2016) Autologous stem cell transplantation in patients with idiopathic premature ovarian failure. J Tissue Sci Eng. https://doi.org/10.4172/2157-7552.C1.030. Proceedings of Global Congress on Tissue Engineering, Regenerative & Precision Medicine, December 1–2, 2016 | San Antonio, USA
78. Herraiz S, Romeu M, Buigues A, Martínez S, Díaz-García C, Gómez-Seguí I, Martínez J, Pellicer N, Pellicer A (2018) Autologous stem cell ovarian transplantation to increase reproductive potential in patients who are poor responders. Fertil Steril 110:496–505. https://doi.org/10.1016/j.fertnstert.2018.04.025
79. Herraiz S, Pellicer N, Romeu M, Pellicer A (2019) Treatment potential of bone marrow-derived stem cells in women with diminished ovarian reserves and premature ovarian failure. Curr Opin Obstet Gynecol 31:156–162. https://doi.org/10.1097/GCO.0000000000000531
80. Ding L, Yan G, Wang B, Xu L, Gu Y, Ru T, Cui X, Lei L, Liu J, Sheng X, Wang B, Zhang C, Yang Y, Jiang R, Zhou J, Kong N, Lu F, Zhou H, Zhao Y, Chen B, Hu Y, Dai J, Sun H (2018) Transplantation of UC-MSCs on collagen scaffold activates follicles in dormant ovaries of POF patients with long history of infertility. Sci China Life Sci 61:1554–1565. https://doi.org/10.1007/s11427-017-9272-2
81. Yan L, Wu Y, Li L, Wu J, Zhao F, Gao Z, Liu W, Li T, Fan Y, Hao J, Liu J, Wang H (2020) Clinical analysis of human umbilical cord mesenchymal stem cell allotransplantation in patients with premature ovarian insufficiency. Cell Prolif 2020:e12938. https://doi.org/10.1111/cpr.12938
82. Igboeli P, El Andaloussi A, Sheikh U, Takala H, ElSharoud A, McHugh A, Gavrilova-Jordan L, Levy S, Al-Hendy A (2020) Intraovarian injection of autologous human mesenchymal stem cells increases estrogen production and reduces menopausal symptoms in women with premature ovarian failure: two case reports and a review of the literature. J Med Case Rep 14:108. https://doi.org/10.1186/s13256-020-02426-5
83. Pellicer De Castellvi N, Herraiz S, Romeu M et al (2020) Bone marrow derived stem cells restore ovarian function and fertility in premature ovarian insufficiency women. Interim report of a randomized trial: mobilization versus ovarian injection. In: Abstracts of the 36th virtual annual meeting of the European Society of Human Reproduction and Embryology, O-088, i38, ESHRE Journals, Virtual meeting, 5–8 July 2020
84. Tauchmanovà L, Selleri C, De Rosa G, Sammartino A, Di Carlo C, Musella T, Martorelli C, Lombardi G, Rotoli B, Nappi C, Colaoet A (2007) Estrogen-progestin therapy in women after stem cell transplant: our experience and literature review. Menopause 14:320–330. https://doi.org/10.1097/01.gme.0000232032.84788.8c
85. Xia X, Yin T, Yan J, Yan L, Jin C, Lu C, Wang T, Zhu X, Zhi X, Wang J, Tian L, Liu J, Li R, Qiao J (2015) Mesenchymal stem cells enhance angiogenesis and follicle survival in human cryopreserved ovarian cortex transplantation. Cell Transplant 24:1999–2000. https://doi.org/10.3727/096368914X685267
86. Zhang Y, Xia X, Yan J, Yan L, Lu C, Zhu X, Wang T, Yin T, Li R, Chang H-M, Qiao J (2017) Mesenchymal stem cell-derived angiogenin promotes primodial follicle survival and angiogenesis in transplanted human ovarian tissue. Reprod Biol Endocrinol 15:18. https://doi.org/10.1186/s12958-017-0235-8
87. Cacciottola L, Courtoy GE, Nguyen TYT, Hossay C, Donnez J, Dolmans MM (2021) Adipose tissue-derived stem cells protect the primordial follicle pool from both direct follicle death and abnormal activation after ovarian tissue transplantation. J Assist Reprod Genet 38:151–161. https://doi.org/10.1007/s10815-020-02005-z
88. Stimpfel M, Cerkovnik P, Novakovic S, Maver A, Virant-Klun I (2014) Putative mesenchymal stem cells isolated from adult human ovaries. J Assist Reprod Genet 31:959–974. https://doi.org/10.1007/s10815-014-0254-8
89. Kossowska-Tomaszczuk K, De Geyter C, De Geyter M, Martin I, Holzgreve W, Scherberich A, Zhang H (2009) The multipotency of luteinizing granulosa cells collected from mature ovarian follicles. Stem Cells 27:210–219. https://doi.org/10.1634/stemcells.2008-0233
90. Dalman A, Totonchi M, Valojerdi MR (2018) Establishment and characterization of human theca stem cells and their differentiation into theca progenitor cells. J Cell Biochem 119:9853–9865. https://doi.org/10.1002/jcb.27306

91. Dalman A, Totonchi M, Rezazadeh Valojerdi M (2019) Human ovarian theca-derived multipotent stem cells have the potential to differentiate into oocyte-like cells in vitro. Cell J 20:527–536. https://doi.org/10.22074/cellj.2019.5651
92. Chandramohan Y, Jeganathan K, Sivanesan S, Koka P, Amritha TMS, Vimalraj S, Dhanasekaran A (2021) Assessment of human ovarian follicular fluid derived mesenchymal stem cells in chitosan/PCL/Zn scaffold for bone tissue regeneration. Life Sci 264:118502. https://doi.org/10.1016/j.lfs.2020.118502
93. Virant-Klun I, Omejec S, Stimpfel M, Skerl P, Novakovic S, Jancar N, Vrtacnik-Bokal E (2019) Female age affects the mesenchymal stem cell characteristics of aspirated follicular cells in the in vitro fertilization programme. Stem Cell Rev Rep 15:543–557. https://doi.org/10.1007/s12015-019-09889-0
94. Riva F, Omes C, Bassani R, Nappi RE, Mazzini G, Icaro Cornaglia A, Casasco A (2014) In-vitro culture system for mesenchymal progenitor cells derived from waste human ovarian follicular fluid. Reprod Biomed Online 29:457–469. https://doi.org/10.1016/j.rbmo.2014.06.006
95. Dzafic E, Stimpfel M, Novakovic S, Cerkovnik P, Virant-Klun I (2014) Expression of mesenchymal stem cells-related genes and plasticity of aspirated follicular cells obtained from infertile women. Biomed Res Int 2014:508216. https://doi.org/10.1155/2014/508216
96. Silvestris E, Cafforio P, D'Oronzo S, Felici C, Silvestris F, Loverro G (2018) In vitro differentiation of human oocyte-like cells from oogonial stem cells: single-cell isolation and molecular characterization. Hum Reprod 33:464–473. https://doi.org/10.1093/humrep/dex377
97. Parte S, Bhartiya D, Telang J, Daithankar V, Salvi V, Zaveri K, Hinduja I (2011) Detection, characterization, and spontaneous differentiation in vitro of very small embryonic-like putative stem cells in adult mammalian ovary. Stem Cells Dev 20:1451–1464. https://doi.org/10.1089/scd.2010.0461
98. Virant-Klun I, Zech N, Rozman P, Vogler A, Cvjeticanin B, Klemenc P, Malicev E, Meden-Vrtovec H (2008) Putative stem cells with an embryonic character isolated from the ovarian surface epithelium of women with no naturally present follicles and oocytes. Differentiation 76:843–856. https://doi.org/10.1111/j.1432-0436.2008.00268.x
99. Virant-Klun I, Skutella T, Kubista M, Vogler A, Sinkovec J, Meden-Vrtovec H (2013) Expression of pluripotency and oocyte-related genes in single putative stem cells from human adult ovarian surface epithelium cultured in vitro in the presence of follicular fluid. Biomed Res Int 2013:861460. https://doi.org/10.1155/2013/861460
100. Virant-Klun I (2018) Functional testing of primitive oocyte-like cells developed in ovarian surface epithelium cell culture from small VSEL-like stem cells: can they be fertilized one day? Stem Cell Rev Rep 14:715–721. https://doi.org/10.1007/s12015-018-9832-y
101. Virant-Klun I, Skerl P, Novakovic S, Vrtacnik-Bokal E, Smrkolj S (2019) Similar population of $CD133^{+}$ and $DDX4^{+}$ VSEL-like stem cells sorted from human embryonic stem cell, ovarian, and ovarian cancer ascites cell cultures: the real embryonic stem cells? Cells 8:706. https://doi.org/10.3390/cells8070706
102. Virant-Klun I, Skutella T, Hren M, Gruden K, Cvjeticanin B, Vogler A, Sinkovec J (2013) Isolation of small SSEA-4-positive putative stem cells from the ovarian surface epithelium of adult human ovaries by two different methods. Biomed Res Int 2013:690415. https://doi.org/10.1155/2013/690415
103. Virant-Klun I, Stimpfel M (2016) Novel population of small tumour-initiating stem cells in the ovaries of women with borderline ovarian cancer. Sci Rep 6:34730. https://doi.org/10.1038/srep34730
104. Virant-Klun I, Kenda-Suster N, Smrkolj S (2016) Small putative NANOG, SOX2, and SSEA-4-positive stem cells resembling very small embryonic-like stem cells in sections of ovarian tissue in patients with ovarian cancer. J Ovarian Res 9:12. https://doi.org/10.1186/s13048-016-0221-3
105. Raghavan S, Snyder CS, Wang A, McLean K, Zamarin D, Buckanovich RJ, Mehta G (2020) Carcinoma-associated mesenchymal stem cells promote chemoresistance in ovarian cancer stem cells via PDGF signaling. Cancers (Basel) 12:2063. https://doi.org/10.3390/cancers12082063

106. Chu Y, You M, Zhang J, Gao G, Han R, Luo W, Liu T, Zuo J, Wang F (2019) Adipose-derived mesenchymal stem cells enhance ovarian cancer growth and metastasis by increasing thymosin beta 4X-linked expression. Stem Cells Int 2019:9037197. https://doi.org/10.1155/2019/9037197
107. Luo X, Huang S, He N, Liu C, Chen Y, Liu Y, Mi X, Li N, Sun P, Li Z, Xiang R, Su W (2018) Inflammatory human umbilical cord-derived mesenchymal stem cells promote stem cell-like characteristics of cancer cells in an IL-1beta-dependent manner. Biomed Res Int 2018:7096707. https://doi.org/10.1155/2018/7096707
108. Zhang Y, Nowicka A, Solley TN, Wei C, Parikh A, Court L, Burks JK, Andreeff M, Woodward WA, Dadbin A, Kolonin MG, Lu KH, Klopp AH (2015) Stromal cells derived from visceral and obese adipose tissue promote growth of ovarian cancers. PLoS ONE 10:e0136361. https://doi.org/10.1371/journal.pone.0136361
109. Bu S, Wang Q, Zhang Q, Sun J, He B, Xiang C, Liu Z, Lai D (2016) Human endometrial mesenchymal stem cells exhibit intrinsic anti-tumor properties on human epithelial ovarian cancer cells. Sci Rep 6:37019. https://doi.org/10.1038/srep37019

Chapter 3
In Vitro Activation of Follicles for Treatment of Ovarian Insufficiency

Kazuhiro Kawamura and Tuyen Kim Cat Vo

Abstract

Introduction: Due to delayed childbearing being the current norm, ovarian dysfunction is rapidly emerging in women. Premature ovarian insufficiency (POI) shows severe ovarian dysfunction exhibiting infertility and symptoms caused by estrogen deficiency. In contrast to the recent development of assisted reproductive technology, POI and diminished ovarian reserve (DOR) are still difficult to have their genetically linked offspring, and oocyte donation is the only established approach for successful pregnancies. This article aims to introduce the procedures of in vitro activation (IVA) and drug-free IVA as potential new infertility treatments for ovarian dysfunction patients.

Methods: Based on the findings demonstrating the activation of dormant primordial follicles by the stimulation of Akt signaling pathway in both mice and human ovaries, our group developed a new infertility treatment, in vitro activation (IVA). The IVA includes laparoscopic ovary removal and subsequent ovarian cortical tissue culture with Akt stimulator drugs followed by autotransplantation of activated tissues. During the development of IVA, we also found that ovarian fragmentation disrupted Hippo signaling pathway to increase the expression of downstream growth factors, leading to secondary follicle growth, and then recently developed the drug-free IVA omitting the tissue culture step, which enables DOR patients to conceive using their own eggs by stimulation of residual early follicles by Hippo signal disruption only.

Results: With the combination of ovarian fragmentation and Akt stimulation, the IVA approach has resulted in successful follicle growth and live births in POI patients. Also, the drug-free IVA succeeded in achieving a higher number of antral follicle numbers as well as a higher number of retrieved oocytes as compared with previous in vitro fertilization treatment outcomes before the IVA approach, leading to successful pregnancies and deliveries.

Conclusions: The implementation of VA and drug-free IVA could increase the potential to conceive genetically related children for patients with ovarian dysfunction.

K. Kawamura (✉) · T. K. C. Vo
Advanced Reproduction Research Center, Department of Obstetrics and Gynecology, International University of Health and Welfare Graduate School of Medicine, Narita Chiba 286-8686, Japan

I. Virant-Klun (ed.), *Stem Cells in Reproductive Tissues and Organs*, Stem Cell Biology and Regenerative Medicine 70, https://doi.org/10.1007/978-3-030-90111-0_3

Keywords Akt stimulation · Diminished ovarian reserve · Drug-free IVA · Hippo signal · In vitro activation (IVA) · Ovarian fragmentation · Premature ovarian insufficiency

Abbreviations

AMH	Anti-Müllerian hormone
BIRC	Baculoviral inhibitors of apoptosis repeat containing
BMP4	Bone morphogenetic protein 4
BMP7	Bone morphogenetic protein 7
ET	Embryo transfer
bFGF	Basic fibroblast growth factor (bFGF)
CCN	Cellular communication network factor
c-kit	Tyrosine kinase receptor
FOXO3	Forkhead box O3
FSH	Follicle-stimulating hormone
GnRH	Gonadotropin-releasing hormone
hCG	Human chorionic gonadotropin
ICSI	Intracytoplasmic sperm injection
IVA	In vitro Activation
IVF	In vitro Fertilization
IVF-ET	In vitro Fertilization and embryo transfer
KGF	Keratinocyte growth factor
LATS1	Large tumor suppressor kinase 1
LH	Luteinizing hormone
LIF	Leukemia inhibitory factor
MST1	Macrophage stimulating 1
mTOR	Mammalian target of rapamycin
PDK1	Phosphatidylinositol-dependent kinase 1
PFs	Primordial follicles
PI3K	Phosphatidylinositol-3-kinase
PIP2	Phosphatidylinositol (4, 5) bisphosphate
PIP3	Phosphatidylinositol (3, 4, 5) triphosphate
POI	Premature ovarian insufficiency
POR-DOR	Poor ovarian response with diminishing ovarian reserve
PTEN	Phosphatase and tensin homolog
SAV1	Salvador family WW domain-containing protein 1
TAZ	Tafazzin PDZ-binding motif
TEAD	Transcription factors containing the TEA/ATTS DNA binding domain
TSC1	Tuberous sclerosis complex 1
TSC2	Tuberous sclerosis complex 2
VEGF	Vascular endothelial growth factor
YAP	Yes-associated transcriptional regulator protein

Introduction

In current modern society, professional pursuing and seeking better living condition desires have urged women to delay their childbearing, and ovarian dysfunction is quickly emerging as a common manifestation. Recent demographic survey data showed a significant rise in the mean age of having the first child of women [1, 2], and it has increased to 31 years of age in Japan. This change is associated with ovarian aging, resulting in ovarian dysfunction caused by a decline in both quantity and quality of oocytes. Premature ovarian insufficiency (POI) is an infertile status characterized by a rapid decline in ovarian follicles in women younger than 40 years and has risen from 1 to 2% in the last few decades [3, 4]. Although recent advances in assisted reproductive technologies have yielded a higher successful pregnancy rate in infertile patients [5], the treatment outcomes for patients with ovarian dysfunction including POI and poor ovarian response with diminishing ovarian reserve (POR-DOR) have limited success. Oocyte donation followed by in vitro fertilization and embryo transfer (IVF-ET) or adoption is the only established option to bear children for these patients. However, most couples would like to have their genetically linked offspring. Because conventional gonadotropin treatment usually demonstrates limited effects on the follicular growth in POI and POR-DOR patients, implementation therapeutic strategies for women with a low ovarian reserve are required to have genetically related offspring.

The ovarian dysfunction as well as POI is caused by autoimmune ovarian diseases or genetic aberrations involving the X chromosome, autosomes, or some specific genes [6]. Moreover, iatrogenic interventions, including ovarian surgery and chemo and/or radiation therapies could also induce ovarian dysfunction. Recently, in cancer patients becoming POI after gonadotoxic therapies, cryopreservation of ovarian tissues, mature oocytes, and/or preimplantation embryos before the cancer treatments are becoming available for potential options to preserve their fertility [7]. However, the most effective infertility treatments for other types of POI patients involve IVF-ET with the use of donor eggs from young healthy women [8]. Despite the effectiveness to treat these patients, several considerations need to be addressed with oocyte donation approach. The principal issue is the fact that patients cannot have their genetically related offspring, leading to personal and ethical concerns. Furthermore, in some countries, especially the majority of Islamic countries, oocyte donation is prohibited due to mainly religious reasons. In addition, pregnancy after oocyte donation is related to high-risk pregnancies due to immune incompatibility between fetus and mother [9, 10]. Moreover, a recent meta-analysis demonstrated that oocyte donation is associated with an increased risk of preeclampsia in singleton pregnancies [11].

To develop preovulatory follicles containing mature oocytes, a number of small primordial follicles (PFs) are periodically activated from the pool of PFs to undergo folliculogenesis. In ovarian physiology, if the number of residual follicles decreases until lower than 1,000 by aging or pathophysiological reasons, there would be neither regular normal activations of dormant PFs nor recruitment of developing follicles, resulting in anovulation and amenorrhea. Similarly, in ovarian dysfunction patients with low number of residual PFs, spontaneous activation of dormant follicles becomes irregular. As the decline of ovarian follicles is progressive, the activation of dormant follicles eventually arrests despite still having varying amounts of residual dormant PFs in ovaries. Although this is common nature between aging women and POI/POR-DOR patients, the decreasing speed of residual follicles is very fast in most POI/POR-DOR cases. Because of the limited number of growing follicles, these patients exhibit poor ovarian response and thus unlikely conceive with their own eggs [12]. To overcome the issue of ovulation disorder and poor response to ovarian stimulation in patients with ovarian dysfunction, we recently developed new infertility treatments and named them as in vitro activation (IVA) and drug-free IVA, which enable POI and POR-DOR patients to conceive using their own genetic eggs by artificial activation of residual dormant PFs and stimulate early gonadotropin-independent stage follicles. In this chapter, we summarize the potential use of these IVA approaches as a new infertility therapy in patients with POI and POR-DOR.

In Vitro Activation (IVA) of Ovarian Follicles

In mammalian ovary, the activation of dormant PFs is a complicated process with the participation of numerous intraovarian growth factors, including kit ligand, neurotrophins, vascular endothelial growth factor (VEGF), bone morphogenetic protein 4 (BMP4), bone morphogenetic protein 7 (BMP7), leukemia inhibitory factor (LIF), basic fibroblast growth factor (bFGF), and keratinocyte growth factor (KGF) [13]. Although the exact molecular mechanisms of the PFs' activation are currently not fully investigated, several major intracellular signaling systems involving this activation have been clarified [14].

Among these signaling pathways, transgenic animal studies revealed the critical roles of the phosphatidylinositol-3-kinase (PI3K)-Akt-forkhead box O3 (FOXO3) pathway in PFs' activation. After binding a cognate tyrosine kinase receptor, c-kit, Kit ligand stimulates PI3K, inducing the conversion of the lipid second messenger PIP2 [phosphatidylinositol (4, 5) bisphosphate] into PIP3 [phosphatidylinositol (3, 4, 5) triphosphate]. Subsequently, PIP3 modulates phosphatidylinositol-dependent kinase 1 (PDK1), leading to activation of AKT. Activated AKT then translocates to nucleus and inhibits the activity of a transcriptional factor, FOXO3. Phosphatase and tensin homolog (PTEN) suppresses this signaling pathway by dephosphorylating PIP3 to PIP2 [15].

In FOXO3 deficient mice, there was a spontaneous activation of all dormant PFs at early neonatal stage accompanied by the depletion of ovarian follicles during early life, resembling the phenotype of POI ovaries [16]. In another experiment in mice, depletion of oocyte-specific PTEN resulted in similar phenotypes and shrunken ovaries [17]. Moreover, adult mice with inducible deletion of PTEN in the oocyte also demonstrated the enhancement of AKT phosphorylation and nuclear export of FOXO3 proteins, leading to the activation of PFs [18]. Mammalian target of rapamycin (mTOR) is the catalytic subunit of two structurally different complexes, mTORC1 and mTORC2. The rapamycin-responsive mTORC1 enhances cell growth and proliferation, whereas tuberous sclerosis complexes (Tsc) 1 and 2 inhibit the activity of mTORC1 [19]. Mice acquired deficiency of Tsc1 or Tsc2 in oocytes also manifested spontaneous activation of dormant PFs [20, 21]. Moreover, different mTOR activators (phosphatidic acid, propranolol, and MHY1485) were also reported to activate dormant PFs [22] and to induce secondary follicle growth [23]. Furthermore, double deletion of Tsc1 and PTEN exhibits a synergistical effect to enhance PFs' activation as compared with singly mutated mice [21]. These findings support the indispensable and cooperative roles of PI3K-AKT and mTOR signaling pathways in regulating activation of dormant PFs [24].

Based on the thorough understandings of PI3K-Akt-FOXO3 pathway in PFs' activation, we attempted to activate PFs pharmacologically through manipulating the intracellular signaling of the PI3K-Akt-FOXO3 pathway. Twenty-four-hour culture of neonatal murine ovaries with a PTEN inhibitor, bpV(Hopic) and a PI3K-stimulating phosphopeptide, 740YP followed by additional 24 h culture with 740YP only could activate dormant PFs by increasing AKT activity, leading to nuclear exclusion of FOXO3 in oocytes in primordial follicles. Following grafting of the activated ovaries under kidney capsules of ovariectomized adult host mice with daily follicle-stimulating hormone (FSH) injection, the activated PFs developed to the preovulatory ones, and mature MII stage oocytes exhibiting normal epigenetic marks of imprinted genes were yielded. After IVF-ET using oocytes obtained from the ovaries with the PTEN inhibitor and the PI3K activator, normal implantation and live offspring rates were obtained, and healthy pups were delivered [25]. With long-term follow-up, we could confirm the normality of the progeny in appearance, behavior, reproductive activities as well as the absence of chronic diseases [26], confirming the safety and effectiveness of in vitro pharmacological activation of dormant PFs in producing mature oocytes in animal models. In translational studies using human ovarian samples, culture of ovarian cortical fragments with the PTEN inhibitor also could activate dormant PFs. After xenotransplantation to severe combined immunodeficiency mice for 6 months with FSH injection every second day, large antral follicles were developed at 6 months after transplantation and could obtain mature human MII stage oocytes with expansion of cumulus cells in response to human chorionic gonadotropin (hCG) stimulation for oocyte maturation [25].

In POI patients, spontaneous successful pregnancy seldom happens due to exhaustion of ovarian follicles. In controlled clinical trials, the pregnancy rate was reported as low as 1.5% [27]. Furthermore, an extensive study of a cohort of 358 young patients diagnosed with POI (26.6 ± 7.9 years of age at the time of diagnosis) demonstrated a spontaneous pregnancy rate of 4.4% during 13 years of observation [28]. Therefore, egg donation has been commonly used for infertility treatment in POI patients. Although diverse hormonal treatments and ovulation induction therapies have been attempted for infertility treatment of POI patients, these treatments showed limited success in inducing follicle growth. Thus, the establishment of a novel method for treating infertility that would enable POI patients to conceive using their own genetic oocytes has been much anticipated.

To challenge the development of a new infertility treatment approach, we performed a series of basic and translational studies. In our studies, treatment with the PTEN inhibitor and the PI3K activator stimulated murine and human dormant PFs by stimulating the PI3K-AKT-FOXO3 signaling pathway [25, 29]. Furthermore, fragmentation of ovarian cortical tissues induced secondary follicle growth in mice and human by disrupting the Hippo signaling pathway as described below [29]. These findings could provide the basis to develop the IVA approach, a new infertility therapy by activation of dormant primordial and promotion of restrained secondary/preantral follicle growth in POI patients. After confirmation of the effectiveness and safety of IVA by our [17] and other [18] groups, we performed a clinical trial of this approach followed by ovarian stimulation and IVF-ET in POI patients with approval from the ethical committee and written informed consent from the patients (Fig. 3.1) [29]. Because PFs are activated by AKT stimulators, whereas secondary follicle growth is enhanced with ovarian fragmentation, we combined AKT stimulator treatment and ovarian fragmentation to activate residual dormant follicles and stimulate early stage of follicles in ovaries of POI patients [29].

The IVA approach is manipulated under laparoscopic surgery. In detail, one or both ovaries were removed with a prudence avoiding the damage to Fallopian tubes

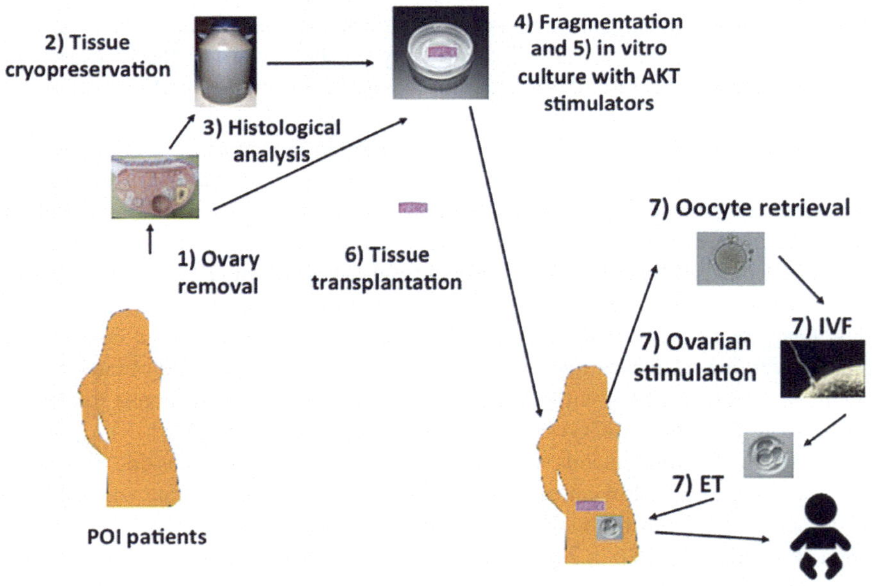

Fig. 3.1 The original in vitro activation (IVA) for patients with premature ovarian insufficiency (POI). This procedure consists of (1) removing one ovary under the first laparoscopic surgery, (2) with or without cryopreservation, (3) histological analysis, (4) fragmentation of cortices to disrupt ovarian Hippo signaling, (5) in vitro culture of ovarian cubes with AKT stimulators for 2 days, (6) grafting ovarian cubes under the second laparoscopic surgery, and (7) ovarian stimulation followed by in vitro fertilization and embryo transfer (IVF-ET)

for later use as grafting sites of cultured ovarian cortical fragments. Because of poor vascularization around and inside POI ovaries, it is rarely required extensive hemostasis by electrocoagulation during the ovariectomy. Since residual follicles at primordial, primary, and secondary stages are located within 2 mm thickness from the surface of ovarian cortex and absence of large antral follicles in POI ovaries [30], ovarian medulla was immediately removed to prepare ovarian cortical tissue where residual follicles were located. Then, the ovarian cortical tissue was cut into small strips (0.5–1.0 × 0.5–1.0 cm, 1–2 mm thickness) [29]. After being prepared, cortical strips were cryopreserved for subsequent tissue culture and autografting procedures using a vitrification method [31]. Before cryopreservation, approximately 10% volumes of the cortices from each cortical strip were dissected and subjected to histological examination to confirm the presence of residual follicles.

Because the patients without residual follicles based on this histological analysis showed no follicle growth during 1 year of follow-up under ovarian stimulation after ovarian cortical tissue grafting in our previous study [31], this histological evaluation of residual follicles was effective to predict successful follicle growth of IVA treatment in POI patients. Although the cryopreservation step of the ovarian cortical tissue is optional for IVA approach, this process provides the following benefits. In ovarian dysfunction patients, the number of residual follicles in ovaries continues to diminish with age [32]. If one ovary is cryopreserved at PCR-DOR or an early stage of POI, patients could undergo additional less invasive infertility treatments before the decision for second IVA surgery for ovarian tissue transplantation. Furthermore, the cryopreservation process allows sufficient time for histological analyses to completely clarifying the presence of residual follicles before a decision for ovarian tissue grafting.

After obtaining the results of histological analyses, cryopreserved ovarian cortical strips were thawed and further fragmented into 1–2 mm of small cubes. These ovarian cortical cubes were placed on the membrane of cell culture insert and covered with medium containing AKT activators. Then the ovarian cortical cubes were cultured for 2 days to activate dormant PFs [29]. After culture, the ovarian tissues were thoroughly washed to avoid the contamination of AKT stimulating drugs in grafting ovarian tissues and autologous transplantations beneath the serosa of Fallopian tubes were conducted under laparoscopic surgery. Saline was preinjected to swell the sites beneath the serosa of Fallopian tubes, followed by cutting the serosa and making a pouch between serosa and Fallopian tube to insert ovarian cortical cubes. In general, we could achieve around 20–80 ovarian cubes, which were grafted beneath the serosa of the two Fallopian tubes at several sites. These grafting sites are subsequently sutured or covered by oxidized regeneration cellulose to avoid cube loss [29, 31]. Fallopian tubes are preferred sites for transplantation because of their high vascularization as well as easier accessibility for later monitoring and oocyte retrieval via transvaginal ultrasound.

After autotransplantation procedure, we monitored the follicular growth by measuring serum estrogen and gonadotrophin levels, and conducting transvaginal ultrasound to detect growing antral follicles for more than 2 weeks. The presence of growing follicles could be clarified by elevation of estrogen levels and following images on ultrasound after a further increase in serum estrogen levels [29, 31]. Furthermore, it is important to suppress elevated endogenous gonadotropins during the induction of follicle growth in POI patients likely by restoring the responsiveness of residual follicles to exogenous gonadotrophin stimulation and by inhibiting early luteinization during follicle growth [32]. For that purpose, patients were treated with estrogen together with GnRH agonist during exogenous gonadotrophin stimulation [29, 31]. Follicle growth was stimulated by daily injection of recombinant FSH or urinary FSH. In cases with difficulty to maintain low LH level by the GnRH agonist treatment, GnRH antagonist was used to avoid the premature LH surge. When growing follicles were reached the preovulatory stage, oocyte maturation was triggered by hCG injection. At 36 h later, oocytes retrieval was conducted under transvaginal ultrasound for later conventional IVF or intracytoplasmic sperm injection (ICSI) [29, 31]. In POI patients, follicle growth was found in about 50% of patients with residual follicles after IVA. Although it takes 4–6 months to develop from primordial to preovulatory stage in human ovarian follicles, preovulatory follicles were detected within a few months after grafting in some cases, suggesting the growth of residual secondary follicles in response to disruption of Hippo signaling pathway [29]. In contrast, some preovulatory follicles were developed after 6 months, likely due to the activation of dormant PFs [25]. After IVF, embryos were cryopreserved at day 2 cleavage stage and then ET was performed under a hormone replacement cycle.

The effectiveness and safety of this treatment have been confirmed by several clinical data. We first declared the successful pregnancy of this procedure, 27 POI patients in their late 30 s with a long duration of amenorrhea (6.8 ± 2.1 years). After the first laparoscopic surgery to remove their ovaries, histological results indicated that 13 out of 27 patients had residual follicles. The IVA procedure and ovarian stimulation were conducted as described previously. Eight out of 13 patients manifested the follicular growth identified by estrogen concentration increase and ultrasound imaging. We could successfully retrieve mature oocytes in five patients for ICSI using the husband's sperm, resulting in successful conception after frozen-thawed ETs in three patients. At the time of publication, a healthy male baby was delivered at term without physical abnormalities and several accumulated cryopreserved embryos in other patients [29]. Our group also reported a subsequent study using the same procedure with adding 10 more patients ($n = 37$ POI patients). Based on the histological analyses, 54% (20/37) of these patients were found to have residual follicles. After ovarian stimulation, follicle growth was observed in 9 out of 20 POI patients, leading to 24 retrieved oocytes in 6 patients. The IVF-ET was then conducted in four patients, yielding three clinical pregnancies, followed by one miscarriage (supposed caused by the advanced patient age) and two healthy live births [31]. At the moment, both babies become more than 7 years of age without exhibiting any developmental abnormalities. Two other pregnancies by two other hospitals have also been achieved [33]. A Chinese group performed a case series study of IVA treatment in 14 POI patients with a mean duration since the last menses of 3.8 years. Among 14 patients, 11 patients showed undetectable AMH levels. One year after the IVA surgical intervention, a total of 15 follicle developmental waves were achieved in 6 out of the 14 patients, resulting in six oocytes from 4 patients. The following oocyte retrieval and IVF-ET resulted in four embryos at day 2 cleavage stage. Following fresh embryo transfer in one patient, the patient delivered a healthy baby boy. Three other patients were waiting for ET at the time of publication [34].

These results suggested the better outcomes that the IVA approach could bring to POI patients. Of note, POI patients in our study were in their late 30 s (mean age: 37.6 ± 4.6), and some cases suffered a long time of menopause. The pregnancy rate was nearly 15.0% (3/20) after the IVA procedure within a short period (<1 year), and two more patients had cryopreserved embryos pending ET [31]. This pregnancy rate is encouraging compared with other data. For instance, it is reported that around 4.4% of POI patients at the mean age of 26.6 ± 7.9 years could be pregnant spontaneously during follow-up as long as 13 years [28]. In another report, the pregnancy rates ranged from 3.6 to 6.8% after hormonal therapy with or without assisted reproductive treatments in young POI patients (median age: 30.4–32.5 years of age) [35–36].

Drug-Free IVA

In IVA procedures, ovarian cortical tissues are fragmented into small cubes to allow ovarian tissue culture with AKT activators. For surgical approach in polycystic ovarian syndrome patients, ovarian wedge resection and laparoscopic ovarian drilling have been demonstrated to enhance the growth of arrested growing antral follicles. Similarly, in oncological patients undergoing sterilizing treatments, fragmentation of ovarian cortices improved freezing and grafting outcomes for fertility preservation along with a higher follicular growth after grafting. A common intervention is involved in these procedures, namely, damaging of ovarian cortices, suggesting induction of follicle growth by alterations of mechanical tensions in ovarian cortices. To assess this hypothesis, rodent ovaries were cut into three pieces, followed by transplantation under kidney capsules. After harvesting, the weights of dissected ovaries were found to be increased in the fragmented ovaries as compared with paired intact ovaries due to stimulation of secondary follicle growth [29].

The Hippo signaling pathway is known as major intracellular signaling in regulating cell proliferation and determining organ size and is well conserved in all metazoan animals [36–38]. In Hippo signaling system, several negative growth regulators act in a serine/threonine kinase cascade, which eventually phosphorylates and inactivates two key transcriptional coactivators, Yes-associated protein (YAP) and transcriptional coactivator with PDZ-binding motif (TAZ) by their nuclear export. Disrupting Hippo signaling decreases YAP phosphorylation followed by an increase in nuclear YAP levels. Along with TEAD (transcription factors containing the TEA/ATTS DNA binding domain) transcriptional factors, nuclear YAP promotes the production of several downstream factors including CCN growth factors and baculoviral inhibitors of apoptosis repeat-containing (BIRC) apoptosis inhibitors. These factors, in turn, modulate cell growth, survival, and proliferation [37]. A network of upstream components involved in regulating cell adhesion, shape, and polarity mainly regulates the Hippo signaling pathway [39]. Actin, a major cytoskeleton, acts as a multifunctional protein that forms microfilaments maintaining many important cellular processes. Rapid polymerization changes in globular actin (G-actin) to form filamentous actin (F-actin) mediate cell adhesion, shape maintenance, and locomotion. F-actin formation in the stress fiber has been demonstrated to disrupt the Hippo signaling pathway leading to accumulation of nuclear YAP [40].

The expression of key Hippo signaling genes (*YAP*, *TAZ*, *MST1/2*, *SAV1*, and *LATS1/2*) was detected in follicles of both murine's and human ovaries at different developmental stages [29]. In our published report, the ovarian fragmentation stimulated a transient increase in the polymerization of G-actin to F-actin. The F-actin disrupted Hippo signaling pathway to decrease phospho-YAP levels and increase nuclear localization of YAP, resulting in the upregulation of downstream CCN growth factors and BIRC apoptosis inhibitors [29]. Mediation of Hippo signal disruption in follicle growth induction was further confirmed the blockade of the follicle growth by using CCN2 antibodies and verteporfin [29], a small molecule capable of suppressing interactions between YAP and TEAD transcriptional factors [41]. In humans, CCN growth factors were also increased in ovarian cortices by cutting thawed human cortical strips into small cubes [29], indicating stimulatory actions of CCN growth factors in ovarian follicle growth. Murine ovaries treated with Jasplakinolide, an actin polymerization-promoting cyclic peptide, or sphingosine-1-phosphate, a follicular fluid constituent known to promote actin polymerization demonstrated enhanced conversion of G-actin to the F-actin and subsequent nuclear translocation of YAP as well as increased expression of downstream CCN growth factors, finally resulted in follicle growth [42].

In ovaries of patients with the early stage of POI and POR-DOR, PFs can be activated spontaneously. However, due to the limited number of residual follicles, the number of growing follicles is very low. Because Hippo signaling disruption alone is effective in stimulation of early follicle growth, we simplified the procedure of IVA to develop drug-free IVA approach to promote early follicle growth to obtain a higher number of mature follicles [13]. In this approach, ovarian cortex was fragmented into small cubes (1–2 mm) to disrupt ovarian Hippo signaling followed by the grafting back into remaining ovaries and beneath serosa of Fallopian tubes without tissue culture [43] (Fig. 3.2). Drug-free IVA procedures allowed our ovarian dysfunction patients to gain successful pregnancies and deliveries [43, 44].

A case series study from our group in 11 POR-DOR women with nearly undetectable AMH levels (median 0.04 ng/ml, min–max, 0–0.8) and extremely low antral follicle count (median 1, min–max, 0–4) revealed that this procedure increased the number of antral follicles under FSH stimulation after the surgical interventions and the number of retrieved mature oocytes per cycle. The rates of fertilization and high-quality embryo were 68.7% and 56.9%, respectively. In consequence, five patients achieved successful pregnancies, resulting in one live birth, two ongoing pregnancies, and one miscarriage. Moreover, three patients and the miscarriage patient could achieve cryopreserved embryos for future transfer [43]. We also reported the clinical outcome of a case series study by adding four more patients in our previous drug-free IVA study. Among 15 POR-DOR patients, 13 patients had a higher number of antral

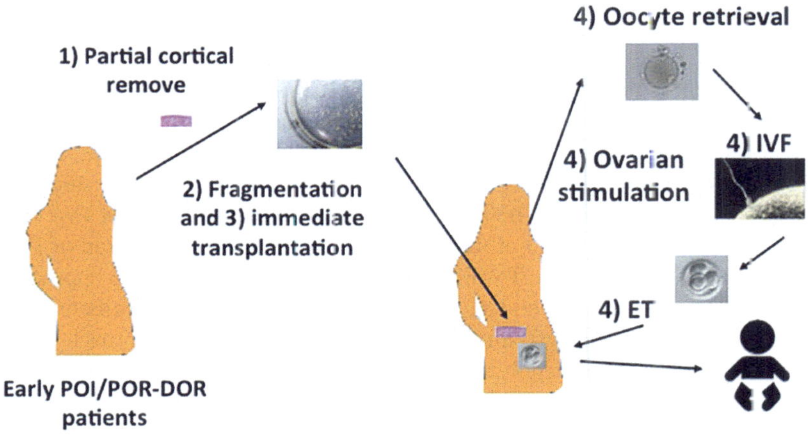

Fig. 3.2 The drug-free IVA for patients with the early stage of POI and poor ovarian response with diminishing ovarian reserve (POR-DOR). This procedure consists of (1) partial ovarian cortical removal under laparoscopic surgery, (2) fragmentation of cortices to disrupt ovarian Hippo signaling, (3) grafting ovarian cubes under the same laparoscopic surgery, and (4) ovarian stimulation followed by IVF-ET

follicles detected by transvaginal ultrasound monitors as well as a higher number of retrieved mature oocytes for IVF as compared with previous IVF outcomes before the drug-free IVA treatment. One spontaneous pregnancy and embryo transfers allowed four live births and one ongoing pregnancy. Five additional patients and one miscarriage patient had cryopreserved embryos for future transfer [44]. In other fertility centers, the drug-free IVA was also declared to be successful in treating POI patients at the early stage. Several studies reported successful outcomes by using the drug-free IVA approach. In a case report, a 32-year-old POI patient with undetectable AMH levels (<0.02 ng/ml) could achieve pregnancy after the drug-free IVA procedure followed by 20 days of ovarian stimulation and IVF-ET [45]. Another study enrolled 14 POI patients who were at the early stage of POI with median age of 33 years (29–36 years), a median length of amenorrhoea of 1.5 years (1–11 years), and a median AMH level of 0.02 ng/ml (0.01–0.1 ng/ml). Under the drug-free IVA procedure, seven patients manifested follicle growth, and oocyte retrieval was successful in five patients, resulting in six embryo transfers and subsequent four pregnancies [46]. In addition to the drug-free IVA, a group reported the clinical outcome of the disruption of Hippo signaling pathway by ovarian biopsy and scratching in 80 POI women. The resumption of ovarian function or follicle growth after FSH stimulation occurred in 13.75% (11/80) patients. Among 10 women undergoing oocyte retrieval and IVF, one could achieve a healthy baby [47].

Conclusion

The development of the original and drug-free IVA approaches increased the chance to conceive genetically related children in ovarian dysfunction patients including POI and POR-DOR, leading to the expansion of modern infertility treatment. Because published studies involve a small group of patients, better designed studies are essential to further investigate before wide clinical implementation of IVA therapies. However, the requirement of the control group with sham operation of IVA is ethically unjustified to conduct. To minimize the invasiveness of the IVA approaches, developing a less invasive method to predict the presence of residual follicles before the surgical intervention is essential. In addition, an alternative approach such as transvaginal injection of actin polymerization-enhancing reagents to ovaries has been considered to disrupt Hippo signaling pathway instead of ovarian fragmentation.

References

1. Martin JA, Hamilton BE, Osterman MJK, Driscoll AK, Drake P (2018) Births: final data for 2016. Natl Vital Stat Rep Cent Dis Control Prev Natl Cent Health Stat Natl Vital Stat Syst 67:1–55
2. Mathews TJ, Hamilton BE (2016) Mean age of mothers is on the rise: United States, 2000–2014. NCHS Data Brief 232:1–8
3. Coulam CB, Adamson SC, Annegers JF (1986) Incidence of premature ovarian failure. Obstet Gynecol 67:604–606
4. Lagergren K, Hammar M, Nedstrand E, Bladh M, Sydsjö G (2018) The prevalence of primary ovarian insufficiency in Sweden; a national register study. BMC Womens Health 18:175. https://doi.org/10.1186/s12905-018-0665-2
5. Niederberger C, Pellicer A, Cohen J, Gardner DK, Palermo GD, O'Neill CL, Chow S, Rosenwaks Z, Cobo A, Swain JE, Schoolcraft WB, Frydman R, Bishop LA, Aharon D, Gordon C, New E, Decherney A, Tan SL, Paulson RJ, Goldfarb JM, Brännström M, Donnez J, Silber S, Dolmans M-M, Simpson JL, Handyside AH, Munné S, Eguizabal C, Montserrat N, Izpisua Belmonte JC, Trounson A, Simon C, Tulandi T, Giudice LC, Norman RJ, Hsueh AJ, Sun Y, Laufer N, Kochman R, Eldar-Geva T, Lunenfeld B, Ezcurra D, D'Hooghe T, Fauser BCJM, Tarlatzis BC, Meldrum DR, Casper RF, Fatemi HM, Devroey P, Galliano D, Wikland M, Sigman M, Schoor RA, Goldstein M, Lipshultz LI, Schlegel PN, Hussein A, Oates RD, Brannigan RE, Ross HE, Pennings G, Klock SC, Brown S, Van Steirteghem A, Rebar RW, LaBarbera AR (2018) Forty years of IVF. Fertil Steril 110:185-324.e5. https://doi.org/10.1016/j.fertnstert.2018.06.005
6. De Vos M, Devroey P, Fauser BCJM (2010) Primary ovarian insufficiency. Lancet 376:911–921. https://doi.org/10.1016/S0140-6736(10)60355-8
7. Donnez J, Dolmans M-M (2013) Fertility preservation in women. Nat Rev Endocrinol 9:735–749. https://doi.org/10.1038/nrendo.2013.205
8. Sauer MV, Paulson RJ, Lobo RA (1990) A preliminary report on oocyte donation extending reproductive potential to women over 40. N Engl J Med 323:1157–1160. https://doi.org/10.1056/NEJM199010253231702
9. van Egmond A, van der Keur C, Swings GMJS, van Beelen E, van Zijl L, Scherjon SA, Claas FHJ (2013) Preservation of human placenta facilitates multicenter studies on the local immune response in normal and aberrant pregnancies. J Reprod Immunol 98:29–38. https://doi.org/10.1016/j.jri.2013.03.001

10. Stoop D, Baumgarten M, Haentjens P, Polyzos NP, De Vos M, Verheyen G, Camus M, Devroey P (2012) Obstetric outcome in donor oocyte pregnancies: a matched-pair analysis. Reprod Biol Endocrinol RBE 10:42. https://doi.org/10.1186/1477-7827-10-42
11. Schwarze JE, Borda P, Vásquez P, Ortega C, Villa S, Crosby JA, Pommer R (2018) Is the risk of preeclampsia higher in donor oocyte pregnancies? a systematic review and meta-analysis. JBRA Assist Reprod 22:15–19. https://doi.org/10.5935/1518-0557.20180001
12. McGee EA, Hsueh AJ (2000) Initial and cyclic recruitment of ovarian follicles. Endocr Rev 21:200–214. https://doi.org/10.1210/edrv.21.2.0394
13. Hsueh AJW, Kawamura K, Cheng Y, Fauser BCJM (2015) Intraovarian control of early folliculogenesis. Endocr Rev 36:1–24. https://doi.org/10.1210/er.2014-1020
14. Adhikari D, Liu K (2009) Molecular mechanisms underlying the activation of mammalian primordial follicles. Endocr Rev 30:438–464. https://doi.org/10.1210/er.2008-0048
15. Reddy P, Shen L, Ren C, Boman K, Lundin E, Ottander U, Lindgren P, Liu Y-X, Sun Q-Y, Liu K (2005) Activation of Akt (PKB) and suppression of FKHRL1 in mouse and rat oocytes by stem cell factor during follicular activation and development. Dev Biol 281:160–170. https://doi.org/10.1016/j.ydbio.2005.02.013
16. Castrillon DH, Miao L, Kollipara R, Horner JW, DePinho RA (2003) Suppression of ovarian follicle activation in mice by the transcription factor Foxo3a. Science 301:215–218. https://doi.org/10.1126/science.1086336
17. Reddy P, Liu L, Adhikari D, Jagarlamudi K, Rajareddy S, Shen Y, Du C, Tang W, Hämäläinen T, Peng SL, Lan Z-J, Cooney AJ, Huhtaniemi I, Liu K (2008) Oocyte-specific deletion of Pten causes premature activation of the primordial follicle pool. Science 319:611–613. https://doi.org/10.1126/science.1152257
18. John GB, Gallardo TD, Shirley LJ, Castrillon DH (2008) Foxo3 is a PI3K-dependent molecular switch controlling the initiation of oocyte growth. Dev Biol 321:197–204. https://doi.org/10.1016/j.ydbio.2008.06.017
19. Laplante M, Sabatini DM (2009) mTOR signaling at a glance. J Cell Sci 122:3589–3594. https://doi.org/10.1242/jcs.051011
20. Adhikari D, Flohr G, Gorre N, Shen Y, Yang H, Lundin E, Lan Z, Gambello MJ, Liu K (2009) Disruption of Tsc2 in oocytes leads to overactivation of the entire pool of primordial follicles. Mol Hum Reprod 15:765–770. https://doi.org/10.1093/molehr/gap092
21. Adhikari D, Zheng W, Shen Y, Gorre N, Hämäläinen T, Cooney AJ, Huhtaniemi I, Lan Z-J, Liu K (2010) Tsc/mTORC1 signaling in oocytes governs the quiescence and activation of primordial follicles. Hum Mol Genet 19:397–410. https://doi.org/10.1093/hmg/ddp483
22. Sun X, Su Y, He Y, Zhang J, Liu W, Zhang H, Hou Z, Liu J, Li J (2015) New strategy for in vitro activation of primordial follicles with mTOR and PI3K stimulators. Cell Cycle 14:721–731. https://doi.org/10.1080/15384101.2014.995496
23. Cheng Y, Kim J, Li XX, Hsueh AJ (2015) Promotion of ovarian follicle growth following mTOR activation: synergistic effects of AKT stimulators. PLoS ONE 10. https://doi.org/10.1371/journal.pone.0117769
24. Adhikari D, Liu K (2010) mTOR signaling in the control of activation of primordial follicles. Cell Cycle 9:1673–1674. https://doi.org/10.4161/cc.9.9.11626
25. Li J, Kawamura K, Cheng Y, Liu S, Klein C, Liu S, Duan E-K, Hsueh AJW (2010) Activation of dormant ovarian follicles to generate mature eggs. Proc Natl Acad Sci U S A 107:10280–10284. https://doi.org/10.1073/pnas.1001198107
26. Adhikari D, Gorre N, Risal S, Zhao Z, Zhang H, Shen Y, Liu K (2012) The safe use of a PTEN inhibitor for the activation of dormant mouse primordial follicles and generation of fertilizable eggs. PLoS ONE 7. https://doi.org/10.1371/journal.pone.0039034
27. van Kasteren YM, Schoemaker J (1999) Premature ovarian failure: a systematic review on therapeutic interventions to restore ovarian function and achieve pregnancy. Hum Reprod Update 5:483–492. https://doi.org/10.1093/humupd/5.5.483
28. Bidet M, Bachelot A, Bissauge E, Golmard JL, Gricourt S, Dulon J, Coussieu C, Badachi Y, Touraine P (2011) Resumption of ovarian function and pregnancies in 358 patients with premature ovarian failure. J Clin Endocrinol Metab 96:3864–3872. https://doi.org/10.1210/jc.2011-1038

29. Kawamura K, Cheng Y, Suzuki N, Deguchi M, Sato Y, Takae S, Ho C, Kawamura N, Tamura M, Hashimoto S, Sugishita Y, Morimoto Y, Hosoi Y, Yoshioka N, Ishizuka B, Hsueh AJ (2013) Hippo signaling disruption and Akt stimulation of ovarian follicles for infertility treatment. Proc Natl Acad Sci U S A 110:17474–17479. https://doi.org/10.1073/pnas.1312830110
30. Haino T, Tarumi W, Kawamura K, Harada T, Sugimoto K, Okamoto A, Ikegami M, Suzuki N (2018) Determination of follicular localization in human ovarian cortex for vitrification. J Adolesc Young Adult Oncol 7:46–53. https://doi.org/10.1089/jayao.2017.0028
31. Suzuki N, Yoshioka N, Takae S, Sugishita Y, Tamura M, Hashimoto S, Morimoto Y, Kawamura K (2015) Successful fertility preservation following ovarian tissue vitrification in patients with primary ovarian insufficiency. Hum Reprod 30:608–615. https://doi.org/10.1093/humrep/deu353
32. Nelson LM (2009) Clinical practice. primary ovarian insufficiency. N Engl J Med 360:606–614. https://doi.org/10.1056/NEJMcp0808697
33. Kawamura K, Cheng Y, Sun Y-P, Zhai J, Diaz-Garcia C, Simon C, Pellicer A, Hsueh AJ (2015) Ovary transplantation: to activate or not to activate. Hum Reprod 30:2457–2460. https://doi.org/10.1093/humrep/dev211
34. Zhai J, Yao G, Dong F, Bu Z, Cheng Y, Sato Y, Hu L, Zhang Y, Wang J, Dai S, Li J, Sun J, Hsueh AJ, Kawamura K, Sun Y (2016) In Vitro Activation of follicles and fresh tissue auto-transplantation in primary ovarian insufficiency patients. J Clin Endocrinol Metab 101:4405–4412. https://doi.org/10.1210/jc.2016-1589
35. Zhang P, Shi Y, Gao X, Wang S, Wang J, Chen Z-J (2007) Clinical analysis of Chinese infertility women with premature ovarian failure. Neuro Endocrinol Lett 28:580–584
36. Bachelot A, Nicolas C, Bidet M, Dulon J, Leban M, Golmard JL, Polak M, Touraine P (2017) Long-term outcome of ovarian function in women with intermittent premature ovarian insufficiency. Clin Endocrinol (Oxf) 86:223–228. https://doi.org/10.1111/cen.13105
37. Meden-Vrtovec H, Gersak K, Franić D (2011) Distribution of etiological factors of hypergonadotropic amenorrhea. Clin Exp Obstet Gynecol 38:369–372
38. Chen X, Chen S-L, Ye D-S, Liu Y-D, He Y-X, Tian X-L, Xu L-J, Tao T (2016) Retrospective analysis of reproductive outcomes in women with primary ovarian insufficiency showing intermittent follicular development. Reprod Biomed Online 32:427–433. https://doi.org/10.1016/j.rbmo.2015.12.011
39. Johnson R, Halder G (2014) The two faces of Hippo: targeting the Hippo pathway for regenerative medicine and cancer treatment. Nat Rev Drug Discov 13:63–79. https://doi.org/10.1038/nrd4161
40. Wada K-I, Itoga K, Okano T, Yonemura S, Sasaki H (2011) Hippo pathway regulation by cell morphology and stress fibers. Dev Camb Engl 138:3907–3914. https://doi.org/10.1242/dev.070987
41. Halder G, Johnson RL (2011) Hippo signaling: growth control and beyond. Development 138:9–22. https://doi.org/10.1242/dev.045500
42. Cheng Y, Feng Y, Jansson L, Sato Y, Deguchi M, Kawamura K, Hsueh AJ (2015) Actin polymerization-enhancing drugs promote ovarian follicle growth mediated by the Hippo signaling effector YAP. FASEB J Off Publ Fed Am Soc Exp Biol 29:2423–2430. https://doi.org/10.1096/fj.14-267856
43. Kawamura K, Ishizuka B, Hsueh AJW (2020) Drug-free in-vitro activation of follicles for infertility treatment in poor ovarian response patients with decreased ovarian reserve. Reprod Biomed Online 40:245–253. https://doi.org/10.1016/j.rbmo.2019.09.007
44. Tanaka Y, Hsueh AJ, Kawamura K (2020) Surgical approaches of drug-free in vitro activation and laparoscopic ovarian incision to treat patients with ovarian infertility. Fertil Steril 114:1355–1357. https://doi.org/10.1016/j.fertnstert.2020.07.029
45. Fabregues F, Ferreri J, Calafell JM, Moreno V, Borrás A, Manau D, Carmona F (2018) Pregnancy after drug-free in vitro activation of follicles and fresh tissue autotransplantation in primary ovarian insufficiency patient: a case report and literature review. J Ovarian Res 11:76. https://doi.org/10.1186/s13048-018-0447-3

46. Ferreri J, Fàbregues F, Calafell JM, Solernou R, Borrás A, Saco A, Manau D, Carmona F (2020) Drug-free in-vitro activation of follicles and fresh tissue autotransplantation as a therapeutic option in patients with primary ovarian insufficiency. Reprod Biomed Online 40:254–260. https://doi.org/10.1016/j.rbmo.2019.11.009
47. Zhang X, Han T, Yan L, Jiao X, Qin Y, Chen Z-J (2019) Resumption of ovarian function after ovarian biopsy/scratch in patients with premature ovarian insufficiency. Reprod Sci 26:207–213. https://doi.org/10.1177/1933719118818906

Chapter 4
Ovarian Cancer Stem Cells and Their Regulatory Mechanisms: Potential Targets for Therapy

Seema C. Parte, Moorthy P. Ponnusamy, Surinder K. Batra, and Sham S. Kakar

Abstract

Introduction: Ovarian cancer is one of the most lethal gynecological cancers. Aspects related to the stemness and cancer stem cell (CSC) related compartments and other components constituting the tumor microenvironment and their dynamics across the trajectory of an individual patients' life during the course of the treatment regimen and the subtle differences of the cellular, molecular parameters, and constituents between chemo-naïve and sensitive ones may provide a multitude of information and novel insights into greater prognostic significance.
Methods: Ovarian cancer stem cells and their regulatory mechanisms were reviewed as potential targets for therapy.
Conclusions: Exploration of novel oncogenes and their regulatory mechanisms in a patient and treatment-specific manner and their combination with conventional oncotherapy may mark a new era in personalized and precision medicine to effectively manage ovarian cancer thus improving patient quality of life and prolonging the progression-free and overall survival post-remission.
Results: Hence, a comprehensive understanding of ovarian CSCs at the primary site and in ascites fluid and the molecular mechanisms governing their malignant potential driving ovarian cancer progression toward relapse and incurable stage are the essential needs of the hour.

Keywords Cancer · Cancer stem cells · Ovarian cancer · Ovary · PTTG1 · Targets · Therapy

S. C. Parte · M. P. Ponnusamy · S. K. Batra
Department of Biochemistry and Molecular Biology, University of Nebraska Medical Center, Omaha, NE 68198, USA

S. S. Kakar (✉)
Department of Physiology and Brown Cancer Center, University of Louisville, Louisville, KY 40202, USA
e-mail: sham.kakar@louisville.edu

I. Virant-Klun (ed.), *Stem Cells in Reproductive Tissues and Organs*, Stem Cell Biology and Regenerative Medicine 70, https://doi.org/10.1007/978-3-030-90111-0_4

Abbreviations

ABCG2	ATP Binding Cassette Subfamily G Member 2 (Junior Blood Group)
ABO	ABO, Alpha 1-3-N-acetylgalactosaminyltransferase and alpha 1-3-galactosyltransferase
AD	Adherent
ALDH1	Aldehyde dehydrogenase 1
ALDH1A1	Aldehyde dehydrogenase 1 family member A1
BKM120	Buparlisib
BMI1	BMI1 proto-oncogene, polycomb ringer finger
BL	Borderline tumor
BN	Benign tumor
BRCA1	BRCA1 DNA repair associated
BRCA2	BRCA2 DNA repair associated
CA125	Cancer antigen 125, ovarian cancer biomarker
CAOV3	Human ovarian cancer cell line from adenocarcinoma
CBS	Cystathionine beta-synthase
CD	Cluster of differentiation
CDK4	Cyclin-dependent kinase 4
CHEK2	Checkpoint kinase 2
CN	Chemo-naïve
CR	Chemo-resistant
CSC	Cancer stem cell
DAPI	4′,6-Diamidino-2-phenylindole
DDX4	DEAD-box helicase 4 (VASA)
DNA	Deoxyribonucleic acid
EMC	Epithelial-mesenchymal transition
EOC	Epithelial ovarian cancer
EpCAM	Epithelial cell adhesion molecule
GLI	GLI family zinc finger
GRO	=CXCL1, C-X-C motif chemokine ligand 1
GSK3β	Glycogen synthase 3 beta
HES1	Hes Family BHLH Transcription Factor 1
HEY1	Hes-related family BHLH transcription factor with YRPW motif 1
HG	High-grade tumor
HIPEC	Hyperthermic intraperitoneal chemotherapy
ICAM1	Intercellular adhesion molecule 1
IL	Interleukin
JAK	Janus kinase
KLF4	Kruppel like factor 4
LEF	Lymphoid enhancer binding factor
LGR5	Leucine rich repeat-containing G protein-coupled receptor 5
LRG6	Leucine rich repeat-containing G protein-coupled receptor 6
LY6A	Lymphocyte antigen 6 family member A

MCP1	=CCL2, C–C motif chemokine ligand 2
MIF	Macrophage migration inhibitory factor
MMP9	Matrix metallopeptidase 9
mTOR	*Mammalian target of rapamycin*
MYC	MYC proto-oncogene, BHLH transcription factor
NAD	Non-adherent
NANOG	Nanog homeobox
NAP2	Nucleosome assembly protein 2
NEK2	NIMA related kinase 2
NOTCH1	Notch receptor 1
OC	Ovarian cancer
OCT4	Octamer-binding transcription factor 4 (POU5F1)
OPG	=TRANSF11B, TNF receptor superfamily member 11b
OSE	Ovarian surface epithelium
PFI	Platinum-free interval
PI3K	Phosphoinositide 3-kinase
PTCH1	Patched 1
PTCH2	Patched 2
PTEN	Phosphatase and tensin homolog
PTTG1	PTTG1 Regulator of sister chromatid separation, securin
RAD51	RAD51 recombinase
RANTES	=CCL5, C–C motif chemokine ligand 5
RNA	*Ribonucleic acid*
mRNA	*Messenger ribonucleic acid*
RT-PCR	*Reverse transcription polymerase chain reaction*
SHH	Sonic hedgehog signaling molecule
SLUG	=SNAI2, snail family transcriptional repressor 2
SMO	Smoothened, frizzled class receptor
SNAIL	=SNAI1, Snail family transcriptional repressor 1
SOX2	SRY-box transcription factor 2
SSEA4	Stage-specific embryonic antigen-4
STAT3	Signal transducer and activator of transcription 3
SUFU	SUFU negative regulator of hedgehog signaling
TCF4	Transcription factor 4
TGFB	Transforming growth factor beta
TERT	Telomerase reverse transcriptase
TIMP2	TIMP metallopeptidase inhibitor 2
TNF	Tumor Necrosis Factor (TNF-alpha)
TNM	Tumor, node, and metastasis
WNT4	Wnt family member 4
ZEB1	Zinc finger E-box binding homeobox 1

Introduction

Ovarian cancer is one of the most lethal gynecological cancers with an estimated new diagnosis of ovarian cancer cases of about 21,750 in the USA of which 13,940 women would succumb to the disease in 2020 [1]. The majority of the cases are epithelial ovarian cancer, of which high-grade serous ovarian tumors reflect worst prognosis. Although incidence rate and mortality trends reveal a decreasing trend of 1.6% and 2.3%, respectively, per year but the risk and predisposition factors (genetic background, family history, age, health, smoking, and alcohol consumption) entitle 38 out of 100 women in the USA to be diagnosed with ovarian cancer once in their lifetime [2]. Globally, approximately 240,000 women will be diagnosed with ovarian cancer of which 150,000 ultimately die. Ovarian cancer is the seventh most common cancer in women and ranks the eighth most common cause of cancer-related deaths. Hence, the significance of early screening and detection is evident from the fact that 5-year survival rate is 29% versus 92% for late and early detection, respectively [3, 4].

Ovarian Cancer Screening, Diagnosis and Treatment

Efforts (randomized trials) are ongoing to establish effective strategies for systematic multimodal annual screening with serum cancer antigen such as CA125, annual trans-vaginal ultrasound screening, coupled with other biomarkers like human epididymis protein 4, a glycoprotein secreted by the Mullerian epithelium of the female reproductive tract. However, these efforts met with but little success so far in terms of reduction of mortality. Hence, genomic-based screening is looked upon as the next promising strategy under consideration. Being a collection of various diseases in terms of its histopathological and patho-physiological symptoms and other complexities based upon distinct precursor lesions, cells of origin, ovarian and especially epithelial ovarian cancer (EOC, 90% of ovarian cancer cases) pose as very enigmatic disease due to varied and random symptoms associated such as abdominal bloating, early satiety, nausea, abdominal distension, change in bowel function, urinary symptoms, back pain, fatigue, and loss of weight. EOC includes other subtypes such as low-grade serous, clear cell, endometrioid, and mucinous based upon histological features [4]. Besides initial screening, EOC staging is followed by the International Federation of Gynecology and Obstetrics (FIGO) staging of ovarian cancer or with a tumor, node, and metastasis (TNM) classifications given by the American Joint Committee on Cancer. Treatment of ovarian cancer includes primary de-bulking of tumors by the surgical process including total abdominal hysterectomy, bilateral salpingo-oophorectomy, omentectomy, and peritoneal examination with aid of biopsy and removal of doubtful region, and para-aortic and pelvic lymph node dissection and ultimately chemotherapy. Clinical trials introducing interval debulking, neo-adjuvant chemotherapy, hyperthermic intraperitoneal chemotherapy (HIPEC), etc. have been studied/are under study, thus pronouncing the need for selection of patients with FIGO stage IIIC and IV for testing the appropriate combination of the treatment regimen.

Higher parity use of oral contraceptives and tubal ligation significantly reduces the risk of ovarian cancer. Several factors have been identified to increase the risk of ovarian cancer such as family history of ovarian or breast cancer (with a mutation in BRCA1, BRCA2, RAD51, and also CHEK2, WNT4, TERT, ABO, etc. genes), higher menopausal age, hormone therapy, obesity, endometriosis, and smoking. Ovarian cancer or cancer per se can be controlled only after the causative agents and reasons are identified. A prediction of risk-based effective screening is performed thus providing an opportunity to intervene and alleviate the reasons and reduce the rate of occurrence and/or recurrence [3].

Despite the above factors under consideration and active trials being undertaken to tailor treatment regimen appropriate for the patients based upon patient tumor status-histology, stage, disease complexity; a high rate (75%) of recurrence is observed in advanced stage (III and IV) [5] with ascites fluid build-up, whereas the initial phase is asymptomatic [4]. The patients initially respond to the chemotherapeutic treatment regimen, but later manifest a refractory phase due to chemoresistance arising out of a shorter duration (6 months) of platinum-free interval (PFI) termed as 'platinum resistant'. Patients with >6 months PFI are platinum sensitive or partially platinum sensitive, hence receive multi-rounds of platinum-based chemotherapy, i.e. intravenous administration of carboplatin and paclitaxel, which, in turn, leads the patients toward recalcitrance and hence the vicious circle sets in [6]. Thus oncotherapy resistance is considered as a major factor responsible for poor survival rate in ovarian cancer patients thus necessitating research efforts toward a broader understanding of the underlying mechanism imparting chemoresistance in patients and toward establishing precision medicine/personalized treatment regimen in accordance with patient's secondary response to chemotherapy.

Collectively it is construed that a special type of cancer cell compartment which appears in minuscule proportion (0.1–3%) but possesses special cellular properties of quiescent state characterized by slow cycling cells, self-renewal with stemness related cell surface markers and with the ability to maintain tumor in vivo known as cancer stem cells (CSCs) are the root cause of chemoresistance, tumor initiation, progression, metastasis and ultimate tumor relapse [7, 8]. Similar counterpart CSCs are detected in malignant ascites fluid as well.

Patient ascites derived transformed clones with typical stem cell characteristics, i.e. tumor stem cells/CSCs clones were first derived by Bapat et al. [9] and found to differentiate in vitro into anchorage-independent spheroids. These tumorigenic clones upon serial transplantation could develop into tumors thus laying the foundation for the concept of ovarian stem/progenitor cell transformation into CSCs. We elaborate upon ascites-derived CSCs further. Subsequently, studies both in the fields of ovarian stem cells [10–15] and CSCs [16, 17] have put forth interesting findings based upon in vivo and in vitro work in mice models and human clinical specimen, acknowledging plasticity of the surface epithelial cell layer in repair in normal physiology and the putative neoplastic transformation of these OSE cells thus, suggesting the possibility of stem cell/CSC origin. Similarities between the fundamental properties of normal stem cells and CSC counterparts, a parallel function of regeneration and growth and the common site of origin of stem and CSCs being the OSE layer, this compartment has been explored in the context of ovarian cancer and CSCs and much remains to be explored yet. Recently, Protein C Receptor (Procr)+ and Lgr5+ cells with distinct cell lineage tracing behavior reported by Wang et al. [18] were implicated in the maintenance of OSE homeostasis during ovulatory rupture and repair process, thus highlighting the dynamics of adult stem cells in the context of tissue injury.

Oncogenic Transformation of Stem Cells to Cancer Stem Cells (CSCs)

Interestingly, as much as their origin, the dysregulation of CSCs remains an unresolved mystery to cancer biologists and oncologists per se and ultimately poses a serious bottleneck in the complete annihilation of cancer and prevention of relapse. CSCs appear to be at the epicenter with exceptional cellular features including cellular reprogramming such as quiescence, the ability for self-renewal, proliferation, and differentiation under the influence of appropriate micro-environmental cues and developmentally high plastic nature thus responsible for tumor initiation, progression, and metastasis. They exist/persist as sub-populations with fundamentally varying metabolic, mechanistic, functional, and hierarchical heterogeneity. Interestingly, recently glycan modification of CSC markers (CD133, CD24, CD44, EpCAM/ESA) has been implicated to play significant role in CSC specific function in tumor cell adhesion, survival, invasiveness, metastasis, pluripotency, stemness, drug resistance, and apoptosis. Thus exclusive signatures of glycome from normal stem cells, CSCs, and non-CSCs and its exclusivity/dynamism associated with respect to oncotherapy may open up new avenues to target cancer cells and CSCs both by combination of conventional cyto-toxic drugs with altered glycan/glycopeptide-based vaccine [19].

Several oncogenic programs govern CSC regulation, and they, in turn, interact with the tumor bulk cells and surrounding elements of the tumor micro-environment thus manifesting as a tumor. Lgr5+ rare adult stem cells have been reported in small intestine and colon and other tissues such as eye, brain, hair follicle, mammary gland, stomach, and reproductive organs. Based on this information, it is speculated that the malignant Lgr5+ cells in various cancers might be the transformed remnants of these stem cell counterparts [20, 21]. Similarly, OSE being the site of repetitive ovulation, it is proposed to undergo a malignant transformation on account of aberrant rupture-repair process. Also, OSE cells on prolonged culture are known to undergo spontaneous transformation with tumorigenic potential in vitro [22]. In addition, the primary human OSE cells upon transfection with SV40, human papillomavirus E6/E7, Ras, hTERT, cyclin D1, CDK4, or hTERT, deletion of p53 with subsequent c-Myc and k-Ras overexpression or overexpression of ß-hCG are known to undergo transformation into cancerous and immortal state. These events may be associated with distinct changes in the genetic composition with respect to genes involved in cell proliferation, metabolism, and cytoskeleton re-organization [23, 24]. All the above events described may be responsible for developing a transformed phenotype of the OSE layer with the resident stem cells and thus implicated in the development of CSCs per se. Besides ovarian CSCs, malignant ascites fluid reveals heterogeneous populations of tumor cells with metastatic potential.

Ascites-Derived Cancer Dissemination and CSCs

Ascites fluid is generally detected in approximately one-third of the ovarian cancer cases and almost in all patients with relapse, serving as a major factor contributing toward patient morbidity and mortality via metastatic spread and peritoneal dissemination of ovarian cancer to surrounding visceral organs. This, in turn, leads to chemoresistance due to its pro-inflammatory and tumor conducive environment and is also associated with poor prognosis [25, 26]. Sufficient understanding of ascites CSCs and their interaction with other components of tumor microenvironment may expedite the discovery of targeted treatments aiming both at CSCs and those belonging to the niche and their complex cross-talk ultimately meant for prolonging progression-free survival and overall survival [25]. Ascitic fluid comprises a number of tumor and non-tumor cells, such as fibroblasts, adipocytes, mesothelial, endothelial and inflammatory cells, cell-free DNA and numerous signaling molecules that control cell behavior and tumor-promoting cytokines, chemokines, growth factors, proteinases and extracellular matrix components contributing to cell growth, tumor invasion and resistance to TNF-related apoptosis-inducing ligand [26]. Multiplex profiling of ascites-derived cytokines from epithelial ovarian cancer patients revealed enhanced expression of several factors including angiogenin, angiopoietin, GRO, ICAM-1, IL-6, IL-6R, IL-8, IL10, leptin, MCP-1. MIF, NAP-2, osteoprotegerin (OPG), RANTES, TIMP-2, and urokinase plasminogen activator receptor (uPAR) [27]. Mechanism: Tumor cells from the primary ovarian tumor capsule escape by shedding process in the peritoneal cavity and migrate to distant sites with the aid of proliferation and epithelial mesenchymal transition (EMT) followed by homing onto the mesothelial extra cellular matrix by MET process where they assume a single cell or multicellular spheroidal aggregate like structure. This process aids the development of secondary lesions and metastasis to distant organs [25].

Putative Ovarian Stem Cell Connections with CSCs

As per the central dogma of ovarian biology, the ovary is supposed to harbor only a limited number of germ cells/follicles that never undergo de novo renewal and eventually get depleted due to aging process culminating into menopause. This theory was contradicted by studies by Johnson et al. [10] who demonstrated the presence of ovarian germline stem cells for the first time, and subsequently several other researchers have contributed to our understanding of the existence of multi- and/or pluri-potent marker expressing or label-retaining cells in mice, germline stem cells in rat and human as well as other mammalian ovaries [11–15, 28]. Extensive studies by Virant-Klun et al. [17, 29, 30] have reported a putative connection of ovarian stem-like cells in the pathophysiology of ovarian cancer by studying several stem cell-based markers (Nanog, Sox2, SSEA4) and showed EMT phenomenon in OSE layer of normal and ovarian tumors. Oncogenic/neoplastic transformation of stem cells residing in the ovaries is suggested to cause high-grade serous adenocarcinomas in mice [16].

Ovarian CSCs

The putative CSC populations are isolated by several methods including clonal selection, FACS-based sorting for cell surface markers (*ALDH1/2, EpCAM, CD24, CD34, CD44, CD133, CD117, LGR5, LY6A*), other marker independent methods such as Side Population analysis by Hoechst dye 33,342 exclusion method and auto-fluorescence [31–33]. Typically, CSCs express stem cell-specific markers such as Oct4, Nanog, Bmi1, ABCG2, and in vivo and in vitro functional assays characterize their self-renewal capacity (asymmetric cell division) as well as differentiation potential to produce differentiated cell progeny contributing toward cellular heterogeneity of the tumor bulk cells. The differentiated tumor cells may differ in terms of the genetic, phenotypic, and epigenetic composition resulting in heterogeneity [31, 34]. Extensive information in the literature suggests extremely complex dynamics occurring within CSC populations in ovarian cancer considering the genetic and epigenetic mutational landscape. Notably, recurrent and late-stage epithelial ovarian cancers reflect a higher expression of stemness markers versus early-stage tumors. Hence, targeted therapies against CSCs mainly depend upon the knowledge of the stemness-related populations that co-exist within the tumors and those that evolve post-therapy contributing to heterogeneity and thus determine patient prognosis [25].

Recently, both linear and parallel forms of clonal evolution have been identified in ovarian cancer patients based on the metastatic status, sites, and genetic/epigenetic profile of the tumor clones and implicated in intra-tumor heterogeneity, which poses as a serious clinical challenge. In addition, the ambiguity encircling the origin of ovarian cancer might be explained with the aid of parallel evolution theory, which suggests the migration and homing of particular clones/populations in early stage and then diverging further bringing about complex metastasis at multiple distant sites of the OSE from fallopian tube epithelium; however, a consensus fails in this regard [35, 36]. In addition, the CSC theory proposes tumor-initiating cells, i.e. CSCs as the origin of cancer.

Identification of Ovarian CSCs

Several combinations of markers have been employed to isolate/identify ovarian CSCs, including CD133+, CD133+/ALDH+, CD44+/MYD88+, CD44+/CD117+, CD44+/CD24+. In addition, ALDH^{hi}/CD44^{hi}/CD24^{low} (breast cancer), CD133^{hi}/CD44hi/Nestin^{hi} (glioblastoma), CD44^{hi}/Lgr5^{hi}/CD133^{hi} (colon cancer), and other solid tumors have been reported besides ovarian tumors [37]. Subsequently, other researchers reported ovarian CSCs with different cellular properties and morphologies such as epithelial, mesenchymal [38, 39]. A research group led by Virant-Klun, extensively studied various stem cell-specific (NANOG, SSEA4, and SOX2) and EMT (Vimentin) specific markers in borderline ovarian cancer and serous ovarian carcinoma, thus proposing the resident ovarian stem cells in the OSE layer as the putative candidates of origin [17, 30]. Virant-Klun and Stimpfel [29] reported the presence of small stem-like cells expressing stemness and germinal lineage, which developed into oocyte-like structures in vitro. The gene expression profiling revealed differentially expressed genes that were distinct in normal ovaries and first reported novel tumor-initiating populations in borderline ovarian tumors. In another recent study from the same group [40], they proposed the presence of small embryonic-like stem cells in human ovaries, ovarian cancer ascites, and human embryonic stem cells and implicated a stem cell population residing in the embryonic stage, in adult ovaries and responsible for tumor initiation. Parte et al. [41] exclusively studied the various CSC populations within the tumors at different stages and normal ovaries in the ovarian surface epithelium (OSE) and cortical portion in normal ovary, and benign, borderline, and high-grade metastatic ovarian tumors by immunohistochemistry, confocal microscopy, and RT-PCR analysis. Diverse and distinct populations of OCT4, SSEA4, ALDH1/2, CD44, and LGR5 co-expressing proliferation marker (Ki67) in HG tumors revealed dual positive proliferating stem and CSCs, few non-proliferating stem/CSC (SSEA4+/Ki67− and ALDH1/2+/Ki67−) and only Ki67+ cells in the cortex, revealing different sizes, i.e. smaller spherical ($\leq 5\ \mu m$) and larger spindle/elliptical shaped (~10 μm) cell populations with high nucleo-cytoplasmic ratio were reported across different tumor stages. Similarly, the same group reported an exclusive germline stem cell marker VASA/DDX4 co-expressed with pluripotent stem cell markers (OCT4, NANOG, SOX2, SSEA1, SSEA4) and CSC (CD44, LGR5) markers within distinct populations identified in the OSE layer and cortical region of normal ovaries compared with benign, borderline and high ovarian grade tumors [42]. Cells with variable dimensions [smaller, spherical ($\leq 5\ \mu m$) and larger, spindle/elliptical shaped (~10 μm)] and combination of markers were identified attributed to variable roles in the initiation and progression of the disease. A summary of various CSC-specific markers implicated in tumor initiation has been provided in a review article recently by Zuber et al. [43], which explains the putative role of each marker, and most of them are described in the context of ovarian CSCs.

PTTG1 a Novel Oncogene in Regulation of Ovarian Cancer and Ascites-Derived Cancer Stem Cells (CSCs)

Ascites accumulation is highly debilitating, and hence elimination of this fluid and adverse symptoms associated with it may help patients with a better quality of life. However, there exists no specific treatment against it, except for some traditional therapies. Catumaxomab is one such medicine approved for treating intraperitoneal malignant ascites in Epcam+ovarian carcinoma patients [44]. Hence, extensive profiling and characterization of ascites CSCs in chemoresistant and chemonaive ovarian cancer patients is required. Ascites-derived CSCs have been identified and studied extensively by Kakar and group [41, 42, 45] whereby they initially characterized various ovarian tumors at different stages (BN, BL, and HG) (Fig. 4.1) and ascites-derived CSCs (Fig. 4.2) (in singlets and spheroidal structures) in comparison with normal ovaries for expression of CSC specific markers (ALDH1, CD34,

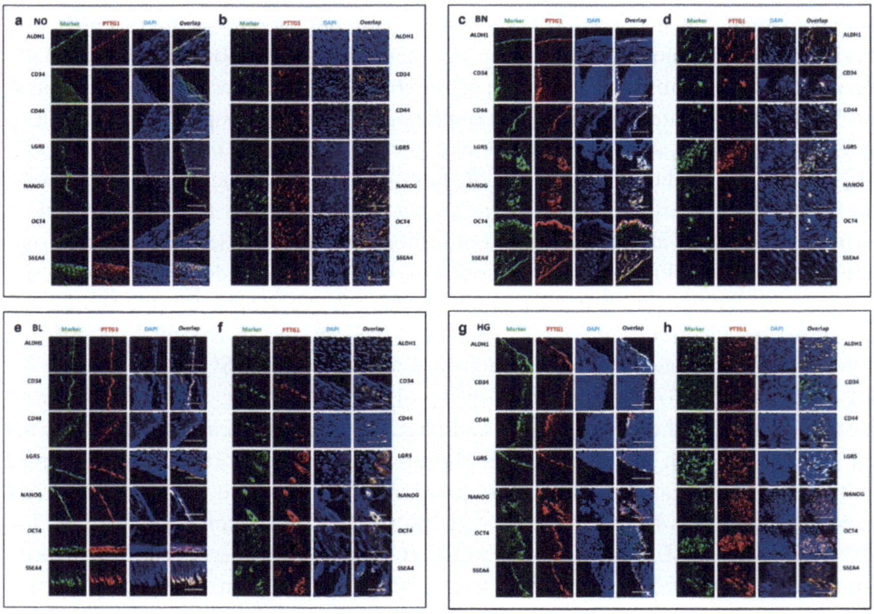

Fig. 4.1 Co-localization of PTTG1/securin with stem cell/cancer stem cell markers in the normal ovary (NO), benign tumor (BN), borderline tumor (BL), and high-grade tumor HG). Anti-ALDH1, CD34, CD44, LGR5, NANOG, OCT4, and SSEA4-specific antibodies (green color) were co-localized with PTTG1 (red color) in OSE (ovarian surface epithelium) and cortex in normal ovarian and tumor tissue sections. Sections were counter-stained with nuclear specific dye DAPI (blue). Overlap of green, red, and blue represents merged image. Alexafluor-labeled secondary antibody (anti-rabbit or anti-mouse) was employed for detection. Scale bar = 50 μm. **a** and **b** = normal ovary (NO), **c** and **d** = benign tumor (BN), **e** and **f** = borderline tumor (BL), and **g** and **h** = high-grade tumor (HG). **a**, **c**, **e**, and **g** = ovarian surface epithelium (OSE), and **b**, **d**, **f**, and **h** = cortex

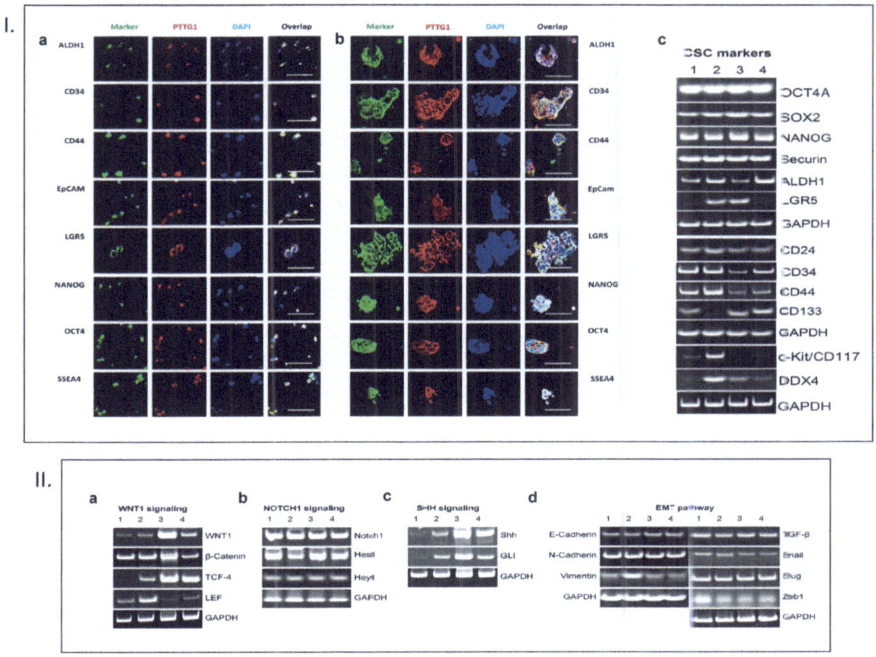

Fig. 4.2 Co-localization of PTTG1/securin with CSCs markers in ascites fluid. The ascites was collected from patients with ovarian cancer at the time of surgery. Specific antibodies for PTTG1 and CSCs markers were used in dual immunofluorescence. **I** (a and b) PTTG1/securin was co-localized with most of the CSCs in single cells and in clusters. (c) RT-PCR amplification of CSC-specific mRNA transcripts. Ascites from patients showed variable levels of expression of mRNA for CSCs. **II** Various self-renewal mechanisms and EMT-specific pathway-related genes, amplified from the ascites fluid-derived CSCs, and a. WNT1 (B-Catenin, TCF-4, LEF), b. NOTCH1 (Notch1, Hes1, Hey1), c. SHH (Shh, Gli), d. Epithelial Mesenchymal Transition (E-Cadherin, N-Cadherin, Vimentin, TGF-B, Snail, Slug, Zeb1). GAPDH was used as a house-keeping gene. Adopted from Parte et al. [45]

CD44, LGR5, NANOG, OCT3/4, SSEA4). These were co-localized with an oncogenic marker extensively studied by them and others earlier and implicated in ovarian cancer known as Securin/PTTG1 [46–51]. The study highlighted the significance of PTTG1 as a novel stem and CSC regulator in ovaries and ovarian cancer with the aid of self-renewal (Wnt1, Notch1, Shh) and EMT-specific pathways (Fig. 4.2) thus correlating the origin of epithelial cancer and CSCs. Several tumor stage-specific expressions of multiple CSC and germline stem cell markers were observed. Unique germline stem cell markers (DDX4/VASA, IFITM3/FRAGILIS) were also studied in relation to PTTG1 down- and up-regulation in ovarian cancer cell line A2780, thus providing a unique link of ovarian stem cells in the etiology of ovarian cancer (tumors and ascites) [45].

Basically, PTTG1 is associated with pleiotropic roles in cell physiology and tumor biology such as facilitating sister chromatid separation in metaphase of mitosis [50], pivotal role in spermatogenesis (differentiation and survival of germ cells), expression in pre-cancerous neoplastic lesions, and in neoplastic angiogenesis. Overexpression of PTTG1 is associated with induction of aneuploidy, and thus associated with tumor cell infiltration, migration, invasion, tumor aggressiveness, and metastatic onset. Recently, PTTG1 co-expression with KLF4, OCT4 is reported to demarcate a sub-population (neoplastic stem cells) in testicular seminoma [52]. Thus, PTTG1 may serve as an excellent potential target for CSC therapy and information regarding its clinical scenario, i.e. its mechanistic regulation based upon therapy and patient response dynamics warrants further exploration.

Similar to the Ki67+ populations of CSCs identified in ovarian tumors at different stages of progression [45], ascites-derived CSCs are reported to produce lactate dehydrogenase suggesting a cohort of tumor cells with CSC properties expressing rapid proliferative and malignant potential. E-Cadherin+ cells in ovarian carcinoma spheroids disseminating into the mesothelial layer exhibit low proliferation and hence least affected by chemotherapy targeting highly proliferating cells and also provide physical barrier for drug penetrance and hence manifest as relapse [25]. Latifi et al. [53] extensively isolated and reported the presence of heterogeneous population of adherent (AD) with mesenchymal properties and non-adherent (NAD) cell types with epithelial properties in both chemo-naive (CN) and chemo-resistant (CR) patient ascites and classified them as epithelial tumorigenic and mesenchymal non-tumorigenic populations. A remarkable heterogeneity of singlets and spheroidal clusters among the ascites CSCs were also observed. The AD type in CR samples expressed Oct4, CD44, MMP9, whereas the NAD counter type showed E-Cadherin, EpCAM, STAT3, Oct4 expressing cells in ascites. Further classification of ascites tumor cells (CA125+EpCAM+/STAT3+) and stromal cells (CD44+/Oct4+) in CR patient samples indicate their vulnerability to the targeting of STAT3 signaling followed by augmentation of cisplatin effect as demonstrated in mouse xenograft model by Kandala and Srivastava [54].

Signal Transduction Pathways Involved in Ovarian Cancer

Genetic instability causes faulty operation of several signal transduction pathways and resultant aberrant protein–protein interactions with compromised cellular functions, thus manifesting the cancer phenotype. Two chief cellular functions—self-renewal and quiescence of normal adult stem cells are impacted whereby they possibly transform into CSCs and undergo high rate of proliferation independent of the transcription factor-mediated signaling pathways obsolete of any external cues such as growth factors, etc. [55]. Following signaling cascades are involved in the pathophysiology and etiology of ovarian cancer, and some are linked to the stem cell compartment (Fig. 4.3).

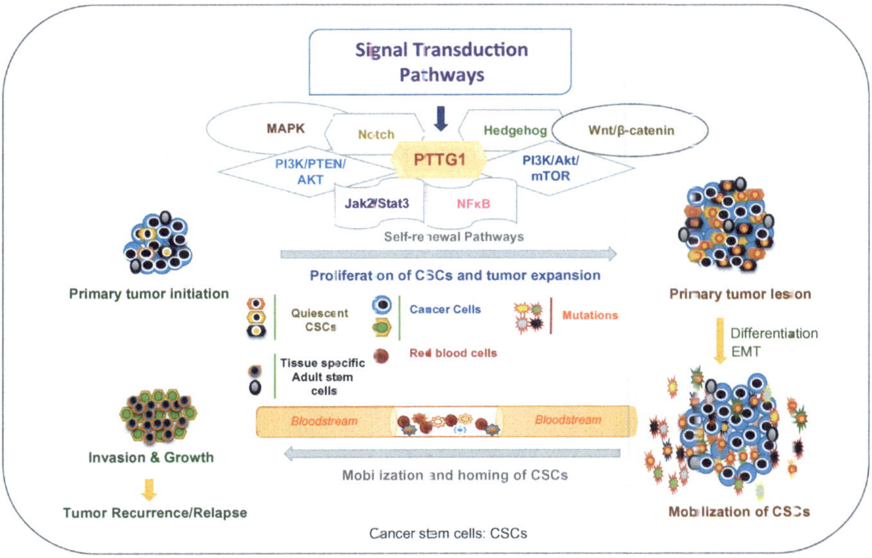

Fig. 4.3 Potential role of various signaling pathways in tumor and CSCs initiation, progression, and metastasis. Adopted from Zuber et al. [43]

PI3K/Akt/mTOR Pathway

This pathway is responsible for cell proliferation, and dysregulation of these pathway components due to mutations in ovarian cancer cells that leads to tumorigenesis, progression, and chemoresistance. Receptor tyrosine kinases located in the cell membrane interact with the extra-cellular cues viz growth factor causing RTK phosphorylation and the activation of PI3K/Akt/mTOR pathway. SKOV3 ovarian CSCs upon treatment with PI3K inhibitor, LY294002 caused suppression of stem cell markers, while suppression of the stem cell signaling molecules may increase sensitivity to a chemotherapy drug, thus suggesting the role of PI3K/Akt/mTOR in CSC formation, imparting drug resistance to CSCs and their maintenance [56]. At present, PI3K/PTEN/AKT inhibitors such as BKM120, Everdimus, and Perifosine are used to treat cancer patients [36]. Future efforts for patient stratification to distinguish patients benefitting from PI3K/PTEN/AKT inhibition will be required for this therapy efficacy in ovarian cancer patients.

MAPK Pathway

MAPK pathway mitogen-activated protein kinase is involved in cell proliferation, survival, and apoptosis in normal cellular physiology. ERK1/ERK2 are activated via phosphorylation through the Ras/RAF pathway, however, in ovarian cancer, the cells possess Ras gene mutation, which leads to constitutive activation of MAPK activity and, thus, cancer. Contrary to this inhibition of cell growth or apoptosis during chemotherapy is controlled by the JNK and p38 MAPK cascade, thus implicating inhibition/decreased activity of these molecules with chemoresistance [57, 58] attempted to target chemo-resistant primary ovarian CD44+/MyD88+ CSCs by Akt inhibitor NV-128 and observed that ROS-dependent activation of ERK could initiate mitochondrial-mediated apoptosis in the cells, thus implicating ERK in inhibiting the drug-induced apoptosis in ovarian CSCs.

Wnt/β-Catenin Pathway

Wnt/β-Catenin pathway is concerned with the embryogenesis and cell fate in the normal development of ovarian and fallopian tube stem cells as well as ovarian cancer initiation and progression by regulation of cell proliferation, apoptosis, and EMT via the Frizzled (Fzd) receptor. EMT and acquisition of stem-like cell properties are interlinked thus imparting drug resistance and mediate dissemination and metastasis [59]. Endometroid ovarian carcinoma reveals β-catenin mutations and other histotypes show variation in regulation of this pathway. Despite the absence of mutations in this pathway per se in various types of ovarian cancer, it is heavily implicated in tumorigenesis [6]. The Fzd receptor elevation in ovarian carcinoma leads to up-regulation of Wnt signaling pathway and linked with poor patient survival [60]. CD117 (c-kit) is known to regulate pI3K-AKT and JAK/STAT pathways and may also alter the Wnt pathway (by phosphorylation of GSK3β and nuclear expression of β-catenin) leading to the gain of stem cell properties in cancer cells and linked with chemoresistance in tumor-initiating cells. Also, the leucine-rich repeat-containing G protein-coupled receptors LGR5 and LGR6 assume an important position in Wnt pathway and are expressed in stem cells and implicated and up-regulated in ovarian tumors [61]. Noncanonical Wnt pathway, Wnt5A is highly expressed in ovarian tumors and ascites. Also, receptor tyrosine kinase-like orphan receptor 1 (ROR1) (a pseudokinase and receptor for Wnt5A) is expressed in ovarian cancer and is associated with poor patient outcome in terms of progression-free and overall survival. ROR+ patient-derived xenografts reveal ALDH+ CSCs, spheroid formation and increased tumorigenicity. Wnt inhibitors targeting Disheveled (NSC668036 and FJ9), Frizzled receptor antibody, Thiazolidinedione (targeting β-catenin reverse transport), and Sulindac (unknown action but potentially effecting β-catenin proteasomal degradation) are under scrutiny for clinical use [36].

Notch Signaling Pathway

This pathway is concerned with cell survival, proliferation, and maintenance of somatic stem cells, and hence any disturbance in this pathway is implicated in stemness properties of CSCs [62, 63]. Notch3 overexpression in >20% of high-grade serous adenocarcinoma is linked to aggressiveness and poor prognostic value and expression of stemness markers Oct4, Nanog, Sox2, etc. [64]. It plays a crucial role in maintenance, differentiation, proliferation, communication, and apoptosis of progenitor cells, and hence elevated expression of ABCG1 results in chemoresistance against platinum and carboplatin. Notch signaling is carried out with the aid of Notch ligands (Jagged and Delta) binding to Notch receptors and via Υ-secretase through a cascade of proteolytic cleavages [65]. Expression of Notch3 ligand Jagged1 in mesothelial cells hint toward the Notch pathway involvement in cell adhesion and tumor proliferation [63, 64]. Moreover, relapsed HG tumors show high Notch3 expression thus serving as a prognostic indicator for recurrent disease. At present, experimental scale γ-secretase inhibitors, γ-secretase modifiers, Notch soluble decoys, and negative regulatory region monoclonal antibodies are already being developed for treating ovarian cancer aiming at the CSC population [36].

Hedgehog Signaling Pathway

This pathway is directly involved in tissue polarity and patterning and stem cell development, maintenance, and regulation of CSC self-renewal, differentiation, tumor initiation. Ligand such as Sonic hedgehog (Shh) and Hh receptors such as Patched (PTCH1, PTCH2) and Smoothened (SMO), and Gli transcription factors 1, 2, 3 or other members such as SUFU show dysregulation by either constitutive expression or mutations [66]. Intracellular expression of Gli is linked with the increased spheroid formation of CSCs and has a role in the expression of ABC transporters ABCG1, ABCG2 by directly binding to promoter regions and thus establishing resistance to Cisplatin and Paclitaxel in ovarian cancer spheroids [67, 68]. Increased expression of SMO and Gli in borderline and malignant ovarian tumors than the normal counterparts implicates their role in ovarian cancer progression [69, 70]. Overexpression of Gli and/or PTCH in ovarian cancer patients correlates with poor survival and, hence, this pathway is heavily implicated as a cardinal regulator of stemness, proliferation, and therapy resistance. Currently, Gli1 antagonists (HPI 1–4), drugs targeting PTCH might benefit patients with activated Hh. Cyclopamine shows significant efficacy in targeting spheroid forming cells by affecting the Gli expression directly impacting the stemness, whereas anti-Gli-shRNA treated ovarian cancer cells exhibited impaired DNA damage response following cisplatin treatment, thus highlighting the role of this pathway in crucial aspects of tumor annihilation in the resistant state [36].

JAK/STAT Signaling Pathway

JAK/STAT signaling pathway is an intracellular signal transduction pathway involved in cell proliferation, differentiation, apoptosis, and immune regulation especially stem cells and CSCs specifically. In ovarian cancer, especially in HG tumors, the JAK/STAT pathway is constitutively active and lacks any inhibition/regulation [71]. During activation of signaling JAK and STAT molecules assemble into homo, hetero-dimers, and multimers [72]. JAK/STAT pathway activation is implicated in various aspects of tumor biology such as tumorigenesis, metastasis, transition of CSCs, chemoresistance, and EMT [73]. CD24+ ovarian CSCs were injected in mice followed by administration of only Cisplatin and Cisplatin + TG101209 (Jak2 inhibitor), and this latter group exhibited better survival in mice and almost no metastasis, thus elucidating the significance of JAK in CSC maintenance. Even ascites-derived cancer cells and Hey8 cells reveal JAK2/STAT3 pathway activation and elevation of stemness genes in response to paclitaxel treatment, thus justifying a clinical scenario of CSC enrichment in response to therapy thus leading to resistance and relapse [36].

NFkB Signaling Pathway

This pathway plays a critical role in normal cellular physiology such as cell survival, proliferation, and cell death. In cancer, this pathway is linked with metastasis, invasion, and chemoresistance [74]. Both canonical and noncanonical NFκB pathways are implicated in imparting stemness, chemoresistance and CSC maintenance in ovarian cancer by the fact that CD44+ CSCs from patient ascites and SKOV3 cells exhibit constitutive activation/overexpression of NFκB pathway (members RelA, RelB, IKKα) with evidence of spheroid formation ability in vitro and tumor formation in vivo. CD44+ CSCs treated with TNFα showed resistance to TNFα induced cell death and increased activity of NFκB pathway, thus highlighting its role in CSC survival during oncotherapy and recalcitrance in ovarian cancer patients [38]. Eriocalyxin B, an inhibitor of NFκB induces apoptosis in ovarian CSCs thus sensitizing CSCs to therapy [75] and reflecting tremendous potential to cater to relapse cases. In ovarian cancer cell line models, overexpression of RelB in ovarian CSC populations such as CD44, ALDH, and CD133 is observed in response to carboplatin treatment. Abrogating expression of RelB in cell lines and xenograft models led to reduction in CSC populations, increased sensitivity to Carboplatin, a reduction in stemness markers per se, and reduced tumorigenic potential in vitro and in vivo, thus providing evidence for RelB's role in stemness and chemoresistance [76]. NFκB pathway may serve as a crucial target for CSC-based therapy for treating ovarian cancer patients [36].

Targeting of CSCs: Recent Developments

A recent article threw light upon a novel approach whereby patient biopsied tumor tissue, or malignant fluid aspirates are used to isolate tumor bulk cells and CSCs and predict the effectiveness of a chemotherapeutic agent or combination. This article provides information regarding a novel high cell kill chemotherapy targeting CSCs (cytotoxicity assay/Chemo ID), which was associated with a better response rate of patients, while the overall and progression-free survival was improved after the annihilation of CSCs. This is a pioneering approach of personalized therapy based upon the CSCs drug response assay performed in vitro to predict alternate treatment regimens for platinum-based relapsed patients based upon their responsiveness/refractoriness [77]. Another report by Uddin et al. [78] addressed the issue of cisplatin resistance in ovarian cancer while exploring the CSC/progenitor cell connection in ovarian cancer cell line A2780 and its Cisplatin resistant version A2780-Cp and a supra resistant version SKOV3-Cp upon Cisplatin treatment. In response to the treatment, both resistant and supra-resistant cells showed increased self-renewal, significant resistance to apoptosis and sub-G1 arrest. ALDEFLUOR assays and SP analysis revealed increased activity and enriched populations in the resistant cell lines compared with their parental counterparts. ALDH1A1 and KLF4 have significantly overexpressed in both resistant/supra resistant cells. Silencing of ALDH1A1 population depleted the stem cells and rendered the resistant cells sensitive to Cisplatin, and reduction in mRNA transcript and protein expression of downstream target NEK2. Silencing expression of either ALDH1A1/NEK-2 led to down-regulation of expression of ABC transporters, while the reverse held for ABC transporters upon overexpression of NEK-2. Collectively, the current study delineated the involvement of ALDH1A1 gene in regulating cisplatin resistance by over-expressing NEK-2 in ovarian cancer. In another study by Zhang et al. [79] discuss the novel therapeutic strategy for targeting ovarian cancer stemness feature by modulating the small nucleolar RNAs (snoRNAs). snoRNAs upon over-expression regulate the self-renewal and migration potential in OVCAR-3 spheroids (OS) and CAOV-3 spheroids (CS) with the ovarian CSCs characteristics and up-regulate stemness markers Nanog, Oct4, and CD133 and mRNA transcripts and protein expression of Notch1 and c-Myc. Silencing of snoRNAs reversed the expression of all stemness-related markers mentioned above and confirmed their findings.

Conclusive Remarks

Aspects related to the stemness and CSC-related compartments and other components constituting the tumor microenvironment and their dynamics across the trajectory of an individual patients' life during the course of the treatment regimen and the subtle differences of the cellular, molecular parameters, and constituents between chemo-naïve and sensitive ones coupled with mutational status/burden may provide a multitude of information and novel insights into greater prognostic significance. Hence, a comprehensive understanding of ovarian CSCs at the primary site and in ascites fluid and the molecular mechanisms governing their malignant potential driving ovarian cancer progression toward relapse and incurable stage are the essential needs of the hour. Exploration of novel oncogenes and their regulatory mechanisms in a patient and treatment-specific manner and their combination with conventional oncotherapy may mark a new era in personalized and precision medicine to effectively manage ovarian cancer thus improving patient quality of life and prolonging the progression-free and overall survival post-remission.

Funding This study was funded by grants from NIH P01CA217798 (SKB, SSK) and T32HL134644 (SSK).

References

1. Siegel RL, Miller KD, Jemal A (2020) Cancer statistics, 2020. CA Cancer J Clin 70:7–30. https://doi.org/10.3322/caac.21590
2. American Cancer Society (2018) Cancer facts & figures 2018. American Cancer Society, Atlanta
3. Webb PM, Jordan SJ (2017) Epidemiology of epithelial ovarian cancer. Best Pract Res Clin Obstet Gynaecol 41:3–14. https://doi.org/10.1016/j.bpobgyn.2016.08.006
4. Lheureux S, Gourley C, Vergote I, Oza AM (2019) Epithelial ovarian cancer. Lancet 393:1240–1253. https://doi.org/10.1016/S0140-6736(18)32552-2
5. Ottevanger PB (2017) Ovarian cancer stem cells more questions than answers. Semin Cancer Biol 44:67–71. https://doi.org/10.1016/j.semcancer.2017.04.009
6. Pieterse Z, Amaya-Padilla MA, Singomat T, Binju M, Madjid BD, Yu Y, Kaur P (2019) Ovarian cancer stem cells and their role in drug resistance. Int J Biochem Cell Biol 106:117–126. https://doi.org/10.1016/j.biocel.2018.11.012
7. Ponnusamy MP, Batra SK (2008) Ovarian cancer: emerging concept on cancer stem cells. J Ovarian Res 1:4. https://doi.org/10.1186/1757-2215-1-4
8. Nimmakayala RK, Batra SK, Ponnusamy MP (2019) Unraveling the journey of cancer stem cells from origin to metastasis. Biochim Biophys Acta Rev Cancer 1871:50–63. https://doi.org/10.1016/j.bbcan.2018.10.006
9. Bapat SA, Mali AM, Koppikar CB, Kurrey NK (2005) Stem and progenitor-like cells contribute to the aggressive behavior of human epithelial ovarian cancer. Cancer Res 65:3025–3029. https://doi.org/10.1158/0008-5472.CAN-04-3931
10. Johnson J, Canning J, Kaneko T, Pru JK, Tilly JL (2004) Germline stem cells and follicular renewal in the postnatal mammalian ovary. Nature 428:145–150. https://doi.org/10.1038/nature02316
11. Szotek PP, Chang HL, Brennand K, Fujino A, Pieretti-Vanmarcke R, Lo Celso C, Dombkowski D, Preffer F, Cohen KS, Teixeira J, Donahoe PK (2008) Normal ovarian surface epithelial label-retaining cells exhibit stem/progenitor cell characteristics. Proc Natl Acad Sci USA 105:12469–12473. https://doi.org/10.1073/pnas.0805012105
12. Flesken-Nikitin A, Hwang C-I, Cheng C-Y, Michurina TV, Enikolopov G, Nikitin AY (2013) Ovarian surface epithelium at the junction area contains a cancer-prone stem cell niche. Nature 495:241–245. https://doi.org/10.1038/nature11979

13. Virant-Klun I, Zech N, Rozman P, Vogler A, Cvjeticanin B, Klemenc P, Malicev E, Meden-Vrtovec H (2008) Putative stem cells with an embryonic character isolated from the ovarian surface epithelium of women with no naturally present follicles and oocytes. Differ Res Biol Divers 76:843–856. https://doi.org/10.1111/j.1432-0436.2008.00268.x
14. Virant-Klun I, Rozman P, Cvjeticanin B, Vrtacnik-Bokal E, Novakovic S, Rülicke T, Dovc P, Meden-Vrtovec H (2009) Parthenogenetic embryo-like structures in the human ovarian surface epithelium cell culture in postmenopausal women with no naturally present follicles and oocytes. Stem Cells Dev 18:137–149. https://doi.org/10.1089/scd.2007.0238
15. Parte S, Bhartiya D, Telang J, Daithankar V, Salvi V, Zaveri K, Hinduja I (2011) Detection, characterization, and spontaneous differentiation in vitro of very small embryonic-like putative stem cells in adult mammalian ovary. Stem Cells Dev 20:1451–1464. https://doi.org/10.1089/scd.2010.0461
16. Ng A, Barker N (2015) Ovary and fimbrial stem cells: biology, niche and cancer origins. Nat Rev Mol Cell Biol 16:625–638. https://doi.org/10.1038/nrm4056
17. Virant-Klun I, Kenda-Suster N, Smrkolj S (2016) Small putative NANOG, SOX2, and SSEA-4-positive stem cells resembling very small embryonic-like stem cells in sections of ovarian tissue in patients with ovarian cancer. J Ovarian Res 9:12. https://doi.org/10.1186/s13048-016-0221-3
18. Wang J, Wang D, Chu K, Li W, Zeng YA (2019) Procr-expressing progenitor cells are responsible for murine ovulatory rupture repair of ovarian surface epithelium. Nat Commun 10:4966. https://doi.org/10.1038/s41467-019-12935-7
19. Barkeer S, Chugh S, Batra SK, Ponnusamy MP (2018) Glycosylation of cancer stem cells: function in stemness, tumorigenesis and metastasis. Neoplasia 20:813–825. https://doi.org/10.1016/j.neo.2018.06.001
20. Barker N, van Es JH, Kuipers J, Kujala P, van den Born M, Cozijnsen M, Haegebarth A, Korving J, Begthel H, Peters PJ, Clevers H (2007) Identification of stem cells in small intestine and colon by marker gene Lgr5. Nature 449:1003–1007. https://doi.org/10.1038/nature06196
21. Barker N et al (2008) Lgr5-positive cells might be adult stem cells that become cancer stem cells in tumors. Nat Clin Pract Gastroenterol Hepatol 5:63–64. https://doi.org/10.1038/ncpgasthep1015
22. Urzúa U, Ampuero S, Roby KF, Owens GA, Munroe DJ (2016) Dysregulation of mitotic machinery genes precedes genome instability during spontaneous pre-malignant transformation of mouse ovarian surface epithelial cells. BMC Genomics 17:728. https://doi.org/10.1186/s12864-016-3068-5
23. Sasaki R, Narisawa-Saito M, Yugawa T, Fujita M, Tashiro H, Katabuchi H, Kiyono T (2009) Oncogenic transformation of human ovarian surface epithelial cells with defined cellular oncogenes. Carcinogenesis 30:423–431. https://doi.org/10.1093/carcin/bgp007
24. Roberts PC, Schmelz EM (2013) In vitro model of spontaneous mouse OSE transformation. Methods Mol Biol Clifton NJ 1049:393–408. https://doi.org/10.1007/978-1-62703-547-7_30
25. Ahmed N, Stenvers KL (2013) Getting to know ovarian cancer ascites: opportunities for targeted therapy-based translational research. Front Oncol 3:256. https://doi.org/10.3389/fonc.2013.00256
26. Ford CE, Werner B, Hacker NF, Warton K (2020) The untapped potential of ascites in ovarian cancer research and treatment. Br J Cancer 123:9–16. https://doi.org/10.1038/s41416-020-0875-x
27. Matte I, Lane D, Laplante C, Rancourt C, Piché A (2012) Profiling of cytokines in human epithelial ovarian cancer ascites. Am J Cancer Res 2:566–580
28. Silvestris E, Cafforio P, D'Oronzo S, Felici C, Silvestris F, Loverro G (2018) In vitro differentiation of human oocyte-like cells from oogonial stem cells: single-cell isolation and molecular characterization. Hum Reprod Oxf Engl 33:464–473. https://doi.org/10.1093/humrep/dex377
29. Virant-Klun I, Stimpfel M (2016) Novel population of small tumour-initiating stem cells in the ovaries of women with borderline ovarian cancer. Sci Rep 6:34730. https://doi.org/10.1038/srep34730

30. Kenda Suster N, Smrkolj S, Virant-Klun I (2017) Putative stem cells and epithelial-mesenchymal transition revealed in sections of ovarian tumor in patients with serous ovarian carcinoma using immunohistochemistry for vimentin and pluripotency-related markers. J Ovarian Res 10:11. https://doi.org/10.1186/s13048-017-0306-7
31. Garson K, Vanderhyden BC (2015) Epithelial ovarian cancer stem cells: underlying complexity of a simple paradigm. Reprod 149:R59-70. https://doi.org/10.1530/REP-14-0234
32. Miranda-Lorenzo I, Dorado J, Lonardo E, Alcala S, Serrano AG, Clausell-Tormos J, Cioffi M, Megias D, Zagorac S, Balic A, Hidalgo M, Erkan M, Kleeff J, Scarpa A, Sainz B, Heeschen C (2014) Intracellular autofluorescence: a biomarker for epithelial cancer stem cells. Nat Methods 11:1161–1169. https://doi.org/10.1038/nmeth.3112
33. Vaz AP, Ponnusamy MP, Seshacharyulu P, Batra SK (2014) A concise review on the current understanding of pancreatic cancer stem cells. J Cancer Stem Cell Res 2:e1004. https://doi.org/10.14343/jcscr.2014.2e1004
34. Kreso A, Dick JE (2014) Evolution of the cancer stem cell model. Cell Stem Cell 14:275–291. https://doi.org/10.1016/j.stem.2014.02.006
35. Roberts CM, Cardenas C, Tedja R (2019) The role of intra-tumoral heterogeneity and its clinical relevance in epithelial ovarian cancer recurrence and metastasis. Cancers 11:1083. https://doi.org/10.3390/cancers11081083
36. Roy L, Cowden Dahl KD (2018) Can Stemness and chemoresistance Be therapeutically targeted via signaling pathways in ovarian cancer? Cancers 10:241. https://doi.org/10.3390/cancers10080241
37. Agliano A, Calvo A, Box C (2017) The challenge of targeting cancer stem cells to halt metastasis. Semin Cancer Biol 44:25–42. https://doi.org/10.1016/j.semcancer.2017.03.003
38. Alvero AB, Chen R, Fu H-H, Montagna M, Schwartz PE, Rutherford T, Silasi D-A, Steffensen KD, Waldstrom M, Visintin I, Mor G (2009) Molecular phenotyping of human ovarian cancer stem cells unravels the mechanisms for repair and chemoresistance. Cell Cycle 8:158–166. https://doi.org/10.4161/cc.8.1.7533
39. Ho MM, Ng AV, Lam S, Hung JY (2007) Side population in human lung cancer cell lines and tumors is enriched with stem-like cancer cells. Cancer Res 67:4827–4833. https://doi.org/10.1158/0008-5472.CAN-06-3557
40. Virant-Klun I, Skerl P, Novakovic S, Vrtacnik-Bokal E, Smrkolj S (2019) Similar population of CD133+ and DDX4+ VSEL-like stem cells sorted from human embryonic stem cell, ovarian, and ovarian cancer ascites cell cultures: the real embryonic stem cells? Cells 8:706. https://doi.org/10.3390/cells8070706
41. Parte SC, Batra SK, Kakar SS (2018) Characterization of stem cell and cancer stem cell populations in ovary and ovarian tumors. J Ovarian Res 11:69. https://doi.org/10.1186/s13048-018-0439-3
42. Parte SC, Smolenkov A, Batra SK, Ratajczak MZ, Kakar SS (2017) Ovarian cancer stem cells: unraveling a germline connection. Stem Cells Dev 26:1781–1803. https://doi.org/10.1089/scd.2017.0153
43. Zuber E, Schweitzer D, Allen D, Parte S, Kakar SS (2020) Stem cells in ovarian cancer and potential therapies. Proc Stem Cell Res Oncog 8:e1001
44. Cerne K, Kobal B (2017) Ascites in advanced ovarian cancer. In: Devaja O, Papadopoulos A (eds) Ovarian cancer: from pathogenesis to treatment. IntechOpen
45. Parte S, Virant-Klun I, Patankar M, Batra SK, Straughn A, Kakar SS (2019) PTTG1: a unique regulator of stem/cancer stem cells in the ovary and ovarian cancer. Stem Cell Rev Rep 15:866–879. https://doi.org/10.1007/s12015-019-09911-5
46. Kakar SS, Jennes L (1999) Molecular cloning and characterization of the tumor transforming gene (TUTR1): a novel gene in human tumorigenesis. Cytogenet Cell Genet 84:211–216. https://doi.org/10.1159/000015261
47. Pei L, Melmed S (1997) Isolation and characterization of a pituitary tumor-transforming gene (PTTG). Mol Endocrinol Baltim Md 11:433–441. https://doi.org/10.1210/mend.11.4.9911
48. Domínguez A, Ramos-Morales F, Romero F, Rios RM, Dreyfus F, Tortolero M, Pintor-Toro JA (1998) hpttg, a human homologue of rat pttg, is overexpressed in hematopoietic neoplasms.

Evidence for a transcriptional activation function of hPTTG. Oncogene 17:2187–2193. https://doi.org/10.1038/sj.onc.1202140
49. Puri R, Tousson A, Chen L, Kakar SS (2001) Molecular cloning of pituitary tumor transforming gene 1 from ovarian tumors and its expression in tumors. Cancer Lett 163:131–139. https://doi.org/10.1016/s0304-3835(00)00688-1
50. Zou H, McGarry TJ, Bernal T, Kirschner MW (1999) Identification of a vertebrate sister-chromatid separation inhibitor involved in transformation and tumorigenesis. Science 285:418–422. https://doi.org/10.1126/science.285.5426.418
51. Hamid T, Malik MT, Kakar SS (2005) Ectopic expression of PTTG1/securin promotes tumorigenesis in human embryonic kidney cells. Mol Cancer 4:3. https://doi.org/10.1186/1476-4598-4-3
52. Grande G, Milardi D, Martini M, Cenci T, Gulino G, Mancini F, Bianchi A, Pontecorvi A, Pierconti F (2019) Protein expression of PTTG-1, OCT-4, and KLF-4 in seminoma: a pilot study. Front Endocrinol 10:619. https://doi.org/10.3389/fendo.2019.00619
53. Latifi A, Luwor RB, Bilandzic M, Nazaretian S, Stenvers K, Pyman J, Zhu H, Thompson EW, Quinn MA, Findlay JK, Ahmed N (2012) Isolation and characterization of tumor cells from the ascites of ovarian cancer patients: molecular phenotype of chemoresistant ovarian tumors. PLoS ONE 7:e46858. https://doi.org/10.1371/journal.pone.0046858
54. Kandala PK, Srivastava SK (2012) Diindolylmethane suppresses ovarian cancer growth and potentiates the effect of cisplatin in tumor mouse model by targeting signal transducer and activator of transcription 3 (STAT3). BMC Med 10:9. https://doi.org/10.1186/1741-7015-10-9
55. Takebe N, Miele L, Harris PJ, Jeong W, Bando H, Kahn M, Yang SX, Ivy SP (2015) Targeting Notch, Hedgehog, and Wnt pathways in cancer stem cells: clinical update. Nat Rev Clin Oncol 12:445–464. https://doi.org/10.1038/nrclinonc.2015.61
56. Ip CKM, Li S-S, Tang MYH, Sy SKH, Ren Y, Shum HC, Wong AST (2016) Stemness and chemoresistance in epithelial ovarian carcinoma cells under shear stress. Sci Rep 6:26788. https://doi.org/10.1038/srep26788
57. Roberts PJ, Der CJ (2007) Targeting the Raf-MEK-ERK mitogen-activated protein kinase cascade for the treatment of cancer. Oncogene 26:3291–3310. https://doi.org/10.1038/sj.onc.1210422
58. Alvero AB, Montagna MK, Holmberg JC, Craveiro V, Brown D, Mor G (2011) Targeting the mitochondria activates two independent cell death pathways in ovarian cancer stem cells. Mol Cancer Ther 10:1385–1393. https://doi.org/10.1158/1535-7163.MCT-11-0023
59. Deng J, Wang L, Chen H, Hao J, Ni J, Chang L, Duan W, Graham P, Li Y (2016) Targeting epithelial-mesenchymal transition and cancer stem cells for chemoresistant ovarian cancer. Oncotarget 7:55771–55788. https://doi.org/10.18632/oncotarget.9908
60. Badiglian Filho L, Oshima CTF, De Oliveira LF, De Oliveira CH, De Sousa DR, Gomes TS, Gonçalves WJ (2009) Canonical and noncanonical Wnt pathway: a comparison among normal ovary, benign ovarian tumor and ovarian cancer. Oncol Rep 21:313–320
61. McClanahan T, Koseoglu S, Smith K, Grein J, Gustafson E, Black S, Kirschmeier P, Samatar AA (2006) Identification of overexpression of orphan G protein-coupled receptor GPR49 in human colon and ovarian primary tumors. Cancer Biol Ther 5:419–426. https://doi.org/10.4161/cbt.5.4.2521
62. Hu W, Liu T, Ivan C, Sun Y, Huang J, Mangala LS, Miyake T, Dalton HJ, Pradeep S, Rupaimoole R, Previs RA, Han HD, Bottsford-Miller J, Zand B, Kang Y, Pecot CV, Nick AM, Wu SY, Lee J-S, Sehgal V, Ram P, Liu J, Tucker SL, Lopez-Berestein G, Baggerly KA, Coleman RL, Sood AK (2014) Notch3 pathway alterations in ovarian cancer. Cancer Res 74:3282–3293. https://doi.org/10.1158/0008-5472.CAN-13-2066
63. Hoarau-Véchot J, Touboul C, Halabi N, Blot-Dupin M, Lis R, Abi Khalil C, Rafii S, Rafii A, Pasquier J (2019) Akt-activated endothelium promotes ovarian cancer proliferation through notch activation. J Transl Med 17:194. https://doi.org/10.1186/s12967-019-1942-z
64. Motohara T, Katabuchi H (2019) Ovarian cancer stemness: biological and clinical implications for metastasis and chemotherapy resistance. Cancers 11:907. https://doi.org/10.3390/cancers11070907

65. Zong X, Nephew KP (2019) Ovarian cancer stem cells: role in metastasis and opportunity for therapeutic targeting. Cancers 11:934. https://doi.org/10.3390/cancers11070934
66. Zhu Q, Shen Y, Chen X, He J, Liu J, Zu X (2020) Self-renewal signalling pathway inhibitors: perspectives on therapeutic approaches for cancer stem cells. OncoTargets Ther 13:525–540. https://doi.org/10.2147/OTT.S224465
67. Chen Y, Bieber MM, Teng NNH (2014) Hedgehog signaling regulates drug sensitivity by targeting ABC transporters ABCB1 and ABCG2 in epithelial ovarian cancer. Mol Carcinog 53:625–634. https://doi.org/10.1002/mc.22015
68. Zeng C, Chen T, Zhang Y, Chen Q (2017) Hedgehog signaling pathway regulates ovarian cancer invasion and migration via adhesion molecule CD24. J Cancer 8:786–792. https://doi.org/10.7150/jca.17712
69. Liao X, Siu MKY, Au CWH, Wong ESY, Chan HY, Ip PPC, Ngan HYS, Cheung ANY (2009) Aberrant activation of hedgehog signaling pathway in ovarian cancers: effect on prognosis, cell invasion and differentiation. Carcinogenesis 30:131–140. https://doi.org/10.1093/carcin/bgn230
70. Chen X, Horiuchi A, Kikuchi N, Osada R, Yoshida J, Shiozawa T, Konishi I (2007) Hedgehog signal pathway is activated in ovarian carcinomas, correlating with cell proliferation: it's inhibition leads to growth suppression and apoptosis. Cancer Sci 98:68–76. https://doi.org/10.1111/j.1349-7006.2006.00353.x
71. Ruan Z, Yang X, Cheng W (2018) OCT4 accelerates tumorigenesis through activating JAK/STAT signaling in ovarian cancer side population cells. Cancer Manag Res 11:389–399. https://doi.org/10.2147/CMAR.S180418
72. Muller R (2019) JAK inhibitors in 2019, synthetic review in 10 points. Eur J Intern Med 66:9–17. https://doi.org/10.1016/j.ejim.2019.05.022
73. Jin W (2020) Role of JAK/STAT3 signaling in the regulation of metastasis, the transition of cancer stem cells, and chemoresistance of cancer by epithelial-mesenchymal transition. Cells 9:217. https://doi.org/10.3390/cells9010217
74. Rinkenbaugh AL, Baldwin AS (2016) The NF-κB pathway and cancer stem cells. Cells 5:16. https://doi.org/10.3390/cells5020016
75. Leizer AL, Alvero AB, Fu HH, Holmberg JC, Cheng Y-C, Silasi D-A, Rutherford T, Mor G (2011) Regulation of inflammation by the NF-κB pathway in ovarian cancer stem cells. Am J Reprod Immunol 65:438–447. https://doi.org/10.1111/j.1600-0897.2010.00914.x
76. House CD, Jordan E, Hernandez L, Ozaki M, James JM, Kim M, Kruhlak MJ, Batchelor E, Elloumi F, Cam MC, Annunziata CM (2017) NFκB promotes ovarian tumorigenesis via classical pathways that support proliferative cancer cells and alternative pathways that support ALDH+ cancer stem-like cells. Cancer Res 77:6927–6940. https://doi.org/10.1158/0008-5472.CAN-17-0366
77. Howard CM, Zgheib NB, Bush S, DeEulis T, Cortese A, Mollo A, Lirette ST, Denning K, Valluri J, Claudio PP (2020) Clinical relevance of cancer stem cell chemotherapeutic assay for recurrent ovarian cancer. Transl Oncol 13:100860. https://doi.org/10.1016/j.tranon.2020.100860
78. Uddin MH, Kim B, Cho U, Azmi AS, Song YS (2020) Association of ALDH1A1-NEK-2 axis in cisplatin resistance in ovarian cancer cells. Heliyon 6:e05442. https://doi.org/10.1016/j.heliyon.2020.e05442
79. Zhang L, Ma R, Gao M, Zhao Y, Lv X, Zhu W, Han L, Su P, Fan Y, Yan Y, Zhao L, Ma H, Wei M, He M (2020) SNORA72 activates the Notch1/c-Myc pathway to promote stemness transformation of ovarian cancer cells. Front Cell Dev Biol 8:583087. https://doi.org/10.3389/fcell.2020.583087

Chapter 5
Ovarian Cancer Stem Cells: Approach to Individualized Medicine

Nataša Kenda Šuster

Abstract

Introduction: Since the first identification of cancer stem cells in leukaemia in 1994, they have been seen as potential therapeutic targets for cancer therapy. It has been proposed that also in ovarian cancer success of treatment depends on ovarian cancer stem cells eradication.
Methods: Ovarian stem cells have been studied from the perspective of individual medicine.
Results: Ovarian cancer stem cells have self-renewal capacity and differentiation potential and contribute to malignancy of ovarian cancer such as heterogeneity, invasiveness, metastasis, multidrug resistance, and recurrence of the disease. The biological activities of cancer stem cells are regulated by various pluripotent transcription factors, intracellular signalling pathways and extracellular factors. Drugs directed toward one or more of those targets would effectively eradicate ovarian cancer stem cells and consequently cure ovarian cancer disease.
Conclusions: According to this knowledge, developing molecules, vaccines, and antibodies targeting cancer stem cells present a key factor to individualized medicine.

Keywords Ovarian cancer · Cancer stem cells · Individualized medicine

Abbreviations

A2780	Human ovarian cancer cell line established from tumour tissue of an untreated patient
ABCB1	ATP binding cassette subfamily B member 1
ABCG2	ATP binding cassette subfamily G member 2
ALDH1	Aldehyde dehydrogenase 1
ATP	Adenosine triphosphate

N. K. Šuster (✉)
Department of Obstetrics and Gynecology, University Medical Center Ljubljana, Ljubljana, Slovenia

I. Virant-Klun (ed.), *Stem Cells in Reproductive Tissues and Organs*, Stem Cell Biology and Regenerative Medicine 70, https://doi.org/10.1007/978-3-030-90111-0_5

CD	Cluster of differentiation
CSC	Cancer stem cell
DNA	Deoxyribonucleic acid
EMT	Epithelial-mesenchymal transition
EpCAM	Epithelial cell adhesion molecule
MDR1	ABCB1, ATP binding cassette subfamily B member 1
NANOG	Nanog homeobox
OC	Ovarian cancer
OCSC	Ovarian cancer stem cell
OCT4	Octamer-binding transcription factor 4
OVCAR3	A high grade serous ovarian cancer stem cell line from adenocarcinoma
ROR1	Receptor tyrosine kinase-like orphan receptor 1
SKOV3	An ovarian cancer cell line derived from the ascites of a 64-year-old Caucasian female with an ovarian serous cystadenocarcinoma
SOX2	SRY-box transcription factor 2
VSEL	Very small embryonic-like stem cell

Ovarian Cancer

Ovarian cancer (OC) is one of the most frequent cancer types in females [1]. It is the leading cause of cancer death among gynaecological tumours affecting a large population of women word wide. According to OC statistics, in the USA, in 2018, there would be approximately 22.240 new cases of OC diagnosed and 14.070 OC deaths [2].

About 90–95% of ovarian malignancies are of epithelial origin [3]. According to histopathology, immunohistochemistry, and molecular genetic analysis, there are five main types of ovarian carcinoma: high-grade serous carcinoma, 70%; low-grade serous carcinoma, <5%; endometrioid carcinoma, 10%; clear cell carcinoma, 10%; and mucinous carcinoma 3% [4]. Different epithelial malignancies have different origins and consequently different morphologies and biological behaviour [5].

The 5-year survival rate of OC varies from 30 to 92%, depending mainly on the dissemination of the disease at the time of diagnosis [1]. As symptoms of OC are nonspecific, such as abdominal discomfort and bloating, 70% of OC patients are diagnosed at advanced disease, with metastatic sites being disseminated widely within the peritoneal cavity, and retroperitoneum, and even in distant organs [6].

Treating disease in its advanced course is demanding and often unsuccessful. The treatment goal is to achieve complete removal of tumour, macroscopic and microscopic lesions. Patients with OC spread beneath ovaries receive radical cytoreductive surgery in combination with neoadjuvant and/or adjuvant platinum and taxane combined chemotherapy [7, 8]. The majority of OC patients initially respond well to combined treatment; however, over 70% of tumour cases eventually relapse, arising chemoresistance and lethal disease [9]. The main reasons for the failure of cancer therapy are tumour residue on microscopic level that remains after surgical debulking and limitations of standard chemotherapy. It is now proposed that those microscopic residues left behind after surgery and chemotherapy are composed of small populations of cancer stem cells (CSCs) being resistant against conventional cancer therapies. For this reason, effective targeted therapies enabling complete eradication of CSCs are urgently needed.

Ovarian Cancer Stem Cells and Their Biomarkers

It is proposed that OC initiation, tumour growth and dissemination are driven by undifferentiated cancer cells with the ability of self-renewal on one hand and differentiation on the other [10]. These so-called CSCs share several biological characteristics with normal stem cells, like self-renewal, resistance to induced apoptosis and ability to undergo differentiation via asymmetric cell division [10]. Furthermore, CSCs could generate a tumour with the hierarchical organization of heterogeneous cells from which only a small part of cells preserve CSCs characteristics.

The first ovarian CSCs were identified from a single cell in the OC patient ascites; this cell produced tumours over several generations in mice [11]. In OC, CSCs have been identified by the expression of different cell surface markers (CD117, CD133, epithelial cell adhesion molecule (EpCAM), CD44, CD24, ROR1 and others), non-surface markers (aldehyde dehydrogenase), transcription factors (NANOG, SOX2, OCT4) and side populations [12–15] CD117 was the first cell surface marker discovered to be significant for ovarian CSCs. Its expression is related to tumour formation, chemoresistance and poor prognosis of disease [16–18]. One of the most reported ovarian CSC surface markers is CD133. It is correlated with a number of stem characteristics, like tumour formation, disease progression, chemoresistance and poor prognosis [19–21]. It was also studied as a target for cancer target therapy [22]. Other commonly used CSC surface markers are CD24, CD44, EpCAM and ROR1. Surface markers can be used for ovarian CSCs identification alone or in combination with other ovarian CSC markers. Enzyme aldehyde dehydrogenase 1 (ALDH1) is also a frequently used marker to identify CSCs in OC. Several studies correlate ALDH1 expression with cell proliferation, migration, chemoresistance and poor survival [23–25]. Expression of ALDH1 alone or in combination with cell surface stem cell markers is frequently used for ovarian CSCs identification. NANOG, OCT4 and SOX2 are transcription factors that play a key role in pluripotency and maintenance of self-renewal in undifferentiated embryonic stem cells [26]. These transcription factors are also commonly expressed in ovarian CSCs [27–29]. Their expression in ovarian CSCs is associated with poor prognosis and chemoresistance [30–33]. CSCs possess the ability to efflux DNA-binding dyes. This characteristic, when using cytometry, results in a side population of cells. CSCs express ATP binding cassette transporters such as MDR1/ABCB1 and ABCG2, which efflux chemotherapeutic agents and so contribute to chemoresistance [34–36].

The exact phenotype of ovarian CSC population and their molecular status have still not been well defined. It is known that CSC phenotype is not uniform among the various cancer types and even amongst tumours of the same histological type [37]. It is believed that OC involves different types of stem cells assembling a complex hierarchical organized process. Heterogeneity of cells within tumours may influence the progression of disease and eventually also its response to treatment in terms of drug resistance [38]. From this point of view, defining the molecular status of CSCs and their exact phenotype is important for planning treatment protocol that would be individualized for each patient.

The Role of Ovarian Cancer Stem Cells in Ovarian Cancer Pathogenesis

The precise mechanism by which normal ovarian cells undergo transformation leading to aggressive cells of OC remains unknown. There are two hypotheses concerning CSC progenitors. The first hypothesis is that CSCs represent the malignant counterparts of normal adult tissue stem cells (Fig. 5.1). There are some promising studies that explain CSC evolution from stem cells of healthy adult human ovary to CSCs in ovarian tumor [13, 39]. In adult human ovaries, similar to other adult tissue and organs, very small embryonic-like (VSEL) stem cells have been identified [40–42]. These cells have been later additionally characterized and differentiated in vitro [40–42]. VSEL stem cells might originate in the embryonic epiblast and probably remain dormant in adult human tissue and organs from this early period of life [43]. The second hypothesis concerning origin of CSCs is that the dedifferentiation of mature cells to a more pluripotent state could be a potential mechanism for the development of cancer cells with stem cell-like features [44]. As stem cells

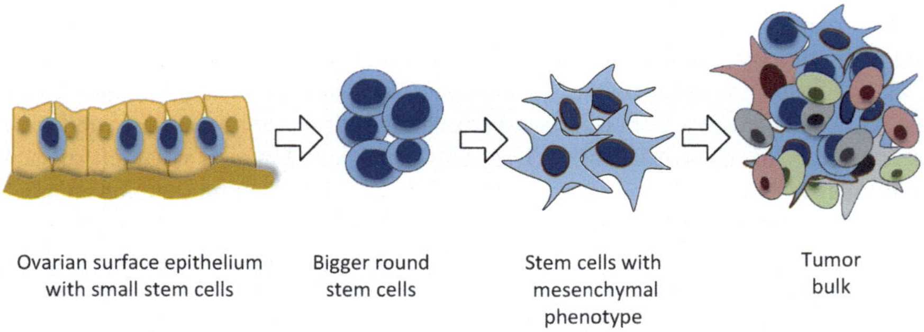

Fig. 5.1 Schematic representation of the role of ovarian cancer stem cells in ovarian cancer pathogenesis. Small, progenitor very small embryonic-like stem cells in the ovarian surface epithelium are developing into bigger round stem cells and gradually undergo transformation into stem cells with mesenchymal-like phenotype

remain in a tissue for a longer period compared with their differentiated progeny, stem cells seem to be more prone to acquire several mutations leading to cancer transformation.

The defining property of CSCs is their potential to generate a tumour of hierarchical organized cells with different rates of differentiation. Subsequent tumour growth, invasion and dissemination are achievable due to CSC biological properties as are self-renewal, asymmetrical division, differentiation and epithelial-mesenchymal transition (EMT). EMT is a physiological cell event active in tissue remodelling during embryogenesis, and during the regeneration of adult tissue in cases of injury. According to many studies, EMT also plays an important role in ovarian carcinogenesis [45] (Fig. 5.2). The so-called EMT programs are supposed to be triggered by different signals of tumour microenvironment [46]. Through EMT cells develop characteristics of mesenchymal cells like epithelial cell polarity and intracellular adhesion, they lose specific cell surface markers and miss epithelial traits. By those changes, tumour cells become more motile, which enhances tumour progression [47]. EMT is also considered a key step in CSC metastases [48]. Acquired mesenchymal properties promote cancer cells invasion into the extracellular matrix

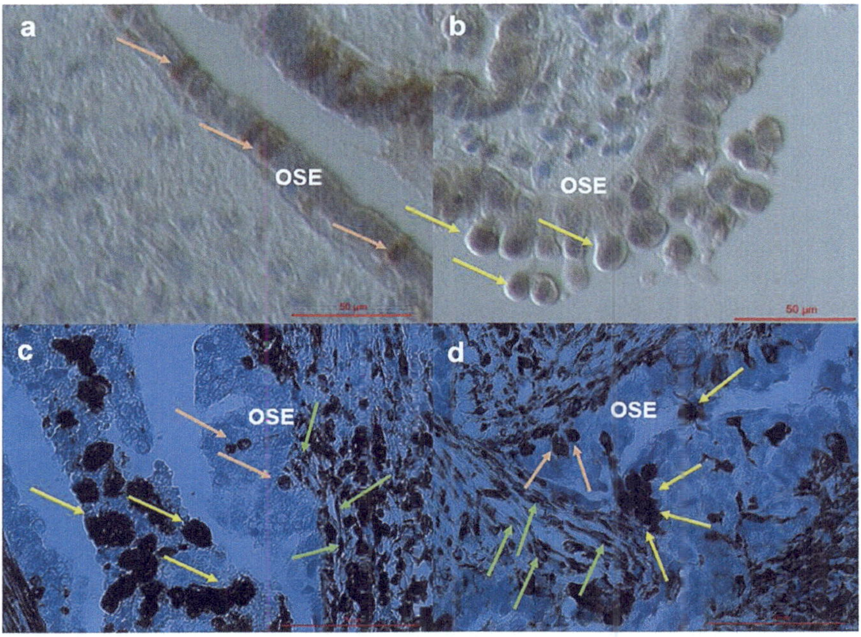

Fig. 5.2 Ovarian cancer tissue sections revealing ovarian cancer stem cells, immunohistochemistry for vimentin. Vimentin-positive small stem cells in ovarian surface epithelium (OSE) (orange arrows) are developing into bigger round stem cells (yellow arrows) and stem cells with mesenchymal-like phenotype (green arrows). Legend: brown colour-positivity for vimentin. Red bar **a**, **b** 50 μm; **c**, **d** 100 μm

and their further dissemination. Transformed tumour cells spread throughout the organism, where they present a reservoir of cancer cell populations [49]. Besides to metastatic progression, vimentin, a major cytoskeletal component of mesenchymal cell, also relates to primary resistance to chemotherapy [50, 51]. Tumour cells undergoing EMT obtain the capability to escape patients antitumor defence, avoid induced apoptosis and resist standard anticancer drugs [51].

An important task for the future is to determine the population of progenitor cells involved in OC initiation and to additionally explore EMT, strongly involved in OC progression and resistance to standard treatment. Both, CSCs progenitors and EMT, might present a key target for cancer target therapy leading to individualized treatment which promises a higher survival rate of OC patients.

The Clinical Significance of Ovarian Cancer Stem Cells

Ovarian cancer, according to its biological characteristics and course of the disease, represents a typical example of CSC-powered disease [10]. It is an extremely aggressive cancer, which invades locally into neighbouring tissues and organs and spreads rapidly through the abdominal cavity causing malignant ascites and later distant metastasis [52]. Malignant ascites, being rich in tumour cells with stem-like properties, represents an important source of CSCs [53–56]. Tumour cell spheroids not only survive and migrate in malignant ascites but also proliferate in it [57, 58]. This characteristic, which represents one of the essential properties of CSCs, is also used in vitro, when assessing cell stemness as formation of cell spheres in suspension culture [59].

Despite aggressive standard treatment of OC, which includes debulking surgery and cytostatic chemotherapy, OC relapses over 70% of all cases [9]. This is because standard treatment reduces tumour size by eliminating only tumour bulk consisting of tumour cells of different differentiation stages, but highly potent subpopulation of ovarian CSCs survives [60] (Fig. 5.3). Standard chemotherapy target actively proliferating cells, so tumour cells displaying a slow-cycling rate stay resistant to cytostatic therapy [61, 62]. As classic OC therapy agents target actively dividing cells in the S or M phase of the cell cycle, cells remaining in the inactive G0 phase, as are CSCs, represent a significant issue in terms of effective treatment.

Resistant dormant ovarian CSCs that survive allegedly successful treatment with cytostatic chemotherapy eventually "wake up" and cause tumour relapse [10]. In time ovarian CSCs repopulate and trigger even more aggressive disease [60]. The arising tumour is richer in CSCs and so presumably more resistant to classic cytotoxic treatment than was primary ovarian tumour [63]. A level of putative CSC marker CD44 was significantly higher in relapsed and metastatic OC in comparison to initial ovarian tumour [63]. Similarly, CD44 was overexpressed in drug-resistant OC cell lines and up-regulated in tumour that reoccurred after chemotherapy treatment [63]. Transcription factor NANOG is another CSC-related marker involved in ovarian carcinoma chemoresistance mechanism [32]. Increased expression of NANOG in

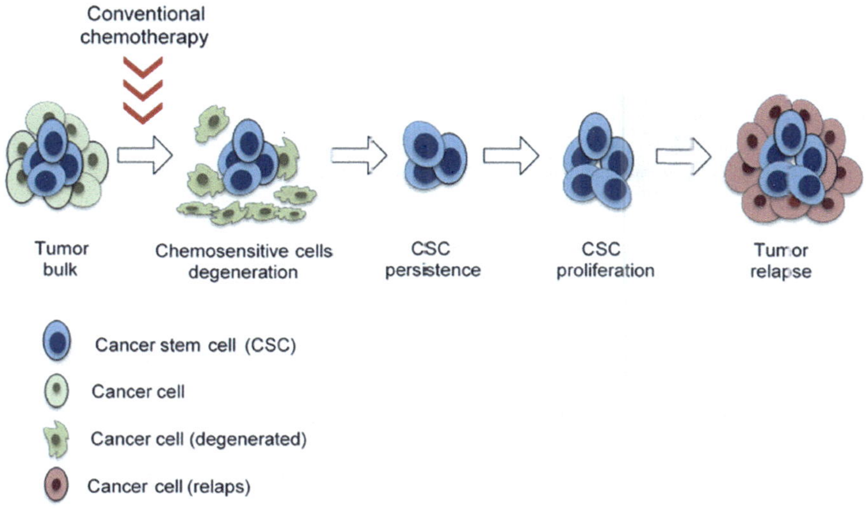

Fig. 5.3 Schematic representation of ovarian cancer stem cells chemoresistance. Cancer cells in ovarian tumour are sensitive to standard chemotherapy leading to cells destruction and reduction of tumour size. Ovarian cancer stem cells are resistant to standard chemotherapy, they persist and further proliferate. Relapsed ovarian tumour is richer in cancer stem cells becoming more resistant to standard chemotherapy

tumour cells was significantly associated with tumour reduced chemosensitivity and consequently OC patients' poor disease-free and overall survival [31].

Besides CSC markers used to identify CSC in various tumours, stemness cells could be also confirmed by functional genomics. Based on gene expression profiles in functional genomic studies, different stemness-associated genes were identified, which, when expressed, relate to OC poor prognosis [64, 65].

Strong implication of ovarian CSCs in course of disease on so many levels can be usefully used in OC treatment.

Ovarian Cancer Stem Cells as Therapeutic Targets

The current focus of anti-cancer therapy is the development of molecular target therapy. It is becoming obvious that CSCs are crucial for chemoresistance and disease recurrence, so targeting these cells seems to be a promising way for efficient treatment of OC. There are different strategies applicable including, targeting stemness markers, signalling pathways, epigenetic features, microenvironment and others (Fig. 5.4).

Stemness markers are not only important for CSC identification and isolation but also for its functionality [66]. Treatment strategy is to recognize CSC-specific antigen and then to either kill the cell with cytotoxic conjugates or to inhibit

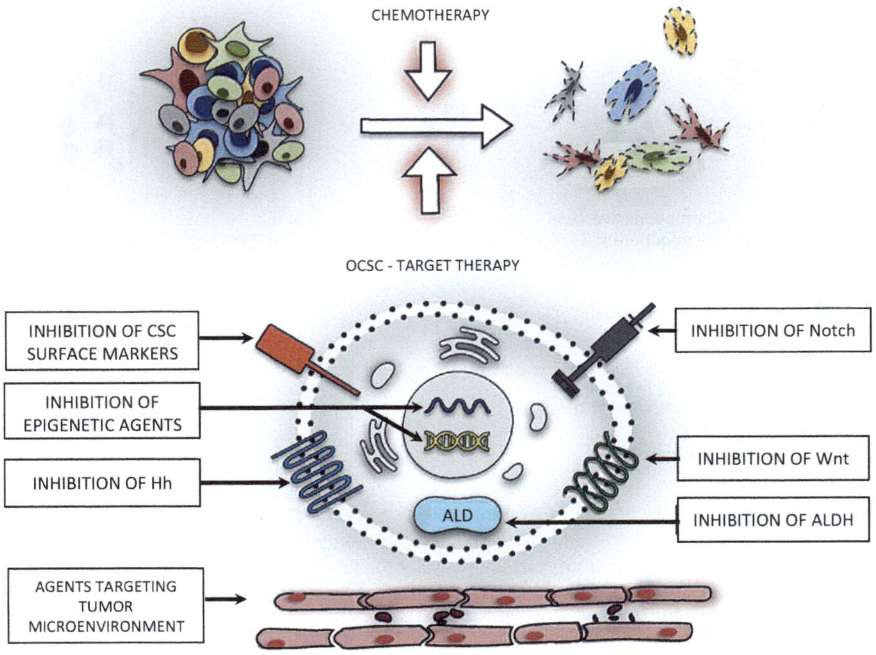

Fig. 5.4 Schematic representation of ovarian cancer stem cells targeting strategies. Combination of standard chemotherapy and target therapy can efficiently eliminate ovarian tumour and residual ovarian cancer stem cells (OCSC)

their biological function. For targeting, surface markers have been created additional approaches, which include antibody–drug conjugates, bispecific antibodies and chimeric antigen receptor T (CAR-T) [67]. It was proven that inhibition of surface marker CD24 decreases cell viability through induction of apoptosis in SKOV3 cell line and restricts tumour growth in nude mice [68]. CD44 also serves as a ligand for hyaluronic acid [69]. The CD44 positive SKOV3 cell lines were targeted by hyaluronic acid-paclitaxel, which caused a decrease in tumour weight and nodules [70]. In another report, CD133 OVCAR5-luc cells were targeted resulting in a considerate decrease in tumour progression [21]. Furthermore, a chimeric antigen receptor-based immunotherapeutic approach pointing at CD133 positive cells successfully removed CSCs from OC cell lines and ascites [22]. Another putative drug for targeting OC is metformin. It was proposed that metformin inhibits CD44 positive and CD177 positive CSCs and EMT process in SKOV3 and A2780 cell lines [71]. It was also shown that metformin reduces ALDH positive CSC populations and decreases angiogenesis [72]. There have been several novel ALDH inhibitors that have been recently evaluated in ovarian CSCs [73–75].

Targeting signalling pathways in CSCs makes an excellent therapeutic option, as dysregulation in signalling pathways is probably one of the key events in CSC initiation and cancer dissemination. There are several key signalling pathways such as wnt, notch and hedgehog (Hh), which are associated with stem cell properties like cell self-renewal, differentiation, proliferation and mesenchymal features [76, 77]. Wnt signalling is involved in dedifferentiation of tumour cells and the implementation of stemness function [78]. Wnt pathway inhibitors that compete with receptors for binding to wnt ligands demonstrate anti-CSC effect and reduce tumour growth [79, 80]. Notch signalling pathway is one of the most intensively studied routes as putative therapeutic targets of CSCs [81, 82]. As part of this, several Notch pathway inhibitors, capable of depleting ovarian CSCs, have been examined [83–86].

Aberrant epigenetic alterations, as are remodelling of chromatin and changes in DNA methylation causing partial or complete loss of epigenetic constraints in cancer cells, are common features of OC [87–91]. These changes allow cancer cells to lose their epithelial phenotype and grow to be invasive, gain metastatic abilities and become chemoresistant. Drugs targeting epigenetic alterations work as differentiation therapy through inducing CSCs to transform from pluripotent to a differentiated state. For the latter activation of differentiation-associated signalling cascades is induced, and re-expression of tumour suppressor genes is stimulated. More differentiated cancer cells are more sensitive to conventional cytotoxic drugs. Promising epigenetic agents are: DNA methyltransferase inhibitors, histone deacetylase inhibitors, histone methyltransferase inhibitors, bromodomain and extraterminal inhibitors and long non-coding RNAs [66].

Hypoxic niches offer suitable environments for OC cells to obtain the properties of cancerous stemness [92, 93]. Therefore, these niches can be also considered as an applicable target for OC target therapy.

Conclusions

In the era of precision oncology, treatment of OC should be tailored to individual patients, but despite progress in understanding ovarian CSCs, developing ovarian CSC-targeted therapeutics remains quite a challenge.

Identification of ovarian CSCs presents an obstacle due to the heterogeneous group of cells in tumour. This challenge could be resolved by a single cell sequencing through which identification and characterization of CSCs from tumour or ascites could be done. Personalized ovarian CSC markers and signalling pathways can be identified, holding a huge promise to the individualized ovarian CSC therapy.

Another barrier in CSC directed target therapy is a lack of specificity of CSC from normal stem cells. There is substantial crosstalk among different signalling pathways involved in cancer and normal stem cells maintenance making efficient targeted therapies difficult to accomplish. To provide target selection and overcome adverse effects, biological differences between cancer and normal stem cells are needed.

According to recent studies, a combination of ovarian CSC-target therapy either before or simultaneously with chemotherapy could be one of the most efficient anti-cancer therapeutic models. To minimize adverse effects and rate of failure of the proposed treatment, protocol evaluations in preclinical and clinical trials are necessary. To achieve this, additional research of ovarian CSC frequency, chemosensitivity, stemness gene signature, genetic/epigenetic background and microenvironment is vital.

To summarize, profound knowledge on ovarian CSCs is crucial for the development of CSC-target therapy and effective combinatorial treatment, which would provide a higher OC disease-free and overall survival rate.

References

1. Torre LA, Trabert B, DeSantis CE (2018) Ovarian cancer statistics. CA: Cancer J Clinicians 68:284–296. https://doi.org/10.3322/caac.21456
2. Siegel RL, Miller KD, Jemal A (2018) Cancer statistics. CA: Cancer J Clincians 68:7–30. https://doi.org/10.3322/caac.21442
3. Zhang XY, Zhang PY (2016) Recent perspectives of epithelial ovarian carcinoma. Oncol Lett 12:3055–3058. https://doi.org/10.3892/ol.2016.5107
4. Prat J (2012) Ovarian carcinomas: five distinct diseases with different origins, genetic alterations, and clinicopathological features. Virchows Arch 460:237–249. https://doi.org/10.1007/s00428-012-1203-5
5. Kurman RJ, Shih Ie M (2016) The dualistic model of ovarian carcinogenesis: revisited, revised, and expanded. Am J Pathol 186:733–747. https://doi.org/10.1016/j.ajpath.2015.11.011
6. Ottevanger PB (2017) Ovarian cancer stem cells more questions than answers. Semin Cancer Biol 44:67–71. https://doi.org/10.1016/j.semcancer.2017.04.009
7. Ozols RF (2005) Treatment goals in ovarian cancer. Int J Gynecol Cancer 15(Suppl) 1:3–11. https://doi.org/10.1111/j.1525-1438.2005.15351.x
8. Schorge JO, McCann C, Del Carmen MG (2010) Surgical debulking of ovarian cancer: what difference does it make? Rev Obstet Gynecol 3:111–117
9. Pieterse Z, Amaya-Padilla MA, Singomat T, Binju M, Madjid BD, Yu Y, Kaur P (2019) Ovarian cancer stem cells and their role in drug resistance. Int J Biochem Cell Biol 106:117–126. https://doi.org/10.1016/j.biocel.2018.11.012
10. Lupia M, Cavallaro U (2017) Ovarian cancer stem cells: still an elusive entity? Mol Cancer 16:64. https://doi.org/10.1186/s12943-017-0638-3
11. Bapat SA, Mali AM, Koppikar CB, Kurrey NK (2005) Stem and progenitor-like cells contribute to the aggressive behavior of human epithelial ovarian cancer. Cancer Res 65:3025–3029. https://doi.org/10.1158/0008-5472.can-04-3931
12. Kenda Suster N, Virant-Klun I, Frkovic Grazio S, Smrkolj S (2016) The significance of the pluripotency and cancer stem cell-related marker nanog in diagnosis and treatment of ovarian carcinoma. Eur J Gynaecol Oncol 37:604–612
13. Virant-Klun I, Kenda-Suster N, Smrkolj S (2016) Small putative nanog, sox2, and ssea-4-positive stem cells resembling very small embryonic-like stem cells in sections of ovarian tissue in patients with ovarian cancer. J Ovarian Res 9:12
14. Kenda Suster N, Virant-Klun I (2019) Presence and role of stem cells in ovarian cancer. World J Stem Cells 11:383–397. https://doi.org/10.4252/wjsc.v11.i7.383
15. Ahmed N, Kadife E, Raza A, Short M, Jubinsky PT, Kannourakis G (2020) Ovarian cancer, cancer stem cells and current treatment strategies: A potential role of magmas in the current treatment methods. Cells 9:719. https://doi.org/10.3390/cells9030719

16. Luo L, Zeng J, Liang B, Zhao Z, Sun L, Cao D, Yang J, Shen K (2011) Ovarian cancer cells with the cd117 phenotype are highly tumorigenic and are related to chemotherapy outcome. Exp Mol Pathol 91:596–602. https://doi.org/10.1016/j.yexmp.2011.06.005
17. Conic I, Stanojevic Z, Jankovic Velickovic L, Stojnev S, Ristic Petrovic A, Krstic M, Stanojevic M, Bogdanovic D, Stefanovic V (2015) Epithelial ovarian cancer with cd117 phenotype is highly aggressive and resistant to chemotherapy. J Obstet Gynaecol Res 41:1630–1637. https://doi.org/10.1111/jog.12758
18. Stemberger-Papic S, Vrdoljak-Mozetic D, Ostojic DV, Rubesa-Mihaljevic R, Krigtofic I, Brncic-Fisher A, Kragevic M, Eminovic S (2015) Expression of cd133 and cd117 in 64 serous ovarian cancer cases. Coll Antropol 39:745–753
19. Curley MD, Therrien VA, Cummings CL, Sergent PA, Koulouris CR, Friel AM, Roberts DJ, Seiden MV, Scadden DT, Rueda BR, Foster R (2009) Cd133 expression defines a tumor initiating cell population in primary human ovarian cancer. Stem Cells (Dayton, Ohio) 27:2875–2883. https://doi.org/10.1002/stem.236
20. Zhang J, Guo X, Chang DY, Rosen DG, Mercado-Uribe I, Liu J (2012) Cd133 expression associated with poor prognosis in ovarian cancer. Mod Pathol 25:456–464. https://doi.org/10.1038/modpathol.2011.170
21. Skubitz AP, Taras EP, Boylan KL, Waldron NN, Oh S, Panoskaltsis-Mortari A, Vallera DA (2013) Targeting cd133 in an in vivo ovarian cancer model reduces ovarian cancer progression. Gynecol Oncol 130:579–587. https://doi.org/10.1016/j.ygyno.2013.05.027
22. Klapdor R, Wang S, Hacker U, Buning H, Morgan M, Dork T, Hillemanns P, Schambach A (2017) Improved killing of ovarian cancer stem cells by combining a novel chimeric antigen receptor-based immunotherapy and chemotherapy. Hum Gene Ther 28:886–896. https://doi.org/10.1089/hum.2017.168
23. Wang Y, Shao F, Chen L (2018) Aldh1a2 suppresses epithelial ovarian cancer cell proliferation and migration by downregulating stat3. Onco Targets Ther 11:599–608. https://doi.org/10.2147/ott.s145864
24. Wang YC, Yo YT, Lee HY, Liao YP, Chao TK, Su PH, Lai HC (2012) Aldh1-bright epithelial ovarian cancer cells are associated with cd44 expression, drug resistance, and poor clinical outcome. Am J Pathol 18:1159–1169. https://doi.org/10.1016/j.ajpath.2011.11.015
25. Januchowski R, Wojtowicz K, Sterzyska K, Sosiska P, Andrzejewska M, Zawierucha P, Nowicki M, Zabel M (2016) Inhibition of aldh1a1 activity decreases expression of drug transporters and reduces chemotherapy resistance in ovarian cancer cell lines. Int J Biochem Cell Biol 78:248–259. https://doi.org/10.1016/j.biocel.2016.07.017
26. Ng PM, Lufkin T (2011) Embryonic stem cells: Protein interaction networks. Biomol Concepts 2:13–25. https://doi.org/10.1515/bmc.2011.008
27. Yu Z, Pestell TG, Lisanti MP, Pestell RG (2012) Cancer stem cells. Int J Biochem Cell Biol 44:2144–2151. https://doi.org/10.1016/j.biocel.2012.08.022
28. Wen Y, Hou Y, Huang Z, Cai J, Wang Z (2017) Sox2 is required to maintain cancer stem cells in ovarian cancer. Cancer Sci 108:719–731. https://doi.org/10.1111/cas.13186
29. Kenda Suster N, Frkovic Grazio S, Virant-Klun I, Verdenik I, Smrkolj S (2017) Cancer stem cell-related marker nanog expression in ovarian serous tumors: a clinicopathological study of 159 cases. Int J Gynecol Cancer 27:2006–2013. https://doi.org/10.1097/igc.0000000000001105]
30. Ruan Z, Yang X, Cheng W (2019) Oct4 accelerates tumorigenesis through activating jak/stat signaling in ovarian cancer side population cells. Cancer Manag Res 11:389–399. https://doi.org/10.2147/cmar.s180418
31. Siu MK, Wong ES, Kong DS, Chan HY, Jiang L, Wong OG, Lam EW, Chan KK, Ngan HY, Le XF, Cheung AN (2013) Stem cell transcription factor nanog controls cell migration and invasion via dysregulation of e-cadherin and foxj1 and contributes to adverse clinical outcome in ovarian cancers. Oncogene 32:3500–3509. https://doi.org/10.1038/onc.2012.363
32. Lee M, Nam EJ, Kim SW, Kim S, Kim JH, Kim YT (2012) Prognostic impact of the cancer stem cell-related marker nanog in ovarian serous carcinoma. Int J Gynecol Cancer 22:1489–1496. https://doi.org/10.1097/IGJ.0b013e3182738307

33. Pan Y, Jiao J, Zhou C, Cheng Q, Hu Y, Chen H (2010) Nanog is highly expressed in ovarian serous cystadenocarcinoma and correlated with clinical stage and pathological grade. Pathobiology 77:283–288. https://doi.org/10.1159/000320866
34. Hu L, McArthur C, Jaffe RB (2010) Ovarian cancer stem-like side-population cells are tumourigenic and chemoresistant. Br J Cancer 102:1276–1283. https://doi.org/10.1038/sj.bjc.6605626
35. Dou J, Jiang C, Wang J, Zhang X, Zhao F, Hu W, He X, Li X, Zou D, Gu N (2011) Using abcg2-molecule-expressing side population cells to identify cancer stem-like cells in a human ovarian cell line. Cell Biol Int 35:227–234. https://doi.org/10.1042/cbi20100347
36. Kobayashi Y, Seino K, Hosonuma S, Ohara T, Itamochi H, Isonishi S, Kita T, Wada H, Kojo S, Kiguchi K (2011) Side population is increased in paclitaxel-resistant ovarian cancer cell lines regardless of resistance to cisplatin. Gynecol Oncol 121:390–394. https://doi.org/10.1016/j.ygyno.2010.12.366
37. Jiang H, Lin X, Liu Y, Gong W, Ma X, Yu Y, Xie Y, Sun X, Feng Y, Janzen V, Chen T (2012) Transformation of epithelial ovarian cancer stemlike cells into mesenchymal lineage via emt results in cellular heterogeneity and supports tumor engraftment. Mol Med 18:1197–1208. https://doi.org/10.2119/molmed.2012.00075
38. Zhou BB, Zhang H, Damelin M, Geles KG, Grindley JC, Dirks PB (2009) Tumour-initiating cells: Challenges and opportunities for anticancer drug discovery. Nat Rev Drug Discov 8:806–823. https://doi.org/10.1038/nrd2137
39. Parte SC, Batra SK, Kakar SS (2018) Characterization of stem cell and cancer stem cell populations in ovary and ovarian tumors. J Ovarian Res 11:69. https://doi.org/10.1186/s13048-018-0439-3
40. Virant-Klun I, Zech N, Rozman P, Vogler A, Cvjeticanin B, Klemenc P, Malicev E, Meden-Vrtovec H (2008) Putative stem cells with an embryonic character isolated from the ovarian surface epithelium of women with no naturally present follicles and oocytes. Differentiation 76:843–856. https://doi.org/10.1111/j.1432-0436.2008.00268.x
41. Parte S, Bhartiya D, Telang J, Daithankar V, Salvi V, Zaveri K, Hinduja I (2011) Detection, characterization, and spontaneous differentiation in vitro of very small embryonic-like putative stem cells in adult mammalian ovary. Stem Cells Dev 20:1451–1464. https://doi.org/10.1089/scd.2010.0461
42. Virant-Klun I, Skutella T, Kubista M, Vogler A, Sinkovec J, Meden-Vrtovec H (2013) Expression of pluripotency and oocyte-related genes in single putative stem cells from human adult ovarian surface epithelium cultured in vitro in the presence of follicular fluid. Biomed Res Int 2013:861460. https://doi.org/10.1155/2013/861460
43. Ratajczak MZ, Zuba-Surma EK, Wysoczynski M, Ratajczak J, Kucia M (2008) Very small embryonic-like stem cells: characterization, developmental origin, and biological significance. Exp Hematol 36:742–751. https://doi.org/10.1016/j.exphem.2008.03.010
44. Liu J (2018) The dualistic origin of human tumors. Semin Cancer Biol 53:1–16. https://doi.org/10.1016/j.semcancer.2018.07.004
45. Kenda Suster N, Smrkolj S, Virant-Klun I (2017) Putative stem cells and epithelial-mesenchymal transition revealed in sections of ovarian tumor in patients with serous ovarian carcinoma using immunohistochemistry for vimentin and pluripotency-related markers. J Ovarian Res 10:11. https://doi.org/10.1186/s13048-017-0306-7
46. Yang J, Weinberg RA (2008) Epithelial-mesenchymal transition: At the crossroads of development and tumor metastasis. Dev Cell 14:818–829. https://doi.org/10.1016/j.devcel.2008.05.009
47. Yan H, Sun Y (2014) Evaluation of the mechanism of epithelial-mesenchymal transition in human ovarian cancer stem cells transfected with a ww domain-containing oxidoreductase gene. Oncol Lett 8:426–430. https://doi.org/10.3892/ol.2014.2063
48. Cioce M, Ciliberto G (2012) On the connections between cancer stem cells and emt. Cell Cycle 11:4301–4302. https://doi.org/10.4161/cc.22809
49. Lam SS, Mak AS, Yam JW, Cheung AN, Ngan HY, Wong AS (2014) Targeting estrogen-related receptor alpha inhibits epithelial-to-mesenchymal transition and stem cell properties of ovarian cancer cells. Mol Ther 22:743–751. https://doi.org/10.1038/mt.2014.1

50. Davidson B, Holth A, Hellesylt E, Tan TZ, Huang RY, Trope C, Nesland JM, Thiery JP (2015) The clinical role of epithelial-mesenchymal transition and stem cell markers in advanced-stage ovarian serous carcinoma effusions. Hum Pathol 46:1–8. https://doi.org/10.1016/j.humpath.2014.10.004
51. Li X, Yang J, Wang X, Li X, Liang J, Xing H (2016) Role of twist2, e-cadherin and vimentin in epithelial ovarian carcinogenesis and prognosis and their interaction in cancer progression. Eur J Gynaecol Oncol 37:100–108
52. Colombo N, Sessa C, du Bois A, Ledermann J, McCluggage WG, McNeish I, Morice P, Pignata S, Ray-Coquard I, Vergote I, Baert T, Belaroussi I, Dashora A, Olbrecht S, Planchamp F, Querleu D (2019) Esmo-esgo consensus conference recommendations on ovarian cancer: Pathology and molecular biology, early and advanced stages, borderline tumours and recurrent disease†. Ann Oncol Off J Eur Soc Med Oncol 30:672–705. https://doi.org/10.1093/annonc/mdz062
53. Ho CM, Chang SF, Hsiao CC, Chien TY, Shih DT (2012) Isolation and characterization of stromal progenitor cells from ascites of patients with epithelial ovarian adenocarcinoma. J Biomed Sci 19:23. https://doi.org/10.1186/1423-0127-19-23
54. Latifi A, Luwor RB, Bilandzic M, Nazaretian S, Stenvers K, Pyman J, Zhu H, Thompson EW, Quinn MA, Findlay JK, Ahmed N (2012) Isolation and characterization of tumor cells from the ascites of ovarian cancer patients: molecular phenotype of chemoresistant ovarian tumors. PLoS ONE 7:e46858. https://doi.org/10.1371/journal.pone.0046858
55. Liao J, Qian F, Tchabo N, Mhawech-Fauceglia P, Beck A, Qian Z, Wang X, Huss WJ, Lele SB, Morrison CD, Odunsi K (2014) Ovarian cancer spheroid cells with stem cell-like properties contribute to tumor generation, metastasis and chemotherapy resistance through hypoxia-resistant metabolism. PLoS One 9:e84941. https://doi.org/10.1371/journal.pone.0084941
56. Mo L, Bachelder RE, Kennedy M, Chen PH, Chi JT, Berchuck A, Cianciolo G, Pizzo SV (2015) Syngeneic murine ovarian cancer model reveals that ascites enriches for ovarian cancer stem-like cells expressing membrane grp78. Mol Cancer Ther 14:747–756. https://doi.org/10.1158/1535-7163.mct-14-0579
57. Tan DSP, Agarwal R, Kaye SB (2006) Mechanisms of transcoelomic metastasis in ovarian cancer. Lancet Oncol 7:925–934. https://doi.org/10.1016/S1470-2045(06)70939-1
58. Kim S, Kim S, Kim J, Kim B, Kim SI, Kim MA, Kwon S, Song YS (2018) Evaluating tumor evolution via genomic profiling of individual tumor spheroids in a malignant ascites. Sci Rep 8:12724. https://doi.org/10.1038/s41598-018-31097-y
59. Dontu G, Abdallah WM, Foley JM, Jackson KW, Clarke MF, Kawamura MJ, Wicha MS (2003) In vitro propagation and transcriptional profiling of human mammary stem/progenitor cells. Genes Dev 17:1253–1270. https://doi.org/10.1101/gad.1061803
60. Clevers H (2011) The cancer stem cell: Premises, promises and challenges. Nat Med 17:313–319. https://doi.org/10.1038/nm.2304
61. Chen W, Dong J, Haiech J, Kilhoffer MC, Zeniou M (2016) Cancer stem cell quiescence and plasticity as major challenges in cancer therapy. Stem Cerlls Int 2016:1740936. https://doi.org/10.1155/2016/1740936
62. Takeishi S, Nakayama KI (2016) To wake up cancer stem cells, or to let them sleep, that is the question. Cancer Sci 107:875–881. https://doi.org/10.1111/cas.12958
63. Raspollini MR, Amunni G, Villanucci A, Baroni G, Taddei A, Taddei GL (2004) C-kit expression and correlation with chemotherapy resistance in ovarian carcinoma: an immunocytochemical study. Ann Oncol Off J Eur Soc Med Oncol 15:594–597
64. Zhang W, Liu Y, Sun N, Wang D, Boyd-Kirkup J, Dou X, Han JD (2013) Integrating genomic, epigenomic, and transcriptomic features reveals modular signatures underlying poor prognosis in ovarian cancer. Cell Rep 4:542–553. https://doi.org/10.1016/j.celrep.2013.07.010
65. Tan TZ, Miow QH, Huang RY, Wong MK, Ye J, Lau JA, Wu MC, Bin Abdul Hadi LH, Soong R, Choolani M, Davidson B, Nesland JM, Wang LZ, Matsumura N, Mandai M, Konishi I, Goh BC, Chang JT, Thiery JP, Mori S (2013) Functional genomics identifies five distinct molecular subtypes with clinical relevance and pathways for growth control in epithelial ovarian cancer. EMBO Mol Med 5:1051–1066. https://doi.org/10.1002/emmm.201201823

66. Zong X, Nephew KP (2019) Ovarian cancer stem cells: role in metastasis and opportunity for therapeutic targeting. Cancers (Basel) 11:934. https://doi.org/10.3390/cancers11070934
67. Saygin C, Matei D, Majeti R, Reizes O, Lathia JD (2019) Targeting cancer stemness in the clinic: from hype to hope. Cell Stem Cell 24:25–40. https://doi.org/10.1016/j.stem.2018.11.017
68. Su D, Deng H, Zhao X, Zhang X, Chen L, Chen X, Li Z, Bai Y, Wang Y, Zhong Q, Yi T, Qian Z, Wei Y (2009) Targeting cd24 for treatment of ovarian cancer by short hairpin rna. Cytotherapy 11:642–652. https://doi.org/10.1080/14653240902878308
69. Chen C, Zhao S, Karnad A, Freeman JW (2018) The biology and role of cd44 in cancer progression: therapeutic implications. J Hematol Oncol 11:64. https://doi.org/10.1186/s13045-018-0605-5
70. Lee SJ, Ghosh SC, Han HD, Stone RL, Bottsford-Miller J, Shen DY, Auzenne EJ, Lopez-Araujo A, Lu C, Nishimura M, Pecot CV, Zand B, Thanapprapasr D, Jennings NB, Kang Y, Huang J, Hu W, Klostergaard J, Sood AK (2012) Metronomic activity of cd44-targeted hyaluronic acid-paclitaxel in ovarian carcinoma. Clin Cancer Res 18:4114–4121. https://doi.org/10.1158/1078-0432.CCR-11-3250
71. Zhang R, Zhang P, Wang H, Hou D, Li W, Xiao G, Li C (2015) Inhibitory effects of metformin at low concentration on epithelial-mesenchymal transition of CD44(+)CD117(+) ovarian cancer stem cells. Stem Cell Res Ther 6:262–262. https://doi.org/10.1186/s13287-015-0249-0
72. Shank JJ, Yang K, Ghannam J, Cabrera L, Johnston CJ, Reynolds RK, Buckanovich RJ (2012) Metformin targets ovarian cancer stem cells in vitro and in vivo. Gynecol Oncol 127:390–397. https://doi.org/10.1016/j.ygyno.2012.07.115
73. Huddle BC, Grimley E, Buchman CD, Chtcherbinine M, Debnath B, Mehta P, Yang K, Morgan CA, Li S, Felton J, Sun D, Mehta G, Neamati N, Buckanovich RJ, Hurley TD, Larsen SD (2018) Structure-based optimization of a novel class of aldehyde dehydrogenase 1a (aldh1a) subfamily-selective inhibitors as potential adjuncts to ovarian cancer chemotherapy. J Med Chem 61:8754–8773. https://doi.org/10.1021/acs.jmedchem.8b00930
74. Chefetz I, Grimley E, Yang K, Hong L, Vinogradova EV, Suciu R, Kovalenko I, Karnak D, Morgan CA, Chtcherbinine M, Buchman C, Huddle B, Barraza S, Morgan M, Bernstein KA, Yoon E, Lombard DB, Bild A, Mehta G, Romero I, Chiang C-Y, Landen C, Cravatt B, Hurley TD, Larsen SD, Buckanovich RJ (2019) A pan-aldh1a inhibitor induces necroptosis in ovarian cancer stem-like cells. Cell Rep 26:3061-3075.e3066. https://doi.org/10.1016/j.celrep.2019.02.032
75. Nwani NG, Condello S, Wang Y, Swetzig WM, Barber E, Hurley T, Matei D (2019) A novel aldh1a1 inhibitor targets cells with stem cell characteristics in ovarian cancer. Cancers (Basel) 11:502. https://doi.org/10.3390/cancers11040502
76. Takebe N, Harris PJ, Warren RQ, Ivy SP (2011) Targeting cancer stem cells by inhibiting wnt, notch, and hedgehog pathways. Nat Rev Clinical Oncol 8:97–106. https://doi.org/10.1038/nrclinonc.2010.196
77. Takebe N, Miele L, Harris PJ, Jeong W, Bando H, Kahn M, Yang SX, Ivy SP (2015) Targeting notch, hedgehog, and wnt pathways in cancer stem cells: Clinical update. Nat Rev Clin Oncol 12:445–464. https://doi.org/10.1038/nrclinonc.2015.61
78. Krishnamurthy N, Kurzrock R (2018) Targeting the wnt/beta-catenin pathway in cancer: update on effectors and inhibitors. Cancer Treat Rev 62:50–60. https://doi.org/10.1016/j.ctrv.2017.11.002
79. Le PN, McDermott JD, Jimeno A (2015) Targeting the wnt pathway in human cancers: therapeutic targeting with a focus on omp-54f28. Pharmacol Ther 146:1–11. https://doi.org/10.1016/j.pharmthera.2014.08.005
80. Jimeno A, Gordon M, Chugh R, Messersmith W, Mendelson D, Dupont J, Stagg R, Kapoun AM, Xu L, Uttamsingh S, Brachmann RK, Smith DC (2017) A first-in-human phase i study of the anticancer stem cell agent ipafricept (omp-54f28), a decoy receptor for wnt ligands, in patients with advanced solid tumors. Clin Cancer Res 23:7490–7497. https://doi.org/10.1158/1078-0432.ccr-17-2157
81. Takebe N, Nguyen D, Yang SX (2014) Targeting notch signaling pathway in cancer: clinical development advances and challenges. Pharmacol Ther 141:140–149. https://doi.org/10.1016/j.pharmthera.2013.09.005

82. Espinoza I, Pochampally R, Xing F, Watabe K, Miele L (2013) Notch signaling: targeting cancer stem cells and epithelial-to-mesenchymal transition. Onco Targets Ther 6:1249–1259. https://doi.org/10.2147/ott.s36162
83. McAuliffe SM, Morgan SL, Wyant GA, Tran LT, Muto KW, Chen YS, Chin KT, Partridge JC, Poole BB, Cheng K-H, Daggett J Jr, Cullen K, Kantoff E, Hasselbatt K Berkowitz J, Muto MG, Berkowitz RS, Aster JC, Matulonis UA, Dinulescu DM (2012) Targeting notch, a key pathway for ovarian cancer stem cells, sensitizes tumors to platinum therapy. Proc Natl Acad Sci U S A 109:E2939–E2948. https://doi.org/10.1073/pnas.1206400109
84. Diaz-Padilla I, Wilson MK, Clarke BA, Hirte HW, Welch SA, Mackay HJ, Biagi JJ, Reedijk M, Weberpals JI, Fleming GF, Wang L, Liu G, Zhou C, Blattler C, Ivy SP, Oza AM (2015) A phase ii study of single-agent ro4929097, a gamma-secretase inhibitor of notch signaling, in patients with recurrent platinum-resistant epithelial ovarian cancer: a study of the princess margaret, chicago and california phase ii consortia. Gynecol Oncol 137:216–222. https://doi.org/10.1016/j.ygyno.2015.03.005
85. Chiorean EG, LoRusso P, Strother RM, Diamond JR, Younger A, Messersmith WA, Adriaens L, Liu L, Kao RJ, DiCioccio AT, Kostic A, Leek R, Harris A, Jimeno A (2015) A phase i first-in-human study of enoticumab (regn421), a fully human delta-like ligand 4 (dll4) monoclonal antibody in patients with advanced solid tumors. Clin Cancer Res 21:2695. https://doi.org/10.1158/1078-0432.CCR-14-2797
86. Chen X, Gong L, Ou R, Zheng Z, Chen J, Xie F, Huang X, Qiu J, Zhang W, Jiang Q, Yang Y, Zhu H, Shi Z, Yan X (2016) Sequential combination therapy of ovarian cancer with cisplatin and γ-secretase inhibitor mk-0752. Gynecol Oncol 140:537–544. https://doi.org/10.1016/j.ygyno.2015.12.011
87. Balch C, Fang F, Matei DE, Huang THM, Nephew KP (2009) Minireview: epigenetic changes in ovarian cancer. Endocrinology 150:4003–4011
88. Balch C, Nephew KP (2010) The role of chromatin, micrornas, and tumor stem cells in ovarian cancer. Cancer Biomark 8:203–221. https://doi.org/10.3233/cbm-2011-0214
89. Mirzaei H, Yazdi F, Salehi R, Mirzaei HR (2016) Sirna and epigenetic aberrations in ovarian cancer. J Cancer Res Ther 12:498–508. https://doi.org/10.4103/0973-1482.153661
90. Smith HJ, Straughn JM, Buchsbaum DJ, Arend RC (2017) Epigenetic therapy for the treatment of epithelial ovarian cancer: a clinical review. Gynecol Oncol Rep 20:81–86. https://doi.org/10.1016/j.gore.2017.03.007
91. Jones BA, Varambally S, Arend RC (2018) Histone methyltransferase ezh2: a therapeutic target for ovarian cancer. Mol Cancer Ther 17:591–602. https://doi.org/10.1158/1535-7163.mct-17-0437
92. Liang D, Ma Y, Liu J, Trope CG, Holm R, Nesland JM, Suo Z (2012) The hypoxic microenvironment upgrades stem-like properties of ovarian cancer cells. BMC Cancer 12:201. https://doi.org/10.1186/1471-2407-12-201
93. Katz E, Skorecki K, Tzukerman M (2009) Niche-dependent tumorigenic capacity of malignant ovarian ascites-derived cancer cell subpopulations. Clinic Cancer Res 15:70. https://doi.org/10.1158/1078-0432.CCR-08-1233

Chapter 6
New Insights in Spermatogonial Stem Cells

Sabine Conrad, Hossein Azizi, Mehdi Amirian, Maryam Hatami, and Thomas Skutella

Abstract

Introduction: Detailed knowledge of the molecular profile of human spermatogonial stem cells (SSCs) is an essential prerequisite to developing germ stem cell therapy strategies, diagnostics, and male infertility therapy and how to handle germ cells in the management of assisted reproduction. It is also essential to better understand testicular cancer manifestation and develop improved and more individualized diagnostic and therapeutic options in the future, including transdifferentiated and reprogrammed SSCs.
Methods: New findings on spermatogonial stem cells are summarized.
Results: Although the molecular tools of single-cell genomics of testicular cells and the internet platform Human Protein Atlas are available, at the present stage, the scientific knowledge about testicular genes and proteins in humans is still insufficiently validated in terms of cell-type-specific localization and function. The establishment of a genome-wide testis-specific molecular atlas of healthy (including all different ages) and pathological stages would be significant for the organization of our knowledge and the increase of molecular understanding of human reproductive biology and disease.
Conclusions: Deeper insights into the molecular regulators and interactomes would potentiate the research related to SSC selection, SSC culture, in vitro spermatogenesis, and SCC plasticity.

Keywords Donor age · Embryonic germ cells · Embryonic stem cells · Germ stem cells · Human adult germ stem cells · Induced pluripotent stem cells · Natural reprogramming · Pluripotency · Primordial germ cells · Small molecules

---

S. Conrad
P.O. Box 1243, 72072 Tübingen, Germany

H. Azizi
Faculty of Biotechnology, Amol University of Special Modern Technologies, Amol, Iran

M. Amirian · M. Hatami · T. Skutella (✉)
Medical Faculty, Institute for Anatomy and Cell Biology, University of Heidelberg, INF 307, 69120 Heidelberg, Germany
e-mail: thomas.skutella@uni-heidelberg.de

I. Virant-Klun (ed.), *Stem Cells in Reproductive Tissues and Organs*, Stem Cell Biology and Regenerative Medicine 70, https://doi.org/10.1007/978-3-030-90111-0_6

Abbreviations

ACVR1B	Activin A receptor type 1B
ACVR2B	Activin A receptor type 2B
AdVac	A-dark spermatogonium with nuclear dilution zone
AdNoVac	A-dark spermatogonium without nuclear vacuole
AMBRA1	Autophagy and beclin 1 regulator 1
ALDOA	Aldolase, fructose-bisphosphate A
BAMBI	BMP and activin membrane-bound inhibitor
BCL2	BCL2 apoptosis regulator
BMPR1B	Bone morphogenetic protein receptor type 1B
C3orf22	Chromosome 3 open reading frame 22
CADH1	Cadherin 1 (CDH1)
CD	Cluster of differentiation
CD9	CD9 molecule
CDK	Cyclin-dependent kinase
CELF4	CUGBP Elav-like family member 4
C-MYC	C-MYC proto-oncogene, BHLH transcription factor
COL1A2	Collagen type I alpha 2 chain
DA	Dopamine, dopaminergic neurons
DAZL	Deleted in azoospermia like
DCAF4L1	DDB1 and CUL4 associated factor 4 like 1
DDX4	DEAD-box helicase 4 (VASA)
DEG	Differentially expressed genes
DMRT1	Doublesex and mab-3-related transcription factor 1
DOCK8	Dedicator of cytokinesis 8
DUSP6	Dual specificity phosphatase 6
DVL1	Dishevelled segment polarity protein 1
EDNRA	Endothelin receptor type A
EGC	Embryonic germ cell
EGR4	Early growth response 4
EIF2	Eukaryotic translation initiation factor 2
EIF4B	Eukaryotic translation initiation factor 4B
EIF4E	Eukaryotic translation initiation factor 4E
EIF4EBP1	Eukaryotic translation initiation factor 4E-binding protein 1
ELAVL2	ELAV-like RNA binding protein 2
ENO2	Enolase 2
ENO3	Enolase 3
EOMES	Eomesodermin
EPHA2	EPH receptor A2
ESC	Embryonic stem cell
ETV5	ETS variant transcription factor 5
FACS	Fluorescence-activated cell sorting
FGF	Fibroblast growth factor

FGFR3	Fibroblast growth factor receptor 3
FSD1	Fibronectin type III and SPRY domain containing 1
Fibs	Fibroblast
FSHR	Follicle-stimulating hormone receptor
FST	Follistatin
GAD1	Glutamate decarboxylase 1
GATA4	GATA binding protein 4
GDF3	Growth differentiation factor
GDNF	Glial cell-derived neurotrophic factor
GFP	Green fluorescent protein
GFRA1	GDNF family receptor alpha 1
GiPSC	GSCs-derived iPSC
GPR125	Adhesion G protein-coupled receptor A3 (ADGRA3)
GSC	Germ stem cell
h	Human
haGSC	Human adult germ stem cell
hFibs	Human fibroblasts
ICA1L	Islet cell autoantigen 1 like
ID4	Inhibitor of DNA binding 4, HLH protein
INTGA6	Integrin A6
iPSC	Induced pluripotent stem cell
KIT	KIT proto-oncogene, receptor tyrosine kinase
KLF4	Kruppel-like factor 4
KLF6	Kruppel-like factor 6
LATS2	Large tumor suppressor kinase 2
L1TD1	LINE1-type transposase domain containing 1
LIF	Leukemia inhibitory factor
LIN28	Lin-28 homolog A
LPPR3	Phospholipid phosphatase-related 3
LY6K	Lymphocyte antigen 6 family member K
MACS	Magnetic-activated cell sorting
MAGEA4	MAGE family member A4
MAGEC1	MAGE family member C1
MCM7	Minichromosome maintenance complex component 7
MEF	Mouse embryonic fibroblast
MKI67	Marker of proliferation Ki-67
MLST8	MTOR-associated protein, LST8 homolog
mRNA	Messenger ribonucleic acid
MSC	Mesenchymal stem cell
mTOR	Mechanistic/mammalian target of rapamycin kinase
NANOG	Nanog homeobox
NANOS2	Nanos C2HC-type zinc finger 2
NANOS3	Nanos C2HC-type zinc finger 3
NOG	Noggin
OCT4	Octamer-binding transcription factor 4

p21	Cyclin-dependent kinase inhibitor 1
p53	Tumor protein P53
PABPC1	Poly(A) binding protein cytoplasmic 1
PASD1	PAS domain-containing repressor 1
PAX7	Paired box 7
PDGFRA	Platelet-derived growth factor receptor alpha
PFKL	Phosphofructokinase, liver type
PGC	Primordial germ cell
PHGDH	Phosphoglycerate dehydrogenase
PIWIL4	Piwi-like RNA-mediated gene silencing 4
PLZF	Zinc finger and BTB domain containing 16 (ZBTB16)
POU5F1	POU class 5 homeobox 1
PPP1R36	Protein phosphatase 1 regulatory subunit 36
PPRC1	PPARG-related coactivator 1
PTPN13	Protein tyrosine phosphatase non-receptor type 13
PVR	PVR cell adhesion molecule
RA	Retinoic acid
RHOX10	Rhox homeobox family member 10
RHOX13	Rhox homeobox family member 13
RHOXF1	Rhox homeobox family member 1
RNA	Ribonucleic acid
RNAi	Ribonucleic acid interference
RPTOR	Regulatory-associated protein of MTOR complex 1
RT-PCR	Reverse transcription polymerase chain reaction
SAGE1	Sarcoma antigen 1
SALL4	Spalt-like transcription factor 4
SCF	Stem cell factor
SCOS	Sertoli-cell-only-syndrome
scRNA	Small conditional RNA
scRNA-seq	Single-cell RNA sequencing
SERINC2	Serine incorporator 2
SIX1	SIX homeobox 1
SOX2	SRY-box transcription factor 2
SPATS2L	Spermatogenesis-associated serine-rich 2 like
SPG	Spermatogonium, spermatogonia
SSC	Spermatogonial stem cell
SSEA1	Stage-specific embryonic antigen-1
SSEA4	Stage-specific embryonic antigen-4
ST3GAL2	ST3 beta-galactoside alpha-2,3-sialyltransferase 2
STRA8	Stimulated by retinoic acid 8
SUSD2	Sushi domain containing 2
SV40T	SV40 large T antigen
TAZ	Tafazzi
TCF3	Transcription factor 3
TCL1	TCL1 family AKT coactivator

TCN1	Transcobalamin 1
TCN2	Transcobalamin 2
TERT	Telomerase reverse transcriptase
TEX15	Testis expressed 15, meiosis and synapsis associated
TGFA	Transforming growth factor alpha
TGFBR	Transforming growth factor beta receptor
TIMP2	TIMP metallopeptidase inhibitor 2
TOP2A	DNA topoisomerase II alpha
TPI1	Triosephosphate isomerase 1
TRA-1-60	T cell receptor alpha locus
TRA-1-81	T cell receptor alpha locus
TSPAN8	Tetraspanin 8
TSPAN33	Tetraspanin 33
TPH1	Tryptophan hydroxylase 1
TRP53	Tumor protein P53
TSPYL1	Testis-specific Y-encoded-like protein 1
TTC14	Tetratricopeptide repeat domain 14
UCHL1	Ubiquitin C-terminal hydrolase L1
USF3	Upstream transcription factor family member 3
UTF1	Undifferentiated embryonic cell transcription factor 1
VASA	Vasa homolog (DDX4)
ZBTB16	Zinc finger and BTB domain containing 16
ZBTB33	Zinc finger and BTB domain containing 33
ZFHX3	Zinc finger homeobox 3
ZNF654	Zinc finger protein 654
YAP	Yes1-associated transcriptional regulator

Introduction

Knowledge of the molecular status of human spermatogonia and other spermatogenic stages (spermatocytes, spermatids, sperm) and supportive Sertoli and Leydig cells is a prerequisite for introducing these cells to clinical practice in the future. Recently, the molecular status of these different spermatogenetic cell types and stages is now better understood, and a precise molecular genome-wide, cell-type-specific mapping is established. One of the leading causes is developing new molecular methods such as single-cell RNA sequencing and bioinformatics, making the molecular characterization of these cells' distinct populations for the first time possible.

Molecular Atlas of Human Spermatogenesis/Human Protein Atlas Platform/The Testis-Specific Proteome/Single-Cell RNA Sequencing

The Human Protein Atlas: Spermatogonial Stem Cells

The Human Protein Atlas is an internet platform launched in 2003. The goal of the platform is to characterize and map all human proteins in cells, tissues, and organs. Various "omics" technologies such as antibody-based imaging, mass spectrometry-based proteomic techniques, transcriptomics, and systems biology methods are used [1–5]. All the knowledge resource data is open access to allow scientists to freely access the information for exploration of the human proteome [1–5]. For the analysis of (SSC) states, limitations include, e.g., missing antibodies, the specificity of the antibody used for the imaging, and no verification of the immunohistochemistry with in situ hybridizations of the same gene. Due to the low occurrence of the different states of SSCs, especially ground states in the testicle tubules, it would be extremely critical to screen routinely large series of testicular sections, which might be at present out of scope for such a platform. Examples of genes/proteins expressed mainly in spermatogonial listed in the Human Protein Atlas include DMRT1, PASD1, SAGE1, MAGEC1, ELAVL2, and C3orf22.

Selection, Transcriptional Profiling, Single-Cell Genomics, and RNA Sequencing: Biomarkers of Spermatogonial Stem Cells

Consistent with the murine data, human germ cells have also been recently analyzed by elaborated transcriptional profiling [6–11]. These generated datasets also enable a much stronger focus on spermatogonial subpopulations' gene expression and protein profiles. These datasets' bioinformatical analysis expands more complete gene profiles of different states of undifferentiated spermatogonia with possible bi-directional transitions between spermatogonial states. Therefore, the primary purpose of these strategies will be to generate a platform for the separation of pure populations of spermatogenic cells (spermatogonia, spermatocytes, round spermatids, sperm) and supporting cells (Sertoli cells, Leydig cells, endothelial cells, and myofibroblasts) from the human testicular tissue and their analysis on the gene expression profile to prepare a molecular atlas of spermatogenesis concerning fertility/infertility and testicular cancer. In particular, the identification, isolation, and cultivation of the basal human SSC population (Fig. 6.1) with apparent SSC reserve functionality could help further studies that intend to select and propagate SSCs to treat male diseases and infertility.

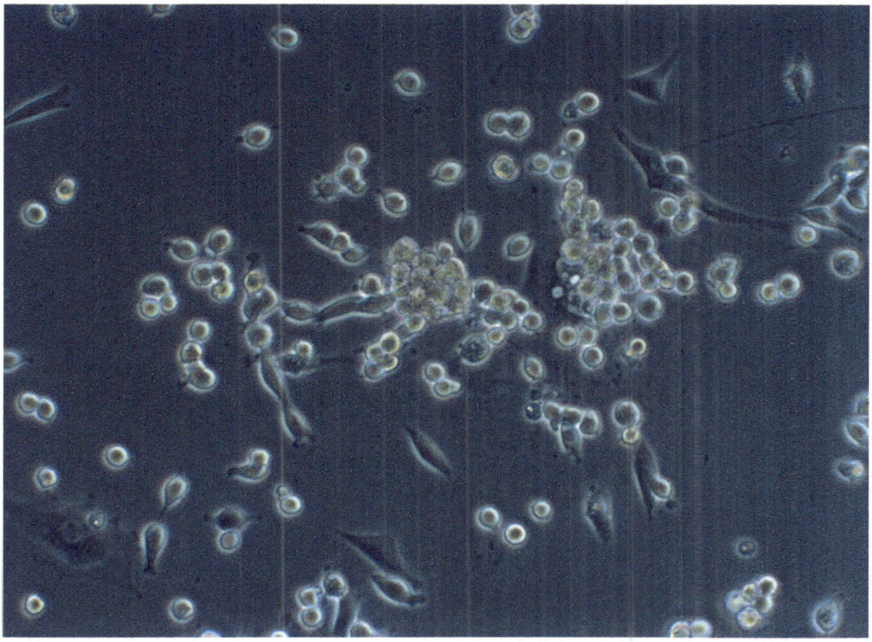

Fig. 6.1 Typical representative morphology of spermatogonia from human testis during culture on CF1 feeder, according to Conrad et al. [18, 21]. Single, pairs, and groups of round cells with a high nucleus/cytoplasm ratio typical of spermatogonia and with diameters of around 10 μm are observable under an inverted microscope (20x objective)

In order to perform single-cell gene expression analyses with RT-PCR, Neuhaus et al. [12] manually separated human spermatogonia from in vitro cultures according to morphological criteria. Their results revealed heterogeneous transcriptional profiles of the cells selected by morphological criteria. Using laser-capture microdissection and subsequent RNA sequencing, Jan et al. [8] used morphological criteria to identify germ cell pools (500 cells from each germ cell subtype) of spermatogonial subtypes (A-dark and A-pale), spermatocytes (leptotin/zygote, early and late pachytene), and round spermatids from testicular tissue of adult males with normal spermatogenesis. In this study, B-spermatogonia and pre-leptotene spermatocytes were excluded because the required number of cells was not reached in these germ cell subtypes. The RNA-Seq data led to the observation that spermatogonia show the highest degree of transcriptional complexity among the germ cell types analyzed, decreasing during more differentiated spermatogenic stages. The spermatogonia of A-dark and A-pale also showed a higher degree of transcriptional heterogeneity than the more differentiated cell types. According to the authors, both types of spermatogonia could not be designated a clear transcriptional expression pattern. It was suggested that the transition from the mitotically inactive to the active state may be regulated by transcriptional changes. It was observed that the genes *glutamate decarboxylase 1* (*GAD1*) and *tryptophan hydroxylase 1* (*TPH1*) were highly expressed in A-dark and low in A-pale spermatogonia. These genes suppress the proliferation of mouse SSCs [13].

Furthermore, it could be observed that spermatogonia already express high numbers of genes that only later become important during spermatogenesis. Caldeira-Brant et al. [14] were able to identify a small group of human reserve SSC AdVac (UTF1+/GFRA1−/MCM7−) in their investigations utilizing morphological studies, kinetics, positioning of the cells concerning intertubular blood vessels, protein expression (UTF1, GFRA1, and KIT) and proliferative activity (MCM7). These cells are a small subpopulation of A-dark with nuclear dilution zone (AdVac) and constitute the human reserve SSC positioned near blood vessels. Within this AdVac population, a small pool (2% of undifferentiated spermatogonia A) of completely dormant cells with high expression of UTF1 and lacking GFRA1 could be identified. In contrast, an actively dividing undifferentiated SSC population of A-dark without nuclear vacuole (AdNoVac) and Apale showed a similar kinetic and high proliferative capacity.

Modern approaches in single-cell genomics allow the complexity of cellular heterogeneity of stem cells, precursor, and terminally differentiated cell types to be decoded with great precision and high resolution. This approach offers the possibility of delineating cellular lineage hierarchies and uncovering molecular cell fate lineages. The biomathematical application of pseudo-time estimation from single-cell gene expression data realizes the extraction of temporal information from otherwise static profiles of single cells. Thus, the modern approaches of single-cell RNA sequencing (scRNA-seq) and its biostatistical elaboration offer unique and powerful tools for studying the development and disease of human tissues and organs. The simultaneous study of gene expression of the cell populations of human spermatogenesis is made possible by the large-scale analysis of many cells of the testicular tissue without involving techniques for initial sorting or other procedures enriching cells [7, 9–11]. The limitations in spatial information obtained from this approach can be compensated by using orthogonal validation approaches such as co-localization of specific RNAs or proteins in fixed tissue via in situ hybridization, immunohistochemistry, or a combination of both [7].

Gua et al. [6] analyzed single-cell transcriptomes of individual human spermatogonia from SSEA4+ cells [6]. Their analysis demonstrated that a portion of the human SSEA4+ germ cells exhibited similarities to mouse SSCs. Others had more in common with progenitor cells and early differentiating spermatogonia. The study results indicate that the human SSEA4+ population includes SSCs, progenitor cells, and differentiating spermatogonia at an early stage. The single-cell transcriptional analysis of SSEA4-sorted cells showed the pluripotency marker genes *OCT4*, *NANOG*, and the *sex-determining region Y-Box 2* (*SOX2*) was not detectable in the selected adult human spermatogonia population. In a further investigation of a limited number of adult human spermatogonia isolated with the surface markers SSEA4 (n = 60 cells) and KIT (n = 32 cells), four different cellular subpopulations were formed concerning their respective transcription profiles [6]. It should be mentioned that the study was based on single testicular tissues that were not characterized in more detail.

In an attempt to generate a human transcriptional cell atlas using three testicular tissues as starting material and performing two technical replicates, Guo et al. [7] identified a new stem cell state called "state 0". Pseudo-time analysis showed a wave-like progression from SSC state 0 to state 4, and clustering analysis defined state-specific gene expression signatures. A striking shift of transcription programs between state 1 and state 2 was observed, which is dominated by the expression of cell cycle/proliferation genes (e.g., *MKI67*). The analyses could identify 490 genes that are either most strongly or expressed explicitly in state 0 (e.g., *PIWIL4*, *EGR4*, *TSPAN33*, *PHGDH*, *PPP1R36*, *ICA1L*) or state 1. The data was validated by sequential mRNA fluorescence in situ hybridization and immunohistochemistry. Using the Human Protein Atlas Resource (http://www.proteinatlas.org/) the pattern of state 0 is verified with 16 candidate markers specific for early SSC. The analysis revealed particular expression in cells in the basal compartment of the testicular tubules. Both the cells of state 0 and 1 co-expressed many important stem cell signaling factors and transcription factors TFs. Zwischen 157 und 158 TFs Transcription factors. The common SSC markers (ID4, FGFR3, TCF3, and UTF1) were expressed in states 0 and 1. The known markers of SSC differentiation (KIT) or proliferation (MKI67) were explicitly expressed during or after state 2 so that states 0 and 1 could contain two different quiescent SSC states. Cells of state 0 probably do not have the SSC surface marker SSEA4 because the SSEA4-catalyzing enzyme ST3GAL2, which catalyzes the formation of SSEA4, is expressed only in small amounts. According to Guo et al. [7], these observations are consistent with the suggestion that although state 0 and state 1 define transcriptionally discrete states, they are likely to represent metastable/heterogeneous cellular phenotypes that give SSCs the ability to adapt to a dynamic niche environment. It is expected that state 0 could represent naive SSC in adults.

Guo et al. [6] were able to identify 4 spermatogonia (SPG) "states" using single-cell RNA sequencing (scRNA-seq). In doing so, markers could be defined to identify the state most likely to be enriched for SSCs. An SSA4 antibody was used to enrich the undifferentiated SPG for these experiments. It should be noted that this is a marker of unclear specificity that introduced a potential bias in SSC characterization so that populations of SSCs may not have been included in their analysis

By computational analysis, von Kopylow and Spiess [15] reanalyzed a plethora of human spermatogonial markers and obtained a robust set of spermatogonia-specific markers. A total of 70 robust SPG markers were obtained by cross-study intersection filtering of all studies. The expression of these markers was compared to the Human Protein Atlas, and 12 markers with exclusive protein expression in SPG are shown. These include PIWIL4, L1TD1, ZBTB33, USF3, TCF3, SIX1, TEX15, AMBRA1, DCAF4L1, ZNF654, PPRC1, and SPATS2L.

Hermann et al. [9] examined the transcriptomes of more than 62,000 single cells of spermatogenesis, ranging from immature stem cells to those of adult testis cells of mice and men. Part of the single-cell transcriptome results was confirmed by RT-PCR. The authors started to analyze pseudo-temporal profiles based on known spermatogonia-specific genes expressed by all spermatogonia (DDX4), as well as genes that distinguish undifferentiated (Gfra1, Id4, and Nanos2) from differentiated spermatogonia (Dmrt1, Kit, Nanos3, and Stra8). It was observed that the expression of genes specific for undifferentiated or differentiated spermatogonia was biased in both unselected and sorted mouse and human spermatogonia, particularly at the beginning and end of trajectories. While Stra8 levels increased in response to the transition from unselected to differentiated spermatogonia in mice, surprisingly, only small amounts of STRA8 mRNA were observable in unselected or differentiated human spermatogonia. The undifferentiated spermatogonia high-expression levels of GFRA1, Id4, and ZBTB16, NANOS2, TCN2, and MAGEA4 were observed. In contrast, in the differentiated spermatogonia higher amounts of Dmrt1, Kit, Nanos3, Rhox13, and Stra8 were detectable. The single-cell transcriptome analysis from Hermann et al. [9] indicated that the expression profiles of SSC self-renewal and progenitor spermatogonia differentiation are different. The authors imply that the SSC state is characterized by a unique signaling and transcriptional regulatory environment that changes upon transition to an intermediate progenitor state. In contrast, when progenitor cells initiate differentiation, they possess increased proliferative activity. The expression of cell cycle genes is increased and a major metabolic shift toward oxidative phosphorylation and activation of protein processing is observable. The prototypical SSC gene signature, which includes the genes Gfra1, Id4, Etv5, Nanos2, Pax7, Tspan8, Rhox10, and Zbtb16, was extended by Hermann et al. [9] to include a number of new genes (Dusp6, Epha2, Ptpn13, Pvr, and Tcl1). Other genes that showed maximal expression at the early to the midpoint of the pseudo-time curve were involved in the control of translation (Eif4e, Eif4ebp1, Pabpc1, and Rptor; EIF2 signaling; mTOR signaling) and the regulation of intracellular cell signaling pathways (F2r, Gnaq, Plce1, Ppp1cb, and Shc1; Plc signaling; thrombin signaling). The authors also observed that human spermatogonia gene cluster 5, which was also located in the middle of the pseudo-time trajectory, showed similar expression of characteristic SSC genes encoding proteins that play a role in translational control (EIF4B, MLST8, and PABPC1; EIF2 signaling; mTOR signaling), glycolysis (ALDOA, ENO3, PFKL, and TPI1), and known SSC genes (ID4 and NANOS2). Interestingly, Hermann et al. [9] also observed that the expression of genes involved in the hepatic stellate cell activation pathway (Bcl2, Ednra, Klf6, Pdgfra, and Tgfa) was increased in human unselected and sorted spermatogonial gene clusters. Activation of this pathway renders the germ cells more sensitive for cytokine signaling [16], suggesting that the transition from quiescent SSCs to proliferative and differentiating male germ cells may involve a response to small proteins important in cell signaling.

The group of Sohni et al. [10] was the next to analyze adult human testes using single-cell RNA sequencing. The protein markers of human SSC identified so far (including ENO2, LIN28, PLZF, SALL4, SSEA4, UCHL1, and UTF1) recognize both undifferentiated and differentiated SPG [17]. Other markers—like ID4 and FGFR3—are relatively specific for undifferentiated SPG [6], but their relative selectivity for human SSCs is unclear. Sohni et al. [10] identified four undifferentiated spermatogonial clusters with specific marker gene expression in the adult testis. These four different cell clusters were designated as (1) SSC-1, (2) SSC cluster-2 (SSC-2), (3) early differentiating (diff) SPG (early diff-SPG), and (4) diff-SPG. The pan-germ cell markers DDX4 and MAGEA4 marked cells in all four clusters. The authors identified marker genes for the primitive SSC state. The differentially expressed genes (DEGs), which are most abundant in the SSC-1 subgroup and thus can be understood as the molecular signature of SSCs, include *CELF4*, *EGR4*, *FGFR3*, *FSD1*, *LPPR3*, *PIWIL4*, and *TSPAN33*. Furthermore, the temporal course of development of male germ cells, neonatal germ cells compared to the differentiating adult SCC stages was characterized. The neonatal germ cell clusters have an expression profile that strongly resembles PAGCs (e.g., expression of pluripotency genes; e.g., *POU5F1* and *NANOG* and were therefore called "PGC-like" (PGCL)).

Data from Sohni et al. [10] map the gene expression profiles of the developing human male germline and the supporting somatic cells. It could be shown that the signaling pathways NOTCH, KIT, HEDGEHOG (Hh), and WNT are used for the maintenance of SSCs in newborns. The PGC-like and SSC markers are candidates for SSC therapy for the treatment. In order to elucidate the nature of the germ cells in the neonatal testis, a clustering analysis was specifically performed only on the germ cells. One cell cluster had an expression profile strongly resembling PAGCs (e.g., expression of pluripotency genes; e.g., POU5F1 and NANOG, and was therefore called "PGC-like" (PGCL)). The second cell cluster expresses genes that mark human pre-SPG, including MAGEA4 and RHOXF1, which is why we have named this cluster "PreSPGs". The PreSPG-1 subcluster expresses several genes that most PreSPG-2 cells express only weakly, including DOCK8, SERINC2, LY6K, MAGEA4, and RHOXF1. Conversely, other genes selectively mark the PreSPG-2 subset, including COL1A2, TIMP2, TTC14, DVL1, and ZFHX3. The analyses of Sohni et al. [10] confirmed that several murine SSC markers (e.g., ID4, EOMES, and ZBTB16) [17] can also be used for the enrichment of human SSCs. The authors observed that human ID4 and ZBTB16 are broadly expressed in human SPG, whereas EOMES is expressed only in a small fraction of primitive undifferentiated human SSC. Furthermore, it was observed that several markers previously proposed for human SSC enrichment—such as GFRA1, FGFR3, and UTF1 [6]—are expressed not only in the primitive SSC-1B subgroup but also in more advanced SPG subgroups, including SSC-2 and/or early diff-SPG. Although these previously defined markers are less selective than the SSC-1B markers described before, they will potentially still be useful as they probably mark SPG precursors and possibly alternative SSC-enriched cell subgroups. Sohni et al. [10] have successfully used antibodies against LPPR3 and TSPAN33 to enrich primitive undifferentiated SPG positively. This strategy could be used in future clinical trials for the enrichment of SSC. Their identification of genes expressed by somatic testicular cells, including those encoding cell–cell signaling factors, could prove useful in developing cocktails containing "testicular niche" factors that might improve the proliferation and extension of human SSCs in vitro for clinical application in vivo.

Guo et al. [18] examined changes in the composition of testicular cells of four boys in puberty with single-cell genomics and compared them with infants and adults. For the spermatogonia, expression patterns of crucial markers were performed. By comparison with earlier work, the germ cells were subordinated to four broad developmental stages. The slowly self-renewing and undifferentiated spermatogonia were labeled with UTF1+ and largely MKI67−. In the 11-year-old sample, a high proportion of cells were still spermatogonia of stages 0 and 1, differentiating spermatogonia and meiotic cells also began to develop. It is noteworthy that in the samples of infants and 7-year-old, the spermatogonia were relatively rare (~3–4% of total testicular cells), while in the 11-year-old and older specimens, the relative proportion of spermatogonia increased considerably and accounted for ~10–15% of whole testicular cells, which is consistent with a spermatogonial amplification and proliferation phase before a differentiation phase. The 13-year-old specimen was mainly similar to the 11-year-old sample (probably reflecting the known age differences at the beginning of puberty). However, the proportion of post-meiotic cells increased proportionally more robust commitment to meiosis. Finally, the composition of germ cells in the 14-year-old sample was similar to that of the adult, indicating almost complete spermatogenesis. The SSCs expressed UTF1, PIWIL4, and TSPAN33, among others. Next, the authors conducted immunofluorescence experiments to confirm their scRNA-seq findings. Undifferentiated UTF1+ spermatogonia (state 0–1) were observed in all analyzed age groups. In contrast, proliferative and differentiated spermatogonia (state 2–3) revealed a strong increase of multiple proliferative markers (e.g., cyclins, CDKs, TOP2A, MKI67, KIT) and MKI67+. Investigation of germ cell-niche interactions during puberty led to the identification of multiple signals due to multiple signaling pathways, including retinoic acid (RA), activin/inhibin, NOTCH, GDNF, FGF, and WNT. The authors assume a crucial role in activin and BMP signaling pathways during puberty. They found that activin receptors (ACVR1B, BMPR1B, and ACVR2B) are expressed in spermatogonia, while key inhibitors of activin signaling (FST, BAMBI, and NOG) are specifically present in undifferentiated spermatogonia (states 0 and 1). The authors argue that the activin signaling pathway is selectively inhibited in slowly self-renewing and undifferentiated spermatogonia but is activated during the proliferation and differentiation of the cells.

The isolation of enriched populations of spermatogenic cells and Sertoli and Leydig cells from testes is one of the crucial steps to address these critical issues. Up to now, various approaches were used to isolate the different testicular cells successfully. Several methods have used magnetic activating (MACS) or fluorescence-activated cell sorting (FACS) to isolate testicular cells [19]. Human spermatogonia were successfully isolated with CD49f, CD9, SSEA4, GPR125- and GFRA1 antibodies [19–24]. Bryant et al. [25] established a velocity sedimentation separation using STA-PUT chambers as another approach to isolate spermatogenic cells. Harichandan et al. [26] demonstrated that a combination of CD164 and SUSD2 enables isolated human SSCs by FACS. Furthermore, they observed that cells positive for Sertoli cell-specific markers FSHR and GATA4 were highly enriched in the CD49f+CD49a−SSEA-4−CD164−population. Chang et al. [27] based on former experiments by Meistrich [28] and Meistrich et al. [29] described a simple approach employing a single protocol to enrich multiple testicular cell populations (Sertoli, Leydig, and several spermatogenic cell populations) from a small amount of tissue. This protocol combines the enzymatic digestion of seminiferous tubules with counter-current centrifugal elutriation, yielding specific testicular cell populations with high purity.

Due to the new characterizations of SSCs using single-cell RNA sequencing, further molecular analyses and comparisons of the SSC populations isolated with the different methods, including antibodies against LPPR3 and TSPAN33 [10], should have to be performed.

Research of spermatogenesis would benefit from a detailed atlas of the interactome of cells involved in spermatogenesis, including the respective stem cell and differentiation compartments. Reports about the potential of two- and three-dimensional in vitro spermatogenesis in rodents, including studies with human SSCs, were reviewed by Richer et al. [30]. It was described that only a limited number of these studies concentrated on building the three-dimensional tubular testicular architecture in vitro [31–35]. While some advances have been made in testicular organoid research in terms of cellular reorganization, none of the described culture systems seems applicable for reconstructing the testicular architecture and in vitro spermatogenesis. It has to be suggested that additional improvements in culture methods and medium composition have to be introduced to initiate testicular tubulogenesis and spermatogenesis in vitro.

Germ Stem Cells and Natural Shift to Pluripotency

By the definition, germ cells are highly specialized unipotent cells for gametes production; however, there are several lines of evidence implying their plasticity, including pluripotential. For example, teratomas are a complex variety of cells and tissues from old germ layers at different maturation stages and almost entirely occur and derive from germ cells in the gonads [36]. Also, the primordial germ cells (PGCs) of 8.5–12.5 days old mouse embryos develop pluripotent cells under appropriate culture conditions [37, 38]. The differentiation features of these embryonic germ cells (EGCs) are like embryonic stem cells (ESCs) isolated from the inner cell mass [39, 40]. As Kanatsu-Shinohara et al. have demonstrated in 2004 [41], mouse neonatal SSCs could be reprogrammed ES-like cells, which are pluripotent and display qualities comparable to ESCs. Later on, Guan et al. and Ko et al. [42–45] themselves reported the derivation of ES-like cells from "adult" germline stem cells (GSCs) from mice up to 7 weeks old. Ko et al. [45] also reported the derivation from mice up to 7 months. However, the data were not shown. It was well-nigh impossible to transfer this reprogramming to true adult GSCs [46]. Besides, mouse EGCs originating from PGCs were successfully established, whereas human EGCs never displayed true pluripotency [47, 48]. While these cells' cellular and molecular characterization proved to be similar to the hESCs, the hEGCs produced no teratomas after transplantation into immunodeficient mice. When human adult germline stem cells (haGSCs) derived from hSSCs were cultured under induction conditions appropriate for the differentiation of hESCs, they differentiated into different types of somatic cells of all three germ layers in vitro and in vivo. But these haGSCs were quite heterogeneous and not fully pluripotent [19, 21]. Most likely, the cells expressing markers of pluripotency in the studies of Mizrak et al. [49], Kossack et al. [50], Chikhovskaya et al. [51], and Gonzalez et al. [52] were generated from mesenchymal stem cells (MSCs) or were similar to MSCs. On the opposite, Stimpfel et al. [53] proved the existence of germ and mesenchymal-derived stem cells originating from human testis biopsies. They were capable of differentiating into cells of all three germ layers. Lim et al. [54] have lately demonstrated the resemblance of haGSCs and hESCs, as they can originate small teratomas.

These findings imply that molecular blocks occur in germline-derived stem cells with increasing age, and humans will have to be identified. Furthermore, strategies will have to be established to remove these molecular blocks to induce fully pluripotent stem cells from germ cells.

Human Primordial Germ Cells and Pluripotency

In contrast to their mouse counterparts, it takes effort to obtain and keep human EGCs over long-term subculture, as very few laboratories demonstrate, including the Gerhaert's group [48, 55–57]. Shamblott et al. [48] postulated that they successfully induced human EGCs to become "pluripotent", closely resembling embryonic germ cells. Embryoid bodies formed by these cells revealed derivatives of all three embryonic germ layers, demonstrating these cells' ability to differentiate into a wide variety of cell types, thus justifying their definition as pluripotent stem cells. These "pluripotent" stem cell lines were generated from the gonadal ridges of human 5–9 weeks postfertilization embryos. Cultures with morphological similarities to mouse ESCs or EGCs were grown on mouse STO fibroblast feeder layers supplemented with human recombinant leukemia inhibitory factor (LIF), human recombinant basic fibroblast growth factor (hrbFGF), and forskolin. Growing multicellular colonies were alkaline phosphatase-, SSEA-1/-3/-4, and TRA-1–60/-81-positive, which could be continuously passaged according to the authors and were found to exhibit a regular and stable karyotype. It had been postulated by Shamblott et al. that it is part of human PGCs' nature to express several key pluripotency controlling factors, such as OCT4, NANOG, and LIN28, although at a much lower level in comparison to hESCs. In contrast to somatic cells, their epigenetic signature was ascertained as similar to stem cells [48, 55–57].

However, later on, the Gerhaert's group described that, unlike mouse EGCs, human EGCs exhibited very low derivation efficiencies and were unstable under the current cell culture methods used. Moreover, human EGCs could be differentiated into all three germ layers in vitro but could not produce teratomas in vivo [48, 55–57]. Due to their characteristics, these human EGCs were more difficult to study and proved to be unprofitable as a study model for the process of PGC reprogramming. Also, it had never been shown that human EGCs display any signs of spontaneous differentiation in vitro into more complex tissue structures such as neural rosettes or contracting cardiomyocyte tissue, as hESCs do after removing growth factors maintaining pluripotency

Recently, it was proven by Bazley et al. [47] that human EGCs are not truly pluripotent, as they gave a detailed account of the direct two-factor reprogramming of human pluripotent stem cells (PSCs) via transfection with SOX2 and OCT4. Zwischen 122+123 muss PSC pluripotent stem cell. As indicated before, magnetic-activated cell sorting with antibodies against the germ cell marker SSEA1 was employed to isolate human PGCs. Concerning human ESCs, PGCs expressed intrinsically lower levels of SOX2, OCT4, and C-MYC mRNA. Bazley et al. [47] utilized an integrative lentiviral of human primordial germ cells to induce pluripotent stem cells (PSCs). These PSCs remained stable for a maximum of 24 subcultures and significantly exceeded the EGCs in derivation efficiency and culture survival rate. The human PGCs' expression of endogenous levels of KLF4 and C-MYC protein was similar to the levels of EGCs. However, their levels of SOX2 and OCT4 were lower. It was necessary to transfect with both SOX2 and OCT4 to induce PGCs to a pluripotent state at an efficiency of 1.71%. The efficiency was increased to 2.33% by a succeeding addition of C-MYC. Immunohistochemical analysis of the SOX2 and OCT4-derived PGC-iPSCs demonstrated the similarity of these cells to ESCs, more than EGCs, concerning colony morphology as well as molecular characterization. LIF's presence integrating an ectopic exposure to C-MYC yielded higher efficiencies, whereas the LIF was redundant for the generation of PGC-iPSCs like EGCs. Besides, the SOX2- and OCT4-derived PGC-iPSCs successfully formed teratomas in vivo and showed differentiation into representative cell types from all three germ layers in vitro. All the five protein levels measured, OCT4, SOX2, NANOG, KLF4, and C-MYC, were significantly expressed higher in ESCs compared to both ECGs and PGCs. The increased heterogeneity of staining within EGC colonies elucidated this higher expression, most likely associated with the human EGCs' tendency to lose their pluripotency and spontaneous culture differentiation. On the contrary, lower OCT4 and C-MYC were expressed by PGCs against EGCs, and no level of SOX2 was discovered. In microarray experiments, the hierarchical analysis of the highest differentially expressed genes in these cell lines was revealed. A similar pattern of gene expression among the SO-derived iPSCs and ESCs was found. This pattern is not identical to the quite comparable pattern of PGCs and EGCs. The mRNA profiles of PGCs and EGCs are resemblant in many features; however, their developmental potential and pluripotent properties are distinguishable. The markers VASA and DAZL were expressed by iPSCs as well. Together with SSEA1 in the iPSCs, these markers' expression indicates the remaining of some of their germ cell originated characteristics after reprogramming.

In conclusion, from the studies of Shamblott et al. and Bazley et al., it becomes evident that human PGCs cannot be naturally converted to a real pluripotent state [47, 48]. At the present stage, the induction with two pluripotency factors OCT4 and SOX2 seems to be a necessary prerequisite. It cannot be ruled out that further enhancement of cell culture conditions and reprogramming yield a more straightforward way of producing human EGCs.

Potent Molecular Barriers of Pluripotency in Germ Cells Depend on Donor Age

Kanatsu-Shinohara et al. [41] described the generation of PSCs from neonatal mouse SSCs and 4–8-week-old p53 knockout GSCs. The colonies were morphologically indistinguishable from ES-like colonies. Although ES-like cells never appeared from fully established wild-type GSCs in PSCs, after long-term culture, the p53 knockout GSCs produced PSCs as long as 6 months after culture initiation.

Using the suppression of DMRT1 and/or TRP53, Tanaka et al. made clear that the potency of the derivation of PSCs from male germline cells declines from PGCs to gonocytes spermatogonia [58]. Tanaka et al. [58] transmitted accumulated PGCs, gonocytes, and spermatogonia with lentiviruses, which expressed short hairpin RNA against DMRT1 and/or TRP53. The scientists detected a great susceptibility of PGCs to reprogramming induction. Merely, the TRP53 depletion was found sufficient for inducing pluripotency. Gonocytes and spermatogonia, on the contrary, showed higher resistance to double-knockdown reprogramming of DMRT1 and TRP53. The PGCs-produced PSCs are partly responsible for chimeras produced by blastocyst injection; however, placenta-only phenotypes were seen in an unspecified number of embryos, implying epigenetic abnormalities of PGC-derived PSCs.

Hence, the different reprogramming potential of PGCs and gonocytes/spermatogonia is evidentiary, and the results suggest that it is not obligatory for fresh and cultured SSCs to share the same properties. Depending on the donor age, this declining potency of germ cells has its analogies in iPSC research. It is of supreme importance that the efficiency of reprogramming was advanced through the removal of various effectors of senescence and that new strategies for a maximum generation of iPSC were implied. The indolence and stochastic nature of reprogramming signify a manifestation of barriers that limit its efficiency, such as the identified senescence.

Various well-known factors contribute to the efficiency of of induced pluripotent stem cell (iPSC) generation. For instance, the starting cell's differentiation state proved to be significant, and the reprogramming efficiency of stem cells is higher than in terminally differentiated cells [59]. Streckfuss-Bomeke et al. [60] found clear proof for the varying efficiency of different somatic cell types originating from the same donor. Furthermore, the effect of cellular senescence on the reprogramming efficiency was evidenced [61–67]. One of the indicators of cellular senescence with advancing age is the cell cycle arrest's irreversibility through the activation of the p53/p21 [63, 66] and p16/Rb pathways [62, 68]. The results imply a determination of the reprogramming efficiency by intrinsic properties of somatic cells. Wang et al. [69] demonstrated that the donor age affects the reprogramming efficiency of murine cells. Somers et al. [70] hinted that donor ages do not interfere with human cells' reprogramming efficiency, in opposition to findings with mice. At the same time, Lapasset et al. [71] demonstrated a successful derivation of iPSC, even from centenarians' fibroblasts. Despite that, the combined effect of age and culture time on human cells' reprogramming efficiency was not reported yet.

The combined negative effect of donor age and time in cultures was investigated by Trokovic et al. [72] in their studies on the reprogramming efficiency of induced pluripotent stem cells. Fibroblasts originating from both young and old donors can result in iPSCs, as their findings prove. Nevertheless, immature and early passage fibroblasts generate the highest reprogramming efficiency. However, the advanced donor age and time in culture leads to a synergistic negative effect, which noticeably decreases the quantity of iPSC colonies. Donor age and time in culture adversely affect the reprogramming efficiency and purport an association with cellular senescence, primarily mediated by the up-regulated cell cycle inhibitory protein p21. Their results validated the potentiation of iPSC generations from late passaged human fibroblasts of old donors employing an inhibition of p21 during the reprogramming process.

Qin et al. [73] evidenced the disclosure of the Hippo pathway as a barrier to reprogramming in mouse PGCs by transcriptional analysis of pluripotency. The transcriptional profiles of mouse germlines were compared to pluripotent and somatic cells, in vivo and in vitro, to detect new barriers to reprogramming pluripotency. The global expression of the transcriptional program for pluripotency in PGCs is noteworthy. Similarities were found for PGC reprogramming to pluripotency and human germ cell tumorigenesis, comprising the loss of LATS2, a tumor suppressor kinase of the Hippo pathway. Enhanced efficiency of induction of pluripotency in human cells was initiated by the knockdown of LATS2. LATS2 RNAi, unlike p53 RNAi, increases the generation of fully reprogrammed iPSCs, in particular, not accelerating cell proliferation. The repression of programming by LATS2 in human cells by post-transcriptionally antagonizing the hippo pathway's two downstream effectors, TAZ but not YAP, was demonstrated. The findings indicate transcriptional parallels between germ cell transformation and the generation of iPSCs and the establishment of a barrier to cellular reprogramming by the Hippo pathway. Senescence can be prevented by over-expressing SV40T antigen or hTERT or by down-regulating p53 or p21, which significantly enhances the efficiency of iPSC generation.

Kimura et al. [74] derived pluripotent stem cells originating from mouse primordial germ cells by the small molecule compounds TGFbR inhibitor and Kenpaullone. Using these two small molecule compounds, pluripotent stem cells under standard embryonic stem cell (ESC) culture conditions lacking hrbFGF and SCF were generated. Especially for the generation of haGSCs from adult men, age and cell senescence become critical factors in cellular reprogramming and have to be redressed.

Feng et al. [75] were able to reprogram GSCs to PSCs with the ectopic expression of NANOG and TET1 concurrent with either Y4F overexpression or p53 knockdown. Interestingly, the reprogramming strategy was successful with the overexpression of NANOG alone in p53 deficient GSCs. The GSCs-derived iPSCs (GiPSCs) exhibited multiple differentiation routes in vitro and in vivo. They could be considered more similar to mouse ESCs than to GSCs based on their transcriptomic profiles

In general, more studies are necessary to study how far GSC reprogramming can be achieved by multiple routes and Y4F overexpression strategy.

Age of the Donor hsas an Impact on PSC Conversion from Mouse SSCs

After repeated experiments generating ESC-like cells from adult mouse SSCs as described by Ko et al. [44, 45], Azizi et al. [46] concluded that the derivation of pluripotent cells from mouse SSCs seems to be age-dependent.

Based on research by Kanatsu-Shinohara et al. [41], Guan et al. [42, 43], and Ko et al. [44, 45], it was still uncertain if the spontaneous conversion of SSCs to ESC-like cells is deficient as of a certain age, considering that SSCs and truly pluripotent ESC-like cells can be segregated from adult mice. The possibility to derive ESC-like cells from mouse SSC cultures before reaching adolescence or up to 7 weeks of age is a proven fact. In 2004, PSCs of mouse testis cells were first generated by Kanatsu-Shinohara et al. [41]. Kanatsu-Shinohara et al. engendered ESC-like cells in an SSC culture from two-day-old pups. He procured these cells 4–7 weeks after culture initiation. Guan et al. [42, 43] obtained ESC-like cells from populations of STRA8-GFP positive cells of 4–7 weeks adult mice. After a repeated induction of pluripotency in 5 weeks old OCT4-GFP positive adult SSCs, Ko et al. [44, 45] detailed the dependency of induction on the initial number of plated SSCs and the length of culture time of OCT4-positive cells without splitting. Additionally, in the later published detailed protocol of converting adult SSCs into pluripotent stem cells, this group employed only SSCs of mice from postnatal day 35 (5 weeks old). The spontaneous conversion of SSCs in ESC-like cells from neonate and nearly adult testis of up to 7-week-old mice could be verified in our experiments. Conversely, the generation of ESC-like cells from mice older than 7 weeks was impossible. Here it would matter to mention how the age of an adult mouse is defined. The NIH criteria classify mice as an adult after reaching 8 weeks of age (http://scienc.education.nih.gov/supplements/nih4/Energy/activities/508/mice-ref/mice_ref.htm). Mice begin to be sexually active when aged 5 and 6 weeks. Flurkey et al. [76] define mice as a mature adult between the ages of 3 and 6 months. However, the impossibility to derive these cells from older, mature adult mice has been referred to and should be corrected.

Since true adult SSCs are unable to shift to pluripotency—which occurs together with a decline in the expression of core pluripotency genes like OCT4, NANOG, and SOX2 in SSCs with age—genes of the spermatogonial differentiation pathway increase at the same time. In the primary cultures of isolated cells from OCT4 transgenic reporter mice, we monitored an expression of the OCT4-GFP signal at a moderate level in the neonate, and up to a low level in older or adult SSCs. While this expression was completely down-regulated during short- and long-term SSC culture, a high-density signal only re-emerged after the conversion to ESC-like cells [77]. As the mRNA expression profiling was validated, the expression of germ-cell-specific genes increased with age, which correlates with the establishment of a full development of a spermatogenic epithelium. Therefore, the expression showed significantly higher in SSCs from 7 to 12-week-old mice in comparison to neonatal SSCs. At the same time, a significantly up-regulated expression of OCT4a and NANOG, and

SOX2 was monitored in neonatal SSCs, and a down-regulated expression in the adult SSCs.

We hence suppose the ideal stoichiometric dose of OCT4 for inducing pluripotency to be higher than the endogenous expression of these cells. The optimum stoichiometry dose for generating iPSCs from mouse embryonic fibroblasts (MEFs) was detailed before, as Thiemann et al. recommended that high OCT4 expression levels outweigh modest levels of SOX2 and KLF4 expression [77]. This predominance would be the most critical factor for reprogramming and reaching a steady condition of pluripotency. The results concerning PGCs are adequate with the discovery, that additional OCT4 is necessary for reaching pluripotency, while stoichiometry was not the object of this study. The self-regulation of OCT4 is a fascinating clarification for this discovery. Pan et al. [78] proved the repression of its own promoter by OCT4 when its expression rises beyond a certain level, whereby it provides negative feedback for its expression. The stable state expression of OCT4 is maintained by this negative feedback loop. As a result, the ES-like undifferentiated state is upheld [79].

Papapetrou et al. [80] demonstrated that the optimal stoichiometry of reprogramming factor expression utilizing human fibroblasts is highly sensitive to OCT4 dosage. They also showed the impact that variances have on the relative ratios of reprogramming factor expression exert on iPSC induction efficiency [80]. The endogenous levels of OCT4 are insufficient for an efficient reprogramming of human PGCs, as revealed by Papapetrou et al. As a member of the POU family of transcription factors, OCT4 is indispensable for pluripotency and successful reprogramming, whereby it seems to control the transcriptional machinery. OCT4 also functions as a nucleocytoplasmic shuttling protein. For instance, a limited potential for cellular reprogramming is displayed by OCT4-mutant ESCs with altered nuclear import/export activity. This fact indicates a nucleocytoplasmic shuttling role of OCT4 in somatic cell reprogramming [81]. OCT4 is expressed in the cytoplasm throughout the cleavage stage in mature human oocytes before their compaction. After compaction, OCT4 can be discovered in the nucleus throughout the blastocyst and in PGCs at the time of their development in the early fetus [82]. In the cytoplasm of dark spermatogonial stem cells of the adult male testis, an OCT4 expression has been reported. These dark SSCs proliferate and initiate lineage-specific differentiation [51]. Compared to the PSCs and iPS-PGCs, no differences were found in the localization of OCT4 staining in the PGCs. Compliant with other reported studies, robust nuclear staining of OCT4 in human PGCs and oogonia were found [83]. In the future, nevertheless, research should be done for the investigation of minor differences that remain in the OCT4 distribution in PGCs, and to search out the role of PSCs in OCT4 levels during PGC reprogramming.

The exact mechanism of the development of ESC-like cells from SSCs remains indeterminate. However, a clearer understanding of the conversion is still missing. A natural conversion of ESC-like cells from mouse testis can only be performed before reaching adolescence, but not after that.

An assay of factors, including small molecules that enhance the probability and the restricted time window of SSC to PSC conversion, would be of particular interest. As evidenced lately, it increases the likelihood of ESC-like cells from SSCs when glycogen synthase kinase-3 inhibitor is added to the testis-derived SC cultures [82]. Bolting factors, such as small molecules, that possibly enhance the probability and determine the SSCs limited time window to PSC conversion would be fascinating.

Human Spermatogonial Stem Cells and Pluripotency

Expression of Pluripotency-Related Markers

Conrad et al. [21] intended to explain the molecular status of single haGSCs and haGSC colonies, which are naturally growing from the CD49f (INTGA6) MACS- and (collagen$^-$/laminin$^+$ binding) matrix-selected fraction of enriched human adult spermatogonia. We performed a Fluidigm BioMark system single-cell transcriptional profiling of a long-term-cultured haGSCs cluster. Thereby, in comparison to hESCs and human fibroblasts (hFibs), features similar to hESCs and a significant difference to somatic hFibs with a characteristic germ pluripotency-associated gene expression profile of haGSCs were disclosed. Furthermore, genome-wide comparisons with microarray analysis show that the haGSC colonies show gene expression heterogeneity with more or less pluripotency. As study results validate, haGSCs are adult stem cells with a specific molecular gene expression profile in vitro. They appear but are not identical with authentic pluripotent stem cells. The haGSC colonies are selectable under ESC conditions and sustain a partly pluripotent state at the molecular level. This level can resemble their cell plasticity and their differentiation capability to split into all germ layers' cells. As a heterogeneous population of cells, the haGSCs differ from hFibs or MSCs and express a lower or higher degree of pluripotency than hESCs. haGSCs arise from the fortified population of CD49f MACS and matrix-selected spermatogonia during the cell culture. They never originate from the negatively selected amount or patients lacking spermatogonia, e.g., Sertoli-Only-Syndrom. Experiments with Sertoli-Cell-Only-Syndrom (SCOS) patients were repeatedly conducted, and it remained impossible to generate haGSC colonies from them.

Detecting the difference of morphology in haGSC colonies and hFibs was easy, as they are much alike the early hESC colonies. These early hESC colonies exemplify through a pivotal cluster with overgrowing epithelial cells. The earliest haGSC colonies/islands in the primary cultures began to manifest 4–6 weeks after the culture of enriched spermatogonia in an hGSC medium. Then followed was a manual selection of denser haGSC aggregations for continued propagation and definition. It was exemplary for haGSC colonies to comprise the central part of the colony and exuberant epithelial cells akin to incipient cell colonies of hESCs. The considerable difference between haGSCs and Fibs was exhibited in single-cell Fluidigm analysis concerning the expression of germ- and pluripotency-associated genes. Some standard features of hFibs were found just with a few outliner hESCs and haGSCs, yet most of them had nothing in common.

The contrasting of single hESCs with haGSCs, hESCs increasingly expressed most pluripotency-related genes such as SOX2, NANOG, LIN28, LIN28B, GDF3, CADH1, OCT4A, TDGF1, and UTF1. Simultaneously, haGSCs more intensely expressed the germ cell-related genes CD9, GFRA1, NANOS, STAT3, TSPYL, GPR125, and MYC. A resemblant profile with lower expression levels was revealed. This signifies that haGSCs only expressed the fundamental pluripotency-related genes, including OCT4a, NANOG, SOX2, and LIN28. These cells still kept a germ cell-related gene expression profile to some extent. Comparing haGSCs to hESCs, their expression of the core pluripotency-related genes was less intense; however, it proved to be substantially higher than in hFibs.

The heterogeneity of haGSC colonies was revealed, as they shared characteristics with an ESC-like state of pluripotency to a greater or lesser extent. Additionally, different haGSC colonies proved to be relatively heterogeneous regarding their expressing of germ- und pluripotency-associated genes in the microarray study. An aggregate analysis of transcriptome and high variance genes in haGSCs compared with hESCs and hFibs displayed the separation of haGSCs from hFibs.

There are current reports about the existence of MSCs in adult human testicles [49–52]. In contrast, we evidenced that most genes that are supposed to be expressed in testicular MSCs lacked expression in the haGSCs [21], like some other researchers did not [49–52]. They demonstrated the isolation of various populations of testicular stem cells from the tissue and cultured them in vitro. It may be possible that different types of adult human stem cells exist in testicles and interact, as they mirror the complexity of this primary reproductive organ.

Furthermore, Chikhovskaya et al. [51] utilized genes that mostly showed a possible derivation of their testis stem cells from MSCs and that were not differentially expressed in haGSCs in comparison to hFibs and hESCs. Moreover, Choi et al. [83] genes CD34 and CD73 for the isolation of fibroblasts originating in testis were not expressed in haGSCs.

The introduction of highly purified SSC populations of state 0 and their stable in vitro cultivation without changing the molecular and cellular phenotype would be another important prerequisite for possible successful programming. For the improvement of programming culture conditions, additional studies are necessary to keep away from the molecular block hindering the haGSCs from fully converting to molecular pluripotent stem cells.

Advantages of Spermatogonial Stem Cells

It can be assumed that autologous hSSC has several advantages over other multipotent/pluripotent stem cells. SSCs and other adult-derived stem cells may be safer to use therapeutically than hESCs or iPSCs cells. These include the lack of immunogenicity after autologous application, a lower probability of tumor occurrence after transplantation, and less complicated ethical regulations than allogeneic pluripotent stem cells. It has been observed that SSCs have the ability of spontaneous transdifferentiation after transplantation. After in vivo application into the corresponding tissues, SSC can transdifferentiate into uterine, prostate, and skin epithelial cells, probably based on SSCs high sensitivity to the tissue-specific microenvironments [54, 84].

Differentiation of Spermatogonial Stem Cells into Other Types of Cells

Boulanger et al. [85] were able to direct the differentiation of testicular stem cells into mammary epithelia, but only under conditions when the enriched SSCs were mixed with mammary epithelial cells and grafted into the mammary fat pad in vivo.

Yang et al. [86] generated functional dopaminergic (DA) neurons by direct transdifferentiation via a simple two-step induction strategy to differentiate hSSCs. The authors achieved a direct conversion of hSSCs into DA neurons displaying the critical morphological, biochemical, and functional characteristics of the midbrain's primary DA neurons. The induction protocol consisted of three weeks of induction with a transdifferentiation efficiency of 21%. hSSCs, from testicular tissues of obstructive azoospermic (OA) patients at the age of 25–45 years who had normal spermatogenesis, were exposed to an improved induction system, which consisted mainly of olfactory ensheathing cell-conditioned culture medium (OECCM) and four candidate small molecules (RA, VPA, SB, forskolin). Growth factors critical for cell specification (SHH, FGF8α, and TGFβ3) were selected for further induction. Morphological changes were observed with the expression of various DA neuron-specific markers (e.g., TUJ1, TH, NURR1, DAT) and several critical pro-DA neurogenesis effectors (e.g., EN1, PITX3, FOXa2, LMX1a, LMX1b, and OTX2). Transcriptome analysis revealed the genetic similarity between the artificially differentiated DA and genuine DA neurons, functional characteristics of converted DA neurons, including synapse formation, dopamine release, electrophysiological activity, and neuron-specific Ca2+ signaling. Finally, hSSCs in the early stage of induction were evaluated for survival, differentiation, migration, tumorigenicity in the mouse striatum, and improved functional deficits in MPTP-induced PD animals.

It is expected that hSSCs can be programmed into specific cell types in vitro without forced expression of transcription factors or reprogramming into the different somatic cell lineages of the human body. Just by their ability to spontaneously transdifferentiate or induced by small molecules, SSCs could be considered another potential source for replacing damaged or degenerated cells in various degenerative diseases. Autologously derived germ stem cells would theoretically avoid any risk of rejection due to host–donor mismatch and may bypass the need for immune suppression post-transplant.

To further clarify these issues, human testicular stem cells from childhood, adolescence, and adulthood should be studied comparatively. It would also be essential for scientific reasons to improve the bottleneck to the access of human testicular tissue. For example, it might make sense to facilitate organ transplantation regulations that the release of testicular tissue to researchers of organ donors would be permitted, and easily carried out in the future for the establishment of stem cell banks from SSC. Improving regulations would significantly facilitate scientific work with human testicular tissue.

Cultivation of Spermatogonial Stem Cells

The cultivation conditions of human spermatogonia, especially when robust cell proliferation is to be achieved, must be improved. Researchers should begin by finding the cellular niche in which the cells spend their entire life in testicular tissue using modern techniques that are more specific to the expression of factors such as single-cell sequencing to identify and analyze them more precisely. Here the question arises: Are there more modern cultural media that can better increase germ stem cells and increase their number during long-term propagation?

Conclusion

In spite of the promise of SSCs as a source of medical regenerative cells, extensive research is required to more accurately identify, propagate, and induce the de/transdifferentiation SSCs into desired cell types. Inherent similarities to ES cells may give the SSCs significant advantages as a starting material for human tissue regeneration compared to adult stem cells. Therefore, more extensive experimental studies should be conducted to investigate the production of tissue-specific target cells of the ectodermal, endodermal, and mesodermal lineages from SSCs. The increasing molecular knowledge characterizing the elementary features of the identifying SSCs ground state and their heterogeneity in interaction with their niche will improve the in vitro propagation and transplantation toward clinical practice. It would be desirable to establish a comprehensive internet platform—SSC atlas—for the molecular and physiological characterization of SSCs. Unipotent SSCs' capability to transdifferentiate into other functional cell types and programming of the cells toward multipotent/pluripotent states offers new, safer perspectives in multipotent/pluripotent stem cell research and regenerative medicine. It would be fascinating to improve culture conditions and apply bolting factors, such as small molecules deterring from the molecular block, which hinders a full germ stem cell or haGSC conversion to molecular pluripotent stem cells. It would also be of great interest to enlarge SSC's limited time window to pluripotent stem cell conversion. The precise future of stem cells in human medicine is unclear. Still, SSCs appear to be a powerful tool that biologists can utilize to regenerate human tissue in various clinical settings.

Acknowledgements None of the authors have any conflicts of interest to disclose, and all authors support the submission to this book.

References

1. Djureinovic D, Fagerberg L, Hallström B, Danielsson A, Lindskog C, Uhlén M, Pontén F (2014) The human testis-specific proteome defined by transcriptomics and antibody-based profiling. Mol Hum Reprod 20:476–488. https://doi.org/10.1093/molehr/gau018
2. Fagerberg L, Hallström BM, Oksvold P, Kampf C, Djureinovic D, Odeberg J, Habuka M, Tahmasebpoor S, Danielsson A, Edlund K, Asplund A, Sjöstedt E, Lundberg E, Szigyarto CA-K, Skogs M, Takanen JO, Berling H, Tegel H, Mulder J, Nilsson P, Schwenk JM, Lindskog C, Danielsson F, Mardinoglu A, Sivertsson A, von Feilitzen K, Forsberg M, Zwahlen M, Olsson I, Navani S, Huss M, Nielsen J, Ponten F, Uhlén M (2014) Analysis of the human tissue-specific expression by genome-wide integration of transcriptomics and antibody-based proteomics. Mol Cell Proteomics 13:397–406. https://doi.org/10.1074/mcp.M113.035600
3. Pineau C, Hikmet F, Zhang C, Oksvold P, Chen S, Fagerberg L, Uhlén M, Lindskog C (2019) Cell type-specific expression of testis elevated genes based on transcriptomics and antibody-based proteomics. J Proteome Res 18:4215–4230. https://doi.org/10.1021/acs.jproteome.9b00351
4. Uhlén M, Fagerberg L, Hallström BM, Lindskog C, Oksvold P, Mardinoglu A, Sivertsson Å, Kampf C, Sjöstedt E, Asplund A, Olsson I, Edlund K, Lundberg E, Navani S, Szigyarto CA-K, Odeberg J, Djureinovic D, Takanen JO, Hober S, Alm T, Edqvist P-H, Berling H, Tegel H, Mulder J, Rockberg J, Nilsson P, Schwenk JM, Hamsten M, von Feilitzen K, Forsberg M, Persson L, Johansson F, Zwahlen M, von Heijne G, Nielsen J, Pontén F (2015) Proteomics. Tissue-based map of the human proteome. Science 347:1260419. https://doi.org/10.1126/science.1260419
5. Yu NY-L, Hallström BM, Fagerberg L, Ponten F, Kawaji H, Carninci P, Forrest ARR, Fantom Consortium, Hayashizaki Y, Uhlén M, Daub CO (2015) Complementing tissue characterization by integrating transcriptome profiling from the Human Protein Atlas and from the FANTOM5 consortium. Nucleic Acids Res 43:6787–6798. https://doi.org/10.1093/nar/gkv608
6. Guo J, Grow EJ, Yi C, Mlcochova H, Maher GJ, Lindskog C, Murphy PJ, Wike CL, Carrell DT, Goriely A, Hotaling JM, Cairns BR (2017) Chromatin and single-cell RNA-seq profiling reveal dynamic signaling and metabolic transitions during human spermatogonial stem cell development. Cell Stem Cell 21:533-546.e6. https://doi.org/10.1016/j.stem.2017.09.003
7. Guo J, Grow EJ, Mlcochova H, Maher GJ, Lindskog C, Nie X, Guo Y, Takei Y, Yun J, Cai L, Kim R, Carrell DT, Goriely A, Hotaling JM, Cairns BR (2018) The adult human testis transcriptional cell atlas. Cell Res 28:1141–1157. https://doi.org/10.1038/s41422-018-0099-2
8. Jan SZ, Vormer TL, Jongejan A, Röling MD, Silber SJ, de Rooij DG, Hamer G, Repping S, van Pelt AMM (2017) Unraveling transcriptome dynamics in human spermatogenesis. Development 144:3659–3673. https://doi.org/10.1242/dev.152413
9. Hermann BP, Cheng K, Singh A, Roa-De La Cruz L, Mutoji KN, Chen I-C, Gildersleeve H, Lehle JD, Mayo M, Westernströer B, Law NC, Oatley MJ, Velte EK, Niedenberger BA, Fritze D, Silber S, Geyer CB, Oatley JM, McCarrey JR (2018) The Mammalian spermatogenesis single-cell transcriptome, from spermatogonial stem cells to spermatids. Cell Rep 25:1650-1667.e8. https://doi.org/10.1016/j.celrep.2018.10.026
10. Sohni A, Tan K, Song H-W, Burow D, de Rooij DG, Laurent L, Hsieh T-C, Rabah R, Hammoud SS, Vicini E, Wilkinson MF (2019) The neonatal and adult human testis defined at the single-cell level. Cell Rep 26:1501-1517.e4. https://doi.org/10.1016/j.celrep.2019.01.045
11. Wang M, Liu X, Chang G, Chen Y, An G, Yan L, Gao S, Xu Y, Cui Y, Dong J, Chen Y, Fan X, Hu Y, Song K, Zhu X, Gao Y, Yao Z, Bian S, Hou Y, Lu J, Wang R, Fan Y, Lian Y, Tang W,

Wang Y, Liu J, Zhao L, Wang L, Liu Z, Yuan R, Shi Y, Hu B, Ren X, Tang F, Zhao X-Y, Qiao J (2018) Single-cell RNA sequencing analysis reveals sequential cell fate transition during human spermatogenesis. Cell Stem Cell 23:599-614.e4. https://doi.org/10.1016/j.stem.2018.08.007
12. Neuhaus N, Yoon J, Terwort N, Kliesch S, Seggewiss J, Huge A, Voss R, Schlatt S, Grindberg RV, Schöler HR (2017) Single-cell gene expression analysis reveals diversity among human spermatogonia. Mol Hum Reprod 23:79–90. https://doi.org/10.1093/molehr/gaw079
13. Du Y, Du Z, Zheng H, Wang D, Li S, Yan Y, Li Y (2013) GABA exists as a negative regulator of cell proliferation in spermatogonial stem cells. Cell Mol Biol Lett 18 149–162. https://doi.org/10.2478/s11658-013-0081-4
14. Caldeira-Brant AL, Martinelli LM, Marques MM, Reis AB, Martello R, Almeida FRCL, Chiarini-Garcia H (2020) A subpopulation of human Adark spermatogonia behaves as the reserve stem cell. Reproduction 159:437–451. https://doi.org/10.1530/REP-19-0254
15. von Kopylow K, Spiess A-N (2017) Human spermatogonial markers. Stem Cell Res 25:300–309. https://doi.org/10.1016/j.scr.2017.11.011
16. Tsuchida T, Friedman SL (2017) Mechanisms of hepatic stellate cell activation. Nat Rev Gastroenterol Hepatol 14:397–411. https://doi.org/10.1038/nrgastro.2017.38
17. Fayomi AP, Orwig KE (2018) Spermatogonial stem cells and spermatogenesis in mice, monkeys and men. Stem Cell Res 29 207–214. https://doi.org/10.1016/j.scr.2018.04.009
18. Guo J, Nie X, Giebler M, Mlcochova H, Wang Y, Grow EJ, DonorConnect KR, Tharmalingam M, Matilionyte G, Lindskog C, Carrell DT, Mitchell RT, Goriely A, Hotaling JM, Cairns BR (2020) The dynamic transcriptional cell atlas of testis development during human puberty. Cell Stem Cell 26:262-276.e4. https://doi.org/10.1016/j.stem.2019.12.005
19. Conrad S, Renninger M, Hennenlotter J, Wiesner T, Just L, Bonin M, Aicher W, Bühring H-J, Mattheus U, Mack A, Wagner H-J, Minger S, Matzkies M, Reppel M, Hescheler J, Sievert K-D, Stenzl A, Skutella T (2008) Generation of pluripotent stem cells from adult human testis. Nature 456:344–349. https://doi.org/10.1038/nature07404
20. Conrad S, Azizi H, Hatami M, Kubista M, Bonin M, Hennenlotter J, Renninger M, Skutella T (2014) Differential gene expression profiling of enriched human spermatogonia after short- and long-term culture. BioMed Res Int 2014:138350. https://doi.org/10.1155/2014/138350
21. Conrad S, Azizi H, Hatami M, Kubista M, Bonin M, Hennenlotter J, Sievert K-D, Skutella T (2016) Expression of genes related to germ cell lineage and pluripotency in single cells and colonies of human adult germ stem cells. Stem Cells Int 2016:8582526. https://doi.org/10.1155/2016/8582526
22. Lim JJ, Sung S-Y, Kim HJ, Song S-H, Hong JY, Yoon TK, Kim JK, Kim K-S, Lee DR (2010) Long-term proliferation and characterization of human spermatogonial stem cells obtained from obstructive and non-obstructive azoospermia under exogenous feeder-free culture conditions. Cell Prolif 43:405–417. https://doi.org/10.1111/j.1365-2184.2010.00691.x
23. Izadyar F, Wong J, Maki C, Pacchiarotti J, Ramos T, Howerton K, Yuen C, Greilach S, Zhao HH, Chow M, Chow Y-C, Rao J, Barritt J, Bar-Chama N, Copperman A (2011) Identification and characterization of repopulating spermatogonial stem cells from the adult human testis. Hum Reprod 26:1296–1306. https://doi.org/10.1093/humrep/der026
24. He Z, Kokkinaki M, Jiang J, Zeng W, Dobrinski I, Dym M (2012) Isolation of human male germ-line stem cells using enzymatic digestion and magnetic-activated cell sorting. Methods Mol Biol 825:45–57. https://doi.org/10.1007/978-1-61779-436-0_4
25. Bryant JM, Meyer-Ficca ML, Dang VM, Berger SL, Meyer RG (2013) Separation of spermatogenic cell types using STA-PUT velocity sedimentation. J Vis Exp 2013:50648. https://doi.org/10.3791/50648
26. Harichandan A, Sivasubramaniyan K, Hennenlotter J, Schwentner C, Stenzl A, Bühring H-J (2013) Isolation of adult human spermatogonial progenitors using novel markers. J Mol Cell Biol 5:351–353. https://doi.org/10.1093/jmcb/mjt029
27. Chang Y-F, Lee-Chang JS, Panneerdoss S, MacLean JA, Rao MK (2011) Isolation of Sertoli, Leydig, and spermatogenic cells from the mouse testis. Biotechniques 51(341–342):344. https://doi.org/10.2144/000113764

28. Meistrich ML (1977) Separation of spermatogenic cells and nuclei from rodent testes. Methods Cell Biol 15:15–54. https://doi.org/10.1016/s0091-679x(08)60207-1
29. Meistrich ML, Longtin J, Brock WA, Grimes SR, Mace ML (1981) Purification of rat spermatogenic cells and preliminary biochemical analysis of these cells. Biol Reprod 25:1065–1077. https://doi.org/10.1095/biolreprod25.5.1065
30. Richer G, Baert Y, Goossens E (2020) In-vitro spermatogenesis through testis modelling: toward the generation of testicular organoids. Andrology 8:879–891. https://doi.org/10.1111/andr.12741
31. Baert Y, Dvorakova-Hortova K, Margaryan H, Goossens E (2019) Mouse in vitro spermatogenesis on alginate-based 3D bioprinted scaffolds. Biofabrication 11:035011. https://doi.org/10.1088/1758-5090/ab1452
32. Pendergraft SS, Sadri-Ardekani H, Atala A, Bishop CE (2017) Three-dimensional testicular organoid: a novel tool for the study of human spermatogenesis and gonadotoxicity in vitro. Biol Reprod 96:720–732. https://doi.org/10.1095/biolreprod.116.143446
33. Perrard M-H, Sereni N, Schluth-Bolard C, Blondet A, D Estaing SG, Plotton I, Morel-Journel N, Lejeune H, David L, Durand P (2016) Complete human and rat ex vivo spermatogenesis from fresh or frozen testicular tissue. Biol Reprod 95:89. https://doi.org/10.1095/biolreprod.116.142802
34. Vermeulen M, Del Vento F, de Michele F, Poels J, Wyns C (2018) Development of a cytocompatible scaffold from pig immature testicular tissue allowing human Sertoli cell attachment, proliferation and functionality. Int J Mol Sci 19:227. https://doi.org/10.3390/ijms19010227
35. von Kopylow K, Schulze W, Salzbrunn A, Schaks M, Schäfer E, Roth B, Schlatt S, Spiess A-N (2018) Dynamics, ultrastructure and gene expression of human in vitro organized testis cells from testicular sperm extraction biopsies. Mol Hum Reprod 24:123–134. https://doi.org/10.1093/molehr/gax070
36. Stevens LC (1984) Spontaneous and experimentally induced testicular teratomas in mice. Cell Differ 15:69–74. https://doi.org/10.1016/0045-6039(84)90054-x
37. Matsui Y, Zsebo K, Hogan BLM (1992) Derivation of pluripotential embryonic stem cells from murine primordial germ cells in culture. Cell 70:841–847. https://doi.org/10.1016/0092-8674(92)90317-6
38. Resnick JL, Bixler LS, Cheng L, Donovan PJ (1992) Long-term proliferation of mouse primordial germ cells in culture. Nature 359:550–551. https://doi.org/10.1038/359550a0
39. Evans MJ, Kaufman MH (1981) Establishment in culture of pluripotential cells from mouse embryos. Nature 292:154–156. https://doi.org/10.1038/292154a0
40. Martin GR (1981) Isolation of a pluripotent cell line from early mouse embryos cultured in medium conditioned by teratocarcinoma stem cells. Proc Natl Acad Sci USA 78:7634–7638. https://doi.org/10.1073/pnas.78.12.7634
41. Kanatsu-Shinohara M, Inoue K, Lee J, Yoshimoto M, Ogonuki N, Miki H, Baba S, Kato T, Kazuki Y, Toyokuni S, Toyoshima M, Niwa O, Oshimura M, Heike T, Nakahata T, Ishino F, Ogura A, Shinohara T (2004) Generation of pluripotent stem cells from neonatal mouse testis. Cell 119:1001–1012. https://doi.org/10.1016/j.cell.2004.11.011
42. Guan K, Nayernia K, Maier LS, Wagner S, Dressel R, Lee JH, Nolte J, Wolf F, Li M, Engel W, Hasenfuss G (2006) Pluripotency of spermatogonial stem cells from adult mouse testis. Nature 440:1199–1203. https://doi.org/10.1038/nature04697
43. Guan K, Wolf F, Becker A, Engel W, Nayernia K, Hasenfuss G (2009) Isolation and cultivation of stem cells from adult mouse testes. Nat Protoc 4:143–154. https://doi.org/10.1038/nprot.2008.242
44. Ko K, Tapia N, Wu G, Kim JB, Bravo MJA, Sasse P, Glaser T, Ruau D, Han DW, Greber B, Hausdörfer K, Sebastiano V, Stehling M, Fleischmann BK, Brüstle O, Zenke M, Schöler HR (2009) Induction of pluripotency in adult unipotent germline stem cells. Cell Stem Cell 5:87–96. https://doi.org/10.1016/j.stem.2009.05.025
45. Ko K, Araúzo-Bravo MJ, Kim J, Stehling M, Schöler HR (2010) Conversion of adult mouse unipotent germline stem cells into pluripotent stem cells. Nat Protoc 5:921–928. https://doi.org/10.1038/nprot.2010.44

46. Azizi H, Conrad S, Hinz U, Asgari B, Nanus D, Peterziel H, Hajizadeh Moghaddam A, Baharvand H, Skutella T (2016) Derivation of pluripotent cells from mouse SSCs seems to be age dependent. Stem Cells Int 2016:8216312. https://doi.org/10.1155/2016/8216312
47. Bazley FA, Liu CF, Yuan X, Hao H, All AH, De Los AA, Zambidis ET, Gearhart JD, Kerr CL (2015) Direct reprogramming of human primordial germ cells into induced pluripotent stem cells: efficient generation of genetically engineered germ cells. Stem Cells Dev 24:2634–2648. https://doi.org/10.1089/scd.2015.0100
48. Shamblott MJ, Axelman J, Wang S, Bugg EM, Littlefield JW, Donovan PJ, Blumenthal PD, Huggins GR, Gearhart JD (1998) Derivation of pluripotent stem cells from cultured human primordial germ cells. Proc Natl Acad Sci USA 95:13726–13731. https://doi.org/10.1073/pnas.95.23.13726
49. Mizrak SC, Chikhovskaya JV, Sadri-Ardekani H, van Daalen S, Korver CM, Hovingh SE, Roepers-Gajadien HL, Raya A, Fluiter K, de Reijke TM, de la Rosette JJMCH, Knegt AC, Belmonte JC, van der Veen F, de Rooij DG, Repping S, van Pelt AMM (2010) Embryonic stem cell-like cells derived from adult human testis. Hum Reprod 25:158–167. https://doi.org/10.1093/humrep/dep354
50. Kossack N, Meneses J, Shefi S, Nguyen HN, Chavez S, Nicholas C, Gromoll J, Turek PJ, Reijo-Pera RA (2009) Isolation and characterization of pluripotent human spermatogonial stem cell-derived cells. Stem Cells 27:138–149. https://doi.org/10.1634/stemcells.2008-0439
51. Chikhovskaya JV, Jonker MJ, Meissner A, Breit TM, Repping S, van Pelt AMM (2012) Human testis-derived embryonic stem cell-like cells are not pluripotent, but possess potential of mesenchymal progenitors. Hum Reprod 27:210–221. https://doi.org/10.1093/humrep/der383
52. Gonzalez R, Griparic L, Vargas V, Burgee K, Santacruz P, Anderson R, Schiewe M, Silva F, Patel A (2009) A putative mesenchymal stem cells population isolated from adult human testes. Biochem Biophys Res Commun 385:570–575. https://doi.org/10.1016/j.bbrc.2009.05.103
53. Stimpfel M, Skutella T, Kubista M, Malicev E, Conrad S, Virant-Klun I (2012) Potential stemness of frozen-thawed testicular biopsies without sperm in infertile men included into the in vitro fertilization programme. J Biomed Biotechnol 2012:291038. https://doi.org/10.1155/2012/291038
54. Lim JJ, Kim H, Kim K-S, Hong J, Lee D (2013) In vitro culture-induced pluripotency of human spermatogonial stem cells. BioMed Res Int 2013:143028. https://doi.org/10.1155/2013/143028
55. Kerr CL, Shamblott MJ, Gearhart JD (2006) Pluripotent stem cells from germ cells. Methods Enzymol 419:400–426. https://doi.org/10.1016/S0076-6879(06)19016-3
56. Pashai N, Hao H, All A, Gupta S, Chaerkady R, De Los Angeles A, Gearhart JD, Kerr CL (2012) Genome-wide profiling of pluripotent cells reveals a unique molecular signature of human embryonic germ cells. PloS One 7:e39088. https://doi.org/10.1371/journal.pone.0039088
57. Turnpenny L, Brickwood S, Spalluto CM, Piper K, Cameron IT, Wilson DI, Hanley NA (2003) Derivation of human embryonic germ cells: an alternative source of pluripotent stem cells. Stem Cells 21:598–609. https://doi.org/10.1634/stemcells.21-5-598
58. Tanaka T, Kanatsu-Shinohara M, Hirose M, Ogura A, Shinohara T (2015) Pluripotent cell derivation from male germline cells by suppression of Dmrt1 and Trp53. J Reprod Dev 61:473–484. https://doi.org/10.1262/jrd.2015-059
59. Eminli S, Foudi A, Stadtfeld M, Maherali N, Ahfeldt T, Mostoslavsky G, Hock H, Hochedlinger K (2009) Differentiation stage determines potential of hematopoietic cells for reprogramming into induced pluripotent stem cells. Nat Genet 41:968–976. https://doi.org/10.1038/ng.428
60. Streckfuss-Bömeke K, Wolf F, Azizian A, Stauske M, Tiburcy M, Wagner S, Hübscher D, Dressel R, Chen S, Jende J, Wulf G, Lorenz V, Schön MP, Maier LS, Zimmermann WH, Hasenfuss G, Guan K (2013) Comparative study of human-induced pluripotent stem cells derived from bone marrow cells, hair keratinocytes, and skin fibroblasts. Eur Heart J 34:2618–2629. https://doi.org/10.1093/eurheartj/ehs203
61. Banito A, Rashid ST, Acosta JC, Li S, Pereira CF, Geti I, Pinho S, Silva JC, Azuara V, Walsh M, Vallier L, Gil J (2009) Senescence impairs successful reprogramming to pluripotent stem cells. Genes Dev 23:2134–2139. https://doi.org/10.1101/gad.1811609

62. Campisi J, d'Adda di Fagagna F (2007) Cellular senescence: when bad things happen to good cells. Nat Rev Mol Cell Biol 8:729–740. https://doi.org/10.1038/nrm2233
63. Kawamura T, Suzuki J, Wang YV, Menendez S, Morera LB, Raya A, Wahl GM, Izpisúa Belmonte JC (2009) Linking the p53 tumour suppressor pathway to somatic cell reprogramming. Nature 460:1140–1144. https://doi.org/10.1038/nature08311
64. Li H, Collado M, Villasante A, Strati K, Ortega S, Cañamero M, Blasco MA, Serrano M (2009) The Ink4/Arf locus is a barrier for iPS cell reprogramming. Nature 460:1136–1139. https://doi.org/10.1038/nature08290
65. Marion RM, Strati K, Li H, Tejera A, Schoeftner S, Ortega S, Serrano M, Blasco MA (2009) Telomeres acquire embryonic stem cell characteristics in induced pluripotent stem cells. Cell Stem Cell 4:141–154. https://doi.org/10.1016/j.stem.2008.12.010
66. Marion RM, Strati K, Li H, Murga M, Blanco R, Ortega S, Fernandez-Capetillo O, Serrano M, Blasco MA (2009) A p53-mediated DNA damage response limits reprogramming to ensure iPS cell genomic integrity. Nature 460:1149–1153. https://doi.org/10.1038/nature08287
67. Utikal J, Polo JM, Stadtfeld M, Maherali N, Kulalert W, Walsh RM, Khalil A, Rheinwald JG, Hochedlinger K (2009) Immortalization eliminates a roadblock during cellular reprogramming into iPS cells. Nature 460:1145–1148. https://doi.org/10.1038/nature08285
68. Narita M, Nũnez S, Heard E, Narita M, Lin AW, Hearn SA, Spector DL, Hannon GJ, Lowe SW (2003) Rb-mediated heterochromatin formation and silencing of E2F target genes during cellular senescence. Cell 113:703–716. https://doi.org/10.1016/s0092-8674(03)00401-x
69. Wang B, Miyagoe-Suzuki Y, Yada E, Ito N, Nishiyama T, Nakamura M, Ono Y, Motohashi N, Segawa M, Masuda S, Takeda S (2011) Reprogramming efficiency and quality of induced Pluripotent Stem Cells (iPSCs) generated from muscle-derived fibroblasts of mdx mice at different ages. PLoS Curr 3:RRN1274. https://doi.org/10.1371/currents.RRN1274
70. Somers A, Jean J-C, Sommer CA, Omari A, Ford CC, Mills JA, Ying L, Sommer AG, Jean JM, Smith BW, Lafyatis R, Demierre M-F, Weiss DJ, French DL, Gadue P, Murphy GJ, Mostoslavsky G, Kotton DN (2010) Generation of transgene-free lung disease-specific human induced pluripotent stem cells using a single excisable lentiviral stem cell cassette. Stem Cells 28:1728–1740. https://doi.org/10.1002/stem.495
71. Lapasset L, Milhavet O, Prieur A, Besnard E, Babled A, Aït-Hamou N, Leschik J, Pellestor F, Ramirez J-M, De Vos J, Lehmann S, Lemaitre J-M (2011) Rejuvenating senescent and centenarian human cells by reprogramming through the pluripotent state. Genes Dev 25:2248–2253. https://doi.org/10.1101/gad.173922.111
72. Trokovic R, Weltner J, Otonkoski T (2015) Generation of iPSC line HEL24.3 from human neonatal foreskin fibroblasts. Stem Cell Res 15:266–268. https://doi.org/10.1016/j.scr.2015.05.012
73. Qin H, Blaschke K, Wei G, Ohi Y, Blouin L, Qi Z, Yu J, Yeh R-F, Hebrok M, Ramalho-Santos M (2012) Transcriptional analysis of pluripotency reveals the Hippo pathway as a barrier to reprogramming. Hum Mol Genet 21:2054–2067. https://doi.org/10.1093/hmg/dds023
74. Kimura T, Kaga Y, Sekita Y, Fujikawa K, Nakatani T, Odamoto M, Funaki S, Ikawa M, Abe K, Nakano T (2015) Pluripotent stem cells derived from mouse primordial germ cells by small molecule compounds. Stem Cells 33:45–55. https://doi.org/10.1002/stem.1838
75. Feng Y, Ning Y, Lin X, Zhang D, Liao S, Zheng C, Chen J, Wang Y, Ma L, Xie D, Han C (2018) Reprogramming p53-deficient germline stem cells into pluripotent state by Nanog. Stem Cells Dev 27:692–703. https://doi.org/10.1089/scd.2018.0047
76. Flurkey K, Brandvain Y, Klebanov S, Austad SN, Miller RA, Yuan R, Harrison DE (2007) PohnB6F1: a cross of wild and domestic mice that is a new model of extended female reproductive life span. J Gerontol A Biol Sci Med Sci 62:1187–1198. https://doi.org/10.1093/gerona/62.11.1187
77. Tiemann U, Sgodda M, Warlich E, Ballmaier M, Schöler HR, Schambach A, Cantz T (2011) Optimal reprogramming factor stoichiometry increases colony numbers and affects molecular characteristics of murine induced pluripotent stem cells. Cytom Part J Int Soc Anal Cytol 79:426–435. https://doi.org/10.1002/cyto.a.21072

78. Pan G, Li J, Zhou Y, Zheng H, Pei D (2006) A negative feedback loop of transcription factors that controls stem cell pluripotency and self-renewal. FASEB J Off Publ Fed Am Soc Exp Biol 20:1730–1732. https://doi.org/10.1096/fj.05-5543fje
79. Pan G, Thomson JA (2007) Nanog and transcriptional networks in embryonic stem cell pluripotency. Cell Res 17:42–49. https://doi.org/10.1038/sj.cr.7310125
80. Papapetrou EP, Tomishima MJ, Chambers SM, Mica Y, Reed E, Menon J, Tabar V, Mo Q, Studer L, Sadelain M (2009) Stoichiometric and temporal requirements of Oct4, Sox2, Klf4, and c-Myc expression for efficient human iPSC induction and differentiation. Proc Natl Acad Sci USA 106:12759–12764. https://doi.org/10.1073/pnas.0904825106
81. Oka M, Moriyama T, Asally M, Kawakami K, Yoneda Y (2013) Differential role for transcription factor Oct4 nucleocytoplasmic dynamics in somatic cell reprogramming and self-renewal of embryonic stem cells. J Biol Chem 288:15085–15097. https://doi.org/10.1074/jbc.M112.448837
82. Moraveji S-F, Attari F, Shahverdi A, Sepehri H, Farrokhi A, Hassani S-N, Fonoudi H, Aghdami N, Baharvand H (2012) Inhibition of glycogen synthase kinase-3 promotes efficient derivation of pluripotent stem cells from neonatal mouse testis. Hum Reprod 27:2312–2324. https://doi.org/10.1093/humrep/des204
83. Choi WY, Jeon HG, Chung Y, Lim JJ, Shin DH, Kim JM, Ki BS, Song S-H, Choi S-J, Park K-H, Shim SH, Moon J, Jung SJ, Kang HM, Park S, Chung HM, Ko JJ, Cha KY, Yoon TK, Kim H, Lee DR (2013) Isolation and characterization of novel, highly proliferative human CD34/CD73-double-positive testis-derived stem cells for cell therapy. Stem Cells Dev 22:2158–2173. https://doi.org/10.1089/scd.2012.0385
84. Simon L, Ekman GC, Kostereva N, Zhang Z, Hess RA, Hofmann M-C, Cooke PS (2009) Direct transdifferentiation of stem/progenitor spermatogonia into reproductive and nonreproductive tissues of all germ layers. Stem Cells 27:1666–1675. https://doi.org/10.1002/stem.93
85. Boulanger CA, Mack DL, Booth BW, Smith GH (2007) Interaction with the mammary microenvironment redirects spermatogenic cell fate in vivo. Proc Natl Acad Sci U S A 104:3871–3876. https://doi.org/10.1073/pnas.0611637104
86. Yang H, Hao D, Liu C, Huang D, Chen B, Fan H, Liu C, Zhang L, Zhang Q, An J, Zhao J (2019) Generation of functional dopaminergic neurons from human spermatogonial stem cells to rescue parkinsonian phenotypes. Stem Cell Res Ther 10:195. https://doi.org/10.1186/s13287-019-1294-x

Chapter 7
Spermatogonial Stem Cell Cryopreservation for Fertility Preservation

Marija Vilaj, Branka Golubić-Ćepulić, and Davor Ježek

Abstract

Introduction: Recent advances in cancer treatment have led to a higher survival rate in children and young adults. However, oncological therapies are related to impaired fertility due to their direct toxic effect on gonadal function, causing the depletion of germ cell number, or they act indirectly by damaging the somatic testicular cells and spermatogonial stem cells (SSCs), which consequently results in later fertility disruption. Therefore, strategies for fertility preservation and subsequent restoration have become an important focus of researchers around the world.
Methods: We summarize new knowledge on freezing spermatogonia stem cells to maintain fertility in humans.
Results: Cryopreservation of spermatozoa is the most established method of fertility preservation before gonadotoxic treatments in male adolescents who have undergone puberty and it is nowadays a clinical routine. Meanwhile, fertility preservation and restoration in prepubertal boys for whom spermatogenesis has not yet started is challenging because they are unable to produce sperm. To date, all known strategies for this purpose are experimental and far from clinical application. The only currently available option of fertility preservation in this group of cancer patients is the cryopreservation of testicular tissue containing SSCs that can be later used for autologous transplantation in order to restore fertility.
Conclusions: Despite the achievements in this field, cryopreserved human testicular tissue containing SSCs has not yet resulted in sperm production and it is, therefore, necessary to improve and standardize methods for testicular tissue and testicular cell cryopreservation, SSCs isolation, their in vitro culture, and propagation. Furthermore, refinement of sorting methods is also required to achieve the efficient enrichment of human spermatogonia and avoid the coincidental introduction of neoplastic cells.

M. Vilaj · B. Golubić-Ćepulić · D. Ježek (✉)
Department for Transfusion Medicine and Transplantation Biology, University of Zagreb, University Hospital "Zagreb", Kišpatićeva 12, 10000 Zagreb, Croatia

D. Ježek
Department of Histology and Embryology, School of Medicine, University of Zagreb, Šalata 3, 10000 Zagreb, Croatia

I. Virant-Klun (ed.), *Stem Cells in Reproductive Tissues and Organs*, Stem Cell Biology and Regenerative Medicine 70, https://doi.org/10.1007/978-3-030-90111-0_7

Keywords Cancer · Cryopreservation · Fertility restoration · Spermatogonia · Spermatogonial stem cells · Testicular tissue

Abbreviations

2D	Two-dimensional
3D	Three-dimensional
DAB	3,3′-Diaminobenzidine
DMSO	Dimethyl sulfoxide
ECM	Extracellular matrix
EEJ	Electroejaculation
FACS	Fluorescence-activated cell sorting
FN	Fine needle
GCNIS	Germ cell neoplasia in situ
HE	Hematoxylin–eosin stain
ICSI	Intracytoplasmic sperm injection
IUI	Intrauterine insemination
IVF	In vitro Fertilization
L	Leydig cell
M	Microlith
MACS	Magnetic-activated cell sorting
MAGEA3/4	MAGE family member A3/4
P	Parenchyma
PLAP	Alkaline phosphatase
S	Scrotum
Sc	Sertoli cell
SSC	Spermatogonial stem cell
ST	Seminiferous tubule
TESA	Testicular sperm aspiration
TESE	Testicular surgical sperm extraction

Introduction

Approximately 300,000 children aged 0–19 years are diagnosed with cancer each year [1]. Since the cause of the vast majority of cancers in children and adolescents is unknown and therefore their occurrence can not be prevented, great effort is being made to promptly determine the correct diagnosis followed by effective therapy. Significant advances in oncological therapies have led to quite high survival rates of more than 80% of patients in the pediatric population [2–5]. However, anticancer therapies, including chemotherapy, radiotherapy, and hematopoietic stem cell transplantation, have a well-known long-term adverse effect on the gonadal function which encouraged the development of fertility preservation techniques for cancer patients. Consequently, fertility preservation options like cryopreservation of gametes and embryos are usually discussed and offered to patients before the gonadotoxic treatment commences. Although the testes of prepubertal patients do not complete spermatogenesis, numerous studies have suggested that cytotoxic treatments used to treat disease in these patients also affect fertility [6–8]. Since the prepubertal patients are unable to produce their gametes, conventional strategies for the preservation of fertility potential are not possible. Therefore it is necessary to develop alternative approaches that allow the preservation and restoration of fertility for those patients with the hope of future parenthood. Furthermore, fertility preservation options for young cancer patients should be considered in multidisciplinary cooperation with oncologists, urologists, andrologists, and fertility specialists, due to the specific challenges in children and adolescents including ethical, practical, and scientific issues.

Fertility Preservation in Adolescents

The availability of fertility preservation techniques depends on the fact whether the child facing anti-cancer therapy is prepubertal or postpubertal as well as on the urgency of cancer treatment initialization. Options for fertility preservation in adolescents who have undergone puberty are broader than those in prepubertal children. In most male adolescents, semen cryopreservation is possible [9] which makes spermatozoa cryopreservation after masturbation the most established method of fertility preservation prior to gonadotoxic treatments [10]. Research by Bahadur et al. [11] has shown that spermatozoa quality in young boys is not affected by a patient's age or diagnosis, wherefore cryopreserved semen from adolescent cancer patients is suitable and potentially useful for assisted reproductive procedures in the future. Adolescents who can not produce sperm sample by masturbation should be offered options of electroejaculation (EEJ) under sedation or anaesthesia [12], testicular aspiration (TESA), testicular sperm extraction (TESE), or retrieval from a post-masturbation urine sample [13].

Electroejaculation (EEJ)

Study of Berookhim and Mulhall [14] showed that successful spermatozoa retrieval with EEJ was obtained in 60% of boys aged 11–19 years. Complications of the technique were rare but the quality of sperm obtained that way was variable. In the case that EEJ is inadequate to collect sufficient volume or density of motile sperm for cryopreservation and in the case of azoospermia, more invasive methods such as TESA or TESE could be taken into consideration.

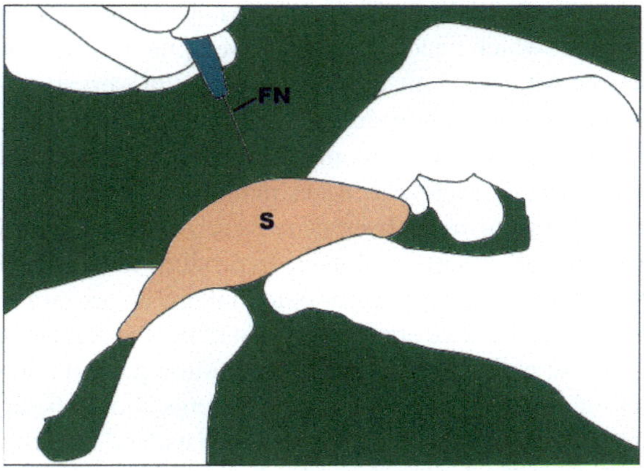

Fig. 7.1 Testicular sperm aspiration (**TESA**). For the TESA purposes, the testis and epididymis must be held in a fixed position, avoiding the puncture of the epididymis. A fine needle (**FN**) is applied directly inside the testicular parenchyma. The whole procedure can be done in an outpatient clinic setting. (**S**, scrotum)

Testicular Sperm Aspiration (TESA)

TESA (Fig. 7.1) is generally performed under local anaesthesia with no incisions needed. This is a simple and routine procedure that usually takes approximately 10 min. However, in some cases, TESA does not provide an adequate yield of spermatozoa, hence the application of the open testis biopsy could be required.

Testicular Surgical Sperm Extraction (TESE)

TESE was safely performed for fertility preservation purposes in adolescents before cancer therapy with a spermatozoa retrieval rate of 50% and was successful even in an 11-year-old boy [14]. Although a scrotal incision (Fig. 7.2) is required to obtain a sizable piece of seminiferous tubules for this approach, it is generally performed under local anaesthesia without any complication. Additionally, an important advantage of the technique is a lack of delay in the initiation of chemotherapy among patients scheduled for urgent cancer treatment. On the other hand, the major shortcoming of TESE is a relatively large volume of testicular parenchyma which is non-selectively removed to obtain sufficient specimen for spermatozoa retrieval.

TESE can also be applied when testicular cancer is diagnosed in an azoospermic patient. If the cancer is detected in an early stage, restricted to a limited area of the testis (Fig. 7.3), there is a possibility to cryopreserve a healthy part of the testicular parenchyma in a form of multiple biopsies. After the orchiectomy (Fig. 7.4a), multiple bioptic samples can be taken from the tumor-free area of the testis (Fig. 7.4b). In our setting, each biopsy is, immediately upon the excision, processed in the laminar flow cabinet (placed inside the operational theatre) and cut into two parts: one part is instantly frozen in a sperm freezing medium for the potential TESE purposes, whereas the other half is fixed in a fixative and subjected to detailed histological analysis

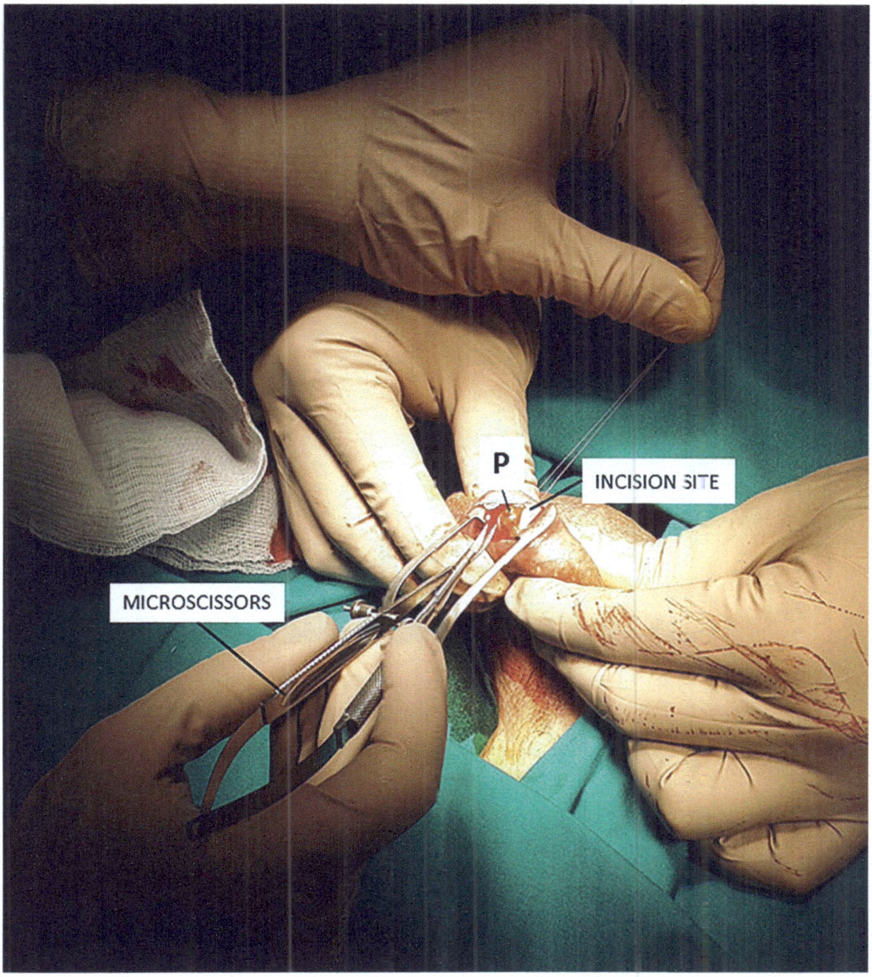

Fig. 7.2 Open biopsy of the testis is usually performed under local or spinal anaesthesia. After an incision of the tunica albuginea, the testicular parenchyma (P) spontaneously protrudes through the incision site and can be easily removed by microscissors

and immunohistochemistry for detection of germ cell neoplasia in situ (GCNIS) (Fig. 7.5a, b). If the histological analysis shows no presence of the tumor, frozen samples can be mobilized from the cryostorage and used for the consequent TESE and intracytoplasmic sperm injection into the oocyte (ICSI) procedure.

In general, in order to achieve pregnancy, fertilization techniques (IVF) are often recommended rather than intrauterine insemination (IUI) due to a limited sperm resource. If a thawed sample contains only a few spermatozoa, fertilization can be yet achieved by performing intracytoplasmic sperm injection (ICSI).

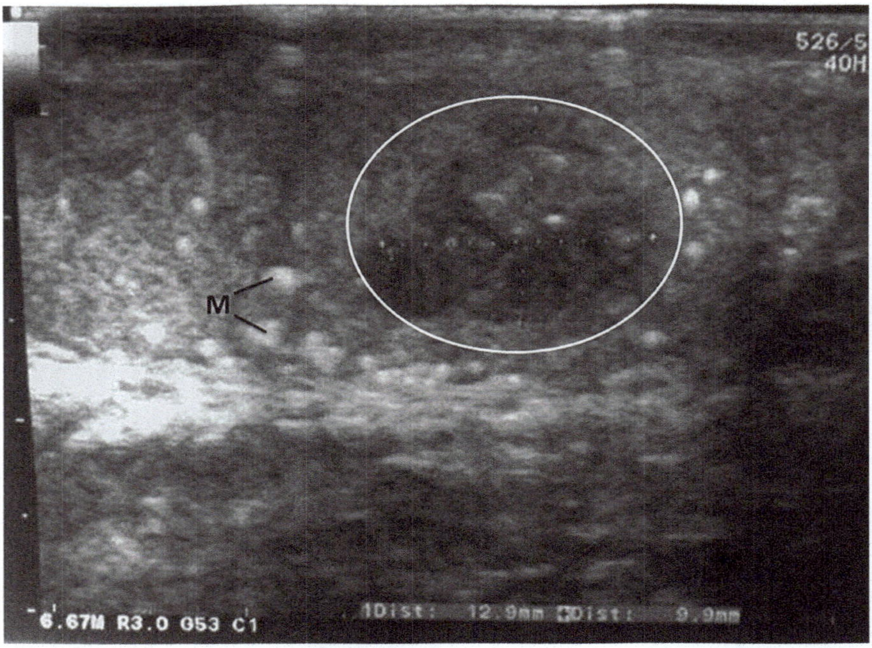

Fig. 7.3 Ultrasound of the testis of an azoospermic patient. Encircled is a limited tumor (12.9 × 9.9 mm) surrounded by the rest of testicular parenchyma with an abundant presence of microliths (**M**)

Cryopreservation of Testicular Tissue

Several different slow programmed freezing protocols for human immature testicular tissue have been described thus far, using cryoprotectants such as 1.5 M ethylene glycol and sucrose [15], 0.7 M DMSO [16] or 0.7 M DMSO and sucrose [17, 18]. Keros et al. [16] reported efficient freezing protocol for prepubertal testicular tissue cryopreservation using 5% DMSO as a cryoprotectant to successfully maintain spermatogonia, Sertoli cells and stromal compartment without clear structural changes during freezing, thawing, and tissue culture. Standardizing a successful slow freezing method for cryopreservation of testicular tissue or testicular cell suspension, with optimized type and concentration of cryoprotectant that would sustain and preserve the quality and viability of the tissue is crucial before clinical application of fertility restoration in prepubertal or adolescent cancer survivors.

On the other hand, vitrification can also be used as an effective method for tissue preservation instead of slow freezing methods [19, 20]. The study conducted by Curaba et al. [19] has demonstrated vitrification as a convenient method for cryopreservation of immature human testicular tissue, with being faster and cheaper than slow freezing. It is still to be determined whether slow freezing and vitrification are equally safe and effective regarding immature testicular tissue cryopreservation or if one of these approaches is predominant in terms of tissue or cells quality preservation.

Since the cryopreserved human testicular tissue or testicular cells have not yet resulted in the spermatozoa production, none of the preservation protocols has been fully validated or been shown to be superior over any other reported method. With the purpose of the future clinical application of fertility restoration, it is necessary to validate, optimize, and standardize cryopreservation protocol for immature testicular tissue or testicular cell suspension. Furthermore, the influence of freezing protocols on genetic and epigenetic modifications in germ cells after cryopreservation must be thoroughly examined.

Fertility Preservation in the Prepubertal Patients

Fertility preservation in prepubertal boys and adolescents, for whom spermatogenesis has not started yet, is challenging due to the fact that testis only contains diploid germ cells (Figs. 7.6 and 7.7), including pro-spermatogonia/SSCs, which makes the use of the conventional approach of sperm cryopreservation impossible. The only option currently available for the future fertility restoration in this group of cancer patients is cryo-banking of testicular tissue containing SSCs [21, 22]. SSCs are a small group of diploid testis cells that are located on the basement membrane of

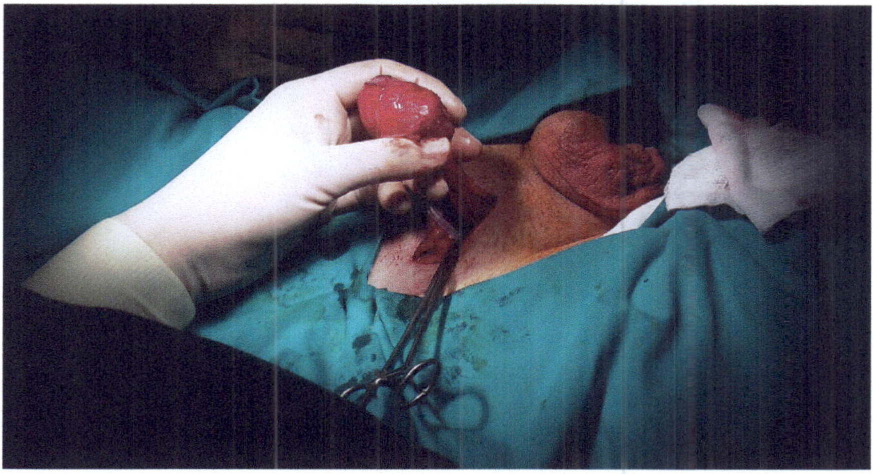

Fig. 7.4 **a** Orchiectomy in an azoospermic patient with a testicular tumor. The testis is removed together with epididymis as well as the majority of the spermatic cord in order to eliminate the neoplasm and isolate the remaining healthy testicular parenchyma. **b** After the orchiectomy and visualization of the tumor area, multisampling of the rest of the testicular parenchyma takes place. Each bioptic piece is divided into two halves: one half for cryopreservation and another for a detailed histological analysis, including immunohistochemistry

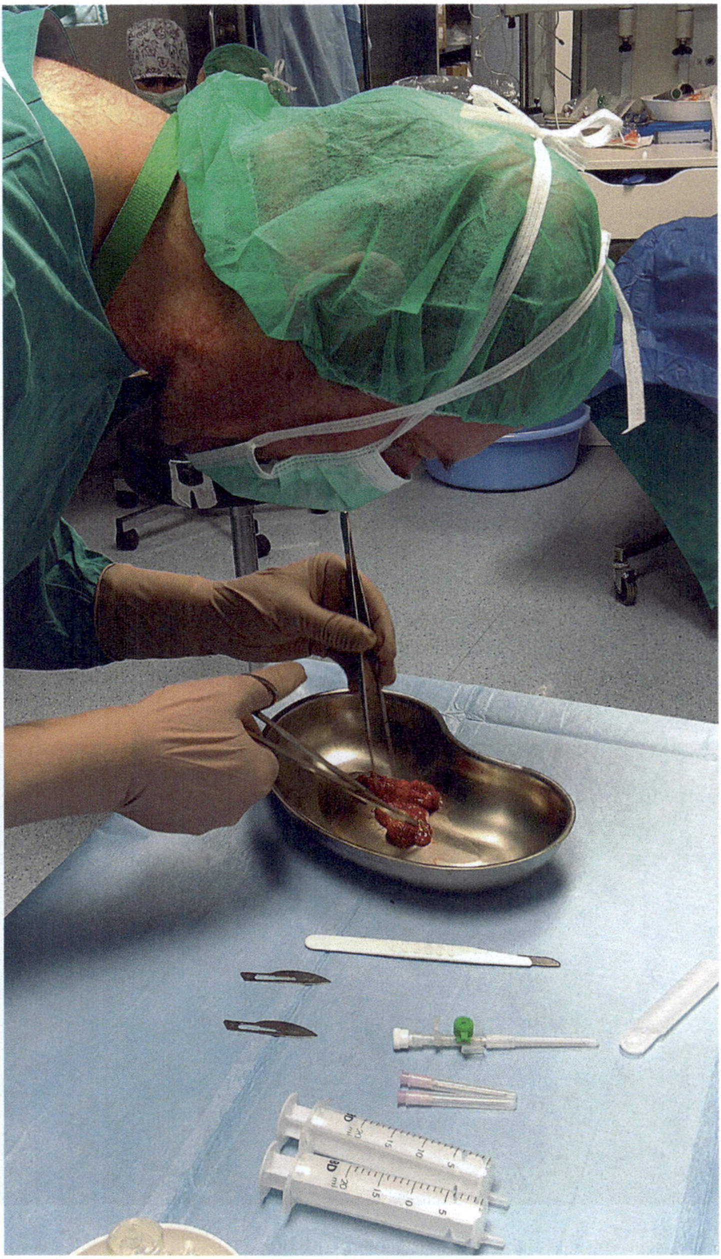

Fig. 7.4 (continued)

seminiferous tubules. These cells have the ability of self-renewal and, thus, they maintain a reserve stem cell pool, but they can also differentiate into other cell types [23]. A fine balance between self-renewal and differentiation of SSCs is determined by their microenvironment, the so-called stem cell niche, which is needed for cell survival, proliferation, and differentiation. Sertoli cells in the seminiferous epithelium of the testis and probably also cells from the interstitial compartment modulate SSCs renewal and differentiation.

SSCs are responsible for the continuous process of spermatogenesis, assuring unlimited sperm production during adult life. Based on morphological studies, three types of spermatogonia in the adult human testis can be distinguished: A_{dark}, A_{pale}, and B spermatogonia, which also differ by their role in spermatogenesis [24]. A_{dark} are known as reserve stem cells. They are mitotically quiescent and become proliferatively active only during the pubertal expansion or in response to spermatogonial depletion as a result of irradiation or toxic exposure. On the other hand, A_{pale} are

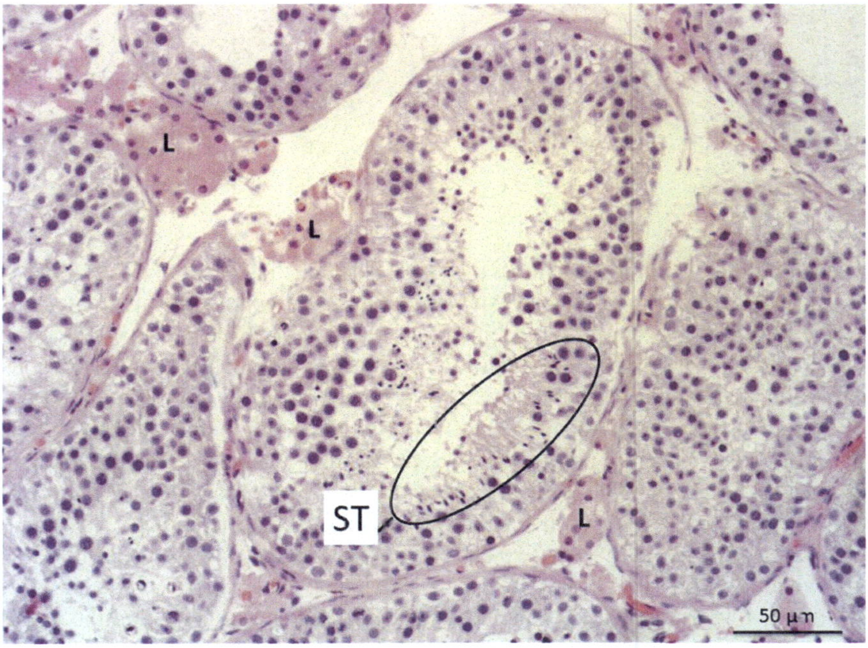

Fig. 7.5 **a** Part of the testicular biopsy of an azoospermic patient with a limited testicular tumor. Within the healthy parenchyma, a seminiferous tubule (**ST**) contains an abundant presence of late spermatids (encircled area). Once released into the lumen, these cells become spermatozoa that can be isolated from the cryopreserved material and used for ICSI. (**L**, Leydig cells) (Hematoxylin eosin /H.E./). **b** Another bioptic piece from the same patient, the expression of placental alkaline phosphatase (PLAP). Within several seminiferous tubules, there is a presence of germ cell neoplasia in situ (GCNIS) (➔). In neighboring tubules, there is an existence of spermatogonia (sp). These cells could be isolated and cultivated for potential spermatozoa production. (3,3,-diaminobenzidine /DAB/ and hemalaun)

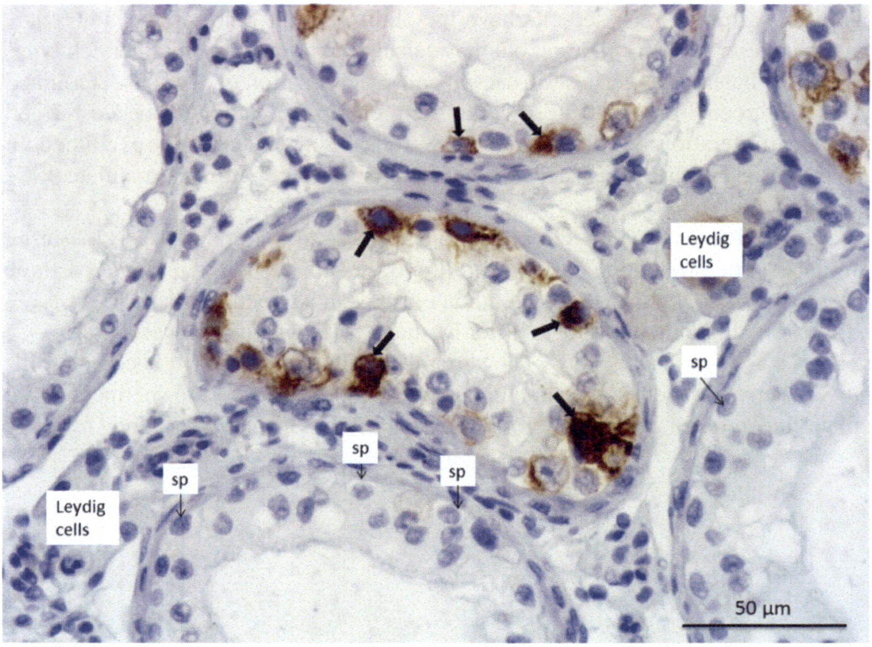

Fig. 7.5 (continued)

considered self-renewing progenitors due to the fact that they are actively dividing to maintain the stem cell pool. Type B spermatogonia differentiate into spermatocytes in which genetic material is recombined and segregated through meiosis, leading to the formation of haploid spermatids. The resulting spermatids undergo various morphological changes (spermiogenesis) in order to transform into spermatozoa.

SSCs are established in the human testis during the first 3 months after birth [25] (Figs. 7.6 and 7.7) and are present in the testis during the whole life unless they are lost due to a disease or some medical treatment [26]. Even the low doses of cytotoxic drugs or irradiation diminish the level of differentiating spermatogonia. This group of testicular cells is highly sensitive due to their rapidly dividing nature, while less sensitive non-proliferating SSCs, as well as spermatocytes and spermatids, survive. Nevertheless, spermatogenesis can be recovered by remaining SSCs but the possibility of reinitiation depends on the type and dose of cytotoxic drug or irradiation used in therapy treatment. Therefore, cryopreservation of testicular tissue pieces containing SSCs or SSCs as a cell suspension is the only available strategy for the fertility preservation in prepubertal boys and adolescents who do not yet produce spermatozoa that could be cryopreserved. Cryopreservation of testicular tissue has been ethically approved, although the restoration of fertility with cryopreserved spermatogonia/testicular tissue containing spermatogonia is still experimental. Therefore, in Europe exist only a few centers offering prepubertal testis cryopreservation (Fig. 7.8), but more to come in the near future.

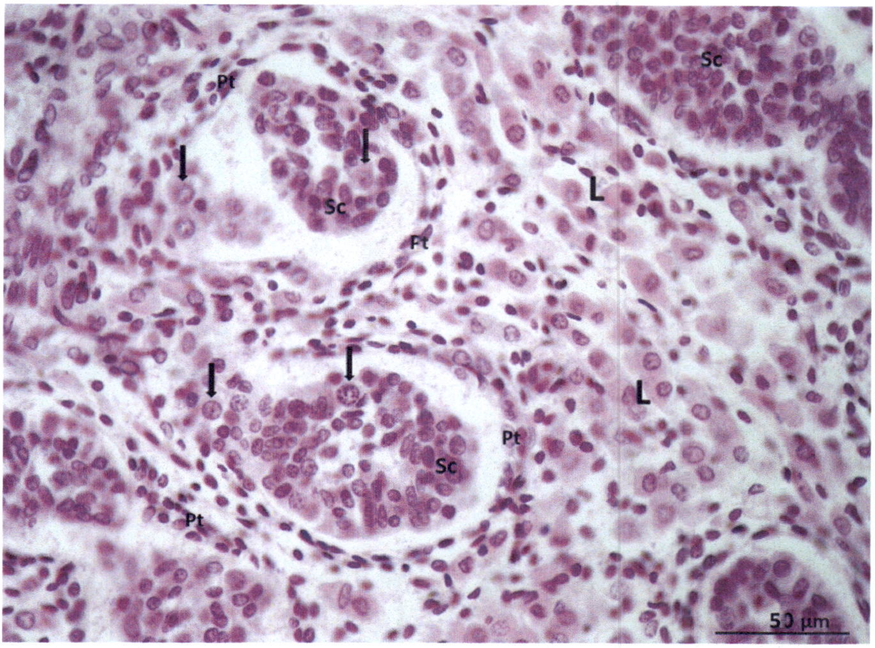

Fig. 7.6 Immature human testis. Sex cords (future seminiferous tubules) do not have a lumen and are lined with pro-spermatogonia (➔) and unripe supporting Sertoli cells (**Sc**). Sex cords are limited by future myoid (peritubular cells, **Pt**) and surrounded by the interstitial compartment rich in Leydig cells (**L**). (H.E.) *(Histology slides collection of Department for Histology and Embryology, School of Medicine University of Zagreb)*

Fertility Restoration

While testicular tissue cryopreservation protocols have been reported [21, 22] and are available to preserve prepubertal testicular tissue and cells, techniques for obtaining mature spermatozoa from the cryopreserved immature testicular tissue are still at the research stage.

Different approaches have been considered for the restoration of fertility using cryopreserved testicular tissue, including autologous transplantation of SSC, testicular tissue grafting or xenografting into immunodeficient mice, de novo testicular morphogenesis, and testicular tissue organ culture and derivation of germ cells from induced pluripotent stem cells [27, 28]. Each one of these strategies has mainly been studied in animals and none of them has yet proven safe and efficient in the human.

Cryopreservation of testicular tissue or cells containing SSC has been imposed as a feasible way of fertility restoration using the autologous transplantation of SSCs (or tissue piece containing SSCs) years after cryo-storage of these same SSCs, due to their ability to regenerate spermatogenesis whether isolated from newborns or adults, freshly obtained or cryopreserved for years.

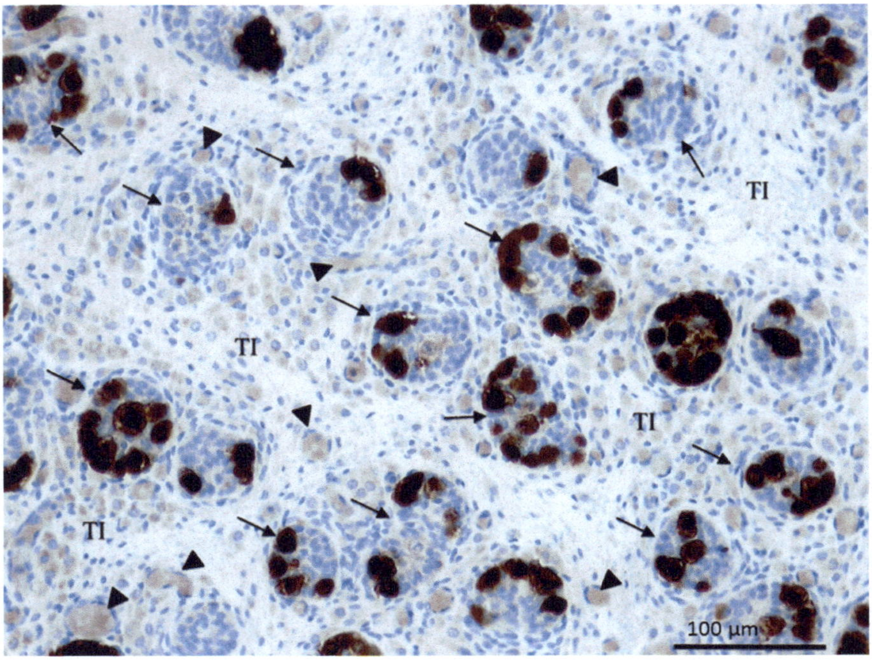

Fig. 7.7 Immature human testis, the expression of MAGE-A3/4 inside the sex cords (➔). Based on the immunohistochemistry reaction, pro-spermatogonia can be easily distinguished from supporting Sertoli cells (negative to MAGE-A3/4). Testis interstitium (**TI**) has an abundant network of blood vessels (▼). (DAB and hemalaun)

In order to preserve the integrity of the stem cell niche, which is the key to normal regeneration of spermatogenesis as well as for the process itself, isolated and cryopreserved biopsies of testicular tissue can be cultured in the form of whole pieces in vitro or in vivo by performing tissue transplantation. If the pieces of testicular tissue are transplanted into a patient to restore normal spermatogenesis in vivo, there is a risk of introducing the cancer cells into the patient. Therefore, this approach of fertility restoration can only be considered if the tissue does not carry the risk of malignant contamination. To overcome this risk, a good alternative strategy is the development of spermatozoa in vitro which can be later on used for ICSI.

Spermatogonial Stem Cell Transplantation for Fertility Restoration

Transplantation of SSCs, although still experimental, is currently considered the most promising strategy for fertility restoration in young cancer patients and might soon be ready for translation into clinical practice. The technique was first described in

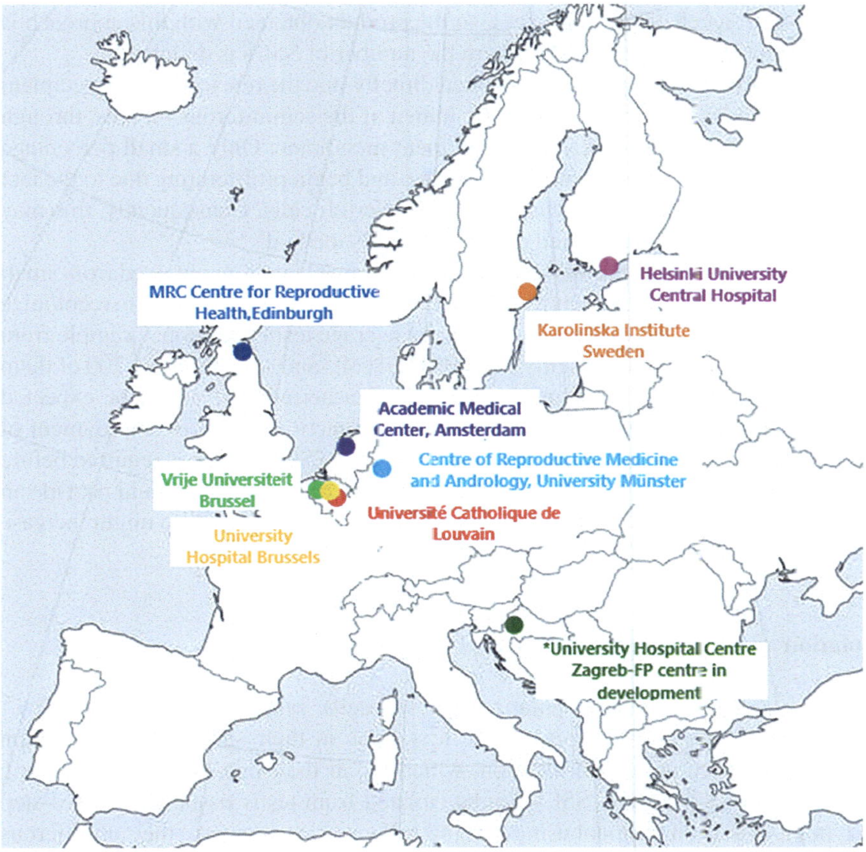

Fig. 7.8 Leading biobanking centers in Europe offering fertility cryopreservation of prepubertal testis (FP)

1994 in mice [29] but has since been replicated in several other mammalian species including human [30, 31]. However, the outcome of this clinical trial performed in 12 male patients with Hodgkin's disease has proven unsuccessful.

Cell suspension of germ cells, including SSCs, can be isolated from human testes and cryopreserved for the future restoration of fertility. After testicular biopsies are obtained, pieces of testicular tissue can be dissociated resulting in isolated testicular cells including SSCs which can be subsequently used in vitro after culture in two-dimensional (2D) or three-dimensional (3D) systems, or in vivo after transplantation. The protocol is straightforward, comprising of mechanical and/or enzymatic digestion of tissue and, thus, can be performed in any assisted reproduction center. However, dissociation of testicular tissue causes impairment of the niche integrity due to the disruption of connections between cells necessary for cell proliferation and differentiation, and also compromise cell survival. Furthermore, the digestion method is not a cell-targeted process, so the product obtained with this approach is a mixed population of testis cells where the number of SSCs is deficient.

Testicular cells are usually transplanted directly into the rete testis of the recipient testis [32]. SSCs must migrate from the lumen of the seminiferous tubules, through the blood–testis barrier to reach the basement membrane. Only a small percentage of the SSCs colonize the basement membrane and begin proliferating due to the fact that most transplanted cells are eliminated by Sertoli cells. Consequently, this may be the reason for overall reduced efficiency of this method.

Another challenge in the application of SSCs transplantation, obtained from small biopsies of prepubertal testes, is an insufficient number of isolated SSCs to recolonize the testicular niche. It has been shown that an average testicular biopsy sample from prepubertal boys provides roughly 390,000 total cells and only about 11,700 of them are considered to be pro-spermatogonia [33]. Furthermore, it would be expected that a much smaller number of them is actually functional, hence, employment of methods that will enable expansion of the number of SSCs in vitro is required before their transplantation [34]. Considering that the stem cell expansion could provide an injection of a high number of pure stem cells, culturing SSCs in vitro might increase the efficiency of the SSCs transplantation.

Isolation and Enrichment of SSCs

SSCs are a rather rare subpopulation of testis cells, ranging from 0.01 to 1% of all testicular cells [35]. Therefore, the first steps in their subsequent application are optimal isolation and purification, which are at the same time major ongoing challenges in SSC culture. SSCs can be isolated from testis tissue with a two-step enzymatic digestion protocol using first the collagenase to disperse the seminiferous tubules followed by an additional enzymatic step in which trypsin is added to obtain a single-cell suspension. Additionally, DNase-I can be added in suspension to prevent adhesion of the resulting cells. The aforementioned protocols for testis tissue digestion have been widely used for the isolation of testicular cells in primates [36] and also in humans [37].

However, digestion protocol results with testis cell suspension composed of SSCs but also of contaminating somatic cells that need to be removed from the culture (or at least their number should be reduced) in order to achieve the efficient enrichment of human spermatogonia. Various strategies for identification and separation of SSCs are used, including approaches that use antibodies, but also methods that do not implement antibodies. Fluorescence-activated cell sorting (FACS) and Magnetic-activated cell sorting (MACS) are methods that use specific antibodies to distinguish SSCs from contaminating cells including malignant cells [38]. Nevertheless, no study has yet been able to determine a single unique marker for SSCs, so the confirmation and subsequent isolation of these cells are based on the identification of multiple markers together.

Besides the sorting methods that are based on the cell's surface antibody recognition, there are enrichment strategies that rely on the distinction in innate cellular properties between SSCs and somatic cell. For instance, Sertoli cells adhere to the culture plate faster than SSCs. Based on this characteristic, the technique of differential planting is used to separate SSCs from somatic cells. Differences in sedimentation velocity or density gradient centrifugation can also be used for separation of somatic cells and SSCs [39].

The major concern of SSCs transplantation is the coincidental introduction of neoplastic cells. Accordingly, sorting methods should also be used to eliminate malignant contamination, considering that it has been shown in animal models that even a very low number of leukemic cells (i.e., just 20 of them) is enough to cause a relapse of the disease [40]. Different sorting techniques that attempt to sort SSCs and exclude neoplastic cells have been reported [41–43]. Nevertheless, complete elimination of cancer cells could not be achieved, neither with magnetic-activated nor fluorescence-activated cell sorting methods. Meanwhile, Sadri-Ardekani et al. [44] have shown that elimination of neoplastic cells can be accomplished when cocultured with testicular cells for 26 days. This indicates that testicular cell culture system could also be used to eliminate containing neoplastic cells before SSCs transplantation, alongside with providing the expansion of SSCs.

In Vitro Culture of SSCs

Establishment of an efficient in vitro culture system that is able to maintain self-renewal as well as proliferation capability of SSCs, and can replicate the process of human spermatogenesis is crucial for SSCs introduction into a clinical application. An important prerequisite for the in vitro spermatogenesis is the reconstruction of stem cell niche in order to establish replacement of dead or differentiating germ cells, and continuous production of in vitro generated germ cells. This replication should resemble natural conditions of the spermatogonial niche in the microenvironment in vivo.

Sadri-Ardekani et al. [34] were first to report a long-term culture of adult human SSCs demonstrating that human SSCs could be maintained in culture for several months and expanded more than 18,000-fold. They used differential plating method for reduction of the testicular somatic cells number and reverse transcriptase polymerase chain reaction as well as immunofluorescence for determination of SSCs presence. The same group in 2011 reported in vitro propagation of SSCs from prepubertal human testes [45]. Since then, more than 20 studies on human SSCs culture have been reported. However, functional proof that these cells are able to re-establish complete spermatogenesis resulting in sperm production as well as the probability of potential cell modification in the culture system, are far from being completely elucidated. Therefore, careful analysis of genetic and epigenetic changes in the testis cell culture is the next challenge to be addressed to ensure their safety for potential clinical application.

Several different research groups have tried to develop an efficient in vitro culture system in humans for SSCs propagation purposes. Three general approaches have been reported to achieve this goal: the 2D culture of testis cell suspension, the 3D culture of testis cell suspension, and organotypic culture of testicular tissue fragments. Organotypic culture of testicular tissue fragments is the latest and newest strategy for in vitro reconstitution of the SSCs niche. Hence, it will be discussed later in the context of the most recent approaches for fertility restoration. The SSC culturing can be performed in the following ways:

- **2D culture systems** of testis cell suspension have been widely used for in vitro proliferation of SSCs as well as the production of haploid male germ cells. To obtain attachment, survival and proliferation in vitro, it has been shown that human spermatogonia require an extracellular matrix (ECM) or co-culture with feeder cells, such as Vero cells, mouse embryonic fibroblasts, mouse endothelial cells, fibroblasts derived from human embryonic stem cells, human Sertoli cells, or human testicular somatic cells. Cremades et al. [46] reported the formation of round spermatids and their differentiation into elongated spermatids when human SSCs were co-cultured with Vero cells isolated from African green monkey. However, due to their animal origin, round spermatids obtained that way are not suitable for clinical purposes. Therefore, another long-term 2D culture system has been developed in which SSCs were co-cultured with Sertoli cells [47], resulting in normal differentiation of late spermatids but with low fertilization potential along with developmental arrest and sex chromosome anomalies in the vast majority of the embryos. Based on all studies regarding the 2D culture of cells that have been conducted so far, it can be concluded that this system has several shortcomings, including the fact that it can not effectively replicate the microenvironment of the testis, with all essential cellular interactions and structural properties of intact seminiferous tubules. The main causes of this drawback are the lack of necessary growth factors and dissociation of the niche integrity.
- **3D culture of testis cell suspension** The 3D culture is an in vitro system that can effectively mimic cell developmental microenvironment in vivo, providing the 3D in situ organization of the seminiferous epithelium. The 3D culture structure consists of a thick layer of soft agar system or the methylcellulose system in which the cells are embedded. This culture technique, in comparison with the 2D system, allows more efficient diffusion of nutrients, oxygen, and other essential molecules and can be used for the assessment of necessary interactions among male germ cells, testicular somatic cells, and the extracellular matrix during spermatogenesis [46, 48]. In the human as well as in the mice and rhesus monkeys, the 3D culture has provided promising results [49–51]. Complete spermatogenesis has been reported after the 3D culturing of juvenile murine and monkey testicular tissue, whereas male germ cells developed to the level of elongating spermatids with the utilization of this technique [49, 52, 53]. It was noticed that expansion of germ cells was prominent when they are co-cultured with Sertoli cells which provide support for developing germ cells in vivo. Stukenborg et al. [54] reported the 3D culture approach using mouse spermatogonia obtained from immature testes. Regardless of the matrix nature, in the presence of gonadotrophins and somatic testicular cells for support, in vitro maturation of male germ cells has been achieved. After 31 days of culture, elongating spermatids and normal spermatozoa were observed.

However, in humans, the 3D culture has not yet been able to complete the entire spermatogenic process.

Further research efforts should be focused on the reproduction of this culture technique in human germ cells and assessment of the fecundability of the spermatozoa obtained in culture. The question also arises if the same results could be obtained with germ cells from human prepubertal tissue.

Testis Tissue Xenografting and Future Directions for Fertility Restoration

Xenografting of Testicular Tissue

Xenografting of testicular human tissue into immunodeficient mice has been explored as a potential alternative to SSCs transplantation for fertility restoration in prepubertal boys [17]. This technique was first used in 1974 [55] and since then has been providing important insight into testicular cells function. Studies performed so far demonstrate that testis xenografts from prepubertal males have better survival and spermatogenesis outcome than that of pubertal and adult donors [56 57]. The goal of this approach is to obtain biopsies from prepubertal patient testes before the gonadotoxic treatment, which can be cryopreserved and stored for future xenografting. When the patient has recovered from the original disease and desires children cryopreserved testis biopsies could be thawed and grafted into immunodeficient mice to obtain fully developed tissue and collect spermatozoa. Collected spermatozoa can be used for ICSI to produce embryos for transfer. However, complete spermatogenesis has not been obtained so far from the cryopreserved human testis tissue transplanted into immunodeficient mice.

The main advantage of this technique over autologous SSCs transplantation is the possibility to avoid malignant contamination with neoplastic cells that may be present in the donor cell suspension at the moment of tissue sampling [58]. Nevertheless, there are serious ethical and safety concerns relating to human testicular tissue xenografting due to the possibility of potentially transferring host pathogens which could have an effect on the developing testis tissue as well as on resulting spermatozoa. Therefore, xenografting of testicular tissue for fertility restoration in prepubertal boys is still in the early experimental stage and extensive safety, feasibility and bioethical evaluation would be required before its application in humans.

The Most Recent Approaches and Future Directions

The most recent approaches and future directions are:

- **Organotypic culture of testicular tissue fragments** Organotypic culture is so far the only technique for in vitro maturation of testicular tissue that is capable to preserve seminiferous tubule structure as well as intact cell interactions within the tissue, wherefore this approach should give a better outcome than the culture of isolated cells. However, in humans, the completion of spermatogenesis in vitro from immature testicular tissue is still not achieved. The main challenge of this technique is the maintenance of tissue viability in the culture for long enough in order to complete in vitro spermatogenesis. This hindrance of testicular tissue culturing arises due to the limited diffusion rates of the tissue as compared to that of monolayer cell cultures. Additionally, a progressive loss of germ cells during the organotypic culture of adult human testicular tissue has been observed [59, 60]. The first study on the culture of immature testicular tissue over 139 days in vitro reported maturation of the prepubertal SSC niche with the progression of Sertoli and Leydig cells from an immature to mature state. Immature testicular tissue was obtained from prepubertal cancer patients between 2 and 12 years of age before undergoing gonadotoxic treatments [61]. However, a loss of spermatogonia over the culture period was noticed and the renewal of spermatogonia and differentiation of haploid germ cells has not been achieved. One year later, the same research group has reported the generation of human haploid germ cells after 16 days of prepubertal testicular tissue culture [62]. Nevertheless, further research should focus on genetic and epigenetic characteristics of germ cells obtained in culture as well as on their ability to completely differentiate into functional sperm before suggesting this technique for potential clinical trials.
- **Organoids** are 3D testis organ-like structures that have been used over the last few years as suitable systems to investigate testicular development, regulation, and function of the SSCs niche, germ cell proliferation, and differentiation in vitro. Compared to 2D systems, 3D organoids provide better and faster testicular cell reorganization leading to initiation and progression of spermatogenesis due to the presence of various testicular somatic cells that are in direct contact with germ cells as well as the 3D organization of seminiferous epithelium (for review see [63]). Pendergraft et al. [64] reported the formation of an organoid from the human adult testicular tissue which produced testosterone and has been maintained in culture for up to 21 days. Furthermore, human testicular organoids were developed using testicular cells from both adult and prepubertal human testes [65]. The latest strategy of 3D organoids using the microwell aggregation approach has been recently described by Sakib et al. [66]. This research group used cryopreserved human prepubertal tissue from 6-month-old and 5-year-old boys for a generation of the organoid which expressed a high level of structural organization similar to that of the testes tissue in vivo, with well-defined interstitial and seminiferous epithelium compartments, separated by a basement membrane. However, the overall tissue architecture was reversed compared to the native structures. This methodology of organoid generations was demonstrated to be applicable across species and provides the organoid that supports the specific function of each cell type it contains.

Organoids are novel strategy intended to explore testicular microenvironment. They could potentially pave the way for the artificial organ creation that would mimic the original, allowing differentiation of germ cells and generation of haploid spermatids for fertility restoration.

Conclusions

The majority of strategies for fertility preservation in prepubertal cancer patients are far experimental and still in development. Although promising results have been obtained with in vitro studies or in animal models, such alternative techniques are not currently translated into clinical practice. To date, none of the studied in vitro systems has enabled full spermatogenesis and resulted in the production of functional spermatozoa. Therefore, further research should be focused on the safety and efficiency of these approaches. Genetic and epigenetic changes potentially occurring in the culture also need to be evaluated. Recently developed organoids and other latest approaches for fertility preservation and restoration are helping researchers to elucidate scientific questions about the spermatogenic process that previous in vitro systems could not help to clarify. Furthermore, more complete insight regarding the regulation of SSCs self-renewal and differentiation could facilitate and accelerate the translation of SSC culture and transplantation into the clinical application.

Patients and their parents should be informed about risks and potential benefits of experimental techniques for fertility preservation in prepubertal boys before cancer treatment, without giving them unrealistic expectations of future fertility. Nevertheless, successful fertility restoration with a cryopreserved ovarian tissue [67] offers hope that the option of fertility restoration will become available for prepubertal cancer survival boys too.

References

1. Steliarova-Foucher E, Colombet M, Ries L et al (2017) International incidence of childhood cancer, 2001–10: a population-based registry study. Lancet Oncol 18:719–731. https://doi.org/10.1016/s1470-2045(17)30186-9
2. Gatta G, Botta L, Rossi S et al (2014) Childhood cancer survival in Europe 1999–2007 results of EUROCARE-5—a population-based study Lancet Oncol 15:35–47. https://doi.org/10.1016/s1470-2045(13)70548-5
3. Smith M, Altekruse S, Adamson P et al (2014) Declining childhood and adolescent cancer mortality. Cancer 120:2497–2506. https://doi.org/10.1002/cncr.28748
4. Anderson C, Smitherman A, Nichols H (2018) Conditional relative survival among long-term survivors of adolescent and young adult cancers. Cancer 124:3037–3043. https://doi.org/10.1002/cncr.31529
5. Howlader N, Krapcho M, Miller D et al (2019) SEER cancer statistics review, 1975–2016. In: SEER. https://seer.cancer.gov/csr/1975_2016/. Accessed 1 Sep 2020

6. Rivkees S, Crawford J (1988) The relationship of gonadal activity and chemotherapy-induced gonadal damage. JAMA J Am Med Assoc 259:2123–2125. https://doi.org/10.1001/jama.259.14.2123
7. Mackie E, Radford M, Shalet S (1996) Gonadal function following chemotherapy for childhood Hodgkin's disease. Med Pediatr Oncol 27:74–78. https://doi.org/10.1002/(sici)1096-911x(199608)27:2%3c74::aid-mpo2%3e3.0.co;2-q
8. Kenney L, Laufer M, Grant F et al (2001) High risk of infertility and long term gonadal damage in males treated with high dose cyclophosphamide for sarcoma during childhood. Cancer 91:613–621. https://doi.org/10.1002/1097-0142(20010201)91:3%3c613::aid-cncr1042%3e3.0.co;2-r
9. Kliesch S, Behre H, Jürgens H, Nieschlag E (1996) Cryopreservation of semen from adolescent patients with malignancies. Med Pediatr Oncol 26:20–27. https://doi.org/10.1002/(sici)1096-911x(199601)26:1%3c20::aid-mpo3%3e3.0.co;2-x
10. Oehninger S (2005) Strategies for fertility preservation in female and male cancer survivors. J Soc Gynecol Investig 12:222–231. https://doi.org/10.1016/j.jsgi.2005.01.026
11. Bahadur G (2002) Semen quality and cryopreservation in adolescent cancer patients. Hum Reprod 17:3157–3161. https://doi.org/10.1093/humrep/17.12.3157
12. Schmiegelow M, Sommer P, Carlsen E et al (1998) Penile vibratory stimulation and electroejaculation before anticancer therapy in two pubertal boys. J Pediatr Hematol Oncol 20:429–430. https://doi.org/10.1097/00043426-199809000-00004
13. Opsahl M, Fugger E, Sherins R, Schulman J (1997) Preservation of reproductive function before therapy for cancer: new options involving sperm and ovary cryopreservation. Cancer J Sci Am 3:189–191
14. Berookhim B, Mulhall J (2014) Outcomes of operative sperm retrieval strategies for fertility preservation among males scheduled to undergo cancer treatment. Fertil Steril 101:805–811. https://doi.org/10.1016/j.fertnstert.2013.11.122
15. Kvist K, Thorup J, Byskov A et al (2006) Cryopreservation of intact testicular tissue from boys with cryptorchidism. Hum Reprod 21:484–491. https://doi.org/10.1093/humrep/dei331
16. Keros V, Hultenby K, Borgström B et al (2007) Methods of cryopreservation of testicular tissue with viable spermatogonia in pre-pubertal boys undergoing gonadotoxic cancer treatment. Hum Reprod 22:1384–1395. https://doi.org/10.1093/humrep/del508
17. Wyns C, Curaba M, Martinez-Madrid B et al (2007) Spermatogonial survival after cryopreservation and short-term orthotopic immature human cryptorchid testicular tissue grafting to immunodeficient mice. Hum Reprod 22:1603–1611. https://doi.org/10.1093/humrep/dem062
18. Poels J, Abou-Ghannam G, Herman S et al (2014) In search of better spermatogonial preservation by supplementation of cryopreserved human immature testicular tissue xenografts with N-acetylcysteine and testosterone. Front Surg 1:47. https://doi.org/10.3389/fsurg.2014.00047
19. Curaba M, Poels J, van Langendonckt A et al (2011) Can prepubertal human testicular tissue be cryopreserved by vitrification? Fertil Steril 95:2123.e9-2123.e12. https://doi.org/10.1016/j.fertnstert.2011.01.014
20. Poels J, Van Langendonckt A, Many M et al (2013) Vitrification preserves proliferation capacity in human spermatogonia. Hum Reprod 28:578–589. https://doi.org/10.1093/humrep/des455
21. Bahadur G, Chatterjee R, Ralph D (2000) Testicular tissue cryopreservation in boys. Ethical and legal issues: case report. Hum Reprod 15:1416–1420. https://doi.org/10.1093/humrep/15.6.1416
22. Pietzak E, Tasian G, Tasian S et al (2015) Histology of testicular biopsies obtained for experimental fertility preservation protocol in boys with cancer. J Urol 194:1420–1424. https://doi.org/10.1016/j.juro.2015.04.117
23. Meachem S (2001) Spermatogonia: stem cells with a great perspective. Reproduction 121:825–834. https://doi.org/10.1530/reprod/121.6.825
24. Clermont Y (1966) Spermatogenesis in man. A study of the spermatogonial population. Fertil Steril 17:705–721. https://doi.org/10.1016/s0015-0282(16)36120-9
25. Hutson J, Li R, Southwell B et al (2013) Germ cell development in the postnatal testis: the key to prevent malignancy in cryptorchidism? Front Endocrinol. https://doi.org/10.3389/fendo.2012.00176

26. Stukenborg J, Jahnukainen K, Hutka M, Mitchell R (2018) Cancer treatment in childhood and testicular function: the importance of the somatic environment. Endocr Connect 7:R69–R87. https://doi.org/10.1530/ec-17-0382
27. Gassei K, Orwig K (2016) Experimental methods to preserve male fertility and treat male factor infertility. Fertil Steril 105:256–266. https://doi.org/10.1016/j.fertnstert.2015.12.020
28. Medrano J, Andrés M, García S et al (2018) Basic and clinical approaches for fertility preservation and restoration in cancer patients. Trends Biotechnol 36:199–215. https://doi.org/10.1016/j.tibtech.2017.10.010
29. Brinster R, Zimmermann J (1994) Spermatogenesis following male germ-cell transplantation. Proc Natl Acad Sci 91:11298–11302. https://doi.org/10.1073/pnas.91.24.11298
30. Radford J, Shalet S, Lieberman B (1999) Fertility after treatment for cancer. BMJ 319:935–936. https://doi.org/10.1136/bmj.319.7215.935
31. Radford J (2003) Restoration of fertility after treatment for cancer. Horm Res Paediatr 59:21–23. https://doi.org/10.1159/000067840
32. Schlatt S, Rosiepen G, Weinbauer G et al (1999) Germ cell transfer into rat, bovine monkey and human testes. Hum Reprod 14:144–150. https://doi.org/10.1093/humrep/14.1.144
33. Wu X, Schmidt J, Avarbock M et al (2009) Prepubertal human spermatogonia and mouse gonocytes share conserved gene expression of germline stem cell regulatory molecules. Proc Natl Acad Sci 106:21672–21677. https://doi.org/10.1073/pnas.0912432106
34. Sadri-Ardekani H, Mizrak S, van Daalen S et al (2009) Propagation of human spermatogonial stem cells in vitro. JAMA 302:2127. https://doi.org/10.1001/jama.2009.1689
35. Tagelenbosch R, de Rooij D (1993) A quantitative study of spermatogonial multiplication and stem cell renewal in the C3H/101 F1 hybrid mouse. Mutat Res/Fundam Mol Mech Mutagen 290:193–200. https://doi.org/10.1016/0027-5107(93)90159-d
36. Maki C, Pacchiarotti J, Ramos T et al (2009) Phenotypic and molecular characterization of spermatogonial stem cells in adult primate testes. Hum Reprod 24:1480–1491. https://doi.org/10.1093/humrep/dep033
37. Kossack N, Meneses J, Shefi S et al (2009) Isolation and characterization of pluripotent human spermatogonial stem cell-derived cells. Stem Cells 27:138–149. https://doi.org/10.1634/stemcells.2008-0439
38. Valli H, Sukhwani M, Dovey S et al (2014) Fluorescence- and magnetic-activated cell sorting strategies to isolate and enrich human spermatogonial stem cells. Fertil Steril 102:566-580.e7. https://doi.org/10.1016/j.fertnstert.2014.04.036
39. Liu S, Tang Z, Xiong T, Tang W (2011) Isolation and characterization of human spermatogonial stem cells. Reprod Biol Endocrinol 9:141. https://doi.org/10.1186/1477-7827-9-141
40. Jahnukainen K, Hou M, Petersen C et al (2001) Intratesticular transplantation of testicular cells from leukemic rats causes transmission of leukemia. Cancer Res 61:706–710
41. Fujita K, Tsujimura A, Miyagawa Y et al (2006) Isolation of germ cells from leukemia and lymphoma cells in a human in vitro model: potential clinical application for restoring human fertility after anticancer therapy. Cancer Res 66:11166–11171. https://doi.org/10.1158/0008-5472.can-06-2326
42. Geens M, Van de Velde H, De Block G et al (2007) The efficiency of magnetic-activated cell sorting and fluorescence-activated cell sorting in the decontamination of testicular cell suspensions in cancer patients. Hum Reprod 22:733–742. https://doi.org/10.1093/humrep/del418
43. Dovey S, Valli H, Hermann B et al (2013) Eliminating malignant contamination from therapeutic human spermatogonial stem cells. J Clin Invest 123:1833–1843. https://doi.org/10.1172/jci65822
44. Sadri-Ardekani H, Homburg C, van Capel T et al (2014) Eliminating acute lymphoblastic leukemia cells from human testicular cell cultures: a pilot study. Fertil Steril 101:1072-1078.e1. https://doi.org/10.1016/j.fertnstert.2014.01.014
45. Sadri-Ardekani H, Akhondi M, van der Veen F et al (2011) In Vitro propagation of human prepubertal spermatogonial stem cells. JAMA 305:2416–2418. https://doi.org/10.1001/jama.2011.791

46. Cremades N, Bernabeu R, Barros A, Sousa M (1999) In-vitro maturation of round spermatids using co-culture on Vero cells. Hum Reprod 14:1287–1293. https://doi.org/10.1093/humrep/14.5.1287
47. Sousa M, Cremades N, Alves C et al (2002) Developmental potential of human spermatogenic cells co-cultured with Sertoli cells. Hum Reprod 17:161–172. https://doi.org/10.1093/humrep/17.1.161
48. Tanaka A, Nagayoshi M, Awata S et al (2003) Completion of meiosis in human primary spermatocytes through in vitro coculture with Vero cells. Fertil Steril 79:795–801. https://doi.org/10.1016/s0015-0282(02)04833-1
49. Stukenborg J, Wistuba J, Luetjens C et al (2008) Coculture of spermatogonia with somatic cells in a novel three-dimensional soft-agar-culture-system. J Androl 29:312–329. https://doi.org/10.2164/jandrol.107.002857
50. Abu Elhija M, Lunenfeld E, Schlatt S, Huleihel M (2012) Differentiation of murine male germ cells to spermatozoa in a soft agar culture system. Asian J Androl 14:285–293. https://doi.org/10.1038/aja.2011.112
51. Huleihel M, Nourashrafeddin S, Plant T (2015) Application of three-dimensional culture systems to study mammalian spermatogenesis, with an emphasis on the rhesus monkey (Macaca mulatta). Asian J Androl 17:972–980. https://doi.org/10.4103/1008-682x.154994
52. Lee D, Kim K, Yang Y et al (2006) Isolation of male germ stem cell-like cells from testicular tissue of non-obstructive azoospermic patients and differentiation into haploid male germ cells in vitro. Hum Reprod 21:471–476. https://doi.org/10.1093/humrep/dei319
53. Lee J, Gye M, Choi K et al (2007) In vitro differentiation of germ cells from nonobstructive azoospermic patients using three-dimensional culture in a collagen gel matrix. Fertil Steril 87:824–833. https://doi.org/10.1016/j.fertnstert.2006.09.015
54. Stukenborg J, Schlatt S, Simoni M et al (2009) New horizons for in vitro spermatogenesis? An update on novel three-dimensional culture systems as tools for meiotic and post-meiotic differentiation of testicular germ cells. Mol Hum Reprod 15:521–529. https://doi.org/10.1093/molehr/gap052
55. Skakkebæk N, Jensen G, Povlsen C, Rygaard J (1974) Heterotransplantation of human foetal testicular and ovarian tissue to the mouse mutant nude: a preliminary study. Acta Obstet Gynecol Scand 53:73–75. https://doi.org/10.3109/00016347409157196
56. Geens M, De Block G, Goossens E et al (2006) Spermatogonial survival after grafting human testicular tissue to immunodeficient mice. Hum Reprod 21:390–396. https://doi.org/10.1093/humrep/dei412
57. Arregui L, Rathi R, Zeng W et al (2008) Xenografting of adult mammalian testis tissue. Anim Reprod Sci 106:65–76. https://doi.org/10.1016/j.anireprosci.2007.03.026
58. Jahnukainen K, Ehmcke J, Söder O, Schlatt S (2006) Clinical potential and putative risks of fertility preservation in children utilizing gonadal tissue or germline stem cells. Pediatr Res 59:40R-47R. https://doi.org/10.1203/01.pdr.0000205153.18494.3b
59. Roulet V, Denis H, Staub C et al (2006) Human testis in organotypic culture: application for basic or clinical research. Hum Reprod 21:1564–1575. https://doi.org/10.1093/humrep/del018
60. Steinberger E (1967) Maintenance of adult human testicular tissue in culture. Anat Rec 157:327–328
61. de Michele F, Poels J, Weerens L et al (2017) Preserved seminiferous tubule integrity with spermatogonial survival and induction of Sertoli and Leydig cell maturation after long-term organotypic culture of prepubertal human testicular tissue. Hum Reprod 32:32–45. https://doi.org/10.1093/humrep/dew300
62. de Michele F, Poels J, Vermeulen M et al (2018) Haploid germ cells generated in organotypic culture of testicular tissue from prepubertal boys. Front Physiol. https://doi.org/10.3389/fphys.2018.01413
63. Alves-Lopes J, Stukenborg J (2018) Testicular organoids: a new model to study the testicular microenvironment in vitro? Hum Reprod Updat 24:176–191. https://doi.org/10.1093/humupd/dmx036

64. Pendergraft S, Sadri-Ardekani H, Atala A, Bishop C (2017) Three-dimensional testicular organoid: a novel tool for the study of human spermatogenesis and gonadotoxicity in vitro. Biol Reprod 96:720–732. https://doi.org/10.1095/biolreprod.116.143446
65. Baert Y, De Kock J, Alves-Lopes J et al (2017) Primary human testicular cells self-organize into organoids with testicular properties. Stem Cell Rep 8:30–38. https://doi.org/10.1016/j.stemcr.2016.11.012
66. Sakib S, Uchida A, Valenzuela-Leon P et al (2019) Formation of organotypic testicular organoids in microwell culture. Biol Reprod 100:1648–1660. https://doi.org/10.1093/biolre/ioz053
67. Donnez J, Dolmans M, Demylle D et al (2004) Livebirth after orthotopic transplantation of cryopreserved ovarian tissue. Lancet 364:1405–1410. https://doi.org/10.1016/s0140-6736(04)17222-x

Chapter 8
Endometrial Stem Cells and Endometriosis

Stefano Canosa, Andrea Roberto Carosso, Marta Sestero, Alberto Revelli, and Benedetta Bussolati

Abstract

Introduction: Endometrial stem/progenitor cells with functional properties of mesenchymal stromal cells have been identified in a perivascular location of human endometrium. There is some evidence that endometrial mesenchymal stromal cells (E-MSCs) may be involved not only in the physiological regeneration of the endometrium after menstruation but also in endometrial proliferative disorders, including endometriosis.
Methods: An association between endometrial stem cells and endometriosis has been established in the literature.
Results: The pathogenesis of early-onset endometriosis has recently been revisited as endometrial stem/progenitor cells that have been identified in menstrual blood, thus coupling the classical theory of retrograde flux of menstrual blood during the reproductive lifespan to neonatal uterine bleeding. In fact, shed endometrial stem/progenitor cells are postulated to migrate in the pelvic cavity where they can survive, proliferate, invade and establish ectopic endometrial lesions characteristic of endometriosis.
Conclusions: Due to the lack of steroid hormone receptors, E-MSCs may be responsible for the high rate of persistence/recurrence of the disease. For this reason, new therapeutic strategies have been investigated to eradicate endometriotic implants, including targeting estrogen receptors, angiogenesis and hypoxia.

Keywords Endometrium · Stem cells · Endometriosis · Pathogenesis

---

S. Canosa · A. R. Carosso · M. Sestero · A. Revelli
Obstetrics and Gynecology 1U, Physiopathology of Reproduction and IVF Unit, Department of Surgical Sciences, S. Anna Hospital, University of Torino, Torino, Italy

B. Bussolati (✉)
Department of Molecular Biotechnology and Health Sciences, University of Torino, Torino, Italy
e-mail: benedetta.bussolati@unito.it

I. Virant-Klun (ed.), *Stem Cells in Reproductive Tissues and Organs*, Stem Cell Biology and Regenerative Medicine 70, https://doi.org/10.1007/978-3-030-90111-0_8

Abbreviations

ABCG2	ATP binding cassette subfamily G member 2 (Junior Blood Group)
BMDSC	Bone marrow-derived stem cells
bmMSC	Bone marrow mesenchymal stem cell
CD	Cluster of differentiation
c-KIT	KIT proto-oncogene, receptor tyrosine kinase (CD117)
DNA	Deoxyribonucleic acid
EGF	Epidermal growth factor
eMSC	Endometrial mesenchymal stem cell
E-MSC	Endometrial mesenchymal stromal cells
EP	Epithelial progenitor
epCAM	Epithelial cell adhesion molecule
ER-alpha	Estrogen receptor alpha
ER-beta	Estrogen receptor beta
GCS-F	Colony-stimulating factor
GnRH	Gonadotropin-releasing hormone
GnRH-a	Gonadotropin-releasing hormone agonist
hES	Human embryonic stem
hESC	Human embryonic stem cell
HGF	Hepatocyte growth factor
HLA-DR	Human leukocyte antigen-DR isotype
HSC	Hematopoietic stem cell
Huvec	Human umbilical vein endothelial cells
MSC	Mesenchymal stem cell
PDGF	Platelet-derived growth *factor*
PDGFRB	Platelet-derived growth factor receptor beta
RAF1	Raf-1 proto-oncogene, serine/threonine kinase
RNA	Ribonucleic acid
mRNA	Messenger ribonucleic acid
SOX9	SRY-box transcription factor 9
SP	Side population
SSEA-1	Stage-specific embryonic antigen-1
SUSD2	Sushi domain containing 2 (W5C5)
UPA	Ulipristal acetate
VEGF	Vascular endothelial growth factor
VEGF-R2	Vascular endothelial growth factor receptor 2
VEGF-R3	Vascular endothelial growth factor receptor 3

Introduction

Human endometrial stem/progenitor cells have been isolated for the first time in 2004. A population of rare clonogenic cells was identified after the purification of human endometrial epithelial and stromal cells from hysterectomy tissue [1]. In the following years, several different populations of stem/progenitor cells have been characterized.

Origin, Characteristics and Isolation of Endometrial Stem Cells

It has been postulated that different subtypes of endometrial stem cells, including endometrial epithelial, stromal and endothelial cells, may contribute to the cyclic regeneration of the uterine wall [2]. In addition, stem cells found in the uterus may also derive from nonuterine sources, such as bone marrow-derived stem cells (BMDSC) and human embryonic stem cells (hESCs). However, these extra-uterine sources of stem cells may have a limited contribution in the natural regenerative processes, particularly involved in providing a suitable endometrial microenvironment via the release of chemotactic factors of other cells [3]. The endometrial tissue comprises two distinctive layers, the functionalis, with luminal and glandular epithelial cells and a basal layer, composed of the vascularized stroma. For these reasons, researchers have focused their attention on the identification of epithelial and mesenchymal stem/progenitor cells [4].

Endometrial Epithelial Progenitors

The human endometrium undergoes a physiological regeneration cycle consisting of proliferation, differentiation and shedding every menses. During this process, high estrogen levels stimulate the epithelial cells to proliferate and regenerate the upper (functionalis) layer of the endometrium whereas the lower (basal) components of the glands remain roughly unchanged. It was postulated that the epithelial progenitor (EP) cell population may reside in the basal portion of the endometrial glands [5–7]. This epithelial population expresses nuclear SOX9 and the cell-surface marker stage-specific embryonic antigen-1 (SSEA-1 or CD15) with a fraction of cells expressing nuclear β-catenin [8]. SSEA-1 is an epitope of a glycoprotein expressed in differentiating human embryonic stem cells (hES) and neutrophils [9]. Cultured SSEA-1+ cells show an increased telomerase activity and longer telomeres, which generate larger spheroids than SSEA-1 cells but express a lower concentration of ER-alfa or progesterone receptors. More recently, N-cadherin surface marker was used to isolate clonogenic, self-renewing human endometrial epithelial progenitor cells with high proliferative potential that can differentiate into cytokeratin+ gland-like structures in vitro. By immunofluorescence and confocal microscopy, it has been shown that these cells localize mainly in the basalis layer adjacent to the myometrium in both premenopausal and postmenopausal endometrium. In addition, N-cadherin+ epithelial cells co-expressed cytokeratin, ER-alfa and E-cadherin. Differently, the proliferation marker Ki-67 was expressed by a limited fraction of N-cadherin+ cells, suggesting a predominant quiescent behavior. Interestingly, N-cadherin+ and SSEA-1+ cells seem to colocalize. SSEA-1 cells are generally more distal to the myometrium and closer to the functionalis than N-cadherin+ cells, and indeed, they have been observed in the deep functionalis as predicted from their lack of ER-alfa. As previously mentioned, endometrial epithelial cells from the basalis express nuclear SOX9 and SSEA-1 [8]. However, SOX9 was rarely found in the N-cadherin+ cells. Based on the different localization of N-cadherin+ and SSEA-1+ cells, a differentiation hierarchy of epithelial cells has been proposed in human endometrial glands within the pseudostratified columnar epithelium. N-cadherin+ epithelial cells appeared to be the most primitive, followed by SSEA-1+ cells and then mature cells that express neither marker [10].

Endometrial Mesenchymal Stromal Cells

Mesenchymal stem cells (MSCs) were isolated from bone marrow biopsies showing fibroblast-like phenotype but have been identified in many organs over the last decade, including the endometrium [6, 11, 12]. The defining characteristics of bone marrow MSCs (bmMSCs) are plastic adherence, clonogenicity, multilineage differentiation in vitro into different types of cells such as osteocytes, chondrocytes and adipocytes and a specific surface markers profile (CD105+, CD146+, CD312, CD29+, CD44+, CD73+, CD90+, CD342 and CD452) compared to hematopoietic stem cells (HSCs) [13]. Endometrial cells from both human [14] and porcine [15] species, as well as the human CD146+ PDGFR-b+ (platelet-derived growth factor receptor beta) and SUSD2+ [sushi domain containing-2 (also W5C5+)] endometrial stromal subpopulation [16, 17] exhibit the same in vitro properties as bmMSCs. Due to the comparable characteristics of endometrial stromal cells to bone marrow stem cells, many authors have focused their attention on the biological role of this endometrial population [18–21]. Endometrial MSCs (eMSCs) are multipotent and self-renewing stromal cells residing in both the basalis and functionalis layers of the endometrium [16]. eMSCs can be detected as CD146+ and PDGFRb+ cells [12] with pericyte-like cells localization in the perivascular space of the blood vessels of endometrial glands in both the basalis and functionalis layer. In addition, SUSD2 was identified as a single marker able to purify eMSCs with functional properties of endometrial stromal tissues regeneration in vivo [16]. As a matter of fact, the number of SUSD2+ cells increases during the proliferative phase of endometrial regeneration, suggesting a direct role in the cyclic growth of functionalis stroma [16]. Perivascular SUSD2+ cells were also isolated from post-menopausal endometrial biopsies showing MSC properties of clonogenicity, multipotency and immunophenotype irrespective of pre-treatment with estrogen to stimulate endometrial regeneration [17], and they were able to regenerate endometrial stroma when xenografted in the mouse model [16]. Interestingly enough, eMSCs can remain dormant for prolonged periods of time and do not require estrogen for survival. eMSCs can undergo multilineage differentiation (osteocytes, adipocytes, smooth muscle cells and chondrocytes) in vitro and are characterized by the following immunophenotypes: CD29+, CD44+, CD73+, CD90+, CD105+, PDGFRB+, CD31−, CD34− and CD45− [22]. Differently from endometrial stromal fibroblasts, eMSCs show high proliferative potential, they may generate large cell clones, do not express ER-alfa and are characterized by a specific gene expression profile [23].

Side Population Cells

The side population (SP) phenotype was proposed as a universal marker of adult stem cells in mammalian species. SP cells have been isolated from a wide variety of mammalian tissues, and in many cases, this cell population has been shown to contain multipotent stem cells [24]. This phenotype is characterized by a high expression of plasma membrane transporters of organic molecules out of cells (e.g., ABCG2), including the DNA-binding dye Hoechst 33,342 [25]. Therefore, the SP is detected as a low-fluorescent phenotype by flow cytometry [26] and has been identified in fresh and cultured human endometrial cell suspensions [27, 28]. The SP population is a heterogeneous mix of epithelial, stromal and endothelial cells, with a predominant presence of endothelial cells. In particular, the SP contained 27% EpCAM+ epithelial cells, 14% CD10+ stromal cells, 51% of endothelial cells and 25% CD146+ endothelial/eMSCs [29]. The relevant contribution of endothelial and eMSCs (CD146+ PDGFRb+) suggests that the endometrial SP may have a role in the angiogenetic process during endometrial regeneration [30]. SP cells are clonogenic and show adult stem cell function in vitro, and in vivo [27, 28], showing typical MSC properties [14] and differentiation capacity into decidual cells [31]. SP cells can also regenerate endometrial tissue in vivo after xenograft in immunocompromised mice, mainly supporting the endothelial vascularization and, to a lesser extent, the epithelial population and surrounding stroma [28, 32]. For this reason, SP cells can act as genuine endometrial stem/progenitor cells, surrounded by endometrial cells providing an appropriate microenvironment. Actually, many authors have suggested that different stem/progenitor cells types may be present in the human endometrial SP population: the vascular wall of capillaries and small vessels in both the functionalis and basalis layers of the human endometrium were demonstrated to be positive with ABCG2 antibodies as a representative marker of SP cells [28] In addition, the endothelial phenotype of the endometrial SP was confirmed by ABCG2 colocalized with CD31. This is also demonstrated by the expression of ER-beta in the endometrial SP as ER-beta was demonstrated to be expressed in endometrial endothelial cells [32].

Role of E-MSCs in the Pathogenesis of Endometriosis

In the last years many theories have been proposed to explain the origin of endometriosis. Among them the endometrial ectopic implantation of endometrial cells derived from retrograde menstruation, coelomic metaplasia, embryonic cell rest, induction followed by lymphatic and vascular dissemination, endocrinological factors, immunological and genetic factors, and abnormalities in the eutopic endometrium [33]. However, there are still some gaps in knowledge around its onset and little is known about the pathological processes involved in the recurrence after treatment [34]. The most accepted theory on endometriosis origin is Sampson's theory, postulating that endometriotic implants arise from retrograde menstruation: a small population of glandular epithelial and mesenchymal cells detached from the endometrium may reach the peritoneal cavity through the fallopian tubes and ectopically implant into the ovary and the adjacent pelvis, where they proliferate, invade and spread to form ectopic lesions [35]. This theory has been confirmed by the finding that endometrial fragments are found in the pouch of Douglas [36], but only a relatively small proportion of women develop endometriosis, besides retrograde menstruation is a very common event, suggesting that other factors may be involved [37, 38]. However, the exact pathogenetic mechanism leading the eutopic endometrium into the abnormal activity is still under investigation [39, 40]. Differently, the "embryonic rest" theory suggests that endometriosis could arise from the presence of Mullerian cells in the peritoneal cavity that may generate endometrial tissue when exposed to specific stimuli [41]. Another hypothesis is represented by the peritoneal metaplasia and the differentiation of mesothelial cells into endometrial cells [42, 43]. However, in the last decade cell populations with markers and functional properties of stem cells have been identified in the endometrium [1, 6, 27, 44], physiologically involved in the cyclic, monthly endometrial regeneration occurring after menstruation. Recently, a new putative pathogenetic mechanism for endometriosis has been proposed based on the direct involvement of the stem cell population of the human endometrium [4]. Through retrograde menstruation, they may migrate into the abdominal cavity, where they proliferate, invade the peritoneum and generate endometriotic implants [4, 45–48]. Endometrial mesenchymal stromal cells (E-MSCs) may play a critical role in all steps of the pathogenesis of the disease: the initial onset, its persistence after hormonal treatment and its recurrence. Previous data have suggested the presence of a small population of adult/progenitor cells with a mesenchymal phenotype deriving from the basal endometrium in the menstrual blood of a healthy woman [49]. Protein analysis of the culture medium of these MSCs has shown the presence of granulocyte–macrophage colony-stimulating factor (GCS-F), platelet-derived growth factor (PDGF) and proangiogenic factors such as angiopoietin 2 and vascular endothelial growth factor (VEGF), stimulating the angiogenetic process supporting the growth of ectopic lesions. Moreover, levels of VEGF, epidermal growth factor (EGF), hepatocyte growth factor (HGF) and angiogenin are significantly upregulated in the peritoneal fluid of patients with endometriosis compared to healthy women, while inhibitors of angiogenesis are downregulated, inducing a chronic inflammatory microenvironment [50–53]. The E-MSCs population may also have a pivotal role in promoting angiogenesis and immunomodulation, which can be involved in the support of ectopic implantation during retrograde menstruation [54].

In addition, E-MSCs isolated from both ovarian and peritoneal endometriotic lesions showed a significantly increased in vitro migration, proliferation and angiogenic properties compared to an endometrial cell from the heathy endometrium [55, 56]. The enhanced invasion and angiogenic abilities were also demonstrated in vivo after xenograft into immunodeficient mice [57]. As a consequence, E-MSCs activity may also be involved in the persistence and/or recurrence of endometriosis after surgical or pharmacological treatment. In fact, currently available treatments do not eradicate all endometriotic implants due to the relatively low sensitivity of ectopic implants to estrogen deprivation as lacking estrogen receptors, which are acquired at later stages, during differentiation, suggesting that their survival and growth is estrogen independent [55]. As a consequence, current available antiestrogenic treatments are effective against differentiated cells only, thus causing the clinical recurrence of the disease.

Novel Strategies for Endometriosis Treatment

The elective treatment of endometriosis is the surgical ablation (burning) and excision (cutting out) of an endometriotic implant under laparoscopic guidance in order to restore the anatomy. However, it is not clear whether laparoscopic surgery may reduce overall pain associated with endometriosis. To date, there is still moderate quality evidence that laparoscopic surgery increases pregnancy chance and no data are reported on live birth [3]. In addition, the current medical treatment for endometriosis is based on the induction of a hypoestrogenic environment producing a chronic anovulation state [58]. This is obtained administrating GnRH agonists (GnRH-a) blocking ovarian estrogen secretion, progestins inducing pseudo-pregnancy or progestins and androgenic progestins inhibiting estrogenic stimulation of the ectopic endometrium [59, 60]. Although medical treatments are effective in relieving pains, they frequently fail to completely eradicate the disease [61–63]. Since estradiol is known to promote proliferation of the endometrium, several studies have tested the selective modulation of estrogen receptor to inhibit the growth of the endometriotic lesion. Raloxifene and bazedoxifene both showed significant regression and inhibition of endometriotic lesions in murine models [64–66]. Also, selective progesterone receptor modulation has been tested. Among progesterone receptor antagonists ulipristal acetate (UPA), asoprisnil and mifepristone have been under investigation. UPA was shown to decrease lesion size, cell proliferation and induce apoptosis in murine endometriosis models [67, 68]. Mifepristone was able to reduce lesion size in both animal models and patients [69, 70].

Table 8.1 Antiangiogenic treatment in endometriosis (modified from Pittatore et al. [71])

Substance type	Compound	Effects
VEGF inhibitor	Bevacizumab Bevacizumab SU5416 and SU6668 Bevacizumab, Sorafenib Sorafenib Anti-VEGF Ab Angiostatin	Reduction in cell proliferation increased apoptosis, decreased vascular density, and reduction in VEGF levels in the peritoneal fluid on nude mice
Endogenous angiogénesis inhibitors Dopamine agonist	Cabergoline	Reduction in the number and size of endometriosis implants on nude mice inhibition of VEGF-R2 phosphorylation, increase in necrotic areas in endometriosis implants and promotion of regression

According to the most recent literature, there is evidence that although endometriosis is a benign disease, it is characterized by some of the adaptive properties of cancer cells, such as deregulated proliferation, invasiveness, and neoangiogenesis [71]. Indeed, the establishment and persistence of endometriotic lesions strictly depend on the formation of new blood vessels, which provide oxygen and nutrient supply to the ectopic tissue [57, 72, 73]. Accordingly, antiangiogenic treatment could represent a possible therapeutic strategy against the disease [74–76] (Table 8.1).

It has been previously demonstrated that the inhibition of the neo-angiogenetic process of the growing implants may lead to a hypoxic state and a consequent growth arrest and necrosis [57, 72, 74, 76]. VEGF was demonstrated to be the main promoter of angiogenesis in endometriosis [77]. Among the different antiangiogenic agents, the tyrosine kinase inhibitor Sorafenib (BAY 43-9006) was shown to target Raf kinases, CD117 (c-Kit), VEGF receptor 2 (VEGF-R2) and VEGF-R3, CD140 (platelet growth factor beta receptor) as well as of the Raf-1 downstream protein Ezrin. This protein was previously shown to be involved in the migration of endometrial stromal cells in patients with endometriosis [78]. Moggio et al. [55] showed that the expression of the activated phosphorylated form of ezrin was higher in ectopic MSCs in respect to eutopic and control lines, indicating possible VEGF-independent targets of Sorafenib in MSCs. Sorafenib treatment reduced the increased proliferation and VEGF release found in ectopic MSC but also inhibited the increased phosphorylation of ezrin in ectopic MSCs, suggesting a limiting action to the increased migration of ectopic MSCs. In addition, recent evidence has demonstrated that E-MSCs may undergo endothelial differentiation, and be potentially involved in the development of endometriotic implants. Canosa et al. [56] observed that a subpopulation of E-MSCs acquired the expression of the endothelial marker CD31 and integrated into tube-like structures when directly in contact with HUVECs. The isolation of CD31+ E-MSCs revealed significant increases in CD31, vascular endothelial growth factor receptor 2, TEK receptor tyrosine kinase and vascular endothelial-cadherin mRNA expression levels with respect to basal and to CD31neg cells. The treatment with Cabergoline, a dopamine receptor-2 agonist, induced a significant reduction of the E-MSC angiogenic potential. The use of the antiangiogenic agents was also tested in vivo in experimental models of endometriosis. Dabrosin et al. [79] reported that angiostatin was able to induce a significant reduction in the number and size of endometriosis implants in mice. In another study, SU5416 and SU6668, two tyrosine-kinase receptor inhibitors targeting VEGF-R2, were administered to rodents with surgically induced endometriosis, obtaining a significant reduction in the microvessel density and the size of lesions [80]. Treatment with Bevacizumab, a recombinant human monoclonal anti-VEGF antibody, induced a significant reduction in cell proliferation, increased apoptosis and decreased vascular density [81]. More recently, Ozer et al. [82] showed that both Bevacizumab and Sorafenib were able to induce a significant decrease in endometriotic implants and microvessels density in rats with induced endometriosis, without compromising follicular integrity within the ovary. Finally, studies conducted on mice with peritoneal endometriosis showed that treatment with daily doses of Cabergoline was able to significantly increase necrotic areas in endometriosis implants and promote their regression [83, 84]. Among the different variables involved in the pathogenesis of endometriosis, hypoxia may play a critical role in modulating molecular pathways in endometriosis, such as cell adhesion, estrogen release and immune functions [85]. Furthermore, hypoxia was demonstrated to promote the epithelial to mesenchymal transition [86], the invasion of endometrial stromal cells via HIF-1α/β-catenin signaling pathway and the expression of target genes, such as VEGF and matrix metalloproteinase 9 [87] but also upregulating the autophagic process [88]. For this reason, hypoxia may represent a promising new therapeutic target for the treatment of endometriosis.

References

1. Chan RWS, Schwab KE, Gargett CE (2004) Clonogenicity of human endometrial epithelial and stromal cells. Biol Reprod 70:1738–1750. https://doi.org/10.1095/biolreprod.103.024109
2. Santamaria X, Mas A, Cervelló I, Taylor H, Simon C (2018) Uterine stem cells: from basic research to advanced cell therapies. Hum Reprod Update 24:673–693. https://doi.org/10.1093/humupd/dmy028
3. Benor A, Gay S, DeCherney A (2020) An update on stem cell therapy for Asherman syndrome. J Assist Reprod Genet 37:1511–1529. https://doi.org/10.1007/s10815-020-01801-x
4. Gargett CE, Nguyen HPT, Ye L (2012) Endometrial regeneration and endometrial stem/progenitor cells. Rev Endocr Metab Disord 13:235–251. https://doi.org/10.1007/s11154-012-9221-9
5. Gargett CE (2007) Uterine stem cells: what is the evidence? Hum Reprod Update 13:87–101. https://doi.org/10.1093/humupd/dml045
6. Gargett CE, Schwab KE, Zillwood RM, Nguyen HPT, Wu D (2009) Isolation and culture of epithelial progenitors and mesenchymal stem cells from human endometrium. Biol Reprod 80:1136–1145. https://doi.org/10.1095/biolreprod.108.075226
7. Schwab KE, Chan RWS, Gargett CE (2005) Putative stem cell activity of human endometrial epithelial and stromal cells during the menstrual cycle. Fertil Steril 84(Suppl 2):1124–1130. https://doi.org/10.1016/j.fertnstert.2005.02.056

8. Valentijn AJ, Palial K, Al-Lamee H, Tempest N, Drury J, Von Zglinicki T, Saretzki G, Murray P, Gargett CE, Hapangama DK (2013) SSEA-1 isolates human endometrial basal glandular epithelial cells: phenotypic and functional characterization and implications in the pathogenesis of endometriosis. Hum Reprod 28:2695–2708. https://doi.org/10.1093/humrep/det285
9. Wright AJ, Andrews PW (2009) Surface marker antigens in the characterization of human embryonic stem cells. Stem Cell Res 3:3–11. https://doi.org/10.1016/j.scr.2009.04.001
10. Nguyen HPT, Xiao L, Deane JA, Tan K-S, Cousins FL, Masuda H, Sprung CN, Rosamilia A, Gargett CE (2017) N-cadherin identifies human endometrial epithelial progenitor cells by in vitro stem cell assays. Hum Reprod 32:2254–2268. https://doi.org/10.1093/humrep/dex289
11. Crisan M, Yap S, Casteilla L, Chen C-W, Corselli M, Park TS, Andriolo G, Sun B, Zheng B, Zhang L, Norotte C, Teng P-N, Traas J, Schugar R, Deasy BM, Badylak S, Buhring H-J, Giacobino J-P, Lazzari L, Huard J, Péault B (2008) A perivascular origin for mesenchymal stem cells in multiple human organs. Cell Stem Cell 3:301–313. https://doi.org/10.1016/j.stem.2008.07.003
12. Schwab KE, Gargett CE (2007) Co-expression of two perivascular cell markers isolates mesenchymal stem-like cells from human endometrium. Hum Reprod 22:2903–2911. https://doi.org/10.1093/humrep/dem265
13. Dominici M, Le Blanc K, Mueller I, Slaper-Cortenbach I, Marini F, Krause D, Deans R, Keating A, Prockop D, Horwitz E (2006) Minimal criteria for defining multipotent mesenchymal stromal cells. The International Society for Cellular Therapy position statement. Cytotherapy 8:315–317. https://doi.org/10.1080/14653240600855905
14. Cervelló I, Gil-Sanchis C, Mas A, Delgado-Rosas F, Martínez-Conejero JA, Galán A, Martínez-Romero A, Martínez S, Navarro I, Ferro J, Horcajadas JA, Esteban FJ, O'Connor JE, Pellicer A, Simón C (2010) Human endometrial side population cells exhibit genotypic, phenotypic and functional features of somatic stem cells. PloS One 5:e10964. https://doi.org/10.1371/journal.pone.0010964
15. Miernik K, Karasinski J (2012) Porcine uterus contains a population of mesenchymal stem cells. Reproduction 143:203–209. https://doi.org/10.1530/REP-11-0202
16. Masuda H, Anwar SS, Bühring H-J, Rao JR, Gargett CE (2012) A novel marker of human endometrial mesenchymal stem-like cells. Cell Transplant 21:2201–2214. https://doi.org/10.3727/096368911X637362
17. Ulrich D, Tan KS, Deane J, Schwab K, Cheong A, Rosamilia A, Gargett CE (2014) Mesenchymal stem/stromal cells in post-menopausal endometrium. Hum Reprod 29:1895–1905. https://doi.org/10.1093/humrep/deu159
18. Hematti P (2012) Mesenchymal stromal cells and fibroblasts: a case of mistaken identity? Cytotherapy 14:516–521. https://doi.org/10.3109/14653249.2012.677822
19. Bianco P, Cao X, Frenette PS, Mao JJ, Robey PG, Simmons PJ, Wang C-Y (2013) The meaning, the sense and the significance: translating the science of mesenchymal stem cells into medicine. Nat Med 19:35–42. https://doi.org/10.1038/nm.3028
20. Phinney DG, Sensebé L (2013) Mesenchymal stromal cells: misconceptions and evolving concepts. Cytotherapy 15:140–145. https://doi.org/10.1016/j.jcyt.2012.11.005
21. Sacchetti B, Funari A, Michienzi S, Di Cesare S, Piersanti S, Saggio I, Tagliafico E, Ferrari S, Robey PG, Riminucci M, Bianco P (2007) Self-renewing osteoprogenitors in bone marrow sinusoids can organize a hematopoietic microenvironment. Cell 131:324–336. https://doi.org/10.1016/j.cell.2007.08.025
22. Darzi S, Werkmeister JA, Deane JA, Gargett CE (2016) Identification and characterization of human endometrial mesenchymal stem/stromal cells and their potential for cellular therapy. Stem Cells Transl Med 5:1127–1132. https://doi.org/10.5966/sctm.2015-0190
23. Barragan F, Irwin JC, Balayan S, Erikson DW, Chen JC, Houshdaran S, Piltonen TT, Spitzer TLB, George A, Rabban JT, Nezhat C, Giudice LC (2016) Human endometrial fibroblasts derived from mesenchymal progenitors inherit progesterone resistance and acquire an inflammatory phenotype in the endometrial niche in endometriosis. Biol Reprod 94:118. https://doi.org/10.1095/biolreprod.115.136010

24. Challen GA, Little MH (2006) A side order of stem cells: the SP phenotype. Stem Cells 24:3–12. https://doi.org/10.1634/stemcells.2005-0116
25. Zhou S, Schuetz JD, Bunting KD, Colapietro AM, Sampath J, Morris JJ, Lagutina I, Grosveld GC, Osawa M, Nakauchi H, Sorrentino BP (2001) The ABC transporter Bcrp1/ABCG2 is expressed in a wide variety of stem cells and is a molecular determinant of the side-population phenotype. Nat Med 7:1028–1034. https://doi.org/10.1038/nm0901-1028
26. Goodell MA, Rosenzweig M, Kim H, Marks DF, DeMaria M, Paradis G, Grupp SA, Sieff CA, Mulligan RC, Johnson RP (1997) Dye efflux studies suggest that hematopoietic stem cells expressing low or undetectable levels of CD34 antigen exist in multiple species. Nat Med 3:1337–1345. https://doi.org/10.1038/nm1297-1337
27. Kato K, Yoshimoto M, Kato K, Adachi S, Yamayoshi A, Arima T, Asanoma K, Kyo S, Nakahata T, Wake N (2007) Characterization of side-population cells in human normal endometrium. Hum Reprod 22:1214–1223. https://doi.org/10.1093/humrep/del514
28. Masuda H, Matsuzaki Y, Hiratsu E, Ono M, Nagashima T, Kajitani T, Arase T, Oda H, Uchida H, Asada H, Ito M, Yoshimura Y, Maruyama T, Okano H (2010) Stem cell-like properties of the endometrial side population: implication in endometrial regeneration. PloS One 5:e10387. https://doi.org/10.1371/journal.pone.0010387
29. Miyazaki K, Maruyama T, Masuda H, Yamasaki A, Uchida S, Oda H, Uchida H, Yoshimura Y (2012) Stem cell-like differentiation potentials of endometrial side population cells as revealed by a newly developed in vivo endometrial stem cell assay. PloS One 7:e50749. https://doi.org/10.1371/journal.pone.0050749
30. Gurung S, Deane JA, Masuda H, Maruyama T, Gargett CE (2015) Stem cells in endometrial physiology. Semin Reprod Med 33:326–332. https://doi.org/10.1055/s-0035-1558405
31. Tsuji S, Yoshimoto M, Takahashi K, Noda Y, Nakahata T, Heike T (2008) Side population cells contribute to the genesis of human endometrium. Fertil Steril 90:1528–1537. https://doi.org/10.1016/j.fertnstert.2007.08.005
32. Critchley HO, Brenner RM, Henderson TA, Williams K, Nayak NR, Slayden OD, Millar MR, Saunders PT (2001) Estrogen receptor beta, but not estrogen receptor alpha, is present in the vascular endothelium of the human and nonhuman primate endometrium. J Clin Endocrinol Metab 86:1370–1378. https://doi.org/10.1210/jcem.86.3.7317
33. Sasson IE, Taylor HS (2008) Stem cells and the pathogenesis of endometriosis. Ann N Y Acad Sci 1127:106–115. https://doi.org/10.1196/annals.1434.014
34. Seli E, Berkkanoglu M, Arici A (2003) Pathogenesis of endometriosis. Obstet Gynecol Clin North Am 30:41–61. https://doi.org/10.1016/s0889-8545(02)00052-9
35. Sampson JA (1927) Peritoneal endometriosis due to the menstrual dissemination of endometrial tissue into the peritoneal cavity. Am J Obstet Gynecol 14:422–469. https://doi.org/10.1016/S0002-9378(15)30003-X
36. Kruitwagen RFPM, Poels LG, Willemsen WNP, Jap PHK, Thomas CMG, Rolland R (1991) Retrograde seeding of endometrial epithelial cells by uterine-tubal flushing. Fertil Steril 56:414–420. https://doi.org/10.1016/S0015-0282(16)54533-6
37. Vigano P, Somigliana E, Vignali M, Busacca M, Blasio AMD (2007) Genetics of endometriosis: current status and prospects. Front Biosci 12:3247–3255. https://doi.org/10.2741/2308
38. Christodoulakos G, Augoulea A, Lambrinoudaki I, Sioulas V, Creatsas G (2007) Pathogenesis of endometriosis: the role of defective "immunosurveillance." Eur J Contracept Reprod Health Care Off J Eur Soc Contracept 12:194–202. https://doi.org/10.1080/13625180701387266
39. Li X, Gong X, Zhu L, Leng J, Fan Q, Sun D, Lang J, Fan Y (2012) Stretch magnitude- and frequency-dependent cyclooxygenase 2 and prostaglandin E2 up-regulation in human endometrial stromal cells: possible implications in endometriosis. Exp Biol Med 237:1350–1358. https://doi.org/10.1258/ebm.2012.012060
40. Sha G, Zhang Y, Zhang C, Wan Y, Zhao Z, Li C, Lang J (2009) Elevated levels of gremlin-1 in eutopic endometrium and peripheral serum in patients with endometriosis. Fertil Steril 91:350–358. https://doi.org/10.1016/j.fertnstert.2007.12.007
41. Signorile PG, Baldi A (2010) Endometriosis: new concepts in the pathogenesis. Int J Biochem Cell Biol 42:778–780. https://doi.org/10.1016/j.biocel.2010.03.008

42. Ferguson BR, Bennington JL, Haber SL (1969) Histochemistry of mucosubstances and histology of mixed müllerian pelvic lymph node glandular inclusions. Evidence for histogenesis by müllerian metaplasia of coelomic epithelium. Obstet Gynecol 33:617–625
43. Matsuura K, Ohtake H, Katabuchi H, Okamura H (1999) Coelomic metaplasia theory of endometriosis: evidence from in vivo studies and an in vitro experimental model. Gynecol Obstet Invest 47(Suppl 1):18–20; discussion 20–22. https://doi.org/10.1159/000052855
44. Dimitrov R, Timeva T, Kyurkchiev D, Stamenova M, Shterev A, Kostova P, Zlatkov V, Kehayov I, Kyurkchiev S (2008) Characterization of clonogenic stromal cells isolated from human endometrium. Reproduction 135:551–558. https://doi.org/10.1530/REP-07-0428
45. Fraser IS (2008) Recognising, understanding and managing endometriosis. J Hum Reprod Sci 1:56–64
46. Leyendecker G, Herbertz M, Kunz G, Mall G (2002) Endometriosis results from the dislocation of basal endometrium. Hum Reprod 17:2725–2736. https://doi.org/10.1093/humrep/17.10.2725
47. Gargett CE, Schwab KE, Deane JA (2016) Endometrial stem/progenitor cells: the first 10 years. Hum Reprod Update 22:137–163. https://doi.org/10.1093/humupd/dmv051
48. Liu Y, Niu R, Yang F, Yan Y, Liang S, Sun Y, Shen P, Lin J (2018) Biological characteristics of human menstrual blood-derived endometrial stem cells. J Cell Mol Med 22:1627–1639. https://doi.org/10.1111/jcmm.13437
49. Herington JL, Bruner-Tran KL, Lucas JA, Osteen KG (2011) Immune interactions in endometriosis. Expert Rev Clin Immunol 7:611–626. https://doi.org/10.1586/eci.11.53
50. Liu Y, Zhang Z, Yang F, Wang H, Liang S, Wang H, Yang J, Lin J (2020) The role of endometrial stem cells in the pathogenesis of endometriosis and their application to its early diagnosis†. Biol Reprod 102:1153–1159. https://doi.org/10.1093/biolre/ioaa011
51. Laschke MW, Menger MD (2018) Basic mechanisms of vascularization in endometriosis and their clinical implications. Hum Reprod Update 24:207–224. https://doi.org/10.1093/humupd/dmy001
52. Rakhila H, Al-Akoum M, Bergeron M-E, Leboeuf M, Lemyre M, Akoum A, Pouliot M (2016) Promotion of angiogenesis and proliferation cytokines patterns in peritoneal fluid from women with endometriosis. J Reprod Immunol 116:1–6. https://doi.org/10.1016/j.jri.2016.01.005
53. Izumi G, Koga K, Takamura M, Makabe T, Satake E, Takeuchi A, Taguchi A, Urata Y, Fujii T, Osuga Y (2018) Involvement of immune cells in the pathogenesis of endometriosis. J Obstet Gynaecol Res 44:191–198. https://doi.org/10.1111/jog.13559
54. Ulrich D, Muralitharan R, Gargett CE (2013) Toward the use of endometrial and menstrual blood mesenchymal stem cells for cell-based therapies. Expert Opin Biol Ther 13:1387–1400. https://doi.org/10.1517/14712598.2013.826187
55. Moggio A, Pittatore G, Cassoni P, Marchino GL, Revelli A, Bussolati B (2012) Sorafenib inhibits growth, migration, and angiogenic potential of ectopic endometrial mesenchymal stem cells derived from patients with endometriosis. Fertil Steril 98:1521-1530.e2. https://doi.org/10.1016/j.fertnstert.2012.08.003
56. Canosa S, Moggio A, Brossa A, Pittatore G, Marchino GL, Leoncini S, Benedetto C, Revelli A, Bussolati B (2017) Angiogenic properties of endometrial mesenchymal stromal cells in endothelial co-culture: an in vitro model of endometriosis. Mol Hum Reprod 23:187–198. https://doi.org/10.1093/molehr/gax006
57. Kao A-P, Wang K-H, Chang C-C, Lee J-N, Long C-Y, Chen H-S, Tsai C-F, Hsieh T-H, Tsai E-M (2011) Comparative study of human eutopic and ectopic endometrial mesenchymal stem cells and the development of an in vivo endometriotic invasion model. Fertil Steril 95:1308-1315.e1. https://doi.org/10.1016/j.fertnstert.2010.09.064
58. Olive DL, Pritts EA (2001) Treatment of endometriosis. N Engl J Med 345:266–275. https://doi.org/10.1056/NEJM200107263450407
59. Lessey BA (2000) Medical management of endometriosis and infertility. Fertil Steril 73:1089–1096. https://doi.org/10.1016/s0015-0282(00)00519-7
60. Valle RF, Sciarra JJ (2003) Endometriosis: treatment strategies. Ann N Y Acad Sci 997:229–239. https://doi.org/10.1196/annals.1290.026

61. Kennedy S, Bergqvist A, Chapron C, D'Hooghe T, Dunselman G, Greb R, Hummelshoj L, Prentice A, Saridogan E, ESHRE Special Interest Group for Endometriosis and Endometrium Guideline Development Group (2005) ESHRE guideline for the diagnosis and treatment of endometriosis. Hum Reprod 20:2698–2704. https://doi.org/10.1093/humrep/dei135
62. Waller KG, Shaw RW (1993) Gonadotropin-releasing hormone analogues for the treatment of endometriosis: long-term follow-up. Fertil Steril 59:511–515. https://doi.org/10.1016/s0015-0282(16)55791-4
63. Revelli A, Modotti M, Ansaldi C, Massobrio M (1995) Recurrent endometriosis: a review of biological and clinical aspects. Obstet Gynecol Surv 50:747–754. https://doi.org/10.1097/00006254-199510000-00022
64. Yao Z, Shen X, Capodanno I, Donnelly M, Fenyk-Melody J, Hausamann J, Nunes C, Strauss J, Vakerich K (2005) Validation of rat endometriosis model by using raloxifene as a positive control for the evaluation of novel SERM compounds. J Investig Surg Off J Acad Surg Res 18:177–183. https://doi.org/10.1080/08941930591004412
65. Kulak J, Fischer C, Komm B, Taylor HS (2011) Treatment with Bazedoxifene, a selective estrogen receptor modulator, causes regression of endometriosis in a mouse model. Endocrinology 152:3226–3232. https://doi.org/10.1210/en.2010-1010
66. Naqvi H, Sakr S, Presti T, Krikun G, Komm B, Taylor HS (2014) Treatment with bazedoxifene and conjugated estrogens results in regression of endometriosis in a murine model. Biol Reprod 90:121. https://doi.org/10.1095/biolreprod.113.114165
67. Huniadi CA, Pop OL, Antal TA, Stamatian F (2013) The effects of ulipristal on Bax/Bcl-2, cytochrome c, Ki-67 and cyclooxygenase-2 expression in a rat model with surgically induced endometriosis. Eur J Obstet Gynecol Reprod Biol 169:360–365. https://doi.org/10.1016/j.ejogrb.2013.03.022
68. Liang B, Wu L, Xu H, Cheung CW, Fung WY, Wong SW, Wang CC (2018) Efficacy, safety and recurrence of new progestins and selective progesterone receptor modulator for the treatment of endometriosis: a comparison study in mice. Reprod Biol Endocrinol 16:32. https://doi.org/10.1186/s12958-018-0347-9
69. Kettel LM, Murphy AA, Morales AJ, Ulmann A, Baulieu EE, Yen SS (1996) Treatment of endometriosis with the antiprogesterone mifepristone (RU486). Fertil Steril 65:23–28. https://doi.org/10.1016/s0015-0282(16)58022-4
70. Mei L, Bao J, Tang L, Zhang C, Wang H, Sun L, Ma G, Huang L, Yang J, Zhang L, Liu K, Song C, Sun H (2010) A novel mifepristone-loaded implant for long-term treatment of endometriosis: in vitro and in vivo studies. Eur J Pharm Sci Off J Eur Fed Pharm Sci 39:421–427. https://doi.org/10.1016/j.ejps.2010.01.012
71. Pittatore G, Moggio A, Benedetto C, Bussolati B, Revelli A (2014) Endometrial adult/progenitor stem cells: pathogenetic theory and new antiangiogenic approach for endometriosis therapy. Reprod Sci 21:296–304. https://doi.org/10.1177/1933719113503405
72. Laschke MW, Menger MD (2007) In vitro and in vivo approaches to study angiogenesis in the pathophysiology and therapy of endometriosis. Hum Reprod Update 13:331–342. https://doi.org/10.1093/humupd/dmm006
73. Taylor RN, Yu J, Torres PB, Schickedanz AC, Park JK, Mueller MD, Sidell N (2009) Mechanistic and therapeutic implications of angiogenesis in endometriosis. Reprod Sci 16:140–146. https://doi.org/10.1177/1933719108324893
74. Hull ML, Charnock-Jones DS, Chan CLK, Bruner-Tran KL, Osteen KG, Tom BDM, Fan T-PD, Smith SK (2003) Antiangiogenic agents are effective inhibitors of endometriosis. J Clin Endocrinol Metab 88:2889–2899. https://doi.org/10.1210/jc.2002-021912
75. Nap AW, Griffioen AW, Dunselman GAJ, Bouma-Ter Steege JCA, Thijssen VLJL, Evers JLH, Groothuis PG (2004) Antiangiogenesis therapy for endometriosis. J Clin Endocrinol Metab 89:1089–1095. https://doi.org/10.1210/jc.2003-031406
76. Van Langendonckt A, Donnez J, Defrère S, Dunselman GAJ, Groothuis PG (2008) Antiangiogenic and vascular-disrupting agents in endometriosis: pitfalls and promises. Mol Hum Reprod 14:259–268. https://doi.org/10.1093/molehr/gan019

77. Soares SR, Martínez-Varea A, Hidalgo-Mora JJ, Pellicer A (2012) Pharmacologic therapies in endometriosis: a systematic review. Fertil Steril 98:529–555. https://doi.org/10.1016/j.fertnstert.2012.07.1120
78. Ornek T, Fadiel A, Tan O, Naftolin F, Arici A (2008) Regulation and activation of ezrin protein in endometriosis. Hum Reprod 23:2104–2112. https://doi.org/10.1093/humrep/den215
79. Dabrosin C, Gyorffy S, Margetts P, Ross C, Gauldie J (2002) Therapeutic effect of angiostatin gene transfer in a murine model of endometriosis. Am J Pathol 161:909–918. https://doi.org/10.1016/S0002-9440(10)64251-4
80. Laschke MW, Elitzsch A, Vollmar B, Vajkoczy P, Menger MD (2006) Combined inhibition of vascular endothelial growth factor (VEGF), fibroblast growth factor and platelet-derived growth factor, but not inhibition of VEGF alone, effectively suppresses angiogenesis and vessel maturation in endometriotic lesions. Hum Reprod 21:262–268. https://doi.org/10.1093/humrep/dei308
81. Ricci AG, Olivares CN, Bilotas MA, Meresman GF, Barañao RI (2011) Effect of vascular endothelial growth factor inhibition on endometrial implant development in a murine model of endometriosis. Reprod Sci 18:614–622. https://doi.org/10.1177/1933719110395406
82. Ozer H, Boztosun A, Açmaz G, Atilgan R, Akkar OB, Kosar MI (2013) The efficacy of bevacizumab, sorafenib, and retinoic acid on rat endometriosis model. Reprod Sci 20:26–32. https://doi.org/10.1177/1933719112452941
83. Novella-Maestre E, Carda C, Noguera I, Ruiz-Saurí A, García-Velasco JA, Simón C, Pellicer A (2009) Dopamine agonist administration causes a reduction in endometrial implants through modulation of angiogenesis in experimentally induced endometriosis. Hum Reprod 24:1025–1035. https://doi.org/10.1093/humrep/den499
84. Novella-Maestre E, Carda C, Ruiz-Sauri A, Garcia-Velasco JA, Simon C, Pellicer A (2010) Identification and quantification of dopamine receptor 2 in human eutopic and ectopic endometrium: a novel molecular target for endometriosis therapy. Biol Reprod 83:866–873. https://doi.org/10.1095/biolreprod.110.084392
85. Li W-N, Wu M-H, Tsai S-J (2021) HYPOXIA AND REPRODUCTIVE HEALTH: the role of hypoxia in the development and progression of endometriosis. Reproduction 161:F19–F31. https://doi.org/10.1530/REP-20-0267
86. Xiong W, Zhang L, Xiong Y, Liu H, Liu Y (2016) hypoxia promotes invasion of endometrial stromal cells via hypoxia-inducible factor 1α upregulation-mediated β-catenin activation in endometriosis. Reprod Sci 23:531–541. https://doi.org/10.1177/1933719115607999
87. Xiong Y, Liu Y, Xiong W, Zhang L, Liu H, Du Y, Li N (2016) Hypoxia-inducible factor 1α-induced epithelial-mesenchymal transition of endometrial epithelial cells may contribute to the development of endometriosis. Hum Reprod 31:1327–1338. https://doi.org/10.1093/humrep/dew081
88. Liu H, Zhang Z, Xiong W, Zhang L, Xiong Y, Li N, He H, Du Y, Liu Y (2017) Hypoxia-inducible factor-1α promotes endometrial stromal cells migration and invasion by upregulating autophagy in endometriosis. Reproduction 153:809–820. https://doi.org/10.1530/REP-16-0643

Chapter 9
The Pathogenesis of Endometriosis: Are Endometrial Stem/Progenitor Cells Involved?

Antonio Simone Laganà and Antoine Naem

Abstract

Introduction: Endometriosis is an enigmatic disease defined by the presence of endometrial-like glands and stroma out the uterine cavity. The pathogenesis of endometriosis was under intensive investigation throughout the last decades and many hypotheses were formulated. In this chapter, we provide a comprehensive review of the current knowledge that concerns the possible involvement of endometrial stem/progenitor cells in the pathogenesis of endometriosis.
Methods: The PubMed/Medline database was searched using the search terms: "Stem, Progenitor, Stemness, Endometriosis, Pathogenesis". All original articles and reviews discussing the role of the embryonic or adult stem cells in the pathogenesis of endometriosis were included.
Results: Stem cells with aberrant stemness markers expression were identified within the ectopic endometrium. These cells exhibited highly proliferative, migratory, and invasive capabilities when cultured in vitro and in vivo. Much of these aggressive characteristics are attributed to the cancer-associated mutations often encountered in endometriosis. Additionally, many post-transcriptional factors could influence the ectopic endometrial stem cells behavior.
Conclusions: The endometrial stem cells appear to have a pivotal role in the pathogenesis of endometriosis. The biological behavior of these cells is supposed to be influenced by genetic, epigenetic, and environmental factors that determine the severity and hormonal responsiveness of endometriosis.

Keywords Endometriosis · Pathogenesis · Progenitor cells · Stem cells

---

A. S. Laganà (✉)
Department of Obstetrics and Gynecology, "Filippo Del Ponte" Hospital, University of Insubria, Piazza Biroldi 1, 21100 Varese, Italy
e-mail: antoniosimone.lagana@uninsubria.it

A. Naem
Faculty of Medicine, Damascus University, Damascus, Syria
e-mail: antoine.naem@gmail.com

I. Virant-Klun (ed.), *Stem Cells in Reproductive Tissues and Organs*, Stem Cell Biology and Regenerative Medicine 70, https://doi.org/10.1007/978-3-030-90111-0_9

Abbreviations

ABCG2	ATP Binding cassette subfamily G member 2 (Junior Blood Group)
AMH	Anti-Müllerian hormone
ARID1A	AT-rich interaction domain 1A
CAM	Cell adhesion molecule
CD	Cluster of differentiation
CD133	Prominin 1
CD146	Melanoma cell adhesion molecule (MCAM)
C-MYC	MYC proto-oncogene, BHLH transcription factor
CXCR-4	C-X-C motif chemokine receptor 4
CXCL-12	C-X-C motif chemokine ligand 12
DDX4	DEAD-box helicase 4
DNA	Deoxyribonucleic acid
EpCAM	Epithelial cell adhesion molecule
ER	Estrogen receptor
HOXA9	Homeobox A9
HOXA10	Homeobox A10
HOXA11	Homeobox A11
HOXA13	Homeobox A13
IFITM3	Interferon induced transmembrane protein 3
K-RAS	KRAS proto-oncogene, GTPase
LGR5	Leucine-rich-repeat-containing G-protein-coupled-receptor 5
MEIS	Myeloid ecotropic viral integration site
MMP9	Matrix metalloproteinase 9
miRNA	MicroRNA
MSI1	Musashi RNA binding protein 1
NANOG	Nanog homeobox
NGS	Next-generation sequencing
NOTCH1	Notch receptor 1
OCT4	Octamer-binding transcription factor 4 (POU5F1)
PBX1	Pre-B-cell leukemia homebox-1
PCNA	Proliferating cell nuclear antigen
PIK3CA	Phosphatidylinositol-4,5-bisphosphate 3-kinase catalytic subunit alpha
POU5F1	POU class 5 homeobox 1
PR	Progesterone receptor
PTEN	Phosphatase and tensin homolog
REX1	REX1 transcription factor
RNA	Ribonucleic acid
SALL4	Spalt like transcription factor 4
SF1	Steroidogenesis factor 1
SOX2	SRY-box transcription factor 2
SOX9	SRY-box transcription factor 9
SSEA-1	Stage-specific embryonic antigen-1

SUD2	Sushi domain containing-2
SYR	Sex determining region
TIAR	Tissue injury and repair
VEGF	Vascular endothelial growth factor

Introduction to Endometriosis and Its Pathogenesis Theories

Endometriosis is a chronic inflammatory, estrogen-dependent, enigmatic disease, defined by the presence of endometrial-like glands and stroma out of the uterine cavity. Its exact prevalence is not determined yet. However, endometriosis was estimated to affect 2–10% of women in the child-bearing age [1], which equates to nearly 190 million patients worldwide [2, 3]. Endometriosis occurrence peaks in the reproductive years due to the estrogenophilic nature of the disease. Nonetheless, endometriosis was encountered in patients of almost all age groups [4], including children, adolescents, and postmenopausal females [4, 5]. Growing evidence suggests that endometriosis is a hereditable disease [6]. The probability of developing endometriosis in first-degree relatives of affected patients is estimated to be 6–9% [7, 8]. This probability increases to 15% for relatives of patients with more severe disease [9, 10]. Endometriosis symptomatology differs among patients and varies in severity from mild symptoms, to severe, impairing illness. The most frequently reported symptoms are chronic pelvic pain, dysmenorrhea, dyspareunia, dyschezia, and subfertility [11, 12]. Moreover, endometriosis was related to adverse pregnancy outcomes, such as pre-eclampsia, preterm birth, and low birth weight [13].

Endometriotic lesions could be found in pelvic and extra-pelvic localizations, with the last being less common [14]. Pelvic endometriosis can take one of either three forms: superficial peritoneal implants, ovarian endometriomas, and deep infiltrating endometriotic nodules [15]. Tubal endometriosis is a more recently introduced subtype of endometriosis since it may have different etiology and specific complications [3]. Extra-pelvic localizations of endometriosis include but not limited to: the skin [16], lungs [17], and brain [18]. As noticed, the wide variation of endometriosis types and localizations further hardens the understanding of the underlying developmental trajectories that result in disease formation. While hormone dysregulation, chronic inflammation, immunotolerance, and angiogenesis are accepted as essential contributors to the disease initiation and progression [19, 20], the exact origin of the endometriotic epithelial and stromal cells is yet to be identified. To date, many hypotheses trying to identify the origin of the ectopic endometrial cells were made. Each hypothesis has its own strengths and weaknesses, but we can say with higher confidence that none of them explains adequately each manifestation of endometriosis.

The retrograde menstruation theory by Sampson is one of the earliest and highly accepted postulations that intended to elucidate the pathogenesis of endometriosis. It states that viable endometrial cells travel backwards through the fallopian tubes to seed within the peritoneal cavity [21]. This hypothesis is further supported by the observation of abnormal uterine contractions, longer, and more intense menses in women suffering from endometriosis [22, 23]. Furthermore, this hypothesis was implicated to explain early onset endometriosis. It was postulated that neonatal menstruation could result in the deposition of less differentiated endometrial cells that could be activated in early puberty [24]. However, male endometriosis [25] and extra-pelvic endometriosis are not explained sufficiently by this theory. The embryonic cell rest theory is another hypothesis that was formulated to elucidate the aforementioned subtypes. It mainly states that the abnormal development or incomplete regression of the Müllerian ducts could result in the deposition of embryonic stem cells of Müllerian origin within the peritoneal cavity. These cells could differentiate into endometrial-like glands, and thus, cause endometriosis in both males and females [26, 27]. Alternatively, abnormal migration of the primordial germ cells from the yolk sac to the urogenital ridges was also suggested as a potential mechanism [28]. This hypothesis explains the existence of endometrial-like islands in the peritoneal cavity of the deceased fetuses. However, a good question to be asked when interpreting the embryonic cell rest theory is "*why different endometriotic lesions could present with various phenotypes in the same adult if they originate from specific and identical embryonic cells?*". Obviously, the question is left unanswered. Bone marrow-derived stem cells involvement in endometriosis pathogenesis was also postulated since these cells were found to regenerate the endometrium in humans and mouse models [29–31]. The coelomic metaplasia is another popular theory that states that the peritoneal mesothelium could undergo metaplastic transformation under certain stimulations and differentiate into any Müllerian-derived cell. This hypothesis is mainly based on previous pathologic findings where transitional cells could be observed between the endometriotic epithelium and the peritoneal mesothelium [15]. Nevertheless, this hypothesis is questionable and the supporting evidence remains scarce.

Despite the discordance between these hypotheses, all of them, explicitly or implicitly, confirm the presence of highly proliferative, regenerative, and multipotential stem cells with invasive and immune-escaping activities within the peritoneal cavity. More recently, the genetic and epigenetic theory was introduced as a supplementary explanation of events that derive all the disorders observed in endometriosis [6]. It states that women with genetic predisposition due to inherited genotypes are at an increased risk of developing endometriosis. Whereas the disease initiation and progression is driven by various and complex epigenetic events [6]. Indeed, the stem cell theory and the genetic and epigenetic theory could be described as "two facets of a single coin". These two hypotheses were supported by later identification of stemness markers in almost all endometriotic lesions [32–34]. Furthermore, genetic polymorphism [35], cancer-associated mutations [36], and post-transcriptional dysregulations [37, 38] were identified in both ectopic and eutopic endometrial stem cells.

In the following chapter, we provide a comprehensive review of the latest evidence concerning the involvement of stem cells in the pathogenesis of endometriosis and discuss the correlation of the genetic and epigenetic abnormalities, with the clinical and experimental observations.

The Potential Role of the Embryonic Stem Cells in Endometriosis Pathogenesis

Genetic Regulation of the Müllerian Ducts Genesis and Differentiation

The mesonephric ducts (Wolffian ducts) and the paramesonephric ducts (Müllerian ducts) are two pairs of genital ducts that are formed at the lateral aspect of the urogenital ridge during the fourth and fifth weeks of embryogenesis. The Wolffian ducts originate within the mesonephros, whereas the Müllerian ducts originate from the invagination of the coelomic epithelium. At this stage, the human embryos continue their development in a sexually indifferent state. By the seventh week, the sex-determining region (SYR) gene on the Y chromosome is activated in male embryos and starts producing the testes determining factor. SYR along with SOX9 binds to the promoter region of the anti-Müllerian hormone (AMH) gene and induces its expression. Additionally, SOX9 and SYR activate the transcription of Steroidogenesis factor 1 (SF1), which in turn stimulates the Leydig cells to secrete AMH, resulting eventually in the regression of the Müllerian ducts [39]. Whereas in female embryos, the Müllerian ducts continue their development to form the fallopian tubes, uterine body, cervix, and the upper one-third of the vagina. The Müllerian ducts development is regulated by various genes and transcription factors. The homebox genes play an essential role in regulating the cell fate and the segmental embryonic formatting [40]. To appropriately function, the HOX genes need to cooperate with specific cofactors that contribute to the binding to the DNA for controlling the expression of target genes [41]. The most important protein cofactors are pre-B-cell leukemia homebox-1 (Pbx-1) and myeloid ecotropic viral integration site (MEIS). Pbx-1 was found to regulate genetic expression of all HOXA family, through the cooperation with other members of the homebox family, such as HOX and MEIS [42]. Pbx-1 is intensively expressed in the mesenchymal tissues of the differentiating urogential ridge [43, 44]. Decreased expression of Pbx-1 leads to reduced differentiation of the urogenital ridge and complete absence of the Müllerian ducts [45]. The HOX genes were found to have the primary role in imparting the positional identity to undifferentiated tissues along the entire body axes [46]. It is noteworthy that each of the HOXA genes was found to be upregulated in a specific part of the evolving Müllerian duct [47]: HOXA-9 is mainly expressed in the fallopian tubes, HOXA-10 and HOXA11 are mainly expressed in the uterus, and finally, HOXA-11 and HOXA-13 are expressed in the cervix and vagina [46, 48]. Therefore, the HOXA genes and associated transcription factors have pivotal role in patterning the female reproductive tract and maintaining the normal differentiation of the Müllerian ducts [49]. In addition, the wingless family, especially Wnt, plays fundamental role in maintaining the normal development of the Müllerian ducts. Wnt4 is essential for the formation of the Müllerian ducts [50], and its down-regulation resulted in complete absence of the Müllerian ducts. Wnt5 was found to be crucial for the development of the genital anterior–posterior axis. Additionally, Wnt7 is involved in the maintenance of HOXA10 and HOXA11 genes [51–53]. Wnt7a was found to be expressed in the stroma, whereas Wnt5a was found to be expressed in the uterine epithelium [54]. Moreover, both of these genes are necessary for appropriate glandular formation [54]. It is noteworthy that β-catenin is the final signal-transducing molecule and multifunctional protein of the Wnt pathways. β-catenin was described as the binding partner of E-cadherin, an important intracellular adhesion molecule and inhibitor of the cells' metastatic activity [23, 55]. Therefore, β-catenin could be regarded an essential protein for epithelial tissue development. After completing the differentiation of the Müllerian ducts, the remnants of the epithelial and mesenchymal stem cells derived from the Müllerian are thought to reside within the basal layer of the adult endometrium, giving rise to the side population cells [56].

Possible Involvement of the Homebox Genes and Wingless Pathway in Endometriosis Pathogenesis

As noticed, the homebox genes and wingless pathway have pivotal influence on the development of the Müllerian ducts. Therefore, it is plausible to think that the abnormal expression of those genes and related pathways would result in Müllerian ducts anomalies and disrupt the normal localization of cells of Müllerian origin [26, 27]. This postulation is further supported by the fact that 46% of patients with uterovaginal anomalies have concomitant endometriosis. On this basis, endometriosis was considered the most common comorbidity in patients with obstructive and non-obstructive symmetric malformations [28, 57]. Another study revealed the coexistence of endometriosis and Mayer-Rokitansky-Küster-Hauser syndrome in 38% of patients [58]. Moreover, Signorile et al. [59, 60] investigated the autopsies of 101 female fetuses and found endometriotic lesions in 9% of them. Interestingly, these lesions were distributed in the rectovaginal space, the pouch of Douglas, and within the myometrium. Therefore, it is attractive to think that abnormal expression of the HOX genes—and HOXA10 specifically—could result in endometrial cells deposited out of the uterine cavity [26]. Furthermore, HOXA10 was found to be expressed within the ectopic endometrial islands. The stromal expression levels of HOXA10 exceed those of the epithelial cells in both eutopic and ectopic endometrium [61]. During the reproductive age, the endometrial expression of HOXA10 was found to peak in the mid-secretory phase in healthy women. However, this elevation in HOXA10 expression was not observed in the eutopic endometrium of patients with endometriosis [62].

The possible involvement of the wingless pathway and β-catenin in the pathogenesis of endometriosis was also investigated. WNT4 and WNT7b were found to be expressed intensively in the human ectopic endometrial cells [63, 64]. Accordingly, increased expression of WNT4 and WNT7b was found in a rat model of endometriosis [65]. In addition, β-catenin expression in eutopic endometrial stromal cells isolated from patients with endometriosis was found to be upregulated significantly under the influence of estrogen in vitro [66]. In addition, β-catenin was found to enhance lesion progression, invasion, and angiogenesis under the influence of local estrogen injection in mouse models of endometriosis. Down-regulation of β-catenin resulted in decreased invasiveness, angiogenesis, and Matrix Metalloproteinase 9 (MMP-9) expression [66]. Additional investigations clarified the relationship between estrogen, β-catenin, and Vascular Endothelial Growth Factor (VEGF). It was demonstrated that a cascade of genes expression is initiated by estrogen resulted in β-catenin up-regulation. β-catenin in turn binds to the promoter region on VEGF gene, and thus, enhances its transcription [67].

On this basis, the homebox genes and wingless pathway could be regarded as important contributors to the endometriosis pathogenesis [27]. Nonetheless, what causes these pathways to malfunction, and whether the aforementioned dysregulations are inherited or stimulated by certain environmental factors, remain unknown.

The Possible Involvement of Primordial Germ Cells in Endometriosis

The primordial germ cells have a relatively long migratory pathway from the yolk sac to the developing urogenital rides through the posterior mesentery of the hindgut [39]. Abnormal migratory pathway, or aberrant regulation of their journey were postulated to cause abnormal localizations of these cells within the coelomic cavity [28]. This postulation was driven by the observation of endometriotic implants in patients with complete Müllerian agenesis [68]. Indeed, while the Müllerian derivatives could be completely absent in patients with endometriosis, the two functional ovaries are always present. This actually led the author to conclude that the eutopic endometrium, ovarian oocytes, and endometriosis may have the same origin, which is the primordial germ cells [57]. This postulation was justified by the isolation of germ cell-specific markers within the endometriotic lesion. Fraunhoffer et al. [69] have carried an interesting study to investigate the possible involvement of the ovarian germ line stem cells in endometriosis pathogenesis. Through immunofluorescence, the authors managed to identify DDX4-positive cytoplasmic stromal cells in all endometriotic samples. They also declared that DDX4-positive stromal cells were encountered more easily in younger patients, and therefore concluded that elevated numbers of DDX4-positive stromal cells are enriched in endometriomas of younger patients. However, this claim is subject to quantifying analysis justification. Moreover, DDX4-expressing cells were found to co-express the nuclear proliferative marker PCNA. In addition, IFITM3, another germ-specific marker, was found to co-localize with DDX4 in the endometriotic stromal cells. None of the previous markers could be observed in eutopic endometrium [69]. These results support the notion of primordial germ cells involvement in endometriosis [28], but obviously, contradicts the common origin of both the endometrium and ovaries.

The Embryonic Cell Rest Theory's Strengths and Weaknesses

The embryonic cell rest theory managed to clarify many questions regarding male endometriosis, endometriosis at extra-pelvic sites, and early onset endometriosis. This hypothesis also explains the reported association between endometriosis and other Müllerian anomalies [58]. It seems logical to postulate that impaired Müllerian ducts regression due to inappropriate function of SOX9 and its related pathway accompanied by genetic dysregulation in the homebox and wingless pathways could lead to the deposition of stem cells of Müllerian origin in males between the prostatic utricle and appendices testes [70]. Moreover, migratory abnormalities mainly driven by HOXA10 and HOXA11 aberrant expression could lead to extra-pelvic deposition of the Müllerian-derived stem cells [26]. This postulation is further enforced with the fact that Müllerian ducts are located in the lower thoracic and lumbar region of the developing embryos [39]. Hereby, the Müllerian-derived stem cells can be left quiescent during childhood and become activated in early puberty under the influence of the hyperestrogenic environment. What derives the activation of such cells in males is not elucidated yet. Alternatively, the presentation of endometriosis at postmenopausal ages without previous history of endometriosis or at least the common symptoms of endometriosis further hardens fully accepting this hypothesis [5]. In other words, it is not clear what induces endometriosis progression in the postmenopausal hypoestrogenic environment and kept the Müllerian-derived stem cells quiescent during all the reproductive years. On the other hand, later genomic studies have demonstrated that the epithelium of the endometriotic lesions is monoclonal. Whereas the endometriotic stromal cells are polyclonal, and therefore, it is not reasonable to postulate that a single cell type is responsible for endometriosis pathogenesis [71, 72]. Moreover, the different endometriotic phenotypes usually encountered in the same patient (for example, deep infiltrating endometriosis accompanied by peritoneal implants) add more confusion to this hypothesis. A question worth asking at this point of the discussion is "*If endometriosis is supposed to originate from identical cells that have the same genotype and share the same microenvironment, then why different phenotypes exist?*". Even if we assumed that the different acquired mutations are the leading cause of these various phenotypes and behaviors, it is not clear why some cells have different mutational profiles, or at least different mutational penetrance that enable them to invade and progress more than the rest. Apparently, the embryonic cell rest theory is unable to give adequate explanation of such observations.

The Eutopic Endometrial Stem Cells, What Do They Tell Us About the Endometriosis Pathogenesis?

Flashback to the Normal Endometrium

The human endometrium is a highly regenerative tissue that undergoes cyclical shedding and regeneration monthly during the reproductive years. A healthy female is estimated to have over 400 menstrual cycle during her life [73]. The human endometrium attains the ability of self-renewal even after extensive iatrogenic procedures, such as endometrial ablation [74, 75]. In each menstrual cycle, the functional and luminal layers of the endometrium are shed, whereas the basal layer is supposed to be left intact. On this basis, the basal layer is considered to be responsible for the endometrium regeneration [76]. Therefore, the existence of endometrial stem cells within the basal was suspected, and heavily investigated throughout the last decades. Indeed, the deeper knowledge we have now regarding the stem cells function, distribution, and genetic profile has enabled us to understand and interpret many of the alterations that are associated with endometriosis development.

At least three types of adult stem cells are known to reside in the basal layer of eutopic endometrium [76, 77]. Mesenchymal stem cells are multipotential, proliferative, regenerative, and self-renewing stem cells found in the basal and functional layer of the endometrium [78]. These can last dormant for a long time, and their survival does not require estrogen [76]. The mesenchymal stem cells are identified mainly through the co-expression of CD146 and Platelets Derived Growth Factor Receptor β (PDGFR β), or the sole expression of Sushi Domain containing-2 (SUD2) [79, 80].

The second type of stem cells located in the endometrium is the endometrial epithelial progenitor. Those are mainly located in the basal layer of the endometrium [76]. Further investigations revealed that clonal expansion of a single epithelial progenitor is capable of regenerating a single gland [81]. In other words, the epithelia of the endometrial glands are monoclonal. These progenitors are identified through the co-expression N-cadherin and Stage-Specific Embryonic Antigen-1 (SSEA-1). Moreover, the endometrial epithelial progenitors' differentiation and maturation were found to occur in hierarchical manner [82]. Epithelial progenitors at the basalis layer are found to be either (N-cadherin$^+$/SSEA$^-$) or (N-cadherin$^+$/SSEA$^+$). The progenitors alter their expression while differentiating, and eventually, give rise to (N-cadherin$^-$/SSEA$^-$) mature epithelium. These are mainly found in the superficial parts of the functional layer and in the luminal layer [82]. Moreover, N-cadherin-positive basalis cells were found to have elevated clonogenicity, proliferative potential, and self-renewal capacity. These characteristics are lost with the loss of N-cadherin expression [19]. Disruption of this highly regulated differentiation process was correlated with endometriosis, as will be discussed later in this chapter

Finally, the side population cells form the third type of endometrial stem cells. The side population cells are a population of epithelial, endothelial, and stromal stem cells that reside in the perivascular spaces in the basal layer of the endometrium [83, 84] and are thought to be the remnants of primordial epithelial stem cells [56].

These cells are identified through the expression of ABCG2 by the means of flow cytometry [85]. The side population cells are correlated to angiogenesis because they are enriched with endothelial and mesenchymal stem cells [86]. In fact, these were found able to regenerate the vascular endothelial tissue of the endometrium in vivo [83, 87].

Aberrant Expression of Stemness Markers in the Eutopic Endometrium of Patients with Endometriosis

In 2002, Leyendecker et al. [88] investigated the expression patterns of estrogen receptor (ER), progesterone receptor (PR), and aromatase in the endometriotic lesions, the eutopic endometrium, and the menstrual blood of women with and without endometriosis. Furthermore, the authors found that patients with endometriosis shed more basal cells during menstruation when compared to healthy women. Accordingly, the ectopic endometrium showed ER and PR expression pattern similar to that of cells in the basal layer of the endometrium. Therefore, they concluded that endometriosis is derived from abnormal eutopic endometrium that shed elevated numbers of basal cells [88]. Indeed, this conclusion was a very useful upgrade to the retrograde menstruation theory that was formulated by Sampson in 1927 [21]. Since then, many stemness markers with their expression pattern in the basal, functional, and luminal layers of the eutopic endometrium were characterized, as we previously discussed.

The loss of the hierarchical differentiation of the epithelial progenitors in the basal layer of the endometrium was related to endometriosis pathogenesis. This conclusion is mainly driven by the observation of enriched "basalis-like" cells within the functional layer of the endometrium in patients with endometriosis [89]. The basalis-like cells show elevated expression of N-cadherin, SSEA, and Sox9 [76, 90]. Moreover, they also showed increased expression of NANOG and OCT-4 genes [89]. The basalis-like epithelial progenitors were able to generate endometrial-like glands similar to those of the ectopic endometrium after 14 days of three dimensional culture [89]. Similar to the basalis-like stem cells, the isolated mesenchymal stem cells from the endometrium of patients with endometriosis showed elevated expression of NANOG and SOX2 [32]. However, the basalis-like epithelial progenitors fail to differentiate into mesodermal cellular lineages [89]. In contrast, the mesenchymal stem cells expressed multipotential characteristics and differentiated into multiple mesodermal cellular lineages, i.e., adipocytes, chondrocytes, and osteocyte when cultured in the respective environment [91]. These cells also showed high colony-forming efficiency when seeded in sub-optimally low dilutions [32]. Cells of the luminal layer of the endometrium also express pivotal stemness markers, including N-cadherin, SSEA, and Leucine-Rich-Repeat-Containing G-Protein-Coupled-Receptor 5 (LGR5) [76, 92]. Cells from the functional and luminal layers are typically shed during menstruation. When taking their stemness features in consideration, it is plausible to postulate that the basalis-like or luminal cells are deposited in the peritoneal cavity through retrograde menstruation. Indeed, the altered uterine peristalsis, longer and heavier menses observed in patients with endometriosis are additional factors that support this postulation [23, 93]. The abnormal uterine contractions could enhance the retrograde flow of menstrual blood, and thus, more progenitors and stem cells are seeded within the peritoneal cavity.

Even though the "modified" retrograde menstruation theory answers many questions and is supported by many valid studies, it remains unable to explain the pathogenesis of endometriosis precisely. In a baboon model, endometriosis induction resulted in aberrant expression of the stemness markers similar to those of the endometrium of patients with endometriosis [89]. Therefore, we are not quite sure whether the observed alterations in the eutopic endometrium of patients with endometriosis are the cause, or rather the result of the disease.

Stemness-Related Markers and Functional Characteristics of Endometriosis

Endometriosis exhibits a biologic behavior very similar to that of the eutopic endometrium in terms of proliferation, shedding, and regeneration [6]. In addition, endometriosis is widely known to be a progressive and highly recurrent disease [94–96]. These characteristics together with our increasing knowledge of the histology of the eutopic endometrium, have derived intensive investigations of the possible involvement and potential functions of stem cells in endometriosis pathogenesis and progression.

Similar to the eutopic endometrium, epithelial and mesenchymal stem cells were isolated and proved to have high clonogenic and self-renewal capacities [91, 97]. The epithelial stem cells expressed EpCAM, cytokeratin, and α6 integrin, whereas the stromal stem cells expressed CD133, Msi-1, and SALL4 [97]. The endometriotic mesenchymal stem cells, similar to their eutopic endometrial counterparts, were found to retain the ability of differentiating to multiple mesodermal lineages, including adipocytes, chondrocytes, myocytes, and osteocytes [97]. The ability of the suspected epithelial progenitors of differentiating into mesodermal cellular lineages was not tested to the best of our knowledge. Even though the eutopic and ectopic mesenchymal stem cells are pretty much similar, many functional studies revealed the presence of essential differences that could explain much of the endometriosis behavior. The ectopic mesenchymal stem cells were proved to have higher growth and proliferation rate than their eutopic counterparts [91]. Additionally, many investigators found increased migration and angiogenesis capacities of ectopic mesenchymal stem cells when compared with the eutopic stem cells [98]. These behavioral differences are attributed mainly to the altered expressions of genes related to growth, proliferation, migration, and self-renewal between these two cell groups.

SSEA and SOX9-expressing mesenchymal stem cells were detected. Additionally, SSEA-expressing endometriotic cells were found to have increased telomerase activity when it was grown in cultures [90].

Endometriosis was found to highly express the pluripotency gene triad: NANOG, SOX2, and OCT-4 [33, 34]. The expression of NANOG and SOX2 was remarkably much higher of their expression in the eutopic endometrium [33, 34]. Whereas OCT-4 was found to be less expressed by endometriosis when compared with eutopic endometrium [33]. Chang et al. [99] also reported the up-regulation of OCT-4 in ectopic lesions and suggested its possible involvement in lesion progression by stimulating the migration of endometrial cells. Moreover, two studies report the up-regulatory influence of estrogen on SOX2 and NANOG [100, 101]. Overexpression of these two genes is suggested to maintain the self-renewal capacity and increase the cell survival in ovarian endometriosis [33].

Notch-1 protein is considered an essential regulatory of cell fate in various cellular types [102, 103]. Therefore, it was postulated that up-regulation of Notch-1 in endometriotic tissues could influence cell proliferation and migration. In an interesting study, He et al. [104] reported the consequences of Notch-1 down-regulating in vitro and in a murine model of endometriosis. The authors reported significant decrease in the expression of NANOG, OCT-4, SOX2, and C-MYC in epithelial and stromal endometriotic cells after silencing the expression of Notch-1. Additionally, there was a significant reduction in the colony-forming efficiency of both epithelial and stromal cells following Notch-1 down regulation. Moreover, the migration of endometriotic stem cells was also impaired after Notch-1 down regulation. Finally, the authors detected significant reduction in the endometriotic lesions' weight and volume in vivo on the 14th and 21st day after the knocking down of Notch-1 [104].

Finally, each of Musashi-1, REX-1, SOX9, and nuclear β-catenin was also found to be expressed in varying intensities in endometriosis [34, 90, 105].

The Mutational Profile of the Eutopic and Ectopic Endometrium

A defective endometrium was suggested to be the origin of endometriosis. A group of Chinese investigators postulated that abnormalities of the endometrial stem cells could result in unfavorable biological behaviors, such as increased proliferation, migration, invasion, and angiogenesis [106, 107]. Indeed, it is plausible to consider genetic alterations as a possible cause of endometriosis, especially when knowing that the hereditability of the disease is reported to be as high as 50% [108]. Moreover, the association between the patient's disease severity, and the probability of developing endometriosis in her first-degree relatives, raises the suspicion of the possible genetic involvement [7–10]. Besides the inherent genetic abnormalities, the endometrium was found to be prone to acquiring new mutations [109], and the mutagenesis risk was reported to be increased by 5% each year [110]. The high proliferative and regenerative activities of the endometrial stem cells is sought to be a key characteristic that derives such genetic abnormalities. As widely known, with every endometrial shedding, the endometrial stem cells in the basal layer of the endometrium get activated, and therefore, undergo mitosis. Even though mitosis is a highly regulated process, and the mutation rate was reported to be in the range between 10^{-8} and 10^{-7} per basepair per cell division [111, 112], mutagenesis appears unavoidable [109]. This is mainly due to the enormous size of the human genome, aging, and the high duplication rate of the endometrial stem cells [113]. It is noteworthy that not all mutations cause altered biological behavior. Some of them remain silent until adequate number of mutations is harvested, or they are simply passenger mutations that do not carry fitness to their bearer [109]. On this basis, it is not surprising anymore to say that high mutational profile was detected in the normal-appearing endometrium, and endometriosis as well [32, 35, 71, 110, 114].

Passenger mutations are random and unique for the harboring cell's progeny. Therefore, passenger mutations are of utmost importance in determining the clonality of the tissue under investigation.

When it comes to endometriosis research, passenger mutations provide invaluable information about the origin of the endometriotic lesions. Similar to the eutopic endometrium, the epithelial component of endometriosis was found to be clonal in origin, whereas the endometriotic stroma was found to be polyclonal [71]. Therefore, it is obvious that the endometriotic lesions do not originate from a single stem cell. Furthermore, Silveira et al. [35] have investigated the clonality of lesions encountered in various anatomical locations in the same patient. Interestingly, the lesions harbored shared genetic alterations in their glandular and stromal components respectively. Therefore, the various endometriotic lesions were thought to originate from a single cell, and thus, clonal in origin. On the other hand, a study that investigated the common origin of paired eutopic and ectopic endometrial glands found that the harbored passenger mutations by the ectopic glandular epithelium differed from those harbored by their counterpart, and thus, the eutopic and ectopic endometrial glands were found to belong to different clones [72]. However, these results should be interpreted carefully since the eutopic endometrial glands are numerous, and only selected clones with specific types of mutations can expand clonally [36].

In contrast, cancer-associated mutations are involved in pathways that are crucial for cell proliferation, survival, and metastatic potential. Detection and identification of the harbored mutations in the endometrium have now improved due to the microdissection techniques and next-generation sequencing technologies (NGS) [109].

Recently, a study investigating the mutational profiles of the endometrium of 110 women having hysterectomy or other medical interventions reported that CAMS were detectable in 51–64% of the included women having normal-appearing endometrium [110]. Another study conducted by Moore et al. [114] reported that 95% of the studied tissue harbored CAMs while they also appear normal. The most frequently reported mutations in normal-appearing endometrium are K-RAS, PTEN, and PIK3CA [32, 110]. Similar to the eutopic endometrium, the endometriotic epithelial was found to harbor the K-RAS mutation. Indeed, a mouse model of endometriosis indicated the strong influence of the oncogenic K-RAS mutation in the *eutopic endometrium* on the development of endometriosis through implantation regulation [115]. Additionally, the increased K-RAS activity was suggested to be essential for lesion survival and contribute to the progesterone treatment-induced apoptosis [19]. Moreover, ARID1A is another frequently reported mutation of the ectopic endometrium. ARID1A is a tumor suppressor gene involved mainly in regulating the DNA methylation and repair [116]. This gene is downregulated in the ectopic and eutopic endometrium, and it was suggested to contribute to the clinical phenotype of endometriosis by increasing the invasiveness. Furthermore, ARID1A down-regulation also affects the Transforming Growth Factor-β pathway and was suggested as a potential contributor to the progesterone resistance [19, 32]. Additionally, PTEN and PIK3CA mutations are also frequently encountered in endometriosis [36, 109]. While the eutopic and ectopic endometrium share many CAMs in their genome, these mutations were shown to be heavily enriched in endometriosis. Moreover, the eutopic endometrium of women with endometriosis was also shown to harbor more CAMs when compared to the endometrium of healthy women [32].

However, all of the previously mentioned mutations are located in the genome of the epithelial component of the ectopic endometrium, whereas the stromal cells did not exhibit any mutational profile [117]. Indeed, this further indicates that the stromal and epithelial components of the endometrial lesions are not derived from the same cell.

Discussion of the Pathogenesis of Endometriosis in Light of the Epigenetic/Genetic Theory

It is now clear that endometriosis contains a population of endometrial mesenchymal stem cells and epithelial progenitors that maintain its progression and regeneration. We have now higher confidence that the endometriotic lesions need at least two cell types to originate, their source remains unknown. The mesenchymal stem cells are mainly involved in the endometriotic stroma formation. On the other hand, the endometrial epithelial progenitors seem to be responsible for the formation of the glandular part of the endometriotic lesions. The mesenchymal stem cells could be also responsible for angiogenesis. Furthermore, the same epithelial progenitor could give rise to multiple lesions at different anatomical locations [35]. Genetic predisposition was confirmed by large twin studies [108]. Therefore, a specific genotype could be regarded as an important predisposing factor for developing endometriosis. Moreover, many acquired CAMs could induce a more aggressive behavior of endometriosis when they are harbored by the stem/progenitor cells. Moreover, these CAMs—although less frequent—are shared with the eutopic endometrium. Some authors interpreted these findings depending on the double-hit theory [6, 109]. In other words, a hereditary predisposing genotype is transmitted to daughter cells at fertilization. These predisposed cell could acquire and accumulate many mutations with aging. When enough mutations are accumulated, then these mutant cells can proliferate, migrate, and invade to form endometriosis. Indeed, this was mainly inspired by the cancer biology, especially when knowing that the eutopic endometrium and the endometrial cancer share the same mutations, but in different frequencies. CAMs in cancer cells are much more than those of the eutopic endometrium [118]. Alternatively, CAMs in endometriosis are also more frequent than the normal-appearing endometrium [32]. However, this seems convincing for the first instance, but an important question should be borne in mind: "*If endometriosis is a genetic disease, then why only 50% of twin develop endometriosis?*". Apparently, the stem cell and genetic theories cannot answer this doubt adequately.

The epigenetic theory state that specific stimulants driven by chronic inflammation and hormone dysregulation could cause the ectopic endometrial cells to express either genetic, or post-transcriptional abnormalities that could lead to the pathogenesis and progression of endometriosis [6]. In fact, this postulation is further supported by the observation of many post-transcriptional dysregulations that were associated with essential characteristics of endometriosis. For example, the endometriotic lesions were found to express lower levels of miRNA-15a-5p and miRNA-34a-5p. These two microRNAs inhibit the transcription of VEGF gene. Thus, when they are downregulated increased angiogenesis would occur [38, 119]. Furthermore, let-7b and miRNA-145 were found to be down regulated in the endometriotic lesions, while miRNA-200b was upregulated. This combination resulted in increased cellular proliferation and endometriosis progression [37]. Furthermore, many authors have explained the increased CAMs frequency in the ectopic endometrium when compared with the eutopic counterpart with the epigenetic theory and the mutagenic factors of the peritoneal milieu [6, 109].

The epigenetic events that cause these alterations are not known yet, but the harsh microenvironment in the peritoneal cavity was suggested to be responsible for many of them. Indeed, continuous retrograde menstruation was also suggested as a potential contributor to the post-transcriptional abnormalities through inducing repeated Tissue Injury And Repair processes (TIAR). The presence of the menstrual debris within the peritoneum could provoke an inflammatory reaction accompanied by increased oxidative stress and iron deposits [6, 20]. In fact, the role of TIAR and chronic inflammation should not be underestimated. Indeed, an initial local inflammatory reaction could be all what it takes to initiate endometriosis through the CXCR-4/CXCL-12 axis. CXCR-4 is a superficial receptor expressed by stem cells, while CXCL-12 is its ligand [120]. CXCL-12 was found to be expressed by injured and inflamed tissues [121]. Moreover, the CXCR-4/CXCL-12 axis was found to induce the migration of the eutopic endometrial cells [122]. In addition, an in vitro study showed that estrogen increased the expression of CXCR-4 in the mouse bone marrow stem cells, while another study reported that progesterone increased the expression of CXCR-12 in the eutopic human endometrium [122, 123]. Indeed, the CXCR-4/CXCL-12 axis could play an important role in lesion initiation, and recruiting the circulating bone marrow stem cells to the stroma [19, 122]. This could also explain the pathogenesis of scar endometriosis and extra-pelvic endometriotic lesions. It would be attractive to think that the circulating endometrial cells stem cells [124] could be attracted through the CXCR-4/CXCL-12 axis activated through local inflammation at any region of the body to form ectopic lesions. Furthermore, this could also explain the formation of stromal endometriosis, since CXCR-4 was found to be highly expressed by the eutopic endometrial stromal cells [122]. Figure 9.1 provides a summary of the suggested events in the pathogenesis of endometriosis.

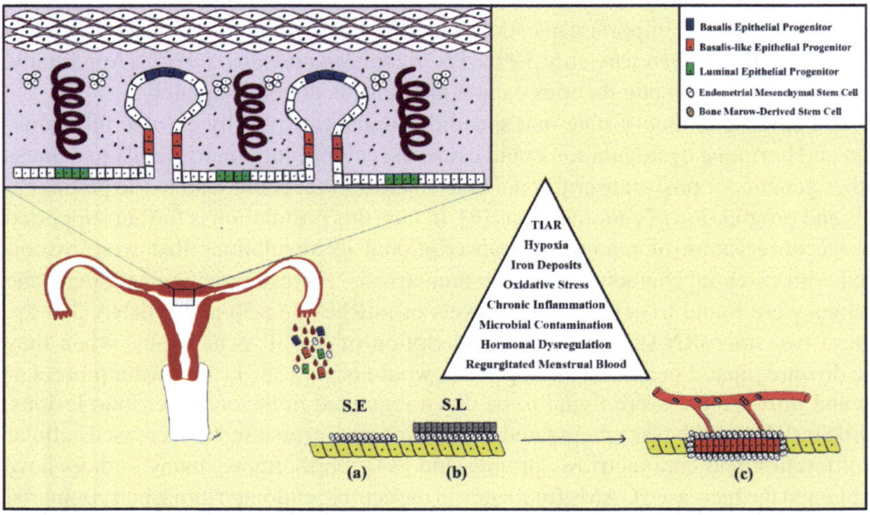

Fig. 9.1 The pathogenesis of endometriosis. **a** Stem cells of the eutopic endometrium. The basalis-like cells of the functional layer (red) and cells from the luminal layer (green) are shed during menstruation. Cells from the basal layer (blue) are also supposed to shed. Each of these cells expresses stemness markers related to self-renewal, proliferation, and clonal expansion. Mesenchymal stem cells (white) also shed during menstruation. All of these cells can be transported to the peritoneal cavity through the retrograde flow. **b** Subtle lesion formation. After reaching the peritoneal cavity, the endometrial mesenchymal stem cells and the epithelial progenitor cells seed in a predisposed peritoneal spot. Peritoneal predisposition could result from repeated TIAR due to continuous regurgitation of the menstrual blood, or other local events that derive an inflammatory reaction. The mesenchymal stem cells with or without the epithelial progenitors are attracted to the predisposed spot under the influence of paracrine factors. The first lesion demonstrated on the left represents stromal endometriosis. Subtle lesions remain quiescent until sufficient stimulations activate endometriosis (endometriotic disease). **c** The endometriotic disease. Harsh environmental factors encountered within the peritoneal milieu provoke additional genetic and epigenetic dysregulations that enable progression, invasion, migration, and angiogenesis. Bone marrow-derived stem cells are recruited and contribute to the endometriotic stroma regeneration. In addition, the endometriotic cells could invade the systemic circulation and spread to other body sites. S.E: stromal endometriosis, S.L: Subtle lesion, TIAR: tissue injury and repair

Conclusion

The pathogenesis of endometriosis is a complicated process. Many hypotheses were made but none of them explain all the manifestations of endometriosis. While we have increased knowledge of many events and molecules involved in the pathogenesis of endometriosis, we still have many more to learn. Strong evidence indicates that stem cells are essential for the development and progression of endometriosis. However, the source of these stem cells is not determined yet. Genetic predisposition paired with driving epigenetic events gives endometriosis much of its aggressive characteristics. Better understanding of the endometriosis pathogenesis is acquired through deep understanding and good correlation between the supposed pathogenesis theories. Each formulated theory forms an important piece that should be puzzled with the rest in order of building a bigger, clearer, and hopefully unifying theory of the endometriosis pathogenesis.

Acknowledgements The authors are grateful to Dr. Bayan Alsaid for assisting in designing Fig. 9.1.

References

1. Meuleman C, Vandenabeele B, Fieuws S, Spiessens C, Timmerman D, D'Hooghe T (2009) High prevalence of endometriosis in infertile women with normal ovulation and normospermic partners. Fertil Steril 92:68–74. https://doi.org/10.1016/j.fertnstert.2008.04.056
2. Shafrir AL, Farland LV, Shah DK, Harris HR, Kvaskoff M, Zondervan K et al (2018) Risk for and consequences of endometriosis: a critical epidemiologic review. Best Pract Res Clin Obst Gynaecol 51:1–15. https://doi.org/10.1016/j.bpobgyn.2018.06.001
3. Hill CJ, Fakhreldin M, Maclean A, Dobson L, Nancarrow L, Bradfield A et al (2020) Endometriosis and the fallopian tubes: theories of origin and clinical implications. J Clin Med 9:1905. https://doi.org/10.3390/jcm9061905
4. Haas D, Chvatal R, Reichert B, Renner S, Shebl O, Binder H et al (2012) Endometriosis: a premenopausal disease? Age pattern in 42,079 patients with endometriosis. Arch Gynecol Obstet 286:667–760. https://doi.org/10.1007/s00404-012-2361-z
5. Naem A, Shamandi A, Al-Shiekh A, Alsaid B (2020) Free large sized intra-abdominal endometrioma in a postmenopausal woman: a case report. BMC Womens Health 20:190. https://doi.org/10.1186/s12905-020-01054-x
6. Koninckx PR, Ussia A, Adamyan L, Wattiez A, Gomel V, Martin DC (2019) Pathogenesis of endometriosis: the genetic/epigenetic theory. Fertil Steril 111:327–340. https://doi.org/10.1016/j.fertnstert.2018.10.013
7. Simpson JL, Elias S, Malinak LR, Buttram VC Jr (1980) Heritable aspects of endometriosis. I. Genetic studies. Am J Obstet Gynecol 137:327–331. https://doi.org/10.1016/0002-9378(80)90917-5
8. Coxhead D, Thomas EJ (1993) Familial inheritance of endometriosis in a British population. A case control study. J Obstet Gynaecol 13:42–44
9. Kennedy S (1998) The genetics of endometriosis. J Reprod Med 43(3 Suppl):263–268
10. Kennedy S, Hadfield R, Westbrook C, Weeks DE, Barlow D, Golding S (1998) Magnetic resonance imaging to assess familial risk in relatives of women with endometriosis. Lancet (London, England) 352:1440–1441. https://doi.org/10.1016/s0140-6736(05)61262-7

11. Carter JE (1994) Combined hysteroscopic and laparoscopic findings in patients with chronic pelvic pain. J Am Assoc Gynecol Laparosc 2:43–47. https://doi.org/10.1016/s1074-3804(05)80830-8
12. Sourial S, Tempest N, Hapangama DK (2014) Theories on the pathogenesis of endometriosis. Int J Reprod Med 2014:179515. https://doi.org/10.1155/2014/179515
13. Koninckx PR, Zupi E, Martin DC (2018) Endometriosis and pregnancy outcome. Fertil Steril 110:406–407. https://doi.org/10.1016/j.fertnstert.2018.06.029
14. Lee HJ, Park YM, Jee BC, Kim YB, Suh CS (2015) Various anatomic locations of surgically proven endometriosis: a single-center experience. Obstet Gynecol Sci 58:53–58. https://doi.org/10.5468/ogs.2015.58.1.53
15. Nisolle M, Donnez J (2019) Reprint of: peritoneal endometriosis, ovarian endometriosis, and adenomyotic nodules of the rectovaginal septum are three different entities. Fertil Steril 112(4 Suppl1):e125–e36. https://doi.org/10.1016/j.fertnstert.2019.08.081
16. Victory R, Diamond MP, Johns DA (2007) Villar's nodule: a case report and systematic literature review of endometriosis externa of the umbilicus. J Min Invas Gynecol 14:23–32. https://doi.org/10.1016/j.jmig.2006.07.014
17. Gates J, Sharma A, Kumar A (2018) Rare case of thoracic endometriosis presenting with lung nodules and pneumothorax. BMJ Case Rep. https://doi.org/10.1136/bcr-2018-224181
18. Maniglio P, Ricciardi E, Meli F, Tomao F, Peiretti M, Caserta D (2018) Complete remission of cerebral endometriosis with dienogest: a case report. Gynecol Endocrinol 34:837–839. https://doi.org/10.1080/09513590.2018.1463362
19. Wang Y, Nicholes K, Shih IM (2020) The origin and pathogenesis of endometriosis. Annu Rev Pathol 15:71–95. https://doi.org/10.1146/annurev-pathmechdis-012419-032654
20. Liu Y, Zhang Z, Yang F, Wang H, Liang S, Wang H et al (2020) The role of endometrial stem cells in the pathogenesis of endometriosis and their application to its early diagnosis†. Biol Reprod 102:1153–1159. https://doi.org/10.1093/biolre/ioaa011
21. Sampson JA (1927) Peritoneal endometriosis due to the menstrual dissemination of endometrial tissue into the peritoneal cavity. Am J Obstet Gynecol 14:422–469
22. D'Hooghe TM, Debrock S (2003) Evidence that endometriosis results from the dislocation of basal endometrium? Hum Reprod (Oxford, England) 18:1130; author reply-1. https://doi.org/10.1093/humrep/deg182
23. Klemmt PAB, Starzinski-Powitz A (2018) Molecular and cellular pathogenesis of endometriosis. Curr Womens Health Rev 14:106–116. https://doi.org/10.2174/1573404813666170306163448
24. Brosens I, Benagiano G (2013) Is neonatal uterine bleeding involved in the pathogenesis of endometriosis as a source of stem cells? Fertil Steril 100:622–623. https://doi.org/10.1016/j.fertnstert.2013.04.046
25. Giannarini G, Scott CA, Moro U, Grossetti B, Pomara G, Selli C (2006) Cystic endometriosis of the epididymis. Urology 68:203.e1–3. https://doi.org/10.1016/j.urology.2006.01.017
26. Zanatta A, Rocha AM, Carvalho FM, Pereira RM, Taylor HS, Motta EL et al (2010) The role of the Hoxa10/HOXA10 gene in the etiology of endometriosis and its related infertility: a review. J Assist Reprod Genet 27:701–710. https://doi.org/10.1007/s10815-010-9471-y
27. Laganà AS, Vitale SG, Salmeri FM, Triolo O, Ban Frangež H, Vrtačnik-Bokal E et al (2017) Unus pro omnibus, omnes pro uno: a novel, evidence-based, unifying theory for the pathogenesis of endometriosis. Med Hypotheses 103:10–20. https://doi.org/10.1016/j.mehy.2017.03.032
28. Makiyan Z (2017) Endometriosis origin from primordial germ cells. Organogenesis 13:95–102. https://doi.org/10.1080/15476278.2017.1323162
29. Taylor HS (2004) Endometrial cells derived from donor stem cells in bone marrow transplant recipients. JAMA 292:81–85. https://doi.org/10.1001/jama.292.1.81
30. Mints M, Jansson M, Sadeghi B, Westgren M, Uzunel M, Hassan M et al (2008) Endometrial endothelial cells are derived from donor stem cells in a bone marrow transplant recipient. Hum Reprod (Oxford, England) 23:139–143. https://doi.org/10.1093/humrep/dem342

31. Ikoma T, Kyo S, Maida Y, Ozaki S, Takakura M, Nakao S et al (2009) Bone marrow-derived cells from male donors can compose endometrial glands in female transplant recipients. Am J Obstet Gynecol 201:608.e1–8. https://doi.org/10.1016/j.ajog.2009.07.026
32. Ponandai-Srinivasan S, Andersson KL, Nister M, Saare M, Hassan HA, Varghese SJ et al (2018) Aberrant expression of genes associated with stemness and cancer in endometria and endometrioma in a subset of women with endometriosis. Hum Reprod (Oxford, England) 33:1924–1938. https://doi.org/10.1093/humrep/dey241
33. Song Y, Xiao L, Fu J, Huang W, Wang Q, Zhang X et al (2014) Increased expression of the pluripotency markers sex-determining region Y-box 2 and Nanog homeobox in ovarian endometriosis. Reprod Biol Endocrinol 12:42. https://doi.org/10.1186/1477-7827-12-42
34. Shariati F, Favaedi R, Ramazanali F, Ghoraeian P, Afsharian P, Aflatoonian B et al (2018) Increased expression of stemness genes REX-1, OCT-4, NANOG, and SOX-2 in women with ovarian endometriosis versus normal endometrium: a case-control study. Int J Reprod Biomed 2016. https://doi.org/10.18502/ijrm.v16i12.3684
35. Silveira CG, Abrão MS, Dias JA Jr, Coudry RA, Soares FA, Drigo SA et al (2012) Common chromosomal imbalances and stemness-related protein expression markers in endometriotic lesions from different anatomical sites: the potential role of stem cells. Hum Reprod (Oxford, England) 27:3187–3197. https://doi.org/10.1093/humrep/des282
36. Kyo S, Sato S, Nakayama K (2020) Cancer-associated mutations in normal human endometrium: surprise or expected? Cancer Sci 111:3458–3467. https://doi.org/10.1111/cas.14571
37. Mashayekhi P, Noruzinia M, Zeinali S, Khodaverdi S (2019) Endometriotic mesenchymal stem cells epigenetic pathogenesis: deregulation of miR-200b, miR-145, and let7b in A functional imbalanced epigenetic disease. Cell J 21:179–185. https://doi.org/10.22074/cellj.2019.5903
38. Liu XJ, Bai XG, Teng YL, Song L, Lu N, Yang RQ (2016) miRNA-15a-5p regulates VEGFA in endometrial mesenchymal stem cells and contributes to the pathogenesis of endometriosis. Eur Rev Medical Pharmacol Sci 20:3319–3326
39. Sadler TW, Langman's medical embryology. Lippincott Williams & Wilkins Philadelphia
40. LaRonde-LeBlanc NA, Wolberger C, Structure of HoxA9 and Pbx1 bound to DNA: hox hexapeptide and DNA recognition anterior to posterior. Genes Dev 17:2060–2072. https://doi.org/10.1101/gad.1103303
41. McGinnis W, Krumlauf R (1992) Homeobox genes and axial patterning. Cell 68:283–302. https://doi.org/10.1016/0092-8674(92)90471-n
42. Capellini TD, Zewdu R, Di Giacomo G, Asciutti S, Kugler JE, Di Gregorio A et al (2008) Pbx1/Pbx2 govern axial skeletal development by controlling Polycomb and Hox in mesoderm and Pax1/Pax9 in sclerotome. Dev Biol 321:500–514. https://doi.org/10.1016/j.ydbio.2008.04.005
43. Schnabel CA, Godin RE, Cleary ML (2003) Pbx1 regulates nephrogenesis and ureteric branching in the developing kidney. Dev Biol 254:262–276. https://doi.org/10.1016/s0012-1606(02)00038-6
44. Schnabel CA, Selleri L, Jacobs Y, Warnke R, Cleary ML (2001) Expression of Pbx1b during mammalian organogenesis. Mech Dev 100:131–135. https://doi.org/10.1016/s0925-4773(00)00516-5
45. Schnabel CA, Selleri L, Cleary ML (2003) Pbx1 is essential for adrenal development and urogenital differentiation. Genesis (New York, NY: 2000) 37:123–130. https://doi.org/10.1002/gene.10235
46. Taylor HS, Vanden Heuvel GB, Igarashi P (1997) A conserved Hox axis in the mouse and human female reproductive system: late establishment and persistent adult expression of the Hoxa cluster genes. Biol Reprod 57:1338–1345. https://doi.org/10.1095/biolreprod57.6.1338
47. Cunha GR (1976) Stromal induction and specification of morphogenesis and cytodifferentiation of the epithelia of the Mullerian ducts and urogenital sinus during development of the uterus and vagina in mice. J Exp Zool 196:361–370. https://doi.org/10.1002/jez.1401960310

48. Branford WW, Benson GV, Ma L, Maas RL, Potter SS (2000) Characterization of Hoxa-10/Hoxa-11 transheterozygotes reveals functional redundancy and regulatory interactions. Dev Biol 224:373–387. https://doi.org/10.1006/dbio.2000.9809
49. Warot X, Fromental-Ramain C, Fraulob V, Chambon P, Dollé P (1997) Gene dosage-dependent effects of the Hoxa-13 and Hoxd-13 mutations on morphogenesis of the terminal parts of the digestive and urogenital tracts. Development (Cambridge, England) 124:4781–4791
50. Vainio S, Heikkilä M, Kispert A, Chin N, McMahon AP (1999) Female development in mammals is regulated by Wnt-4 signalling. Nature 397:405–409. https://doi.org/10.1038/17068
51. Miller C, Sassoon DA (1998) Wnt-7a maintains appropriate uterine patterning during the development of the mouse female reproductive tract. Development (Cambridge, England) 125:3201–3211
52. Mericskay M, Kitajewski J, Sassoon D (2004) Wnt5a is required for proper epithelial-mesenchymal interactions in the uterus. Development (Cambridge, England) 131:2061–2072. https://doi.org/10.1242/dev.01090
53. Selleri L, Depew MJ, Jacobs Y, Chanda SK, Tsang KY, Cheah KS et al (2001) Requirement for Pbx1 in skeletal patterning and programming chondrocyte proliferation and differentiation. Development (Cambridge, England) 128:3543–3557
54. Hayashi K, Yoshioka S, Reardon SN, Rucker EB 3rd, Spencer TE, DeMayo FJ et al (2011) WNTs in the neonatal mouse uterus: potential regulation of endometrial gland development. Bio Reprod 84:308–319. https://doi.org/10.1095/biolreprod.110.088161
55. Serrano-Gomez SJ, Maziveyi M, Alahari SK (2016) Regulation of epithelial-mesenchymal transition through epigenetic and post-translational modifications. Mol Cancer 15:18. https://doi.org/10.1186/s12943-016-0502-x
56. Gargett CE (2007) Uterine stem cells: what is the evidence? Hum Reprod Update 13:87–101. https://doi.org/10.1093/humupd/dml045
57. Makiyan Z (2016) New theory of uterovaginal embryogenesis. Organogenesis 12:33–41. https://doi.org/10.1080/15476278.2016.1145317
58. Dabi Y, Canel V, Skalli D, Paniel BJ, Haddad B, Touboul C (2020) Postoperative evaluation of chronic pain in patients with Mayer - Rokitansky - Küster - Hauser (MRKH) syndrome and uterine horn remnant: experience of a tertiary referring gynaecological department. J Gynecol Obstet Hum Reprod 49:101655. https://doi.org/10.1016/j.jogoh.2019.101655
59. Signorile PG, Baldi F, Bussani R, D'Armiento MR, De Falco M, Boccellino M et al (2010) New evidence sustaining the presence of endometriosis in the human foetus. Reprod Biomed Online 21:142–147
60. Signorile PG, Baldi A, Endometriosis: new concepts in the pathogenesis. Int J Bioch Cell Biol 42:778–870. https://doi.org/10.1016/j.biocel.2010.03.008
61. Browne H, Taylor H (2006) HOXA10 expression in ectopic endometrial tissue. Fertil Steril 85:1386–1390. https://doi.org/10.1016/j.fertnstert.2005.10.072
62. Matsuzaki S, Canis M, Darcha C, Pouly JL, Mage G (2009) HOXA-10 expression in the mid-secretory endometrium of infertile patients with either endometriosis, uterine fibromas or unexplained infertility. Hum Reprod (Oxford, England) 24:3180–3187. https://doi.org/10.1093/humrep/dep306
63. Gaetje R, Holtrich U, Engels K, Kissler S, Rody A, Karn T et al (2007) Endometriosis may be generated by mimicking the ontogenetic development of the female genital tract. Fertil Steril 87:651–656. https://doi.org/10.1016/j.fertnstert.2006.07.1533
64. Gaetje R, Holtrich U, Karn T, Cikrit E, Engels K, Rody A et al (2007) Characterization of WNT7A expression in human endometrium and endometriotic lesions. Fertil Steril 88:1534–1540. https://doi.org/10.1016/j.fertnstert.2007.01.128
65. de Mattos RM, Pereira PR, Barros EG, da Silva JH, Palmero CY, da Costa NM et al (2016) Aberrant levels of Wnt/β-catenin pathway components in a rat model of endometriosis. Histol Histopathol 31:933–942. https://doi.org/10.14670/hh-11-730

66. Xiong W, Zhang L, Yu L, Xie W, Man Y, Xiong Y et al (2015) Estradiol promotes cells invasion by activating β-catenin signaling pathway in endometriosis. Reproduction (Cambridge, England) 150:507–516. https://doi.org/10.1530/rep-15-0371
67. Zhang L, Xiong W, Xiong Y, Liu H, Liu Y (2016) 17 β-Estradiol promotes vascular endothelial growth factor expression via the Wnt/β-catenin pathway during the pathogenesis of endometriosis. Mol Hum Reprod 22:235–526. https://doi.org/10.1093/molehr/gaw025
68. Stewart B, Reddington C, Cameron M (2020) Laparoscopic hemihysterectomy for obstructive uterine didelphys with unilateral vaginal hypoplasia. J Minim Invasive Gynecol 27:1225–1227. https://doi.org/10.1016/j.jmig.2019.12.019
69. Fraunhoffer NA, Meilerman Abuelafia A, Stella I, Galliano S, Barrios M, Vitullo AD (2015) Identification of germ cell-specific VASA and IFITM3 proteins in human ovarian endometriosis. J Ovarian Res 8:66. https://doi.org/10.1186/s13048-015-0193-8
70. Simsek G, Bulus H, Tas A, Koklu S, Yilmaz SB, Coskun A (2012) An unusual cause of inguinal hernia in a male patient: endometriosis. Gut Liver 6:284–285. https://doi.org/10.5009/gnl.2012.6.2.284
71. Noë M, Ayhan A, Wang TL, Shih IM (2018) Independent development of endometrial epithelium and stroma within the same endometriosis. J Pathol 245:265–269. https://doi.org/10.1002/path.5082
72. Suda K, Nakaoka H, Yoshihara K, Ishiguro T, Tamura R, Mori Y et al (2018) Clonal expansion and diversification of cancer-associated mutations in endometriosis and normal endometrium. Cell Rep 24:1777–1789. https://doi.org/10.1016/j.celrep.2018.07.037
73. Jabbour HN, Kelly RW, Fraser HM, Critchley HO (2006) Endocrine regulation of menstruation. Endocr Rev 27:17–46. https://doi.org/10.1210/er.2004-0021
74. Tresserra F, Grases P, Ubeda A, Pascual MA, Grases PJ, Labastida R (1999) Morphological changes in hysterectomies after endometrial ablation. Hum Reprod (Oxford, England) 14:1473–1477. https://doi.org/10.1093/humrep/14.6.1473
75. Muller I, van der Palen J, Massop-Helmink D, Vos-de Bruin R, Sikkema JM (2015) Patient satisfaction and amenorrhea rate after endometrial ablation by ThermaChoice III or NovaSure: a retrospective cohort study. J Gynecol Surg 12:81–87
76. Cousins FL, Gargett CE (2018) Endometrial stem/progenitor cells and their role in the pathogenesis of endometriosis. Best Pract Res Clin Obstet Gynaecol 50:27–38. https://doi.org/10.1016/j.bpobgyn.2018.01.011
77. Tempest N, Maclean A, Hapangama DK (2018) Endometrial stem cell markers: current concepts and unresolved questions. Int J Mol Sci 19:3240. https://doi.org/10.3390/ijms19103240
78. Gargett CE, Masuda H (2010) Adult stem cells in the endometrium. Mol Hum Reprod 16:818–834. https://doi.org/10.1093/molehr/gaq061
79. Schwab KE, Gargett CE (2007) Co-expression of two perivascular cell markers isolates mesenchymal stem-like cells from human endometrium. Hum Reprod (Oxford, England). 22:2903–2911. https://doi.org/10.1093/humrep/dem265
80. Masuda H, Anwar SS, Bühring HJ, Rao JR, Gargett CE (2012) A novel marker of human endometrial mesenchymal stem-like cells. Cell Transplant 21:2201–2214. https://doi.org/10.3727/096368911x637362
81. Chan RW, Schwab KE, Gargett CE (2004) Clonogenicity of human endometrial epithelial and stromal cells. Biol Reprod 70:1738–1750. https://doi.org/10.1095/biolreprod.103.024109
82. Nguyen HPT, Xiao L, Deane JA, Tan KS, Cousins FL, Masuda H et al (2017) N-cadherin identifies human endometrial epithelial progenitor cells by in vitro stem cell assays. Hum Reprod (Oxford, England). 32:2254–2268. https://doi.org/10.1093/humrep/dex289
83. Masuda H, Matsuzaki Y, Hiratsu E, Ono M, Nagashima T, Kajitani T et al (2010) Stem cell-like properties of the endometrial side population: implication in endometrial regeneration. PLoS ONE 5:e10387. https://doi.org/10.1371/journal.pone.0010387
84. Miyazaki K, Maruyama T, Masuda H, Yamasaki A, Uchida S, Oda H et al (2012) Stem cell-like differentiation potentials of endometrial side population cells as revealed by a newly developed in vivo endometrial stem cell assay. PLoS ONE 7:e50749. https://doi.org/10.1371/journal.pone.0050749

85. Goodell MA, Rosenzweig M, Kim H, Marks DF, DeMaria M, Paradis G et al (1997) Dye efflux studies suggest that hematopoietic stem cells expressing low or undetectable levels of CD34 antigen exist in multiple species. Nat Med 3:1337–1345. https://doi.org/10.1038/nm1297-1337
86. Gurung S, Deane JA, Masuda H, Maruyama T, Gargett CE (2015) Stem cells in endometrial physiology. Semin Reprod Med 33:326–332. https://doi.org/10.1055/s-0035-1558405
87. Cervelló I, Mas A, Gil-Sanchis C, Peris L, Faus A, Saunders PT et al (2011) Reconstruction of endometrium from human endometrial side population cell lines. PLoS ONE 6:e21221. https://doi.org/10.1371/journal.pone.0021221
88. Leyendecker G, Herbertz M, Kunz G, Mall G (2002) Endometriosis results from the dislocation of basal endometrium. Hum Reprod (Oxford, England) 17:2725–2736. https://doi.org/10.1093/humrep/17.10.2725
89. Hapangama DK, Drury J, Da Silva L, Al-Lamee H, Earp A, Valentijn AJ et al (2019) Abnormally located SSEA1+/SOX9+ endometrial epithelial cells with a basalis-like phenotype in the eutopic functionalis layer may play a role in the pathogenesis of endometriosis. Hum Reprod (Oxford, England) 34:56–68. https://doi.org/10.1093/humrep/dey336
90. Valentijn AJ, Palial K, Al-Lamee H, Tempest N, Drury J, Von Zglinicki T et al (2013) SSEA-1 isolates human endometrial basal glandular epithelial cells: phenotypic and functional characterization and implications in the pathogenesis of endometriosis. Hum Reprod (Oxford, England) 28:2695–2708. https://doi.org/10.1093/humrep/det285
91. Kao AP, Wang KH, Chang CC, Lee JN, Long CY, Chen HS et al (2011) Comparative study of human eutopic and ectopic endometrial mesenchymal stem cells and the development of an in vivo endometriotic invasion model. Fertil Steril 95:1308–1315.e1. https://doi.org/10.1016/j.fertnstert.2010.09.064
92. Tempest N, Baker AM, Wright NA, Hapangama DK (2018) Does human endometrial LGR5 gene expression suggest the existence of another hormonally regulated epithelial stem cell niche? Hum Reprod (Oxford, England) 33:1052–1062. https://doi.org/10.1093/humrep/dey083
93. Leyendecker G, Kunz G, Herbertz M, Beil D, Huppert P, Mall G et al (2004) Uterine peristaltic activity and the development of endometriosis. Ann N Y Acad Sci 1034:338–355. https://doi.org/10.1196/annals.1335.036
94. Vercellini P, Viganò P, Somigliana E, Fedele L (2014) Endometriosis: pathogenesis and treatment. Nat Rev Endocrinol 10:261–275. https://doi.org/10.1038/nrendo.2013.255
95. Guo SW (2009) Recurrence of endometriosis and its control. Hum Reprod Update 15:441–461. https://doi.org/10.1093/humupd/dmp007
96. Gordts S, Koninckx P, Brosens I (2017) Pathogenesis of deep endometriosis. Fertil Steril 108:285–872.e1. https://doi.org/10.1016/j.fertnstert.2017.08.036
97. Chan RW, Ng EH, Yeung WS (2011) Identification of cells with colony-forming activity, self-renewal capacity, and multipotency in ovarian endometriosis. Am J Pathol 178:2832–2844. https://doi.org/10.1016/j.ajpath.2011.02.025
98. Liu Y, Liang S, Yang F, Sun Y, Niu L, Ren Y et al (2020) Biological characteristics of endometriotic mesenchymal stem cells isolated from ectopic lesions of patients with endometriosis. Stem Cell Res Ther 11:346. https://doi.org/10.1186/s13287-020-01856-8
99. Chang JH, Au HK, Lee WC, Chi CC, Ling TY, Wang LM et al (2013) Expression of the pluripotent transcription factor OCT4 promotes cell migration in endometriosis. Fertil Steril 99:1332–1339.e5. https://doi.org/10.1016/j.fertnstert.2012.11.033
100. Zhang Y, Eades G, Yao Y, Li Q, Zhou Q (2012) Estrogen receptor α signaling regulates breast tumor-initiating cells by down-regulating miR-140 which targets the transcription factor SOX2. J Biol Chem 287:41514–41522. https://doi.org/10.1074/jbc.M112.404871
101. Yang L, Luo L, Ji W, Gong C, Wu D, Huang H et al (2013) Effect of low dose bisphenol A on the early differentiation of human embryonic stem cells into mammary epithelial cells. Toxicol Lett 218:187–193. https://doi.org/10.1016/j.toxlet.2013.01.026
102. Banerjee P, Fazleabas AT (2010) Endometrial responses to embryonic signals in the primate. Int J Dev Biol 54:295–302. https://doi.org/10.1387/ijdb.082829pb

103. Bray SJ (2006) Notch signalling: a simple pathway becomes complex. Nature Rev Mol Cell Biol 7:678–689. https://doi.org/10.1038/nrm2009
104. He H, Liu R, Xiong W, Pu D, Wang S, Li T (2016) Lentiviral vector-mediated down-regulation of Notch1 in endometrial stem cells results in proliferation and migration in endometriosis. Mol Cell Endocrinol 434:210–218. https://doi.org/10.1016/j.mce.2016.07.004
105. Götte M, Wolf M, Staebler A, Buchweitz O, Kelsch R, Schüring AN et al (2008) Increased expression of the adult stem cell marker Musashi-1 in endometriosis and endometrial carcinoma. J Pathol 215:317–329. https://doi.org/10.1002/path.2364
106. Li X, Gong X, Zhu L, Leng J, Fan Q, Sun D et al (2012) Stretch magnitude- and frequency-dependent cyclooxygenase 2 and prostaglandin E2 up-regulation in human endometrial stromal cells: possible implications in endometriosis. Exp Biol Med (Maywood, NJ) 237:1350–1358. https://doi.org/10.1258/ebm.2012.012060
107. Sha G, Zhang Y, Zhang C, Wan Y, Zhao Z, Li C et al (2009) Elevated levels of gremlin-1 in eutopic endometrium and peripheral serum in patients with endometriosis. Fertil Sterilit 91:350–358. https://doi.org/10.1016/j.fertnstert.2007.12.007
108. Treloar SA, O'Connor DT, O'Connor VM, Martin NG (1999) Genetic influences on endometriosis in an Australian twin sample. Fertil Steril 71:701–710. https://doi.org/10.1016/s0015-0282(98)00540-8
109. Guo SW (2020) Cancer-associated mutations in endometriosis: shedding light on the pathogenesis and pathophysiology. Hum Reprod Update 26:423–449. https://doi.org/10.1093/humupd/dmz047
110. Lac V, Nazeran TM, Tessier-Cloutier B, Aguirre-Hernandez R, Albert A, Lum A et al (2019) Oncogenic mutations in histologically normal endometrium: the new normal? J Pathol 249:173–181. https://doi.org/10.1002/path.5314
111. Araten DJ, Golde DW, Zhang RH, Thaler HT, Gargiulo L, Notaro R et al (2005) A quantitative measurement of the human somatic mutation rate. Cancer Res 65:8111–8117. https://doi.org/10.1158/0008-5472.can-04-1198
112. Nachman MW, Crowell SL (2000) Estimate of the mutation rate per nucleotide in humans. Genetics 156:297–304
113. Rozhok AI, DeGregori J (2016) The evolution of lifespan and age-dependent cancer risk. Trends Cancer 2:552–560. https://doi.org/10.1016/j.trecan.2016.09.004
114. Moore L, Leongamornlert D, Coorens THH, Sanders MA, Ellis P, Dentro SC et al (2020) The mutational landscape of normal human endometrial epithelium. Nature 580:640–646. https://doi.org/10.1038/s41586-020-2214-z
115. Cheng CW, Licence D, Cook E, Luo F, Arends MJ, Smith SK et al (2011) Activation of mutated K-ras in donor endometrial epithelium and stroma promotes lesion growth in an intact immunocompetent murine model of endometriosis. J Pathol 224:261–269. https://doi.org/10.1002/path.2852
116. Wu RC, Wang TL, Shih Ie M (2014) The emerging roles of ARID1A in tumor suppression. Cancer Biol Ther 15:655–664. https://doi.org/10.4161/cbt.28411
117. Anglesio MS, Papadopoulos N, Ayhan A, Nazeran TM, Noë M, Horlings HM et al (2017) Cancer-associated mutations in endometriosis without cancer. N Engl J Med 376:1835–1848. https://doi.org/10.1056/NEJMoa1614814
118. Kandoth C, Schultz N, Cherniack AD, Akbani R, Liu Y, Shen H et al (2013) Integrated genomic characterization of endometrial carcinoma. Nature 497:67–73. https://doi.org/10.1038/nature12113
119. Ma Y, Huang YX, Chen YY (2017) miRNA-34a-5p downregulation of VEGFA in endometrial stem cells contributes to the pathogenesis of endometriosis. Mol Med Rep 16:8259–8264. https://doi.org/10.3892/mmr.2017.7677
120. Lai CY, Yamazaki S, Okabe M, Suzuki S, Maeyama Y, Iimura Y et al (2014) Stage-specific roles for CXCR4 signaling in murine hematopoietic stem/progenitor cells in the process of bone marrow repopulation. Stem Cells (Dayton, Ohio) 32:1929–1942. https://doi.org/10.1002/stem.1670

121. Hattori K, Heissig B, Rafii S (2003) The regulation of hematopoietic stem cell and progenitor mobilization by chemokine SDF-1. Leuk Lymphoma 44:575–582. https://doi.org/10.1080/1042819021000037985
122. Leconte M, Chouzenoux S, Nicco C, Chéreau C, Arkwright S, Santulli P et al (2014) Role of the CXCL12-CXCR4 axis in the development of deep rectal endometriosis. J Reprod Immunol 103:45–52. https://doi.org/10.1016/j.jri.2013.12.121
123. Moridi I, Mamillapalli R, Cosar E, Ersoy GS, Taylor HS (2017) Bone marrow stem cell chemotactic activity is induced by elevated CXCl12 in endometriosis. Reprod Sci (Thousand Oaks, Calif). 24:526–533. https://doi.org/10.1177/1933719116672587
124. Pospisilova E, Kiss I, Souckova H, Tomes P, Spicka J, Matkowski R et al (2019) Circulating endometrial cells: a new source of information on endometriosis dynamics. J Clin Med 8:1938. https://doi.org/10.3390/jcm8111938

Chapter 10
Stem Cell Transplantation for Endometrial Regeneration in Humans

Lucía de Miguel Gómez, Antonio Pellicer, and Irene Cervelló

Abstract

Introduction: Patients with a pathologic endometrium, either due to endometrial atrophy and/or Asherman's syndrome, have a diminished probability of achieving a healthy pregnancy. Although different therapeutic alternatives have been tested over the years, stem cell therapy is the most promising.

Methods: Google Scholar and the PubMed and Web of Science databases were used to select studies for this chapter.

Results: Various sources of stem cells, either autologous (bone marrow, adipose tissue, menstrual blood, or endometrium) or allogenic (umbilical cord), have been used for transplantation in women with endometrial pathologies compromising fertility. Of all stem cell sources, bone marrow has been the most explored, probably due to the suggested contribution of bone marrow-derived stem cells to the endometrial stem cell niche.

Conclusions: The stem cell approach has emerged as an effective therapy for treating endometrial pathologies and achieving healthy pregnancies. However, in recent years, these treatments have been redirected to less invasive options. This trend has encouraged the use of stem cell derivatives, such as the secretome or exosomes, that can be used alone or in combination with other approaches like platelet-rich plasma or bioengineering solutions. This chapter presents various alternatives that stem cell therapy offers to treat specific pathologies affecting the endometrium and how these therapeutic approaches have evolved.

Keywords Endometrium · Human · Regeneration · Stem cell · Transplantation

---

L. de Miguel Gómez · I. Cervelló (✉)
Fundación Instituto Valenciano de Infertilidad (FIVI), La Fe Health Research Institute, La Fe University Hospital, Avenida Fernando Abril Martorell, 106., Tower A, 1st Floor, Valencia 46026, Spain
e-mail: Irene.cervello@ivirma.com

L. de Miguel Gómez · A. Pellicer
University of Valencia, 46010 Valencia, Spain

A. Pellicer
IVIRMA Rome, 00197 Rome, Italy

I. Virant-Klun (ed.), *Stem Cells in Reproductive Tissues and Organs*, Stem Cell Biology and Regenerative Medicine 70, https://doi.org/10.1007/978-3-030-90111-0_10

Abbreviations

ADMSC	Adipose-derived mesenchymal stem cell
ART	Assisted reproduction techniques
AS	Asherman's syndrome
BCL-2	B-cell lymphoma 2
BMDSC	Bone marrow-derived stem cell
G-CSF	Granulocyte-colony stimulating factor
BrdU	5-Bromo-2′-deoxyuridine
CD	Cluster of differentiation
CD9	CD9 antigen
CD34	CD34 antigen
CD73	CD73 antigen or 5'-nucleotidase ecto
CD90	CD90 antigen or Thy-1 membrane glycoprotein
CD105	CD105 antigen or endoglin
CD133	CD133 antigen or prominin 1
c-KIT	KIT Proto-oncogene, receptor tyrosine kinase
DNA	Deoxyribonucleic acid
DSC	Derived stem cell
EA	Endometrial atrophy
ECM	Extracellular matrix
EPCAM	Epithelial cell adhesion molecule
ETh	Endometrial thickness
FGF	Fibroblast growth factor
HGF	Hepatocyte growth factor
HSC	Hematopoietic stem cell
ICAM1	Intercellular adhesion molecule 1
IGF1	Insulin-like growth factor 1
IL1β	Interleukin 1β
IL	Interleukin 10
IUA	Intrauterine adhesion
LGR5	Leucine-rich repeat-containing G-protein coupled receptor 5
LIF	Leukemia inhibitory factor
MenMSC	Menstrual blood-derived mesenchymal stem cell
MSC	Mesenchymal stem cell
NK	Natural killer
NOD/SCID	Non-obese diabetic/severe combined immune deficiency (mouse)
OCT-4	Octamer-binding transcription factor 4
PGS scaffold	Poly (glycerol sebacate) scaffold
PRP	Platelet-rich plasma
SDF1α	Stromal cell derived factor 1α
SSEA1	Stage-specific embryonic antigen 1
SP	Side population
TNFα	Tumor necrosis factor α

UCMSC	Umbilical cord derived-mesenchymal stem cell
VEGF	Vascular endothelial growth factor
vWF	von Willebrand factor

Introduction

The human female reproductive system consists of three main internal organs: ovaries, fallopian tubes, and the uterus. The ovaries, or female gonads, house oocytes and produce and secrete female sex hormones (mainly estrogens and progesterone, but also others such as inhibin or relaxin). The fallopian tubes transport oocytes from the ovaries to the uterine cavity, thus they are also called oviducts. The uterus, specifically the endometrial layer, is responsible for accommodating the embryo and subsequent fetal development until birth [1].

The uterus is a hollow, muscular, and inverted pear-shaped organ with thick walls, located intrapelvically between the bladder and the rectum. This large organ—measuring on average 8 cm × 5 cm × 2.5 cm—can be differentiated into three main parts: the fundus, corpus, and cervix. The fundus is located above the fallopian tube entries and, in conjunction with the corpus, constitutes two-thirds of the uterus. The lower part, the cervix, communicates with the vagina and is separated from the corpus by a short isthmus [2]. The corpus uterine wall consists of three different layers (from outside to inside): perimetrium, the external or serous layer; myometrium, the middle muscular and thickest layer; and endometrium, the internal or mucous layer. The latter is absolutely essential in the reproductive process.

The Human Endometrium

The endometrium plays a fundamental role during embryo implantation. Sufficient morphology and thickness and, ultimately, appropriate functionality of this tissue is crucial to achieve a successful pregnancy [1]. Therefore, any problem or complication related to the endometrium could cause fertility difficulties [3].

At structural and morphological levels, the human endometrium comprises three main compartments—the epithelium, stroma, and a vascular network—apart from a population of resident immune cells [4] (Fig. 10.1a).

- The epithelium, responsible for embryo implantation, consists of a monolayer of cuboidal polarized cells covering the uterine cavity that is organized in luminal and glandular compartments. In response to estrogen levels, which increase during the proliferative phase, luminal epithelial cells develop *microvilli* that disappear during the secretory phase. While glandular epithelial cells proliferate during the early secretory phase, they become longer and convoluted as the phase progresses [4].

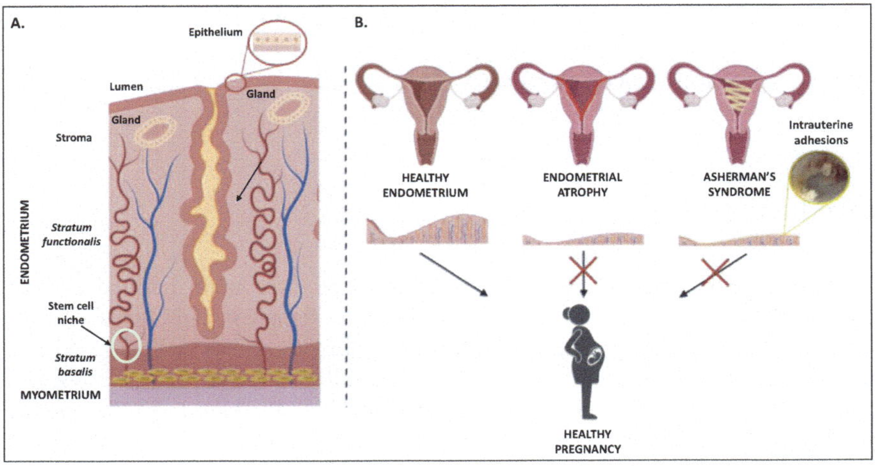

Fig. 10.1 The human endometrium morphology and endometrial pathologies. **a** Sectional view of endometrial tissue showing organization of the two main layers: *stratum basalis*, the layer closest to the myometrium and a putative endometrial stem cell niche; and *stratum functionalis*. **b** Graphical representation of two main endometrial pathologies affecting fertility: endometrial atrophy, characterized by an atrophic endometrium with an insufficient thickness for embryo implantation; and Asherman's syndrome, characterized by the presence of intrauterine adhesions (right upper yellow circle) (image adapted with permission from www.drjennycook.com). Image created with www.BioRender.com

- The stroma consists of different cells, mainly fibroblasts, and extracellular matrix (ECM). These fibroblasts are implicated in remodeling the ECM along the menstrual cycle, favoring stromal decidualization in response to ovarian hormones, which in turn causes morphological changes in the fibroblasts. The initial small and compact cells become large and polygonal, with a well-defined nucleus, developed Golgi apparatus, and parallel laminins of the endoplasmic reticulum, which secrete prolactin, insulin-like growth factor-binding protein 1, and other molecules [4].
- The intrauterine vascular network initiates in the myometrium, forming arcuate arteries that become radial along the myometrium thickness, turning into basal arteries at the endometrium. In this layer, they form a network from which the spiral arteries emerge and maintain the *stratus basalis* and then ramify through the *stratus functionalis*. The importance of this network is on angiogenic[1] events induced in the endometrium during menstruation, to regenerate the vascular network; in the proliferative phase, when tissue growths and arteries elongate; and during the secretory phase, when the spiral arteries grow the most and

[1] Angiogenesis is the growth of blood vessels from existing vasculature.

experiment with intussusception.[2] Additional angiogenic processes occur during embryo implantation and placentation [4].

- The immune components, mainly natural killer (NK) cells, macrophages, and T lymphocytes, are essential to the endometrium for protecting the tissue against pathogens and for avoiding immunological rejection during embryo implantation. NK cells predominate during the proliferative phase (preimplantation), while T lymphocytes and macrophages are present after ovulation.

These four components are organized into two main endometrial layers or compartments, the *stratum basalis* and *stratus functionalis* (Fig. 10.1a). The *stratum basalis*, a thin basal layer adjacent to the myometrium, remains relatively constant along the menstrual cycle. This layer putatively contains a permanent somatic stem cell population, thought to be involved in cyclic regeneration of the functional endometrial layer, as further discussed in this chapter [1, 5]. The *stratum functionalis*, a thick superficial layer adjacent to the endometrial cavity, expands and vascularizes during the menstrual cycle and finally sloughs off during menstruation.

Thus, endometrial tissue has an intrinsic capacity for self-regeneration, thought to be due to the niche of the somatic stem cell population located in the *stratum basalis* (Fig. 10.1a). This cell niche has been postulated to contain not only endogenous endometrial stem cells but also cells derived from an exogenous source, such as bone marrow, that may migrate in response to endometrial stimulation such as damage [6]. Hence, in-depth study of this stem cell population (e.g., cell organization, location, and behavior; influence of other stem cell populations) is essential for a complete understanding of different endometrial pathologies and future effective treatments based on cell therapies.

Understanding Endometrial Pathologies

Absolute uterine factor infertility is attributable to uterine absence or an abnormality that prevents a successful pregnancy [7]. The main cause is hysterectomy, or surgical removal of the uterus, in fertile-age women. Hysterectomy may be performed due to malignant uterine neoplasias, leiomyomas (uterine fibroids), or adenomyosis (ectopic endometrial glands located in the myometrium). Further, congenital conditions can lead to complete absence of the organ, as in Mayer-Rokitansky-Kuster-Hauser syndrome, which is characterized by utero-vaginal atresia and usually occurs in females with a normal phenotype and karyotype (46XX) [8].

Other causes can impair pregnancy but do not necessarily imply absolute infertility. These include leiomyomas or adenomyosis that do not require a hysterectomy and the majority of Müllerian duct anomalies (septate, unicornuate, or bicornuate uterus; uterus didelphys). Endometriosis can also affect reproductive success. This gynecological disorder is defined as the presence of endometrial tissue (endometrial

[2] Process by which a cylindrical microstructure appears that spans the lumen of existing small blood vessels.

glands and stroma) in ectopic locations such as other reproductive organs or even in the abdominal wall, bowel, or lungs [9]. Among pathologies directly related to the endometrial factor, there is differentiation among those that affect the endometrial lining, such as endometrial atrophy or thin endometrium, and those implying intrauterine adhesions (IUAs), such as Asherman's syndrome (AS) [3]. Endometrial hyperplasia or endometritis, which is systemic inflammation of the endometrial tissue, can also compromise fertility.

Asherman's Syndrome

Asherman's syndrome is considered a rare disease and is characterized by the presence of IUAs. It occurs when the normal endometrium is replaced by fibrotic tissue inside the uterine cavity and/or endocervix (Fig. 10.1b). The most common symptoms are menstrual abnormalities, infertility, pelvic pain, recurrent miscarriage, abnormal placentation, and attendant psychological distress [10]. There are different methods to categorize AS patients, but the classification system described by the American Fertility Society in 1988 is probably most commonly used. This system accounts for extensiveness and morphology of IUAs as evaluated by hysteroscopy or hysterosalpingography as well as the menstrual pattern. The method classifies AS patients in three groups: stage I or mild, with few filmy IUAs involving less than a third of the uterine cavity and normal menses or sometimes hypomenorrhea; stage II or moderate, with filmy and dense IUAs involving one-third to two-thirds of the cavity and hypomenorrhea; and stage III or severe, with dense IUAs involving more than two-thirds of the cavity and amenorrhea [11].

Several risk factors contribute to development of AS, usually related to an iatrogenic trauma such as miscarriage, postpartum curettage, or hysteroscopic surgery. Other causes can be a Müllerian duct malformation, uterine artery embolization, or, less frequently, genital infections, caesarean sections, or even insertion of an intrauterine device [12].

Endometrial Atrophy

Endometrial thickness, usually measured by transvaginal sonography, is defined as the minimal distance between echogenic interfaces at the junction of endometrium and myometrium (Fig. 10.1b). This distance varies along the menstrual cycle and usually ranges 1–4 mm during menstruation, 4–8 mm in the middle proliferative phase, 8–14 mm in the late proliferative phase, and 7–14 mm in the secretory phase. Various studies indicate that embryo implantation cannot be achieved if the endometrium thickness is below a certain critical cut-off limit, generally 6 mm; however, there is controversy about the exact value [13].

Diminished endometrial thickness is the cause of endometrial atrophy (EA), also referred to as thin or refractory endometrium. This pathology has an incidence of around 2.4% among patients undergoing in vitro fertilization treatment but is most common in older women, reaching 25% in women >40 years old [14]. Although EA can result from various factors, the most common causes are inflammatory and iatrogenic origins. Inflammatory causes such as acute or chronic infections can lead to destruction of the *stratum basalis*. Iatrogenic causes can be surgical, due to repeated or vigorous curettages or myomectomies, or medical, due to indiscriminate use of drugs such as clomiphene citrate. EA can also have idiopathic causes, such as those derived from intrinsic properties of the endometrium or uterine architecture [15].

Regardless of the cause, an atrophic or thin endometrium is characterized by poor vascularity. Patients have deficient growth of the glandular epithelium, poor vascular development, and decreased vascular endothelial growth factor (VEGF) expression [16].

Stem-Cell Based Therapeutic Strategies for Endometrial Regeneration

Throughout this chapter, we will go along the different alternatives that the stem cell therapy offers to deal with specific pathologies affecting the endometrium and how these therapeutic approaches have evolved during the years.[3]

Regenerative Medicine in Human Reproduction

New Regenerative Field: Stem Cells

Regenerative medicine seeks to develop therapeutic approaches to regenerate, repair, or replace cells, tissues, or organs damaged by injury or disease. This discipline includes generation and use of therapeutic stem cells and scaffold-based strategies, such as tissue engineering or the production of artificial organs.

Stem cells are usually defined as undifferentiated cells with the potential to perpetuate themselves for long periods through self-renewal by cell division. They can divide into more stem cells (symmetric division) or, under certain physiologic or experimental conditions, can be induced to become specific differentiated/mature cells with special functions (asymmetric division or differentiation) [17]. Briefly, stem cells can be classified in five groups according to the number of cells that they can differentiate into totipotent, pluripotent, multipotent, oligopotent, and unipotent. Also, based on their origin, stem cells can be divided in six groups following the

[3] Google Scholar, PubMed, and Web of Science databases were used to select studies (only in English) published through November 2020 assessing the use of different types of stem cells for treating endometrial pathologies. The following search terms were used: "adipose tissue", "amnion", "Asherman", "bone marrow", "endometrial atrophy", "endometrium", "menstrual stem cells", "platelet-rich plasma", "stem cells", "stem cell therapy", "thin endometrium", and "umbilical cord". We reviewed human studies as well as studies from murine models.

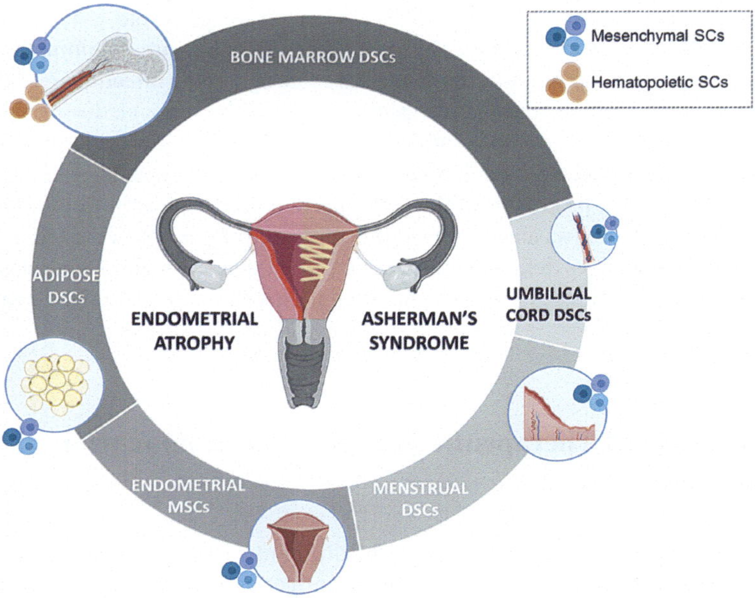

Fig. 10.2 Representative diagram of stem cells used to treat endometrial pathologies. Diagram depicting five types or sources of stem cells used to treat endometrial atrophy and/or Asherman's syndrome. Proportion in the circle diagram corresponds to the number of published works described in human studies. Bone marrow, umbilical cord, menstrual blood, endometrial, and adipose stem cells are included in this summary. DSCs: derived stem cells; MSCs: mesenchymal stem cells. Image created with www.BioRender.com

classification system proposed by Ilic and Polak in 2011 [18]: primordial, embryonic, fetal, perinatal (from umbilical cord, placenta, and amnion), adult (also referred to as somatic stem cells), and induced pluripotent stem cells.

From all existing types of stem cells, the focus here is on perinatal and adult stem cells, which are most commonly used to treat endometrial pathologies (Fig. 10.2). Perinatal stem cells can be obtained from the placenta, umbilical cord, amnion, and amniotic fluid. Epithelial, endothelial progenitors, mesenchymal, and hematopoietic stem cells can be isolated from the cord [19], while mesenchymal, amniotic, chorionic, and epithelial stem cells can be isolated from the rest of perinatal fluids, placenta [20], amnion, and amniotic fluid [21]. Adult stem cells, with an important role in tissue regeneration and self-renewal, are present in many tissues but are most often isolated from bone marrow and adipose tissue. For identification and classification of different stem cell types, cluster of differentiation (CD) antigens are cellular hallmarks [22]. CD73, CD90, and CD105 [23] are the gold-standard antigens for identifying mesenchymal stem cells (MSCs), while CD34, CD45, and CD133 [24] are used to identify hematopoietic stem cells (HSCs).

In 1978, Schofield [25] first postulated that stem cells were located in defined protective and dynamic environments in specific body tissues, called niches. The niche ensures the stem cell population as well as supplies the tissue with differentiated cells; thus, the niche maintains tissue homeostasis.

Stem Cells in the Endometrium

Adult stem cells have long been suspected to be present in the endometrium, as this tissue is highly regenerative [26]. However, the first evidence came in 2004 by Chan et al. [27], who reported the existence of putative endometrial epithelial and stromal stem cells by demonstrating the presence of rare clonogenic populations with high proliferative potential. In light of this evidence, other research groups also postulated the presence of these endometrial stem cells in murine models. These studies used a labeling retention method with 5-bromo-2′-deoxyuridine (BrdU), a thymidine analog that is incorporated de novo into synthesized DNA of dividing cells during mitosis. The cell capacity to retain this chemical label is higher in slow-cycling cells, such as adult stem cells, than in fast-cycling cells. Chan and Gargett in 2006 as well as Cervelló's group in 2007 detected the presence of slow-cycling BrdU-labeled cells in the stratum basalis of the stromal cell fraction in murine endometrial tissue [28, 29]. These labeled cells were positive for stem cell markers such as c-KIT and OCT-4, reinforcing the existence of a stem cell niche in the endometrium.

Despite this evidence, there is still no consensus about specific endometrial stem cell markers. However, studies suggest using markers previously described in adult stem cells from other tissues, such as Musashi-1 [30], B-cell lymphoma 2 (BCL-2) and CD34 [31], leucine-rich repeat-containing G-protein coupled receptor 5 (LGR5) [32], epithelial cell adhesion molecule (EPCAM) [33], stage-specific embryonic antigen 1 (SSEA1) [34], and intercellular adhesion molecule 1 (ICAM1) [35]. Somatic stem cells in the endometrium were historically detected using functional assays based on cell cytometry and cell sorting.

Another procedure to detect adult stem cells is the side population (SP) approach. The SP was first described by Goodell et al. in 1996 [36] as a stem cell subpopulation located in the bone marrow. They described these cells as having high plasticity and the ability to repopulate and self-renew. SP cells can extrude the vital dye Hoechst 33,342 more quickly than other cell populations because they have a higher concentration of ABC transmembrane transporters. Thus, efflux of the vital dye gives SP cells a specific fluorescence pattern that can be distinguished by flow cytometry. Several groups identified SP cells in the endometrium [37–39], reinforcing the presence of a somatic stem cell population in this tissue. Further, in the absence of specific cell markers, stem cell properties have been corroborated in endometrial stem cells, including clonogenicity, in vivo reconstruction, multilineage differentiation potential, and long-term culturing capacity [10].

The population of stem cells in the endometrium is postulated to have a dual origin consisting of tissue-specific endogenous stem cells as well as bone marrow-derived stem cells (BMDSCs) in the endometrial stem cell niche.

Endogenous Origin of Stem Cells in the Endometrium

Endogenous endometrial stem cells were described as composed of epithelial, endothelial, and stromal cells [6]. However, other studies reported that these tissue-specific cells have either a mesenchymal or mesodermal origin. A mesenchymal origin is substantiated by reports that these cells express the CD90 MSC marker but are negative for CD34 and CD45, which are hematopoietic markers [38]. A mesodermal origin is based on studies demonstrating that these cells can differentiate into osteocytes, adipocytes [38, 39], or even myocytes [40]. A possible origin from remaining fetal cells has also been suggested, probably from Müllerian ducts [41, 42].

Exogenous Origin of Stem Cells in the Endometrium

The presence of BMDSCs in tissues other than bone marrow has been described for the liver [43] and heart [44], among others. When these cells arrive at an exogenous location, they can remain in an undifferentiated state or differentiate themselves [45]. Thus, the same may be true for the endometrium.

That BMDSCs contribute to the endometrial stem cell population and, consequently, to the endometrial regeneration process that occurs during each menstrual cycle was first hypothesized after findings of extra-pelvis endometriosis [46]. The authors hypothesized that BMDSCs, which can differentiate into non-hematopoietic cells, may migrate toward the endometrium and differentiate into endometrial stem cells. However, there is still no evidence for the degree of contribution of these BMDSCs to the endometrium. It is unknown if these BMDSCs migrate toward the endometrium in each endometrial cycle, only after a bone marrow transplant, or in specific injury contexts. In a 2012 study of BMDSCs in the endometrium of patients receiving a bone marrow transplant from a male donor, Cervelló et al. [47] concluded that these cells are an exogenous source of stem cells that can migrate to the uterus in certain situations but that do not constitute the endometrial stem cell niche.

Thus, endometrial adult stem cells could be hypothesized to have two different origins: endogenous, constituted by resident endometrial somatic stem cells, and exogenous, constituted by BMDSCs that migrate and reach the endometrial tissue in response to certain stimuli.

Stem Cells Therapies for Endometrial Pathologies

Stem Cell Transplantation

The BMDSC hypothesis prompted the proposed use of granulocyte-colony stimulating factor (G-CSF) to treat endometrial pathologies. This glycoprotein promotes production of progenitor and stem cells by bone marrow and their subsequent release into the bloodstream. By stimulating BMDSC recruitment to the endometrium, the endometrial tissue could regenerate more easily. However, the effectiveness of G-CSF remains controversial. Two independent clinical trials [48, 49] reported that this factor does not significantly increase endometrial thickness. In contrast, a 2017 meta-analysis of 11 different studies [50] suggested that intrauterine perfusion of G-CSF could improve endometrial thickness and clinical pregnancy and implantation rates in patients with a thin endometrium.

Moreover, stem cell therapy is proposed as a promising therapeutic option to induce endometrial tissue regeneration. For a complete understanding of the therapeutic mechanism of stem cells and how the scientific community arrived at the existence of a somatic stem cell population in the endometrium, we next discuss some animal model studies related to this topic. All human studies using stem cells to treat endometrial pathology reviewed in this section are summarized in Table 10.1.

Bone Marrow Stem Cells

The first steps toward use of stem cell therapy for treating endometrial pathologies were introduced by Taylor in 2004 [46]. Donor-derived endometrial cells were detected in endometrium from women who received a bone marrow transplant to treat leukemia from HLA-mismatched female donors. These results introduced bone marrow as an extrauterine source of stem cells with potential to regenerate the endometrium. Taylor's group reinforced these results with a later murine study from 2007, reporting that BMDSCs from male donors were detected in the endometrium of female recipients [51]. Similar results were reported by Ikoma et al. in 2009 [52].

Another group reported in 2008 that bone marrow transplant also promoted the presence of donor endothelial cells in the endometrium, which promoted angiogenesis [53]. They corroborated these results in a woman with aplastic anemia receiving a bone marrow transplant as well as in a murine model. Similarly, preliminary results from Kearns et al. in 1982 indicated that decidual cells could derive from endothelial progenitors from bone marrow [54]. All these studies indicate that engraftment of BMDSCs in the endometrium could directly relate to de novo tissue generation. However, a human study in 2012 qualified these results, reporting that while male BMDSCs are detected in the endometrium of human endometrial biopsies, the male cells do not arrive at the stem cell niche and presumably consist of SP cells [47].

Together, such findings have led to the use of BMDSCs for treating endometrial pathologies. In 2011, Nagori et al. [55] presented a case report in which a woman with AS was treated with autologous angiogenic BMDSCs (isolated using CD9, CD90, and CD133 antibodies) and achieved sufficient endometrial thickness, good vascularity, and ultimately a biochemical pregnancy after embryo transfer. Later studies reported encouraging results using $CD133^+$ BMDSCs from human females in a murine model [56] and also in a pilot study of patients with AS/EA [57]. Cervelló et al. [56] reported that after injecting $CD133^+$ BMDSCs in NOD/SCID

Table 10.1 Human studies using stem cell therapies for endometrial pathologies. ART: assisted reproduction techniques; AS: Asherman's syndrome; BMDSCs: bone marrow-derived stem cells; EA: endometrial atrophy; ERα: estrogen receptor alpha; ETh: endometrial thickness; IUAs: intrauterine adhesions; MenMSCs: menstrual mesenchymal stem cells; MSCs: mesenchymal stem cells; UCMSCs: umbilical cord-derived stem cells; vWF: von Willebrand factor

Authors	Stem cells (markers)	Endometrial pathologies	Population size	Administration method	Main outcomes
Nagori et al. (2011)	Autologous angiogenic BMDSCs (CD9, CD44, CD90, CD133)	AS	1 (case report)	Injection in fundus of uterus	↑ ETh; multilayered endometrium pregnancy achieved by ART
Zhao et al. (2013)	Autologous BMMSCs	AS	1 (case report)	Instillation into uterine cavity with implantation at end of adhesiolysis procedure	Pregnancy achieved by intercourse based on monitoring ovulation
Sing et al. (2014)	Autologous $CD34^{+}$ BMDSCs	AS	6	Injection into sub-endometrial zone by ovum pickup needle	↑ ETh (maintained up to 9 months) restored menstruation
Santamaría et al. (2016)	Autologous $CD133^{+}$ BMDSCs	AS and EA	16	Intra-arterial catheterization	↑ ETh; ↑ mature vessel density restored menstruation 3/16 spontaneous pregnancies 7/16 pregnancies by ART
Tan et al. (2016)	Autologous MenMSCs (CD44, CD73, CD90, CD105)	AS	7	Instillation into fundus of uterus with internal canula	↑ ETh; ↑ endometrial morphology 1/7 spontaneous pregnancy 2/7 pregnancies by ART

(continued)

Table 10.1 (continued)

Authors	Stem cells (markers)	Endometrial pathologies	Population size	Administration method	Main outcomes
Cao et al. (2018)	UCMSCs embedded in collagen scaffold	AS	26	Insertion of SC/collagen scaffold into uterine cavity using Foley catheter after adhesioolysis procedure	↑ ETh; ↑ endometrial markers of good functionality (ERα, Ki67, vWF) milder IUAs 7/26 spontaneous pregnancies 3/26 pregnancies by ART 8/26 newborns
Sudoma et al. (2019)	Autologous ADMSCs (CD34, CD45)	Thin endometrium	25	Introduction into sub-endometrial zone by intracavital or transmyometrial injection using needles for amniocentesis or oocyte aspiration	20/25 ↑ ETh 13/25 pregnancies by ART 9/25 newborns
Lee et al. (2020)	Autologous ADMSCs	AS	6	Instillation into fundus of uterus with embryo transfer canula	↑ ETh restored menstruation ↑ endometrial morphology
Ma et al. (2020)	Autologous MenMSCs (CD73, CD90, CD105, CD90)	AS	12	Injection in fundus of uterus	↑ ETh ↑ menstruation duration 1/12 spontaneous pregnancies 4/12 pregnancies by ART
Sapozhak et al. (2020)	Endometrial MSCs (CD73, CD90, CD105)	Thin endometrium	1 (case report)	Sub-mucosal administration	Twin pregnancy by ART 1 healthy newborn

(continued)

Table 10.1 (continued)

Authors	Stem cells (markers)	Endometrial pathologies	Population size	Administration method	Main outcomes
Tersoglio et al. (2020)	Endometrial MSCs (CD73, CD90, CD105)	Thin endometrium	29	Instillation into fundus of uterus with embryo transfer catheter	↑ ETh 23/29 pregnancies by ART 10/29 newborns

mice, either intrauterine or through the tail vein, the cells localized around endometrial blood vessels, inducing proliferation in surrounding cells as well as regulation of the paracrine factors thrombospondin 1 and insulin-like growth factor 1 (IGF1). Stem cells used by Cervelló et al. came from a pilot study of 16 patients with AS/EA in which $CD133^+$ BMDSCs were mobilized from patients after G-CSF injection, isolated via peripheral blood aphaeresis, and catheterized into uterine spiral arterioles. These results suggest that autologous $CD133^+$ BMDSCs have an effective, although transient (maximum effect was three months after injection and was diminished after six months), effect on endometrial repair in terms of menstruation restoration, endometrial morphology and thickness, neoangiogenesis, and pregnancy rates. A later study to investigate this paracrine effect concluded that stem cells create an immunomodulatory environment that precedes the pro-regenerative effects (proliferation, angiogenesis) that are intrinsic to healing [58].

In addition to CD133, CD34 also has been used to isolate BMDSCs for treating endometrial injury. In 2014, Singh et al. [59] used CD34 to isolate mononuclear cells from bone marrow to treat six women with AS. These patients received stem cell injections in the sub-endometrial zone transmyometrially; after monitoring up to 9 months, all of the patients had significantly increased endometrial thickness, and five of them recovered from amenorrhea. It is also worth mentioning a murine study published by Alawhadi et al. in 2014 [60], in which tail-vein injection of the whole fraction of bone marrow cells from male mice resulted in reduced fibrotic area as well as improved fertility rates compared to untreated control mice.

Bone Marrow Mesenchymal Stem Cells

Another subpopulation of BMDSCs is bone marrow mesenchymal stem cells (BMMSCs). A case study from 2013 reported that administration of autologous BMMSCs after an adhesiolysis surgical procedure in a woman with severe AS resulted in a spontaneous pregnancy [61]. In addition to this single study, several animal studies also reported the effectiveness of BMMSCs in murine models of endometrial injury [62–64]. These studies showed that stem cell therapy increased endometrial thickness and the number of endometrial glands; upregulated factors related to normal endometrial function, such as leukemia inhibitory factor (LIF) or tumor necrosis factor α (TNFα); and reduced fibrotic area as well as pro-inflammatory cytokines such as interleukin (IL) 1β.

BMMSCs have also been tested in combination with bioengineering approaches and less invasive regenerative solutions expected to strengthen the cells' regenerative action. For example, collagen scaffolds are three-dimensional porous biomaterials that can mimic characteristics of the natural endometrial matrix. These structures provide a template for tissue regeneration, promoting cell proliferation, attachment, and migration, as well as ECM restoration and other pro-regenerative events [65]. In 2014, Ding et al. [66] loaded BMMSCs into a collagen scaffold and transplanted the construct in the basal endometrial layer of rats with an injured uterus. The construct resulted in increased expression of growth factors in the tissue, increased vascularity and endometrial thickness, and improved implantation and pregnancy rates, suggesting it promoted endometrial regeneration. In 2019, Xiao et al. [67] also reported a comparison of different types of scaffolds [made from lactic-co-glycolic acid, collagen, and poly (glycerol sebacate) (PGS)] in combination with BMMSCs, concluding that PGS scaffolds best enhanced the regenerative action of stem cells in an AS rat model of endometrial injury.

Another bioengineering approach to regenerate the endometrium is the use of hydrogels. These materials are hydrophilic gels consisting of networks of crosslinked polymer chains [68]. Yang et al. [69] published a study in 2017 implementing this technology to treat endometrial damage in a rat model. To promote the effect of stem cells, the authors applied them in combination with vitamin C and a commercial hydrogel and reported endometrial regeneration in terms of morphology, fibrosis, and expression of cytokeratin (structural protein used as a marker of endometrial epithelial cells), von Willebrand factor (vWF), and IL1β. However, these relatively new combined strategies need further testing in human approaches

Other Sources of Stem Cells: Adipose Tissue

The relatively easy access and collection of adipose tissue make adipose-derived mesenchymal stem cells (ADMSCs) great adult stem cell candidates for regenerative treatments in various medical fields [70], including gynecology. Several studies have reported the effectiveness of ADMSCs for restoring fertility in AS murine models in terms of endometrial morphology and reduced fibrosis [71], increased expression of pro-regenerative factors such as VEGF or IGF1 [72], or improved embryo implantation [73]. Further, Han et al. [74] showed in an AS rat model that ADMSCs in combination with an acellular human amniotic membrane significantly improved angiogenesis, an essentially biological process for tissue regeneration, in injured endometrium.

Concerning human studies, Sudoma et al. [75] presented promising work in 2019 in which they locally injected ADMSCs into the sub-endometrial area by intracavital or transmyometrial injection in 25 women with a thin endometrium (<5 mm). They reported that 20 patients increased their endometrial thickness, 13 patients became pregnant, and nine patients had a healthy delivery. However, this study had a small population size as well as variability among patients (mainly in onset and duration of the therapeutic effect of stem cells) that prevented determination of which etiology of a thin endometrium is most suitable for ADMSC treatment. Lee et al. [76] reported similar results from an unblinded pilot study of six patients diagnosed with AS in which they administered ADMSCs into the endometrial cavity using an embryo transfer device. All patients showed increased endometrial thickness and recovered, at least partially, endometrial morphology and menstrual patterns. One patient even obtained pregnancy, although she had a miscarriage after nine weeks.

These two studies suggest that ADMSCs may be a better option to treat endometrial pathologies, mainly because of the ease and reduced invasiveness of obtaining ADMSCs compared to stem cells obtained from bone marrow. However, further studies including more patients are needed to reinforce and corroborate this hypothesis.

Other Sources of Stem Cells: Umbilical Cord

The umbilical cord has also emerged as a promising stem cell source, not only for treating damaged endometrium but also in other pathologies. Concerning endometrial tissue, the effectiveness of umbilical cord derived-mesenchymal stem cells (UCMSCs) has been more widely studied in AS murine models [77, 78], which demonstrate improved endometrial morphology and expression of regeneration-related factors compared to human studies. The most relevant study is probably that by Cao et al. [79] in 2018, which demonstrated the therapeutic effect of UCMSCs in combination with collagen scaffolds (see section "Bone Marrow Mesenchymal Stem Cells"). In a phase I clinical trial, patients with IUAs were treated with UCMSCs loaded onto a collagen scaffold that was introduced into the uterine cavity just after an adhesiolysis procedure. This novel therapeutic approach resulted in increased endometrial thickness while reducing IUA formation. This study also assessed endometrial biopsies before and after stem cell therapy and reported that expression of estrogen receptors, vimentin (a structural protein usually used as a marker of endometrial stroma cells), Ki67 (a cell proliferation marker), and vWF were upregulated after treatment, while ΔNP63 oncogene, a member of the p53 family that is upregulated in AS patients [80], was downregulated. Again, the results of this combined therapy are consistent with murine studies supporting its effectiveness [81, 82].

Other Sources of Stem Cells: Menstrual Blood, Amniotic Membrane, and More

In 2016, results of a non-controlled clinical trial were reported for seven patients with AS treated with autologous menstrual blood-derived mesenchymal stem cells (MenMSCs). All patients exhibited an increase in endometrial thickness that, in some cases, resulted in successful pregnancies [83]. To increase the regenerative potential of MenMSCs, this group later combined delivery of the cells with administration of platelet-rich plasma (PRP). PRP promoted MenMSC proliferation and differentiation processes in vitro [84], and the cells could regenerate the uterine horns in an IUA murine model [85]. The next step would be corroboration of the effectiveness of this combined therapy in patients with endometrial pathologies.

In 2020, another group reported results of a study of patients with AS, showing that administration of autologous MenMSCs improved endometrial thickness measurements, restored menstruation, and increased pregnancy rates [86]. Several murine studies corroborated the beneficial effect of MenMSCs on an injured endometrium in terms of endometrial thickness and expression of regenerative and endometrial markers [87, 88].

Notably, MenMSCs are endometrial stem cells isolated from menstrual blood [89], which is possible because the uterus contains autologous adult stem cells. A recent case report shows that injection of autologous MSCs from the endometrium resulted in pregnancy in a patient with thin endometrium and repeated but unsuccessful attempts of assisted reproduction cycles [90]. Tersoglio et al. [91] also performed a longitudinal study of patients with thin endometrium who were injected with endometrial MSCs, reporting increased endometrial thickness and pregnancy outcomes.

Intriguingly, Gan et al. [92] reported results in a rat model using MSCs obtained from an amniotic membrane. Stem cell therapy decreased expression of pro-inflammatory cytokines while increasing anti-inflammatory molecules. Likewise, Saho et al. [93] and Li et al. [94] in 2019 as well as Ouyang et al. [95] and Bai et al. [96] in 2020 showed positive results for endometrial regeneration after applying human amnion epithelial cells in a murine model of IUAs.

Finally, it is interesting to note a recent study from Ji et al. [97] reporting a preliminary study in which a human induced pluripotent stem cell-derived MSC-loaded hydrogel scaffold (constructed with a 3D bioprinting device) can regenerate the endometrium in an AS mouse model.

Other Therapies Derived from Stem Cells

Conclusions drawn from works highlighted in this chapter and studies in other fields [98–100] prompted the stem cell paracrine hypothesis. Evidence suggests that stem cells exert a paracrine action when executing their regenerative function. This characteristic aligns with the fact that, despite the effective therapeutic action of stem cell therapy, this approach sometimes entails low retention of the cells and also some risk of teratoma development [101]. Consequently, other therapeutic approaches derived from stem cells have been tested for their ability to repair the endometrium, such as using the stem cell secretome or extracellular vesicles containing active paracrine molecules with high potential in regenerative medicine, also known as exosomes [102].

In this line, Liu et al. [103] reported in 2019 that use of a combined approach composed of the secretome from human BMMSCs and a hyaluronic acid hydrogel could repair a damaged endometrium in an AS murine model and also improve reproductive outcomes. This study measured the presence of different factors in the secretome directly related to tissue regeneration, and specifically in the endometrium, such as fibroblast growth factor 9 (FGF9) and IGF1. A year later, in 2020, a group reported that exosomes derived from ADMSCs could also promote endometrial regeneration (in terms of morphology, percentage of fibrotic tissue, and expression of integrin β3, LIF, and VEGF factors) and restore fertility (in terms of time to conceive, implanted embryos, and pregnancy rate) in a rat model [104]. These results were corroborated in a rabbit model by Yao et al. [105] using exosomes derived from BMMSCs, which showed better endometrial morphology and increased expression of regeneration-related factors as well as decreased expression of factors related to fibrosis events. The authors confirmed the comparable regenerative effect of exosomes to the use of stem cells per se.

Xin et al. [106] reported the use of exosomes derived from UCMSCs in a rat model of IUAs, in which exosomes were administered in combination with collagen scaffolds. This work confirmed the therapeutic potential of exosomes for endometrial regeneration as well as the benefit of administering in combination with a collagen scaffold. Finally, Saribas et al. [107] showed that using exosomes from uterus-derived MSCs was effective at triggering cell proliferation and vascularization while reducing fibrosis in a murine model of AS. Again, these new strategies still need corroboration in human studies, including women with endometrial pathologies.

To further harness paracrine functions, other studies focused on specific molecules that the cells secrete (e.g., cytokines, growth factors) and could potentially serve as effectors of their regenerative action [108–110]. G-CSF, which promotes production of progenitors and stem cells, has been extensively tested in human studies. However, results are controversial, with some studies reporting ineffective treatment [48, 49] while others support its use [50]. As far back as the 2000s, several growth factors were reported to have a positive effect on endometrial tissue in vitro, including hepatocyte growth factor (HGF) [111], transforming growth factor β [112], platelet-derived growth factor [113], epidermal growth factor, and even basic FGF [114]. This increasing interest in growth factors has probably encouraged use of PRP in the assisted reproduction field.

PRP is a blood derivative that has an above-average platelet concentration [115] and potential in tissue regeneration [116–118] including for gynecology. PRP contains α-granules located inside the platelets that contain and excrete (after the degranulation process, also called activation) bioactive molecules, including cytokines, growth factors [e.g., HGF, IGF1, stromal cell-derived factor 1α (SDF1α)], and other proteins involved in tissue repair [119]. PRP's regenerative effect on the endometrium was demonstrated in vitro [120, 121], increasing cell proliferation and migration among other biological processes. Its effect was also demonstrated in vivo in murine models [122–124], improving parameters such as endometrial thickness and implantation rate and decreasing fibrosis and other related factors.

PRP has also been tested in women with endometrial pathologies, including thin endometrium from EA [125–129] and AS [130]. However, most studies lack sufficient population size or adequate study design—only two randomized clinical trials have used PRP to treat women with endometrial pathologies. The first trial was performed in women with EA, in which the authors described improved endometrial thickness as well as implantation and pregnancy rates in the treatment group [131]. The second trial, performed in AS patients, reported no improvement in terms of restoration of menstrual patterns or decreased severity of IUAs [132].

PRP also has been used to enhance the regenerative effect of stem cells. Recent work from 2020 shows that PRP can promote the effect of BMMSC transplant in a murine model of endometrial injury [133]. Addition of PRP could enhance stem cell differentiation and expression of factors related to endometrial regeneration (e.g., IGF1, IL10) as well as restore endometrial thickness, in comparison with addition of BMMSCs alone. Similarly, Yi et al. [134] showed in 2019 that administration of SDF1α, a single paracrine factor and chemoattractant of stem cells had comparable regenerative power to BMMSCs on injured endometrium of mice.

Conclusions and Future Directions

Therapy with stem cells, from the bone marrow or other sources, is an effective approach for treating endometrial pathologies compromising fertility. Nonetheless, the suggested paracrine action of stem cells, together with the disadvantages of stem cell treatment, has directed management of endometrial pathologies to less invasive techniques, such as the use of single factors or other approaches like PRP or bioengineering solutions. However, these relatively recent therapeutic solutions need further corroboration in well-designed clinical trials. Will these emerging approaches be the most promising future direction to treat endometrial pathologies affecting fertility?

Acknowledgements This work was supported by the Carlos III Health Institute, grants CPI19/00149 and PI17/01039 to I.C., and the Regional Valencian Ministry of Education, PROMETEO/2018/137 to L.dM-G., A.P., and I.C.

References

1. Cunningham F, Leveno K, Bloom S, et al (2015) Williams Obstetricia, 24th ed. Mc Graw Hill Education
2. Vaamonde D, du Plessis SS, Agarwal A (2016) Exercise and human reproduction: Induced fertility disorders and possible therapies. Exerc Hum Reprod Induc Fertil Disord Possible Ther 1–351. https://doi.org/10.1007/978-1-4939-3402-7
3. Galliano D, Bellver J, Díaz-García C et al (2015) Art and uterine pathology: how relevant is the maternal side for implantation? Hum Reprod Update 21:13–38. https://doi.org/10.1093/humupd/dmu047
4. Simón C, Horcajadas JA, García-Velasco J, Pellicer A (2009) El Endometrio Humano: desde la investigación a la clínica, 1st ed. Editorial Médica Panamericana S.A.
5. Cervelló I, Simón C (2009) Somatic stem cells in the endometrium. Reprod Sci 16:200–205. https://doi.org/10.1177/1933719108329955
6. Gargett CE, Masuda H (2010) Adult stem cells in the endometrium. Mol Hum Reprod 16:818–834. https://doi.org/10.1093/molehr/gaq061
7. Brännström M (2019) Introduction: uterus transplantation. Fertil Steril 112:1–2. https://doi.org/10.1016/j.fertnstert.2019.05.032
8. Bombard DS, Mousa SA (2014) Mayer-Rokitansky-Kuster-Hauser syndrome: complications, diagnosis and possible treatment options: A review. Gynecol Endocrinol 30:618–623. https://doi.org/10.3109/09513590.2014.927855
9. Kim JH, Han E (2018) Endometriosis and female pelvic pain. Semin Reprod Med 36:143–151. https://doi.org/10.1055/s-0038-1676103
10. Santamaria X, Mas A, Cervelló I et al (2018) Uterine stem cells: From basic research to advanced cell therapies. Hum Reprod Update 24:673–693. https://doi.org/10.1093/humupd/dmy028
11. Smikle C, Yarrarapu S, Khetarpal S Asherman (2020) Syndrome - StatPearls - NCBI Bookshelf. https://www.ncbi.nlm.nih.gov/books/NBK448088/. Accessed 16 Oct 2020
12. Conforti A, Alviggi C, Mollo A et al (2013) The management of Asherman syndrome: a review of literature. Reprod Biol Endocrinol 11:118. https://doi.org/10.1186/1477-7827-11-118
13. Senturk LM, Erel CT (2008) Thin endometrium in assisted reproductive technology. Curr Opin Obstet Gynecol 20:221–228. https://doi.org/10.1097/GCO.0b013e328302143c
14. Kasius A, Smit JG, Torrance HL et al (2014) Endometrial thickness and pregnancy rates after IVF: a systematic review and meta-analysis. 20:530–541. https://doi.org/10.1093/humupd/dmu011
15. Mahajan N, Sharma S (2016) The endometrium in assisted reproductive technology: how thin is thin? J Hum Reprod Sci 9:3–8
16. Miwa I, Tamura H, Takasaki A et al (2009) Pathophysiologic features of "thin" endometrium. Fertil Steril 91:998–1004. https://doi.org/10.1016/j.fertnstert.2008.01.029
17. Alberts B, Johnson A, Lewis J et al (2002) Molecular biology of the cell, 4th edn. Garland Science, New York
18. Ilic D, Polak JM (2011) Stem cells in regenerative medicine: introduction. Br Med Bull 98:117–126. https://doi.org/10.1093/bmb/ldr012
19. Alatyyat SM, Alasmari HM, Aleid OA et al (2020) Umbilical cord stem cells: background, processing and applications. Tissue Cell 65. https://doi.org/10.1016/j.tice.2020.101351
20. Antoniadou E, David AL (2016) Placental stem cells. Best Pract Res Clin Obstet Gynaecol 31:13–29. https://doi.org/10.1016/j.bpobgyn.2015.08.014
21. Rennie K, Gruslin A, Hengstschläger M et al (2012) Applications of amniotic membrane and fluid in stem cell biology and regenerative medicine. Stem Cells Int 2012. https://doi.org/10.1155/2012/721538
22. Tarnok A, Ulrich H, Bocsi J (2010) Phenotypes of stem cells from diverse origin. Cytom Part A 77:6–10. https://doi.org/10.1002/cyto.a.20844
23. Lin C-S, Xin Z-C, Dai J, Lue TF (2013) Commonly used mesenchymal stem cell markers and tracking labels: limitations and challenges. Histol Histopathol 28:1109–1116

24. Wognum AW, Eaves C, Thomas TE (2004) Identification and isolation of hematopoietic stem cells. Arch Med Res 34:461–475
25. Schofield R (1978) The relationship between the spleen colony-forming cell and the haemopoietic stem cell. Blood Cells 4:7–25
26. Prianishnikov V (1978) Model of the structure of the epithelium normal, hyperplastic, and malignant human endometrium. Review 428:420–428
27. Chan RWS, Schwab KE, Gargett CE (2004) Clonogenicity of human endometrial epithelial and stromal cells. Biol Reprod 70:1738–1750. https://doi.org/10.1095/biolreprod.103.024109
28. Chan RWS, Gargett CE (2006) Identification of label-retaining cells in mouse endometrium. Stem Cells 24:1529–1538. https://doi.org/10.1634/stemcells.2005-0411
29. Cervelló I, Martínez-Conejero JA, Horcajadas JA et al (2007) Identification, characterization and co-localization of label-retaining cell population in mouse endometrium with typical undifferentiated markers. Hum Reprod 22:45–51. https://doi.org/10.1093/humrep/del332
30. Götte M, Wolf M, Staebler A et al (2008) Increased expression of the adult stem cell marker Musashi-1 in endometriosis and endometrial carcinoma. J Pathol 215:317–329. https://doi.org/10.1002/path.2364
31. Cho NH, Park YK, Kim YT et al (2004) Lifetime expression of stem cell markers in the uterine endometrium. Fertil Steril 81:403–407. https://doi.org/10.1016/j.fertnstert.2003.07.015
32. Cervelló I, Gil-Sanchis C, Santamaría X et al (2017) Leucine-rich repeat–containing G-protein–coupled receptor 5–positive cells in the endometrial stem cell niche. Fertil Steril 107:510-519.e3. https://doi.org/10.1016/j.fertnstert.2016.10.021
33. Gargett CE, Schwab KE, Zillwood RM et al (2009) Isolation and culture of epithelial progenitors and mesenchymal stem cells from human endometrium. Biol Reprod 80:1136–1145. https://doi.org/10.1095/biolreprod.108.075226
34. Valentijn AJ, Palial K, Al-Lamee H et al (2013) SSEA-1 isolates human endometrial basal glandular epithelial cells: phenotypic and functional characterization and implications in the pathogenesis of endometriosis. Hum Reprod 28:2695–2708. https://doi.org/10.1093/humrep/det285
35. López-Pérez N, Gil-Sanchis C, Ferrero H et al (2019) Human endometrial reconstitution from somatic stem cells: the importance of niche-like cells. Reprod Sci 26:77–87. https://doi.org/10.1177/1933719118766251
36. Goodell M, Brose K, Paradis G et al (1996) Isolation and functional properties of murine hematopoietic stem cells that are replicating in vivo. J Exp Med 183:1797–1806. https://doi.org/10.1084/jem.183.4.1797
37. Kato K, Yoshimoto M, Kato K et al (2007) Characterization of side-population cells in human normal endometrium. Hum Reprod 22:1214–1223. https://doi.org/10.1093/humrep/del514
38. Cervelló I, Gil-Sanchis C, Mas A et al (2010) Human endometrial side population cells exhibit genotypic, phenotypic and functional features of somatic stem cells. PLoS ONE 5. https://doi.org/10.1371/journal.pone.0010964
39. Cervelló I, Mas A, Gil-Sanchis C et al (2011) Reconstruction of endometrium from human endometrial side population cell lines. PLoS ONE 6:1–11. https://doi.org/10.1371/journal.pone.0021221
40. Shoae-Hassani A, Sharif S, Seifalian AM et al (2013) Endometrial stem cell differentiation into smooth muscle cell: A novel approach for bladder tissue engineering in women. BJU Int 112:854–863. https://doi.org/10.1111/bju.12195
41. Snyder EY, Loring JF (2005) A role for stem cell biology in the physiological and pathological aspects of aging. J Am Geriatr Soc 53:287–291. https://doi.org/10.1111/j.1532-5415.2005.53491.x
42. Lu X, Lin F, Fang H et al (2011) Expression of a putative stem cell marker Musashi-1 in endometrium. Histol Histopathol 26:1127–1133. https://doi.org/10.14670/HH-26.1127
43. Theise ND, Nimmakayalu M, Gardner R et al (2000) Liver from bone marrow in humans. Hepatology 32:11–16. https://doi.org/10.1053/jhep.2000.9124
44. Orlic D, Kajstura J, Chimenti S et al (2001) Bone marrow cells regenerate infarcted myocardium. Nature 410:701–705. https://doi.org/10.1038/35070587

45. Blau HM, Brazelton TR, Weimann JM (2001) The evolving concept of a stem cell: entity or function? Cell 105:829–841. https://doi.org/10.1016/S0092-8674(01)00409-3
46. Taylor HS (2004) Endometrial cells derived from donor stem cells in bone marrow transplant recipients. JAMA 292:81–85
47. Cervelló I, Gil-Sanchis C, Mas A et al (2012) Bone marrow-derived cells from male donors do not contribute to the endometrial side population of the recipient. PLoS ONE 7. https://doi.org/10.1371/journal.pone.0030260
48. Eftekhar M, Sayadi M, Arabjahvani F (2014) Transvaginal perfusion of G-CSF for infertile women with thin endometrium in frozen ET program: a non-randomized clinical trial. Iran J Reprod Med 12:661–666
49. Barad DH, Yu Y, Ph D et al (2014) A randomized clinical trial of endometrial perfusion with factor in in vitro fertilization cycles: impact on endometrial thickness and clinical pregnancy rates. Fertil Steril 101:710–715. https://doi.org/10.1016/j.fertnstert.2013.12.016
50. Xie Y, Tian Z, Zhang J et al (2017) Efficacy of intrauterine perfusion of granulocyte stimulating factor (GCSF) for Infertile women with thin endometrium: a systematic review and meta-analysis. Am J Reprod Immunol 78:1–10. https://doi.org/10.1111/aji.12701
51. Du H, Taylor HS (2007) Contribution of bone marrow-derived stem cells to endometrium and endometriosis. Stem Cells 25:2082–2086. https://doi.org/10.1634/stemcells.2006-0828
52. Ikoma T, Kyo S, Maida Y, Ozaki S (2009) Bone marrow – derived cells from male donors can compose endometrial glands in female transplant recipients. Am J Obstet Gynecholgy 201:608.e1-608.e8. https://doi.org/10.1016/j.ajog.2009.07.026
53. Mints M, Jansson M, Sadeghi B et al (2008) Endometrial endothelial cells are derived from donor stem cells in a bone marrow transplant recipient. Hum Reprod 23:139–143
54. Kearns M, Lala PK (1982) Bone marrow origin of decidual cell precursors uterus in the pseudopregnant mouse uterus. J Exp Med 1155(5):1537–1554
55. Nagori C, Panchal S, Patel H (2011) Endometrial regeneration using autologous adult stem cells followed by conception by in vitro fertilization in a patient of severe Ashermans syndrome. J Hum Reprod Sci 4:43–48. https://doi.org/10.4103/0974-1208.82360
56. Cervelló I, Gil-Sanchis C, Santamaría X et al (2015) Human CD133+ bone marrow-derived stem cells promote endometrial proliferation in a murine model of Asherman syndrome. Fertil Steril 104:1552–1560. https://doi.org/10.1016/j.fertnstert.2015.08.032
57. Santamaria X, Cabanillas S, Cervelló I et al (2016) Autologous cell therapy with CD133+ bone marrow-derived stem cells for refractory Asherman's syndrome and endometrial atrophy: a pilot cohort study. Hum Reprod 31:1087–1096. https://doi.org/10.1093/humrep/dew042
58. de Miguel-Gómez L, Ferrero H, López-Martínez S et al (2019) Stem cell paracrine actions in tissue regeneration and potential therapeutic effect in human endometrium: a retrospective study. BJOG 127:551–560. https://doi.org/10.1111/1471-0528.16078
59. Singh N, Mohanty S, Seth T et al (2014) Autologous stem cell transplantation in refractory Asherman's syndrome: a novel cell based therapy. J Hum Reprod Sci 7:93–98. https://doi.org/10.4103/0974-1208.138864
60. Alawadhi F, Du H, Cakmak H, Taylor HS (2014) Bone Marrow-Derived Stem Cell (BMDSC) transplantation improves fertility in a murine model of Asherman' s syndrome. PLoS ONE 9:1–6. https://doi.org/10.1371/journal.pone.0096662
61. Zhao Y, Wang A, Tang X et al (2013) Intrauterine transplantation of autologous bone marrow derived mesenchymal stem cells followed by conception in a patient of severe intrauterine adhesions. Open J Obstet Gynecol 3:377–380. https://doi.org/10.4236/ojog.2013.33069
62. Jing Z, Qiong Z, Yonggang W, Yanping L (2014) Rat bone marrow mesenchymal stem cells improve regeneration of thin endometrium in rat. Fertil Steril 101:587–594. https://doi.org/10.1016/j.fertnstert.2013.10.053
63. Wang J, Ju B, Pan C et al (2016) Application of bone marrow-derived mesenchymal stem cells in the treatment of intrauterine adhesions in rats. Cell Physiol Biochem 39:1553–1560. https://doi.org/10.1159/000447857
64. Gao L, Huang Z, Lin H et al (2019) Bone Marrow Mesenchymal Stem Cells (BMSCs) restore functional endometrium in the rat model for severe Asherman syndrome. Reprod Sci 26:436–444. https://doi.org/10.1177/1933719118799201

65. Nikolova MP, Chavali MS (2019) Recent advances in biomaterials for 3D scaffolds: a review. Bioactive Materials 4:271–292. https://doi.org/10.1016/j.bioactmat.2019.10.005
66. Ding L, Sun H, Su J et al (2014) Biomaterials Transplantation of bone marrow mesenchymal stem cells on collagen scaffolds for the functional regeneration of injured rat uterus. Biomaterials 35:4888–4900. https://doi.org/10.1016/j.biomaterials.2014.02.046
67. Xiao B, Yang W, Lei D et al (2019) PGS scaffolds promote the in vivo survival and directional differentiation of bone marrow mesenchymal stem cells restoring the morphology and function of wounded rat uterus. Adv Healthc Mater 8:1–14. https://doi.org/10.1002/adhm.201801455
68. Ahmed EM (2015) Hydrogel : Preparation, characterization, and applications: a review. J Adv Res 6:105–121. https://doi.org/10.1016/j.jare.2013.07.006
69. Yang H, Wu S, Feng R et al (2017) Vitamin C plus hydrogel facilitates bone marrow stromal cell-mediated endometrium regeneration in rats. Stem Cell Res Ther 8. https://doi.org/10.1186/s13287-017-0718-8
70. Miana VV, Prieto González EA (2018) Adipose tissue stem cells in regenerative medicine. Ecancermedicalscience 12:1–14. https://doi.org/10.3332/ecancer.2018.822
71. Kilic S, Yuksel B, Pinarli F, Albayrak A (2014) Effect of stem cell application on Asherman syndrome, an experimental rat model. J Assist Reprod Genet 31:975–982. https://doi.org/10.1007/s10815-014-0268-2
72. Çil N, Yaka M, Ünal MS et al (2020) Adipose derived mesenchymal stem cell treatment in experimental Asherman syndrome induced rats. Mol Biol Rep 47:4541–4552. https://doi.org/10.1007/s11033-020-05505-4
73. Yotsumoto F, Iwaguro H, Harada Y et al (2020) Adipose tissue-derived regenerative cells improve implantation of fertilized eggs in thin endometrium. Regen Med 15:1891–1904. https://doi.org/10.2217/rme-2020-0037
74. Han X, Ma Y, Lu X et al (2020) Transplantation of human adipose stem cells using acellular human amniotic membrane improves angiogenesis in injured endometrial tissue in a rat intrauterine adhesion model. Cell Transplant 29:1–11. https://doi.org/10.1177/0963689720952055
75. Sudoma I, Pylyp L, Kremenska Y (2019) Application of autologous adipose- derived stem cells for thin endometrium treatment in patients with failed ART programs. J Stem Cell Ther Transplant 3:001–008. https://doi.org/10.29328/journal.jsctt.1001013
76. Lee SY, Shin JE, Kwon H et al (2020) Effect of autologous adipose-derived stromal vascular fraction transplantation on endometrial regeneration in patients of Asherman' s syndrome: a pilot study. Reprod Sci 27:561–568. https://doi.org/10.1007/s43032-019-00055-y
77. Tang YQ, Gan L, Xu Q et al (2016) Effects of human umbilical cord mesenchymal stem cells on intrauterine adhesions in a rat model. Int J Clin Exp Pathol 9:12119–12129
78. Zhang L, Li Y, Guan CY et al (2018) Therapeutic effect of human umbilical cord-derived mesenchymal stem cells on injured rat endometrium during its chronic phase. Stem Cell Res Ther 9:1–15. https://doi.org/10.1186/s13287-018-0777-5
79. Cao Y, Sun H, Zhu H et al (2018) Allogeneic cell therapy using umbilical cord MSCs on collagen scaffolds for patients with recurrent uterine adhesion: a phase I clinical trial. Stem Cell Res Ther 9:1–10. https://doi.org/10.1186/s13287-018-0904-3
80. Zhao G, Cao Y, Zhu X et al (2017) Transplantation of collagen scaffold with autologous bone marrow mononuclear cells promotes functional endometrium reconstruction via downregulating ΔNp63 expression in Asherman's syndrome. Sci China Life Sci 60:404–416. https://doi.org/10.1007/s11427-016-0328-y
81. Xu L, Ding L, Wang L et al (2017) Umbilical cord-derived mesenchymal stem cells on scaffolds facilitate collagen degradation via upregulation of MMP-9 in rat uterine scars. Stem Cell Res Ther 8:1–13. https://doi.org/10.1186/s13287-017-0535-0
82. Xin L, Lin X, Pan Y et al (2019) A collagen scaffold loaded with human umbilical cord-derived mesenchymal stem cells facilitates endometrial regeneration and restores fertility. Acta Biomater 92:160–171. https://doi.org/10.1016/j.actbio.2019.05.012
83. Tan J, Li P, Wang Q et al (2016) Autologous menstrual blood-derived stromal cells transplantation for severe Asherman's syndrome. Hum Reprod 31:2723–2729. https://doi.org/10.1093/humrep/dew235

84. Zhang S, Li P, Yuan Z, Tan J (2018) Effects of platelet-rich plasma on the activity of human menstrual blood-derived stromal cells in vitro. Stem Cell Res Ther 9:48. https://doi.org/10.1186/s13287-018-0795-3
85. Zhang S, Li P, Yuan Z, Tan J (2019) Platelet-rich plasma improves therapeutic effects of menstrual blood-derived stromal cells in rat model of intrauterine adhesion. Stem Cell Res Ther 10:1–12. https://doi.org/10.1186/s13287-019-1155-7
86. Ma H, Liu M, Li Y et al (2020) Intrauterine transplantation of autologous menstrual blood stem cells increases endometrial thickness and pregnancy potential in patients with refractory intrauterine adhesion. J Obstet Gynaecol Res 46:2347–2355. https://doi.org/10.1111/jog.14449
87. Hu J, Song K, Zhang J et al (2019) Effects of menstrual blood-derived stem cells on endometrial injury repair. Mol Med Rep 19:813–820. https://doi.org/10.3892/mmr.2018.9744
88. Chen L, Guo L, Chen F et al (2020) Transplantation of menstrual blood-derived mesenchymal stem cells (MbMSCs) promotes the regeneration of mechanical injuried endometrium. Am J Transl Res 12:4941–4954
89. Meng X, Ichim TE, Zhong J et al (2007) Endometrial regenerative cells: A novel stem cell population. J Transl Med 5:57. https://doi.org/10.1186/1479-5876-5-57
90. Sapozhak IM, Gubar S, Rodnichenko AE, Zlatska AV (2020) Application of autologous endometrial mesenchymal stromal/stem cells increases thin endometrium receptivity: a case report. J Med Case Rep 14:190. https://doi.org/10.1186/s13256-020-02515-5
91. Tersoglio AE, Tersoglio S, Salatino DR et al (2020) Regenerative therapy by endometrial mesenchymal stem cells in thin endometrium with repeated implantation failure. A novel strategy. J Bras Reprod Assist 24:118–127. https://doi.org/10.5935/1518-0557.20190061
92. Gan L, Duan H, Xu Q et al (2017) Human amniotic mesenchymal stromal cell transplantation improves endometrial regeneration in rodent models of intrauterine adhesions. Cytotherapy 19:603–616. https://doi.org/10.1016/j.jcyt.2017.02.003
93. Shao X, Ai G, Wang L et al (2019) Adipose-derived stem cells transplantation improves endometrial injury repair. Zygote 1–8. https://doi.org/10.1017/S096719941900042X
94. Li B, Zhang Q, Sun J, Lai D (2019) Human amniotic epithelial cells improve fertility in an intrauterine adhesion mouse model. Stem Cell Res Ther 10:1–14. https://doi.org/10.1186/s13287-019-1368-9
95. Ouyang X, You S, Zhang Y et al (2020) Transplantation of human amnion epithelial cells improves endometrial regeneration in rat model of intrauterine adhesions. Stem Cells Dev 29:1346–1362. https://doi.org/10.1089/scd.2019.0246
96. Bai X, Liu J, Yuan W et al (2020) Therapeutic effect of human amniotic epithelial cells in rat models of intrauterine adhesions. Cell Transplant 29:1–12. https://doi.org/10.1177/0963689720908495
97. Ji W, Hou B, Lin W et al (2020) 3D Bioprinting a human iPSC-derived MSC-loaded scaffold for repair of the uterine endometrium. Acta Biomater 116:268–284. https://doi.org/10.1016/j.actbio.2020.09.012
98. Tögel F, Weiss K, Yang Y et al (2007) Vasculotropic, paracrine actions of infused mesenchymal stem cells are important to the recovery from acute kidney injury. Am J Physiol Ren Physiol 292:1626–1635. https://doi.org/10.1152/ajprenal.00339.2006
99. Mirotsou M, Jayawardena TM, Schmeckpeper J et al (2011) Paracrine mechanisms of stem cell reparative and regenerative actions in the heart. J Mol Cell Cardiol 50:280–289
100. Gnecchi M, Danieli P, Malpasso G, Ciuffreda MC (2016) Paracrine mechanisms of mesenchymal stem cells in tissue repair. Methods Mol Biol 1416:123–146. https://doi.org/10.1007/978-1-4939-3584-0
101. Musiał-Wysocka A, Kot M, Majka M (2019) The pros and cons of mesenchymal stem cell-based therapies. Cell Transplant 28:801–812
102. Zhang B, Yin Y, Lai RC, Tan SS, Choo AB, Lim SK (2014) Mesenchymal stem cells secrete immunologically active exosomes. Stem Cells Dev 23:1233–1244. https://doi.org/10.1089/scd.2013.0479

103. Liu F, Hu S, Yang H et al (2019) Hyaluronic acid hydrogel integrated with mesenchymal stem cell-secretome to treat endometrial injury in a rat model of Asherman's syndrome. Adv Healthc Mater 8. https://doi.org/10.1002/adhm.201900411
104. Zhao S, Qi W, Zheng J et al (2020) Exosomes derived from adipose mesenchymal stem cells restore functional endometrium in a rat model of intrauterine adhesions. Reprod Sci 27:1266–1275. https://doi.org/10.1007/s43032-019-00112-6
105. Yao Y, Chen R, Wang G et al (2019) Exosomes derived from mesenchymal stem cells reverse EMT via TGF-β1/Smad pathway and promote repair of damaged endometrium. Stem Cell Res Ther 10:1–17. https://doi.org/10.1186/s13287-019-1332-8
106. Xin L, Lin X, Zhou F et al (2020) A scaffold laden with mesenchymal stem cell-derived exosomes for promoting endometrium regeneration and fertility restoration through macrophage immunomodulation. Acta Biomater 113:252–266. https://doi.org/10.1016/j.actbio.2020.06.029
107. Saribas GS, Ozogul C, Tiryaki M et al (2020) Effects of uterus derived mesenchymal stem cells and their exosomes on Asherman's syndrome. Acta Histochem 122. https://doi.org/10.1016/j.acthis.2019.151465
108. Oskowitz A, McFerrin H, Gutschow M et al (2011) Serum-deprived human multipotent mesenchymal stromal cells (MSCs) are highly angiogenic. Stem Cell Res 6:215–225. https://doi.org/10.1016/j.scr.2011.01.004
109. Mussano F, Genova T, Corsalini M et al (2017) Cytokine, chemokine, and growth factor profile characterization of undifferentiated and osteoinduced human adipose-derived stem cells. Stem Cells Int 2017:6202783. https://doi.org/10.1155/2017/6202783
110. An SY, Jang YJ, Lim HJ et al (2017) Milk fat globule-EGF factor 8, secreted by mesenchymal stem cells, protects against liver fibrosis in mice. Gastroenterology 152:1174–1186. https://doi.org/10.1053/j.gastro.2016.12.003
111. Yoshida S, Matsumoto K, Tomioka D et al (2004) Recombinant hepatocyte growth factor accelerates cutaneous wound healing in a diabetic mouse model recombinant hepatocyte growth factor accelerates cutaneous wound healing in a diabetic mouse model. Growth Factors 22:111–119. https://doi.org/10.1080/08977190410001701005
112. Nasu K, Ph D, Nishida M et al (2005) Regulation of proliferation, motility, and contractivity of cultured human endometrial stromal cells by transforming growth factor-beta isoforms. Fertil Steril 84:1114–1123. https://doi.org/10.1016/j.fertnstert.2005.02 055
113. Matsumoto H, Nasu K, Nishida M et al (2005) Regulation of proliferation, motility, and contractility of human endometrial stromal cells by platelet-derived growth factor. J Clin Endocrinol Metab 90:3560–3567. https://doi.org/10.1210/jc.2004-1918
114. Gargett CE, Chan RWS, Schwab KE (2008) Hormone and growth factor signaling in endometrial renewal: role of stem/progenitor cells. Mol Cell Endocrinol 288:22–29. https://doi.org/10.1016/j.mce.2008.02.026
115. Andia I (2016) Clinical indications and treatment protocols with platelet-rich plasma in dermatology. Ediciones Mayo, Barcelona
116. Sommeling CE, Heyneman A, Hoeksema H et al (2013) The use of platelet-rich plasma in plastic surgery: a systematic review. J Plast Reconstr Aesthetic Surg 66:301–311. https://doi.org/10.1016/j.bjps.2012.11.009
117. Bernuzzi G, Petraglia F, Pedrini MF et al (2014) Use of platelet-rich plasma in the care of sports injuries: our experience with ultrasound-guided injection. Blood Transfus 12:s229–s234. https://doi.org/10.2450/2013 0293-12
118. Tsachiridi M, Galyfos G, Andreou A et al (2019) Autologous platelet-rich plasma for nonhealing ulcers: a comparative study. Vasc Spec Int 35:22–27. https://doi.org/10.5758/vsi.2019.35.1.22
119. Mussano F, Genova T, Munaron L et al (2016) Cytokine, chemokine, and growth factor profile of platelet-rich plasma. Platelets 27:467–471. https://doi.org/10.3109/09537104.2016.1143922
120. Anitua E, de la Fuente M, Ferrando M et al (2016) Biological effects of plasma rich in growth factors (PRGF) on human endometrial fibroblasts. Eur J Obstet Gynecol Reprod Biol 206:125–130. https://doi.org/10.1016/j.ejogrb.2016.09.024

121. Aghajanova L, Houshdaran S, Balayan S et al (2018) In vitro evidence that platelet-rich plasma stimulates cellular processes involved in endometrial regeneration. J Assist Reprod Genet 35:757–770. https://doi.org/10.1007/s10815-018-1130-8
122. Jang HY, Myoung SM, Choe JM et al (2017) Effects of autologous platelet-rich plasma on regeneration of damaged endometrium in female rats. Yonsei Med J 58:1195–1203. https://doi.org/10.3349/ymj.2017.58.6.1195
123. Kim JH, Park M, Paek JY et al (2020) Intrauterine infusion of human platelet-rich plasma improves endometrial regeneration and pregnancy outcomes in a murine model of Asherman's syndrome. Front Physiol 11:1–9. https://doi.org/10.3389/fphys.2020.00105
124. de Miguel-Gómez L, López-Martínez S, Campo H et al (2020) Comparison of different sources of platelet-rich plasma as treatment option for infertility-causing endometrial pathologies. Fertil Steril 1–10. https://doi.org/10.1016/j.fertnstert.2020.07.053
125. Nazari L, Salehpour S, Hoseini S et al (2016) Effects of autologous platelet-rich plasma on implantation and pregnancy in repeated implantation failure: a pilot study. Int J Reprod Biomed 14:625–628. https://doi.org/10.29252/ijrm.14.10.625
126. Tandulwadkar S, Naralkar M, Surana A et al (2017) Autologous intrauterine platelet-rich plasma instillation for suboptimal endometrium in frozen embryo transfer cycles: a pilot study. J Hum Reprod Sci 10:208–212. https://doi.org/10.4103/jhrs.JHRS_28_17
127. Molina AM, Sánchez J, Sánchez W, Vielma V (2018) Platelet-rich plasma as an adjuvant in the endometrial preparation of patients with refractory endometrium. J Bras Reprod Assist 22:42–48. https://doi.org/10.5935/1518-0557.20180009
128. Chang Y, Li J, Wei LN et al (2019) Autologous platelet-rich plasma infusion improves clinical pregnancy rate in frozen embryo transfer cycles for women with thin endometrium. Medicine (Baltimore) 98. https://doi.org/10.1097/MD.0000000000014062
129. Kim H, Shin JE, Koo HS et al (2019) Effect of autologous platelet-rich plasma treatment on refractory thin endometrium during the frozen embryo transfer cycle : a pilot study. Front Endocrinol (Lausanne) 10:1–9. https://doi.org/10.3389/fendo.2019.00061
130. Zadehmodarres S, Salehpour S, Saharkhiz N, Nazari L (2017) Treatment of thin endometrium with autologous platelet-rich plasma: a pilot study. J Bras Reprod Assist 21:54–56. https://doi.org/10.5935/1518-0557.20170013
131. Eftekhar M, Tabibnejad N, Alsadat A (2018) The thin endometrium in assisted reproductive technology: an ongoing challenge. Middle East Fertil Soc J 23:1–7. https://doi.org/10.1016/j.mefs.2017.12.006
132. Javaheri A, Kianfar K, Pourmasumi S, Eftekhar M (2020) Platelet-rich plasma in the management of Asherman's syndrome: An RCT. Int J Reprod Biomed 18:113–120. https://doi.org/10.18502/ijrm.v18i2.6423
133. Zhou Y, Shen H, Wu Y et al (2020) Platelet-rich plasma therapy enhances the beneficial effect of bone marrow stem cell transplant on endometrial regeneration. 8:1–11. https://doi.org/10.3389/fcell.2020.00052
134. Yi KW, Mamillapalli R, Sahin C et al (2019) Bone marrow-derived cells or C-X-C motif chemokine 12 (CXCL12) treatment improve thin endometrium in a mouse model. Biol Reprod 100:61–70. https://doi.org/10.1093/biolre/ioy175

Chapter 11
Germinal Origin of Very Small Embryonic-Like Stem Cells (VSELs): Relation to Primordial Germ Cells

Mariusz Z. Ratajczak, Janina Ratajczak, and Magda Kucia

Abstract

Introduction: The germline is immortal from an evolutionary point of view and transfers DNA and mitochondria to the next generation. Mammals including their various stem cell compartments, develop after the fusion of gametes (oocyte and sperm), derived from primordial germ cells (PGCs). During early embryogenesis, PGCs are the first stem cells specified as separate stem cell population in the proximal epiblast. From there, these cells migrate to the extraembryonic mesoderm and subsequently make a turn and enter again into the embryo through primitive streak and migrate to the aorta-gonad-mesonephros region (AGM) to settle down in genital ridges populating developing gonads.
Methods: In this chapter, we will present a challenging concept that PGCs give rise during their migration to a population of very small embryonic likes stem cells (VSELs) that become deposited in developing organs during embryogenesis and play a role as a backup population for monopotent tissue-committed stem cells. We will also present a novel view on the adult stem cell compartment as well as the presence of these early-development cells in postnatal tissues that challenge the current concept on the hierarchy of the adult stem cell compartment.
Results: VSELs express genes characteristic from pluripotent stem cells as well as several markers of late migratory PGCs. VSELs are quiescent because of the erasure of differently methylated loci at paternally imprinted genes. They become activated during stressful situations and are mobilized into circulation. The number of these cells decreases with age.
Conclusions: All experimental results and literature suggest that VSELs develop from primordial germ cells and are preserved in adult tissues and organs.

M. Z. Ratajczak (✉) · J. Ratajczak · M. Kucia
Stem Cell Institute at James Graham Brown Cancer Center, University of Louisville, 500 S. Floyd Street, Rm. 107, Louisville, KY, USA
e-mail: mzrata01@louisville.edu

Department of Regenerative Medicine, Center for Preclinical Research and Technology, Medical University of Warsaw, Warszawa, Poland

I. Virant-Klun (ed.), *Stem Cells in Reproductive Tissues and Organs*, Stem Cell Biology and Regenerative Medicine 70, https://doi.org/10.1007/978-3-030-90111-0_11

Keywords Adult stem cells · Aorta-gonad-mesonephros region · Embryogenesis · Epiblast · Igf2–H19 locus · Imprinted genes · Primordial germ cell · Tissue regeneration · Tumorigenesis · Very small embryonic-like stem cell (VSEL)

Abbreviations

AGM	Aorta-gonad-mesonephros region
BFU-E	Erythroid progenitors
BLIMP1	B-Lymphocyte-induced maturation protein 1 (PRDM1)
BM	Bone marrow
BMP4	Bone morphogenetic protein 4
BrdU	Bromodeoxyuridine
CD	Cluster of differentiation
CD34	CD34 molecule, antigen
CD45	CD45 antigen (PTPRC)
CD133	CD133 antigen
CFU	Colony formation unit
CFU-GM	Granulocyte-monocytic colony formation units
CFU-Meg	Megakaryocytic colony formation unit
CFU-Mix	Multilineage progenitor cell colony formation units
CpG	Cytosine and guanine nucleotide sites
CTCF	CCCTC-binding factor
DAZL	Deleted in azoospermia like
DDX4	DEAD-box helicase 4 (VASA)
DMEM	Dulbecco's modified eagle medium
DMR	Differently methylated region
DNA	Deoxyribonucleic acid
DND1	DND microRNA-mediated repression inhibitor 1
DNMT3L	DNA methyltransferase 3 like
DPPA3	Developmental pluripotency associated 3 (STELLA)
EPC	Endothelial progenitor cell
ESC	Embryonic stem cell
FGF2	Fibroblat growth factor 2
FGF5	Fibroblast growth factor 5
PRDM1	PR/SET domain 1
FSH	Follicle-stimulating hormone
GBX2	Gastrulation brain homeobox 2
DND1	DND microRNA-mediated repression inhibitor 1
GEF	Guanine nucleotide exchange factor
H19	H19 imprinted maternally expressed transcript
HSC	Hematopoietic stem cell
IFITM1	Interferon induced transmembrane protein 1
IGF-1	Insulin-like growth factor 1

IGF-2	Insulin-like growth factor 2
IGF-1R	Insulin-like growth factor 1 receptor
IGF-2R	Insulin-like growth factor 2 receptor
Igf2–H19	Insulin-like growth factor 2–H19 imprinted maternally expressed transcript
iPSC	Induced pluripotent stem cell
KL	Klotho
LIN	Lin-28 homolog A
LH	Luteinizing hormone
MASC	Multipotent adult stem cell
MIAMI	Marrow-isolated adult multilineage-inducible (cells)
MPC	Multipotent progenitor cell
mRNA	Messenger ribonucleic acid
MSC	Mesenchymal stem cell
NANOG	Nanog homeobox
NANOS3	Nanos C2HC-Type Zinc Finger 3
NODAL	Nodal growth differentiation factor
OCT4	Octamer-binding transcription factor 4
PB	Peripheral blood
PGC	Primordial germ cell
POU5F1	POU class 5 homeobox 1
PRDM14	PR/SET domain 14
PRL	Prolactin
RASGRF1	Ras protein specific guanine nucleotide releasing factor 1
RNA	Ribonucleic acid
SCA1	Spinocerebellar ataxia type 1 protein (ATXN1)
SIRT1	Sirtuin 1
STRA8	Stimulated by retinoic acid 8
SSEA-1	Stage-specific embryonic antigen-1
TCSC	Mono-potent tissue-committed stem cells
UCB	Umbilical cord blood
UM177	Small molecule UM177
USSC	Unrestricted somatic stem cell
VSEL	Very small embryonic-like stem cell

Introduction

Regenerative medicine is looking for a potential candidate pluripotent/multipotent stem cell able to differentiate across germ layers and be safely employed in cell therapies to regenerate organs and tissues. Unfortunately, with the exception of hematopoietic stem cells (HSCs) that since 50 years are employed for hematological applications, the current clinical results with stem cells in tissue organ regeneration are somewhat disappointing. The potential clinical applications of the more primitive immortal cell lines established from embryonic stem cells (ESCs) and induced pluripotent stem cells (iPSCs) have so far been discouraging, as both cell types exhibit several problems, including genomic instability, risk of teratoma formation, and the possibility of rejection by the immune system.

Therefore, the only safe stem cells that have so far been employed in regenerative medicine are histocompatible monopotent stem cells, such as the abovementioned HSCs or mesenchymal stem cells (MSCs) that are isolated from postnatal tissues. However, their monopotency, and therefore limited differentiation potential, is a barrier to their broader application in the clinic. Moreover, while HSCs may reconstitute after transplantation hematopoiesis, effects of MSCs are mostly paracrine due to release of soluble factors as well as extracellular microvesicles, that may improve regeneration of damaged organs.

Interestingly, and what is a topic of this chapter, results have accumulated indicating that adult tissues contain rare, early-development stem cells known as very small embryonic-like stem cells (VSELs), which can differentiate into cells from more than one germ layer. Therefore, these cells isolated initially from hematopoietic tissues fulfill some of the criteria of pluripotency. These most primitive, dormant, postnatal tissue-derived stem cells are regulated by epigenetic changes in the expression of certain parentally imprinted genes, a molecular phenomenon previously well described for keeping PGCs in a quiescent state. Specifically, VSELs show erasure of parental imprinting at the important Igf2–H19 locus, which regulates cell proliferation, and this epigenetic modification keeps them in a quiescent state [1]. A similar phenomenon of erasure of parentally imprinted genes occurs also in migrating PGCs [2, 3]. To date, the presence of these cells in adult postnatal tissues have been demonstrated by at least 35 independent laboratories [4–26]. One of the hurdles to employ these cells in the clinic was so far their quiescent state and lack of efficient ex vivo expansion protocol. Our most recent data indicate that VSELs expand in vivo and in vitro after the reestablishment of somatic imprinting at Igf2–H19 locus by nicotinamide treatment in response to stimulation, e.g., by pituitary gonadotrophins (follicle-stimulating factor, luteinizing hormone, and prolactin) and gonadal androgens and estrogens [27, 28]. These small cells could be also successfully expanded ex vivo in the presence of the small molecule UM177 [29]. The expansion has been so far achieved for VSELs isolated from human and murine bone marrow (BM), umbilical cord blood (UCB) as well as VSELs isolated from ovarian surface epithelium and testes. What is important is that it has been nicely demonstrated that VSELs are undergoing asymmetric divisions, a feature that is characteristic for stem cells that allow to keep their number constant. In this review, we will present recent research progress on VSELs isolated from adult tissues and discuss their potential relationship to PGCs.

It is an important topic as these cells as postulated in seminal papers in ovary and testes have been proposed to be precursors of gametes, and perhaps could be employed in future as a source of functional gametes to treat cases of infertility.

Two Major Questions to be Addressed

The hierarchy of stem cell compartment in adult tissues is still incompletely understood [30]. Various types of stem cells residing in postnatal organs that possess more than one germline specification potential have been described This may result because of similar overlapping population of development early cells isolated from postnatal tissues which pending isolation strategy are assigned different names. The examples for such cells include spore-like stem cells [31], multipotent adult stem cells (MASCs) [32], mesenchymal stem cells (MSCs) [33], unrestricted somatic stem cells (USSCs) [34], marrow-isolated adult multilineage-inducible (MIAMI) cells [35, 36], or multipotent progenitor cells (MPCs) [37]. We also postulate that VSELs are at the top of the hierarchy of all of these various overlapping populations of mentioned above stem cells that are endowed with pluri/multipotent differentiation potential, and side by side comparative studies should be performed [38]. Therefore, the undisputed fact that adult tissues contain such rare cells with potential for broader germ layer specification gives rise to two important questions that we will discuss in this chapter.

The first question we will ask in this chapter is related to the well-known fact that the first stem cells specified in the developing embryo in both mice and humans are PGCs. Based on this, one could ask: Are some of the stem cells from the embryonic stage of development related to PGCs deposited into and reside in adult tissues in the quiescent state? This emerging concept suggests a developmental continuum in the stem cell compartment, beginning from the fertilized zygote to adult tissue-committed monopotent stem cells. We will also ask how much germline potential is still present in adult stem cells including the population of VSELs? The second question is: What is a mechanism that keeps these developmentally early stem cells that express markers of pluripotency in a quiescent state in adult tissues, so that they do not form teratomas? To address this question, we demonstrated that VSELs that are deposited in the adult tissues are kept in a quiescent state, similarly as migrating PGCs, by changes in expression of certain paternally imprinted genes, which proper expression of these genes is crucial for initiation of embryogenesis and cell proliferation [1, 39].

We will discuss below all these issues and present evidence that developmentally early VSELs, sharing several markers with PGCs and the epiblast cells, are present in postnatal tissues. We will focus on these cells isolated from BM as evidence accumulated that BM-residing VSELs can be specified e.g., into HSCs, endothelial progenitor cells (EPCs), mesenchymal stem cells (MSCs) and other types of monopotent tissue-committed stem cells. We will also refer to seminal papers showing that VSELs isolated from ovarian surface epithelium or from testes could be potentially specified as precursors of gametes.

Question # 1. Are There Primordial Germ Cell (PGCs) Precursors of Very Small Embryonic-Like Stem Cells (VSELs)?

From a developmental point of view, PGCs are the most important highly specialized stem cells that are precursors of gametes[3]. PGC following meiosis develop in gonads as haploid sperm and eggs that in the process of fertilization and creation of zygote may generate a new organism. These cells are endowed with developmental totipotency and transmit genetic and epigenetic information, as well as mitochondria, to next generations and ensure the reproduction and survival of a species [40]. Based on this, one can postulate that PGCs have evolved to become immortal and create during embryogenesis a mortal soma that helps them to fulfill in an adult life a mission to pass genes to offspring.

These cells become specified early in development at the time shortly before gastrulation in a proximal epiblast of developing embryo as a small group of cells that could be identified by high alkaline phosphatase activity. From there, these PGCs migrate to the extraembryonic mesoderm and then make a turn and re-enter again into the embryo through primitive streak and migrate into the posterior endoderm to the aorta-gonad-mesonephros region (AGM) [41]. At this destination, they settle down in genital ridges in developing gonads. PGCs differentiate at different times in male testes and female ovary development, and recent molecular studies suggest that final determination occurs after these cells colonize the developing gonad. PGCs express markers characteristic for pluripotent stem cells including alkaline phosphatase, Oct4 (POU5F1), Fragilis (IFITM1), Stella (DPPA3), Dazl, and Vasa (DDX4) [2]. The maintenance of pluripotency within this cell population that prevents them from parthenogenesis and teratoma formation arises through epigenetic modifications of parentally imprinted genes such as mentioned above Igf2–H19 tandem gene locus, which erasure on the differently methylated region (DMR) suppresses somatic differentiation programs [2].

VSELs that were initially identified by our group in postnatal murine and human bone marrow [42, 43] express markers of pluripotency and genes characteristic to late migrating PGCs [44]. They are also a highly migratory cell population and may become specified as demonstrated, e.g., into tissue-committed monopotent HSCs and endothelial progenitor cells (EPCs) [6, 45, 46]. Therefore, they somehow fulfill the criteria of hemangioblast, stem cells for both hematopoietic and endothelial lineage [47–50]. What is also important: migratory route of PGCs overlap with time and space when thefirst primitive hematopoietic stem cells and endothelial progenitors are identified in the extraembryonic mesoderm at bottom of the yolk sac, and later on in aorta-gonad-mesonephros (AGM) region where definitive hematopoietic stem cells arise from the area around the dorsal aorta. Thus, the developmental route of PGCs overlaps with the emergence of the first primitive HSCs in time and space—first in the so-called blood islands at the bottom of the yolk sac and later with the emergence of definitive HSCs in the AGM region of the developing embryo proper [41, 51, 52]. It is very likely that in both places in which primitive and definitive hematopoiesis is initiated VSELs that are derived from migrating PGCs and may become specified via hemangioblast into hematopoietic and endothelial precursors (Fig. 11.1). Similarly, at the same time, we envision a population of PGCs-derived VSELs may become deposited in developing organs where they remain into adulthood as the back population for monopotent tissue-committed stem cells. Finally, as demonstrated in several elegant papers, VSELs could also survive into adulthood in gonads where they could be precursor cells for gametes [5, 12, 26, 53, 54]

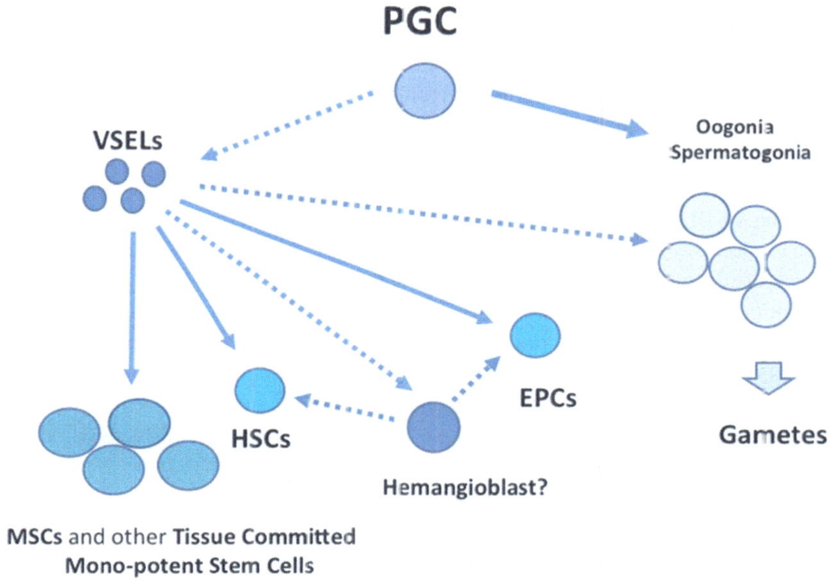

Fig. 11.1 Proposed developmental interrelationship between primordial germ cells (PGCs), very small embryonic-like stem cells (VSELs), hematopoietic stem cells (HSCs), endothelial progenitor cells (EPCs), mesenchymal stem cells (MSCs), and other monopotent tissue-committed stem cells. We propose that migratory PGCs, aside from their major role in establishing gametogenesis, may be a source of VSELs that in bone marrow give rise to HSCs, EPCs, MSCs, and other monopotent tissue-committed stem cells (TCSCs). Specification of VSELs into HSCs and EPCs may involve putative hemangioblast as an intermediate precursor cell. VSELs can also, as postulated in seminal papers, be the direct precursors of gametes at pre-stage of germinal stem cells (gonocytes)—oogonia and spermatogonia. Dotted lines pathways are still under investigation

How Much Germ Line Potential Is on Adult BM—A Proposed Potential Developmental Link Between PGCs, VSELs, and HSCs

The potential developmental relationship between PGCs and VSELs and their deposition in developing bone marrow is supported by several observations indicating a presence of the germline potential in this organ. Accordingly, SSEA-1^{+}Oct-4^{+}Stella^{+}Mvh^{+} cells, isolated from murine BM with anti-SSEA-1 immunomagnetic beads, may differentiate as reported after stimulation by BMP-4 (bone morphogenetic factor 4) into gamete precursors [17]. Similarly, BM-derived Oct-4^{+}Mvh^{+}Dazl^{+}Stella^{+} putative germ cells affect the recurrence of oogenesis in mice sterilized by chemotherapy. In another elegant study, a population of Oct-4^{+}Mvh^{+}Stella^{+} cells that was isolated from the BM of male Stra8–GFP transgenic mice, expressed several molecular markers of spermatogonial stem cells and differentiating spermatogonia [55]. Finally, in a very recent elegant study, murine BM-derived VSELs have been demonstrated to differentiate into germline cells [56]. In our hands, VSELs purified from human umbilical cord blood or bone marrow could be also expanded ex vivo in presence of pituitary sex hormones into cells that highly express germline markers including Stella and MvH (Fig. 11.2). On other hand, in direct mutagenesis studies, it has also been demonstrated that BM cells exposed to a carcinogen that is methylchloranthrene may give rise to germline tumors [57, 58]. In addition to BM and gonads, early-development stem cells with germline characteristics were also isolated from other tissues and organs including skin, heart, liver, brain, pancreas, and kidneys. Nevertheless, their relationship to VSELs purified from hematopoietic tissues requires further comparative studies.

The potential developmental relationship of VSELs with germline cells prompt us to phenotype murine and human VSELs for expression of sex hormone receptors, and we analyzed the expression of receptors for pituitary follicle-stimulating hormone (FSH), luteinizing hormone (LH), prolactin (PRL), and gonadal (androgen, estrogen, and progesterone) in these cells [27]. We observed that all these receptors were expressed by VSELs. More importantly, we found that 10-day administration of each of the pituitary or gonadal sex hormones into mice directly stimulated the proliferation of VSELs, as evaluated by the percentage of cells that incorporated bromodeoxyuridine (BrdU) [27]. Specifically, the percentage of quiescent BrdU positive VSELs increased from 2 to 15–40%. The highest response was observed for LH, FSH, and danazol for both female and male cells and for PRL for female cells. At the same time, we identified the expression of sex hormone receptors in murine and human HSCs and noticed that these hormones stimulate their proliferation. To support this notion, the administration of sex hormones accelerated the recovery of peripheral blood (PB) cell counts in sublethally irradiated mice and slightly mobilized egress of HSCs from BM into circulation. Finally, in direct in vitro clonogenic experiments on purified murine and human progenitor cells, we observed a stimulatory effect of these hormones on clonogenic potential in the presence of suboptimal doses of the colony-stimulating factors to grow mixed lineages (CFU-Mix), erythroid (BFU-E), megakaryocytic (CFU-Meg), and granulocyte-monocytic (CFU-GM) colonies.

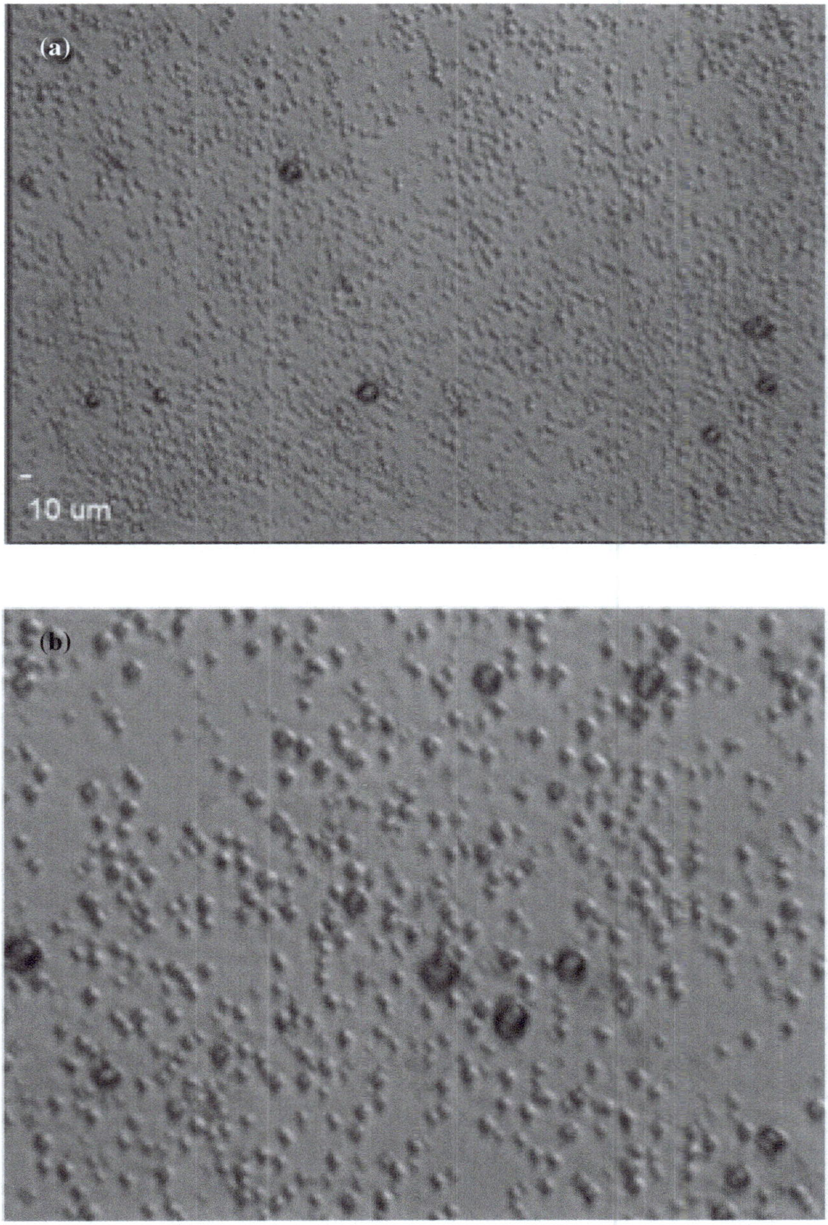

Fig. 11.2 Example of expansion of human umbilical cord blood-derived VSELs. Freshly sorted VSELs (5×10^2) were plated in 0.2 ml of DMEM + 10% artificial serum, supplemented with NMA and a cocktail of two pituitary sex hormones, FSH and LH together with BMP-4, IGF-2, and KL. Cells were cultured for 1 month and half of culture medium has been changed every 7 days

Thus, our data indicate that pituitary—and gonadal—secreted SexHs directly stimulate the expansion of stem cells in BM [27]. On the other hand, this link between PGCs and HSCs is even stronger in light of our data which demonstrated that human and murine germline-derived teratocarcinoma cells lines, as well as ovarian cancer cell lines, express functional erythropoietin receptors, so specific for hematopoiesis, and these cells respond to erythropoietin by chemotaxis, increased adhesion, and phosphorylation of MAPKp42/44 and AKT.

These observations support the presence of PGCs-derived VSELs in postnatal BM and a potential developmental link between PGCs, VSELs, and HSCs in this organ. In addition, as reported, VSELs purified from BM or UCB could be specified into EPCs and MSCs [6, 7]. This data supports the presence of stem cell hierarchy in hematopoietic tissues where at the top are PGCs-derived VSELs and that BM-residing stem cells are responsive to germline mediators.

Question # 2. What Are the Mechanisms Controlling Proliferation of VSELs and Their Specification into Tissue Committed Monopotent Stem Cells?

Two decades ago, it has been proposed that tissue-committed stem cells, such as, HSCs could change their fate and differentiate into stem cells for other lineages, for example, cardiac stem cells [59, 60]. Nevertheless, this concept of the so-called "stem cell plasticity" did not stand up to critical examination [61, 62]. My team from the beginning proposed that stem cell plasticity and explanation for why some degree of chimerism may be sometimes observed in various tissues after transplantation of BM cells could be explained by the fact that the adult BM contains early-development multi/pluripotent stem cells [63]. To support this challenging idea, we succeeded in isolating from adult murine BM cells that were slightly smaller than erythrocytes and that expressed pluripotency markers such as Oct-4 and Nanog and named them VSELs [42]. Similar cells corresponding to BM-derived VSELs were also isolated by other groups from ovarian surface epithelium and testes [53, 64, 65].

Molecular analysis of gene libraries established from murine BM-purified VSELs isolated by multiparameter sorting as small Sca-1^{+}Lin^{-}CD45^{-} cells, revealed that despite a similar small size and primitive morphology characteristic for early-development cells, these cells are heterogeneous. To support this, we found at least three different types of libraries generated from single, sorted VSELs, that exhibited a strong epiblast—PGC-like or mixed—gene expression pattern [44]. In support of PGCs and epiblast connection, we observed that murine BM-derived VSELs express several genes that are characteristic of epiblast SCs (*Gbx2, Fgf5,* and *Nodal*) and, more importantly, of germline specification and migrating PGCs (*Stella, Prdm14, Fragilis, Blimp1, Nanos3,* and *Dnd1*). What is important is the expression of some of these crucial genes that have subsequently been confirmed by demonstrating the presence of transcriptionally active unmethylated promoters in these genes associated with transcription promoting histone modifications [39, 44]. Importantly, as mentioned above, BM-residing murine VSELs and their human UCB-derived counterparts sorted as small $CD133^{+}CD34^{+}Lin^{-}CD45^{-}$ cells respond in vivo to stimulation by the pituitary and gonadal sex hormones as evidenced by in vivo BrdU accumulation assay [27]. Furthermore, gene expression analysis and immunohistochemical staining confirms that these cells express several sex hormone receptors.

As mentioned above, evidence accumulated that VSELs are endowed with broad trans-germ layer differentiation and multi/pluripotential. Nevertheless, there are some both in vitro and in vivo criteria to call a given stem cell "pluripotent". For in vitro criteria, a pluripotent stem cell candidate has to show undifferentiated morphology, the presence of euchromatin in the nucleus, and a high nuclear/cytoplasmic ratio. Pluripotent stem cells should also express markers of pluripotency such as Oct-4, Nanog, and SSEA, and exhibit bivalent domains in the promoters of genes encoding important developmental, homeobox-containing transcription factors. In addition, pluripotent stem cells isolated from females should reactivate the X chromosome [66]. Most important, pluripotent stem cells should differentiate in appropriate culture conditions into cells from all three germ layers (meso-, ecto-, and endoderm). In parallel, there are also proposed in vivo criteria that apply to pluripotent established ESCs and iPSCs. They include their ability to complement blastocyst development and grow teratomas in an in vivo assay after injection into immunodeficient mice.

VSELs purified from hematopoietic tissues fulfill all the in vitro criteria. Nevertheless, since they are highly quiescent in culture, some special assays and procedures are needed to differentiate them into cells for various tissues. However, at the same time, in contrast to established immortalized ESCs and iPSCs cell lines, VSELs do not fulfill the in vivo criteria, as they do not complete blastocyst development and do not grow teratoma. This is a good and important message, as these are normal cells isolated from adult tissues.

The main reason for the quiescent state of VSELs similar to PGCs is the modification of expression of certain paternally imprinted genes [67]. So far there have been identified ~50–100 paternally imprinted genes in the mammalian genome that are expressed either from the maternal or paternal chromosome only [68, 69]. These genes play an important role in embryonal development and some of them, for example, the tandem gene insulin-like growth factor 2 (*Igf2*)–*H19*, is of particular importance for maintaining totipotential state of the zygote, initiating embryogenesis, fetal growth, and pluripotency of developmentally early stem cells [70, 71].

To explain the molecular mechanisms behind the developmental role of paternally imprinted genes, mammalian development requires proper gene dosage of these paternally imprinted genes, which is enabled by their imprinting, so that a single paternal allele (maternal or paternal) is expressed in the cell at proper level. Thus, phenomenon of genomic imprinting is an epigenetic program that ensures the parent-of-origin-specific monoallelic transcription of imprinted genes [72, 73]. This results in intracellular expression of imprinted genes from only one of the two paternal chromosomes—derived either from the mother or the father, which assures "proper dosage of genetic information".

The epigenetics behind regulating the expression of imprinted genes is based on the imposition of epigenetic marks by DNA methylation within differentially methylated regions (DMRs), which are CpG-rich *cis*-elements within their loci. These epigenetic marks imposed on DMRs in the female germline act on the promoters of imprinted genes, which results in the heritable repression of the given genes at maternal chromosomes. By contrast, the imposition of epigenetic marks by methylation of the chromosomes in the male germline does not occur at the promoters but rather within the intergenic regions as seen, for example, at the *Igf2–H19* DMR locus located between gene encoding mRNA for Igf2 and H19 long non coding mRNA [74]. Figure 11.3 panel A shows the location of DMR between Igf2 and H19 loci and a fact that expression of this tandem gene is regulated by a distal enhancer. This Figure also shows differential expression of both loci in normal somatic cells from maternal (M) or paternal (P) chromosomes. Since maternal imprinting at the DMR for this tandem gene is erased at the maternal (M) chromosome (open lollypops), this DMR binds CTCF protein (insulator), which forms a physical barrier between Igf2 and H19 and thereby prevents the distal enhancer from activating transcription of mRNA for Igf2 from the maternal allele. In this situation, distal enhancer promotes transcription of mRNA for long noncoding mRNA from the H19 locus. By contrast, the DMR region at the paternal chromosome (P) is imprinted (methylated) demonstrated by black lollypops. This epigenetic modification prevents CTCF binding to the DMR at the paternal chromosome. As a result of this, the distal enhancer activates transcription of mRNA for Igf2 from the paternal allele. In somatic cells, that have imprinted DMR at the maternal chromosome, it is a proposer transcription balance between mRNA for Igf2 and H19. As a result of this differential regulation of transcription from Igf2–H19 locus translation of mRNA for Igf2, there arises a potent growth factor that promotes proliferation (IGF-2). In contrast, long noncoding mRNA H19 is spliced into several miRNAs that negatively affect cell proliferation interfering with insulin-like growth factors signaling. Nevertheless, as the result of normal, balanced paternal imprinting in somatic cell nuclei, there is a balance in the expression of IGF-2 that promotes proliferation and H19 miRNAs that have opposite effects.

In contrast, erasure of genomic imprints at the DMR at the *Igf2–H19* locus (Fig. 11.3, panel B), both maternal and paternal DMRs bind CTCF insulator protein, and the distal enhancer activates transcription of *H19* from both parental alleles. Cells affected by this epigenetic mechanism do not express proliferation promoting IGF-2 and overexpress noncoding H19 mRNA, thereby negatively affecting via generated inhibitory miRNA cell proliferation. This epigenetic change preventing the expression of *Igf2–H19* locus occurs both in PGCs and VSELs and explains why these cells remain quiescent. This mechanism also prevents PGCs and VSELs from participating in complementation of blastocyst, teratoma formation and what is also noted nuclei from both these cell types do not contribute in clonote formation after nuclear transfer into denucleated oocyte [1, 75, 76]. This mechanism can be, however, reverted with what happened in response to re-methylation of erased DMR loci by de novo DNA

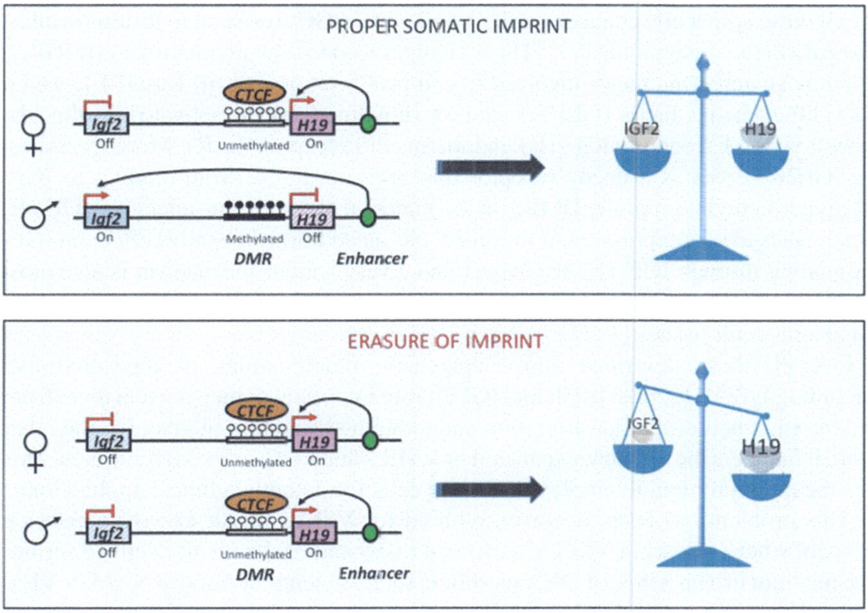

Fig. 11.3 Regulation of expression of the *Igf2–H19* tandem gene. **Panel A**. The *Igf2* and *H19* coding regions are separated by a differentially methylated region (DMR) that is unmethylated (open lollypops) on the maternal chromosome (M) and methylated (filled lollypops) on the paternal chromosome (P). Expression of both genes is regulated by a 3' distal enhancer depicted in green. Since the DMR is unmethylated on the maternal chromosome, it binds CTCF, and this prevents activation of the *Igf2* promoter by the distal enhancer. As a result, only H19 mRNA is transcribed from the maternal chromosome (red arrow). By contrast, methylation of the DMR on the paternal chromosome prevents binding of the CTCF insulator protein and allows activation of the *Igf2* promoter by the distal enhancer and transcription of Igf2 mRNA from the paternal chromosome (red arrow). Normal somatic imprinting observed in all somatic cells results in properly balanced expression of Igf2 from the paternal chromosome and H19 from the maternal chromosome. **Panel B**. Erasure of imprinting at the *Igf2–H19* locus, as seen in PGCs and VSELs, leads to a situation in which DMRs on the maternal and parental chromosomes both bind CTCF, and the 3' distal enhancer activates transcription of only *H19* from both chromosomes. Therefore, erasure of imprinting at the *Igf2–H19* locus leads to overexpression of proliferation-inhibiting H19 mRNA. We noticed that a proper somatic imprinting at this locus is re-established after exposure of VSELs to nicotinamide

methyltransferases. PGCs remethylated erased loci after first meiotic division during gametogenesis, and VSELs remethylated erased loci in response to nicotinamide (NAM) that is a form of vitamin B_3 or to valproic acid (VPA) [28].

To get a full picture of these epigenetic changes and their effects on cell proliferation, erasure of imprinting at the Igf2–H19 locus is not the only epigenetic change. As we demonstrated, murine BM-residing VSELs also erase the paternally methylated imprints at DMRs for RasGrf1 [1]. In parallel, they hypermethylate the maternally methylated DMR for the insulin-like growth factor 2 receptor gene (*IGF2R*) [1].

All these epigenetic changes render VSELs like PGCs resistant to insulin/insulin-like growth factor signaling [77]. These changes lead to downregulation of (i) IGF-2, which is an autocrine factor involved in cell proliferation and (ii) RasGRF1, which is a GTP-exchange factor (GEF) crucial for signaling from the activated insulin-like growth factor 1 receptor (IGF-1R) and the insulin receptor (InsR). Moreover, since the IGF2R serves as a decoy receptor that prevents IGF-2 from binding to IGF-1R, hypermethylation of the DMRs on the maternal chromosome encoding IGF-2R, which leads to overexpression of this gene, has an additional negative effect on IGF-2 signaling through IGF-1R. We expect that a very similar mechanism is also most likely responsible for the quiescent state of human VSELs not only in bone marrow, but also in adult tissues [1, 77].

Overall, these described above epigenetic modifications of imprinted loci (including Igf2–H19, RasGRF1, and IGF2R) are important as they prevent them from undergoing uncontrolled proliferation and teratoma formation in vivo. On the other hand, it hampers the efficient expansion of VSELs and PGCs in ex vivo cultures. This was the main problem to employ these rare cells for potential clinical applications.

This problem has been, however, omitted for VSELs in our ex vivo expansion protocol where quiescent VSELs are forced to expand in chemically defined serum-free medium in a presence of DNA modifiers such as mentioned above NAM or VPA, selected sex hormones (FSH and LH), and a cocktail of growth factors including BMP-4, FGF2, KL, and IGF-1 [27, 28]. By employing this approach, we can expand VSELs up to 3000 times. To explain our expansion approach, both NMA or VPA, are inhibitors of the histone deacetylase Sirt-1 [78–81]. This enzyme inhibits the activity of the de novo DNA methyltransferase DnmT3L, which is crucial for methylation of the regulatory regions of paternally imprinted genes [78, 82, 83]. The fact that Sirt-1 maintains a low intracellular level of DnmT3L explains why it has beneficial effects on longevity by preventing premature depletion of VSELs from adult tissues [78]. By contrast, downregulation of Sirt-1 by nicotinamide or valproic acid in culture promotes ex vivo expansion of these cells. The epigenetic changes during VSEL expansion in the presence of the small molecule UM177 are not clear at this point [29].

Other Implications of VSELs Presence in Adult Tissues

In addition to their role in hematopoiesis and vasculogenesis, VSELs are involved in several other processes. It has been convincingly demonstrated that VSELs can be isolated from the ovarian surface epithelium of young and postmenopausal women as well as from testes [5, 12, 26, 54, 65, 84, 85]. This may have an important implication to treat infertility including patients that were treated with radio/chemotherapy. VSELs are more resistant to such treatment as compared to other stem cell populations [54]. What has been proposed and supported by experimental evidence: VSELs isolated from the gonads may differentiate into gamete-like cells [53]. Recently, it has been reported in an excellent report that ovary-isolated VSELs differentiate into oocyte-like cells in response to sperm cells and release the zona pellucida, which is the first step in fertilization process [12]. This has not been achieved so far with oocyte-like cells derived from ESCs or iPSCs.

VSELs play also an important role in the aging process [86]. Their number correlates with longevity in certain long-living murine strains such as Laron dwarf or Ames dwarf mice that have an undetectable low level of circulating IGF-1 in PB [87]. Their number can be increased in experimental animals as we have demonstrated by caloric restriction, regular exercise, and administration of DNA modifiers such as nicotinamide or valproic acid [88, 89]. These cells are a challenge for modern pharmacology to develop compounds that will increase their survival in adult tissues. The most important target is to ameliorate insulin/insulin-like growth factors signaling. We postulate that VSELs are burned during adult life in a "fire of insulin/insulin-like growth factors signaling". Interestingly, the administration of metformin that regulates insulin effects on cells has a beneficial effect on these cells. By contrast, the opposite effect has exposure of animals to increased insulin/insulin-like growth factor signaling which leads to premature aging and depletion of VSELs from the tissues [86, 87]. Thus, VSELs sensitivity to insulin/insulin-like growth factors signaling, positive effect of caloric restriction, and regular physical activity also has a beneficial effect on human fertility which again straightens a concept that PGCs, VSELs, and gametes have a common developmental origin.

Finally, VSELs have been demonstrated to be beneficial in regeneration in several experimental models of tissue/organ injuries. Several papers have already been published showing a contribution by injected purified VSELs to hematopoiesis, osteogenesis, and angiogenesis as well as to myocardium, liver, bones, and pulmonary alveolar epithelium in appropriate in vivo models [6–8, 90–92]. The well-demonstrated presence of chimerism in several organs after injection of these cells indicates the potential of these cells to differentiate across germ layers. We expect that these effects will be even more pronounced after the application of ex vivo expanded VSELs. We are currently working on this challenging topic.

Conclusions

Evidence has accumulated for the existence of a developmental link between PGCs, VSELs, and tissue-committed stem cells including, for example, HSCs, EPCs, and MSCs shedding new light on the developmental hierarchy of the stem cell compartment in adult BM. A similar important relationship occurs between PGCs, VSELs, and oocyte and sperm stem cells in gonads. Our group has identified VSELs in adult tissues and demonstrated that epigenetic modification of certain imprinted genes in these cells plays a crucial role in controlling their proliferation. On the other hand, reversal of this imprinting mechanism is crucial to employing these cells in regenerative medicine. Our encouraging data indicate that we are able to expand VSELs ~3 $\times$ 10^3 in serum-free medium in the presence of NAM and a cocktail of FSH, LH, BMP-4, and KL. Most important in this chemically defined medium is that

VSELs undergo asymmetric divisions which is an important hallmark of primitive stem cells. In addition, another team was successful in expanding ex vivo quiescent VSELs in presence of small pyrimido-indole derivative molecule UM177 [29]. In sum, to date, the presence of VSELs or VSELs-like cells present in adult postnatal tissues has been demonstrated by at least 35 independent laboratories. These cells, as we believe, are the hope for regenerative medicine because they are free of risk for teratoma formation. It is also a hope that they could become one day a source of gametes to treat infertility.

Acknowledgements This work was supported by NIH grants 2R01 DK074720 and the Stella and Henry Endowment to MZR.

References

1. Shin DM et al (2009) Novel epigenetic mechanisms that control pluripotency and quiescence of adult bone marrow-derived Oct4(+) very small embryonic-like stem cells. Leukemia 23:2042–2051. https://doi.org/10.1038/leu.2009.153
2. Hayashi K, de Sousa Lopes SM, Surani MA (2007) Germ cell specification in mice. Science 316:394–396. https://doi.org/10.1126/science.1137545
3. McLaren A (2003) Primordial germ cells in the mouse. Dev Biol 262:1–15. https://doi.org/10.1016/s0012-1606(03)00214-8
4. Chang YJ, Tien KE, Wen CH, Hsieh TB, Hwang SM (2014) Recovery of CD45(−)/Lin(−)/SSEA-4(+) very small embryonic-like stem cells by cord blood bank standard operating procedures. Cytotherapy 16:560–565. https://doi.org/10.1016/j.jcyt.2013.10.009
5. Ganguly R, Metkari S, Bhartiya D (2018) Dynamics of bone marrow VSELs and HSCs in response to treatment with gonadotropin and steroid hormones, during pregnancy and evidence to support their asymmetric/symmetric cell divisions. Stem Cell Rev Rep 14:110–124. https://doi.org/10.1007/s12015-017-9781-x
6. Guerin CL et al (2015) Bone-marrow-derived very small embryonic-like stem cells in patients with critical leg ischaemia: evidence of vasculogenic potential. Thromb Haemost 113:1084–1094. https://doi.org/10.1160/TH14-09-0748
7. Havens AM et al (2013) Human very small embryonic-like cells generate skeletal structures, in vivo. Stem Cells Dev 22:622–630. https://doi.org/10.1089/scd.2012.0327
8. Kassmer SH et al (2013) Very small embryonic-like stem cells from the murine bone marrow differentiate into epithelial cells of the lung. Stem Cells 31:2759–2766. https://doi.org/10.1002/stem.1413
9. Kuruca SE, Celik DD, Ozerkan D, Erdemir G (2019) Characterization and isolation of very small embryonic-like (VSEL) stem cells obtained from various human hematopoietic cell sources. Stem Cell Rev Rep 15:730–742. https://doi.org/10.1007/s12015-019-09896-1
10. Monti M et al (2017) (2017) A novel method for isolation of pluripotent stem cells from human umbilical cord blood. Stem Cells Dev 26:1258–1269. https://doi.org/10.1089/scd.2017.0012
11. Shaikh A, Nagvenkar P, Pethe P, Hinduja I, Bhartiya D (2015) Molecular and phenotypic characterization of CD133 and SSEA4 enriched very small embryonic-like stem cells in human cord blood. Leukemia 29:1909–1917. https://doi.org/10.1038/leu.2015.100
12. Virant-Klun I (2018) Functional testing of primitive oocyte-like cells developed in ovarian surface epithelium cell culture from small VSEL-like stem cells: can they be fertilized one day? Stem Cell Rev Rep 14:715–721. https://doi.org/10.1007/s12015-018-9832-y

13. McGuckin C, Jurga M, Ali H, Strbad M, Forraz N (2008) Culture of embryonic-like stem cells from human umbilical cord blood and onward differentiation to neural cells in vitro. Nat Protoc 3:1046–1055. https://doi.org/10.1038/nprot.2008.69
14. Dyce PW et al (2011) (2011) In vitro and in vivo germ line potential of stem cells derived from newborn mouse skin. PLoS One 6:e20339. https://doi.org/10.1371/journal pone.0020339
15. Wu JH, Wang HJ, Tan YZ, Li ZH (2012) Characterization of rat very small embryonic-like stem cells and cardiac repair after cell transplantation for myocardial infarction. Stem Cells Dev 21:1367–1379. https://doi.org/10.1089/scd.2011.0280
16. Iskovich S et al (2012) Elutriated stem cells derived from the adult bone marrow differentiate into insulin-producing cells in vivo and reverse chemical diabetes. Stem Cells Dev 21:86–96. https://doi.org/10.1089/scd.2011.0057
17. Shirazi R et al (2012) BMP4 can generate primordial germ cells from bone-marrow-derived pluripotent stem cells. Cell Biol Int 36:1185–1193. https://doi.org/10.1042/CBI20110065
18. Igura K, Okada M, Kim HW, Ashraf M (2013) Identification of small juvenile stem cells in aged bone marrow and their therapeutic potential for repair of the ischemic heart. Am J Physiol Heart Circ Physiol 305:H1354-1362. https://doi.org/10.1152/ajpheart.00379.2013
19. Lee SJ et al (2014) Adult stem cells from the hyaluronic acid-rich node and duct system differentiate into neuronal cells and repair brain injury. Stem Cells Dev 23:2831–2840. https://doi.org/10.1089/scd.2014.0142
20. Hwang S et al (2014) Nonmarrow hematopoiesis occurs in a hyaluronic-acid-rich node and duct system in mice. Stem Cells Dev 23:2661–2671. https://doi.org/10.1089/scd.2014.0075
21. Gounari E et al (2019) Isolation of a novel embryonic stem cell cord blood-derived population with in vitro hematopoietic capacity in the presence of Wharton's jelly-derived mesenchymal stromal cells. Cytotherapy 21:246–259. https://doi.org/10.1016/j.jcyt.2018.11.00
22. Golipoor Z et al (2016) Migration of bone marrow-derived very small embryonic-like stem cells toward an injured spinal cord. Cell J 17:639–647. https://doi.org/10.22074/cellj.2016.3836
23. Nakatsuka R et al (2016) Identification and characterization of lineage(−)CD45(−)Sca-1(+) VSEL phenotypic cells residing in adult mouse bone tissue. Stem Cells Dev 25:27–42. https://doi.org/10.1089/scd.2015.0168
24. Lo Sicco C et al (2015) Identification of a new cell population constitutively circulating in healthy conditions and endowed with a homing ability toward injured sites. Sci Rep 5:16574. https://doi.org/10.1038/srep16574
25. Makar T et al (2018) A subset of mobilized human hematopoietic stem cells express germ layer lineage genes which can be modulated by culture conditions. Stem Cell Res Ther 9:127. https://doi.org/10.1186/s13287-018-0358-5
26. Virant-Klun I, Skerl P, Novakovic S, Vrtacnik-Bokal E, Smrkolj S (2019) Similar population of CD133+ and DDX4+ VSEL-like stem cells sorted from human embryonic stem cell, ovarian, and ovarian cancer ascites cell cultures: the real embryonic stem cells? Cells 8:706. https://doi.org/10.3390/cells8070706
27. Mierzejewska K et al (2015) Hematopoietic stem/progenitor cells express several functional sex hormone receptors-novel evidence for a potential developmental link between hematopoiesis and primordial germ cells. Stem Cells Dev 24:927–937. https://doi.org/10.1089/scd.2014.0546
28. Ratajczak MZ et al (2017) A novel view of the adult stem cell compartment from the perspective of a quiescent population of very small embryonic-like stem cells. Circ Res 120 166–178. https://doi.org/10.1161/CIRCRESAHA.116.309362
29. Lahlil R et al (2018) VSELs Maintain their Pluripotency and Competence to Differentiate after Enhanced Ex Vivo Expansion. Stem Cell Rev Rep 14:510–524. https://doi.org/10.1007/s12015-018-9821-1
30. Ratajczak MZ (2015) A novel view of the adult bone marrow stem cell hierarchy and stem cell trafficking. Leukemia 29:776–782. https://doi.org/10.1038/leu.2014.346
31. Vacanti MP, Roy A, Cortiella J, Bonassar L, Vacanti CA (2001) Identification and initial characterization of spore-like cells in adult mammals. J Cell Biochem 80:455–460

32. Seandel M et al (2007) Generation of functional multipotent adult stem cells from GPR125+ germline progenitors. Nature 449:346–350. https://doi.org/10.1038/nature06129
33. Jiang Y et al (2002) Pluripotency of mesenchymal stem cells derived from adult marrow. Nature 418:41–49. https://doi.org/10.1038/nature00870
34. Kogler G et al (2004) A new human somatic stem cell from placental cord blood with intrinsic pluripotent differentiation potential. J Exp Med 200:123–135. https://doi.org/10.1084/jem.20040440
35. D'Ippolito G et al (2004) Marrow-isolated adult multilineage inducible (MIAMI) cells, a unique population of postnatal young and old human cells with extensive expansion and differentiation potential. J Cell Sci 117:2971–2981. https://doi.org/10.1242/jcs.01103
36. Prockop DJ (1997) Marrow stromal cells as stem cells for nonhematopoietic tissues. Science 276:71–74. https://doi.org/10.1126/science.276.5309.71
37. Howell JC et al (2003) Pluripotent stem cells identified in multiple murine tissues. Ann N Y Acad Sci 996:158–173. https://doi.org/10.1111/j.1749-6632.2003.tb03244.x
38. Ratajczak MZ, Ratajczak J, Kucia M (2019) Very small embryonic-like stem cells (VSELs). Circ Res 124:208–210. https://doi.org/10.1161/CIRCRESAHA.118.314287
39. Shin DM et al (2012) Global gene expression analysis of very small embryonic-like stem cells reveals that the Ezh2-dependent bivalent domain mechanism contributes to their pluripotent state. Stem Cells Dev 21:1639–1652. https://doi.org/10.1089/scd.2011.0389
40. Surani MA (2007) Germ cells: the eternal link between generations. C R Biol 330:474–478. https://doi.org/10.1016/j.crvi.2007.03.009
41. Molyneaux K, Wylie C (2004) Primordial germ cell migration. Int J Dev Biol 48:537–544. https://doi.org/10.1387/ijdb.041833km
42. Kucia M et al (2006) A population of very small embryonic-like (VSEL) CXCR4(+)SSEA-1(+)Oct-4+ stem cells identified in adult bone marrow. Leukemia 20:857–869. https://doi.org/10.1038/sj.leu.2404171
43. Kucia M et al (2007) Morphological and molecular characterization of novel population of CXCR4+ SSEA-4+ Oct-4+ very small embryonic-like cells purified from human cord blood: preliminary report. Leukemia 21:297–303. https://doi.org/10.1038/sj.leu.2404470
44. Shin DM et al (2010) Molecular signature of adult bone marrow-purified very small embryonic-like stem cells supports their developmental epiblast/germ line origin. Leukemia 24:1450–1461. https://doi.org/10.1038/leu.2010.121
45. Ratajczak J et al (2011) Hematopoietic differentiation of umbilical cord blood-derived very small embryonic/epiblast-like stem cells. Leukemia 25:1278–1285. https://doi.org/10.1038/leu.2011.73
46. Ratajczak J et al (2011) Adult murine bone marrow-derived very small embryonic-like stem cells differentiate into the hematopoietic lineage after coculture over OP9 stromal cells. Exp Hematol 39:225–237. https://doi.org/10.1016/j.exphem.2010.10.007
47. Lacaud G, Kouskoff V (2017) Hemangioblast, hemogenic endothelium, and primitive versus definitive hematopoiesis. Exp Hematol 49:19–24. https://doi.org/10.1016/j.exphem.2016.12.009
48. Lis R et al (2017) Conversion of adult endothelium to immunocompetent haematopoietic stem cells. Nature 545:439–445. https://doi.org/10.1038/nature22326
49. Park C, Ma YD, Choi K (2005) Evidence for the hemangioblast. Exp Hematol 33:965–970. https://doi.org/10.1016/j.exphem.2005.06.003
50. Ribatti D (2008) Hemangioblast does exist. Leuk Res 32:850–854. https://doi.org/10.1016/j.leukres.2007.12.001
51. Mizuochi C et al (2012) Intra-aortic clusters undergo endothelial to hematopoietic phenotypic transition during early embryogenesis. PLoS One 7:e35763. https://doi.org/10.1371/journal.pone.0035763
52. Mikkola HK, Orkin SH (2006) The journey of developing hematopoietic stem cells. Development 133:3733–3744. https://doi.org/10.1242/dev.02568
53. Sriraman K, Bhartiya D, Anand S, Bhutda S (2015) Mouse ovarian very small embryonic-like stem cells resist chemotherapy and retain ability to initiate oocyte-specific differentiation. Reprod Sci 22:884–903. https://doi.org/10.1177/1933719115576727

54. Anand S, Patel H, Bhartiya D (2015) Chemoablated mouse seminiferous tubular cells enriched for very small embryonic-like stem cells undergo spontaneous spermatogenesis in vitro. Reprod Biol Endocrinol 13:33. https://doi.org/10.1186/s12958-015-0031-2
55. Nayernia K et al (2006) Derivation of male germ cells from bone marrow stem cells. Lab Invest 86:654–663. https://doi.org/10.1038/labinvest.3700429
56. Johnson J et al (2005) Oocyte generation in adult mammalian ovaries by putative germ cells in bone marrow and peripheral blood. Cell 122:303–315. https://doi.org/10.1016/j.cell.2005.06.031
57. Liu C, Chen Z, Chen Z, Zhang T, Lu Y (2006) Multiple tumor types may originate from bone marrow-derived cells. Neoplasia 8:716–724. https://doi.org/10.1593/neo.06253
58. Cheng L (2004) Establishing a germ cell origin for metastatic tumors using OCT4 immunohistochemistry. Cancer 101:2006–2010. https://doi.org/10.1002/cncr.20566
59. Mezey E, Chandross KJ, Harta G, Maki RA, McKercher SR (2000) Turning blood into brain: cells bearing neuronal antigens generated in vivo from bone marrow. Science 290:1779–1782. https://doi.org/10.1126/science.290.5497.1779
60. Orlic D et al (2001) Bone marrow cells regenerate infarcted myocardium. Nature 410:701–705. https://doi.org/10.1038/35070587
61. Castro RF et al (2002) Failure of bone marrow cells to transdifferentiate into neural cells in vivo. Science 297:1299. https://doi.org/10.1126/science.297.5585.1299
62. Murry CE et al (2004) Haematopoietic stem cells do not transdifferentiate into cardiac myocytes in myocardial infarcts. Nature 428:664–668. https://doi.org/10.1038/nature02446
63. Ratajczak MZ et al (2004) Stem cell plasticity revisited: CXCR4-positive cells expressing mRNA for early muscle, liver and neural cells 'hide out' in the bone marrow. Leukemia 18:29–40. https://doi.org/10.1038/sj.leu.2403184
64. Virant-Klun I et al (2008) Putative stem cells with an embryonic character isolated from the ovarian surface epithelium of women with no naturally present follicles and oocytes. Differentiation 76:843–856. https://doi.org/10.1089/scd.2007.0238
65. Parte S et al (2011) (2011) Detection, characterization, and spontaneous differentiation in vitro of very small embryonic-like putative stem cells in adult mammalian ovary. Stem Cells Dev 20:1451–1464. https://doi.org/10.1089/scd.2010.0461
66. Surani MA, Hayashi K, Hajkova P (2007) Genetic and epigenetic regulators of pluripotency. Cell 128:747–762. https://doi.org/10.1016/j.cell.2007.02.010
67. Maatouk DM et al (2006) DNA methylation is a primary mechanism for silencing postmigratory primordial germ cell genes in both germ cell and somatic cell lineages. Development 133:3411–3418. https://doi.org/10.1242/dev.02500
68. Bartolomei MS, Ferguson-Smith AC (2011) Mammalian genomic imprinting. Cold Spring Harb Perspect Biol 3:a002592. https://doi.org/10.1101/cshperspect.a002592
69. Mann MR, Bartolomei MS (2002) Epigenetic reprogramming in the mammalian embryo: struggle of the clones. Genome Biol 3:REVIEWS1003. https://doi.org/10.1186/gb-2002-3-2-reviews1003
70. Zemel S, Bartolomei MS, Tilghman SM (1992) Physical linkage of two mammalian imprinted genes, H19 and insulin-like growth factor 2. Nat Genet 2:61–65. https://doi.org/10.1038/ng0992-61
71. Kono T et al (2004) Birth of parthenogenetic mice that can develop to adulthood. Nature 428:860–864. https://doi.org/10.1038/nature02402
72. Reik W, Walter J (2001) Genomic imprinting: parental influence on the genome. Nat Rev Genet 2:21–32. https://doi.org/10.1038/35047554
73. Delaval K, Feil R (2004) Epigenetic regulation of mammalian genomic imprinting. Curr Opin Genet Dev 14:188–195. https://doi.org/10.1016/j.gde.2004.01.00
74. Keniry A et al (2012) The H19 lincRNA is a developmental reservoir of miR-675 that suppresses growth and Igf1r. Nat Cell Biol 14:659–665. https://doi.org/10.1038/ncb2521
75. Surani MA, Durcova-Hills G, Hajkova P, Hayashi K, Tee WW (2008) Germ line, stem cells, and epigenetic reprogramming. Cold Spring Harb Symp Quant Biol 73:9–15. https://doi.org/10.1101/sqb.2008.73.015

76. Durcova-Hills G, Tang F, Doody G, Tooze R, Surani MA (2008) Reprogramming primordial germ cells into pluripotent stem cells. PLoS One 3:e3531. https://doi.org/10.1371/journal.pone.0003531
77. Ratajczak MZ, Shin DM, Schneider G, Ratajczak J, Kucia M (2013) Parental imprinting regulates insulin-like growth factor signaling: a Rosetta Stone for understanding the biology of pluripotent stem cells, aging and cancerogenesis. Leukemia 27:773–779. https://doi.org/10.1038/leu.2012.322
78. Heo J et al (2017) Sirt1 regulates DNA methylation and differentiation potential of embryonic stem cells by antagonizing Dnmt3l. Cell Rep 18:1930–1945. https://doi.org/10.1016/j.celrep.2017.01.074
79. Chaurasia P, Gajzer DC, Schaniel C, D'Souza S, Hoffman R (2014) Epigenetic reprogramming induces the expansion of cord blood stem cells. J Clin Invest 124:2378–2395. https://doi.org/10.1172/JCI70313
80. Broxmeyer HE (2014) Inhibiting HDAC for human hematopoietic stem cell expansion. J Clin Invest 124:2365–2368. https://doi.org/10.1172/JCI75803
81. Peled T et al (2012) Nicotinamide, a SIRT1 inhibitor, inhibits differentiation and facilitates expansion of hematopoietic progenitor cells with enhanced bone marrow homing and engraftment. Exp Hematol 40:342-355.e341. https://doi.org/10.1016/j.exphem.2011.12.005
82. Ou X et al (2011) SIRT1 deficiency compromises mouse embryonic stem cell hematopoietic differentiation, and embryonic and adult hematopoiesis in the mouse. Blood 117:440–450. https://doi.org/10.1182/blood-2010-03-273011
83. Singh SK et al (2013) Sirt1 ablation promotes stress-induced loss of epigenetic and genomic hematopoietic stem and progenitor cell maintenance. J Exp Med 210:987–1001. https://doi.org/10.1084/jem.20121608
84. Virant-Klun I, Bui HT, Ratajczak MZ (2016) Challenges in translating germinal stem cell research and therapy. Stem Cells Int 2016:4687378. https://doi.org/10.1155/2016/4687378
85. Virant-Klun I, Stimpfel M, Cvjeticanin B, Vrtacnik-Bokal E, Skutella T (2013) Small SSEA-4-positive cells from human ovarian cell cultures: related to embryonic stem cells and germinal lineage? J Ovarian Res 6:24. https://doi.org/10.1186/1757-2215-6-24
86. Kucia M et al (2011) Reduced number of VSELs in the bone marrow of growth hormone transgenic mice indicates that chronically elevated Igf1 level accelerates age-dependent exhaustion of pluripotent stem cell pool: a novel view on aging. Leukemia 25:1370–1374. https://doi.org/10.1038/leu.2011.98
87. Kucia M et al (2013) The negative effect of prolonged somatotrophic/insulin signaling on an adult bone marrow-residing population of pluripotent very small embryonic-like stem cells (VSELs). Age (Dordr) 35:315–330. https://doi.org/10.1007/s11357-011-9364-8
88. Grymula K et al (2014) Positive effects of prolonged caloric restriction on the population of very small embryonic-like stem cells—hematopoietic and ovarian implications. J Ovarian Res 7:68. https://doi.org/10.1186/1757-2215-7-68
89. Marycz K et al (2016) Endurance exercise mobilizes developmentally early stem cells into peripheral blood and increases their number in bone marrow: implications for tissue regeneration. Stem Cells Int 2016:5756901. https://doi.org/10.1155/2016/5756901
90. Chen ZH et al (2015) Hepatic regenerative potential of mouse bone marrow very small embryonic-like stem cells. J Cell Physiol 230:1852–1861. https://doi.org/10.1002/jcp.24913
91. Wojakowski W et al (2009) Mobilization of bone marrow-derived Oct-4+ SSEA-+ very small embryonic-like stem cells in patients with acute myocardial infarction. J Am Coll Cardiol 53:1–9. https://doi.org/10.1016/j.jacc.2008.09.029
92. Dawn B et al (2008) Transplantation of bone marrow-derived very small embryonic-like stem cells attenuates left ventricular dysfunction and remodeling after myocardial infarction. Stem Cells 26:1646–1655. https://doi.org/10.1634/stemcells.2007-0715

Chapter 12
The Role of Very Small Embryonic-Like Stem Cells (VSELs) in Reproductive Tissues

Deepa Bhartiya, Pushpa Singh, Ankita Kaushik, and Diksha Sharma

Abstract

Introduction: Adult reproductive tissues including testes, ovary, and uterus harbor two populations of stem cells namely very small embryonic-like stem cells (VSELs) and tissue specific progenitors (immediate descendants of VSELs) spermatogonial stem cells (SSCs) in testis, ovarian stem cells (OSCs) in ovary, and endometrial stem cells (EnSCs) in uterus.
Methods: We summarize new knowledge about VSELs in reproductive tissues and organs.
Results: VSELs are small (2–6 μm) in size, pluripotent (express embryonic markers and differentiate into 3 germ layers in vitro), and can be studied with a cell surface profile of LIN-CD45-SCA-1+ in mice and LIN-CD45-CD133+ in humans by flow cytometry. They undergo asymmetrical cell divisions to give rise to slightly bigger progenitors (SSCs, OSCs, and EnSCs) which divide rapidly in a symmetrical manner and undergo clonal expansion prior to initiating differentiation. Being quiescent in nature, VSELs in the ovaries and testes survive oncotherapy and they also survive in atrophied uterus after bilateral ovariectomy. Non-functional gonads as well as the uterus can be rejuvenated by transplanting mesenchymal stromal cells (MSCs) 'paracrine providers', extracellular vesicles, or exosomes secreted by the MSCs that facilitate the tissue-resident stem cells (VSELs) to undergo differentiation and enable regeneration. Stem cells do not pellet down (remain buoyant) when cells suspensions are centrifuged at 200–600 g to collect cells for various studies. This low speed is used to ensure viability of the somatic cells, but the stem cells get discarded unknowingly. Stem cells can be enriched by further centrifuging the supernatant at 1000 g. Negative data reported by various groups on the stem cells by single cell RNAseq in adult tissues is because they spin cells at lower speed for the experiments and unknowingly discard the stem cells.
Conclusions: VSELs serve as a back-up pool for tissue committed progenitors which further differentiate and thus tissue homeostasis is maintained life-long. Any

D. Bhartiya (✉) · P. Singh · A. Kaushik · D. Sharma
Stem Cell Biology Department, ICMR-National Institute for Research in Reproductive Health, Jehangir Merwanji Street, Parel, Mumbai 400 012, India
e-mail: deepa.bhartiya@yahoo.in

I. Virant-Klun (ed.), *Stem Cells in Reproductive Tissues and Organs*, Stem Cell Biology and Regenerative Medicine 70, https://doi.org/10.1007/978-3-030-90111-0_12

disruption to the stem cells compartment results in various pathologies including cancer.

Keywords Cancer · Infertility · Ovary · Stem cells · Testes · Uterus · Very small embryonic-like stem cells (VSELs)

Abbreviations

ABCG2	ATP binding cassette subfamily G member 2 (Junior blood group)
ACD	Asymmetrical cell division
AXIN2	Axin 2
BMI1	BMI1 proto-oncogene, polycomb ring finger
CD	Cluster of differentiation
CD45	CD45 antigen (PTPRC)
CD133	CD133 antigen, prominin 1 (PROM1)
CK18	Cytokeratin 18
c-KIT	KIT proto-oncogene, receptor tyrosine kinase
CXCR4	C-X-C motif chemokine receptor 4
DAZL	Deleted in azoospermia like
DDX4	DEAD-box helicase 4
DES	Diethylstilbestrol
DNA	Deoxyribonucleic acid
E	Estrogen
EnSC	Endometrial stem cell
ER	Estrogen receptor
Erα	Estrogen receptor alpha
ESC	Embryonic stem cell
FGF	Fibroblast growth factor
FGSC	Female germ-line stem cell
FSH	Follicle-stimulating hormone
FSHR	Follicle-stimulating hormone receptor
FSHR1	Follicle-stimulating hormone receptor 1
FSHR3	Follicle-stimulating hormone receptor 3
GFP	Green fluorescent protein
GFRA	GDNF family receptor alpha
GnRH	Gonadotropin-releasing hormone
H&E	Hematoxylin and eosin (staining)
IGF1	Insulin like growth factor 1
iPSC	Induced pluripotent stem cell
LH	Luteinizing hormone
MSC	Mesenchymal stem cell
MVH	Human VASA (DDX4) homolog
NP95	Ubiquitin like with PHD and ring finger domains 1

NANOG	Nanog homeobox
NOD/SCID	Non-obese diabetic/severe combined immunodeficiency (mouse)
NUMB	NUMB endocytic adaptor protein
OCT-4	Octamer-binding transcription factor 4
OSC	Ovarian stem cell
OSE	Ovarian surface epithelium
P	Progesterone
PCNA	Proliferating cell nuclear antigen
PCOS	Polycystic ovary syndrome
PGC	Primordial germ cell
PGCLC	Primordial germ cell-like cell
POF	Premature ovarian failure
RNA	Ribonucleic acid
RNAseq	RNA sequencing
RT-PCR	Reverse transcription polymerase chain reaction
SCA-1	Stem cell antigen-1
SCD	Symmetrical cell division
scRNAseq	Single-cell RNA sequencing
SOX2	SRY-box transcription factor 2
SP	Side population
SSEA-1	Stage-specific embryonic antigen-1
SSEA-4	Stage-specific embryonic antigen-4
SSC	Spermatogonial stem cell
STELLA	Developmental pluripotency associated 3
STRA8	Stimulated by retinoic acid
VSEL	Very small embryonic-like stem cell

Introduction

Tissue resident adult stem cells are crucial for maintaining life-long homeostasis under physiological conditions and to regenerate the tissue after any kind of insult. Stem cells have infinite lifespan and are expected to be relatively quiescent (out of cell cycle and in low metabolic state) in nature, rarely divide in an asymmetrical fashion whereby they divide to self-renew, and give rise to slightly bigger 'progenitors' with finite lifespan that undergo rapid, symmetrical cell divisions, and clonal expansion (cell division with incomplete cytokinesis) prior to initiating differentiation. Quiescent nature of the primitive stem cells is expected to protect them from accumulating mutations during DNA replication whereas the actively dividing progenitors have the ability to rapidly differentiate to maintain homeostasis and replace damaged tissue. This presence of two populations of stem cells in adult tissues was discussed at length about a decade ago [1, 2]. However, it has proved difficult to isolate the quiescent stem cells as a physical entity from adult tissues, and thus it is being argued that rather than the stem cells, mature cells dedifferentiate to stem-like state [3]. Recent single-cell RNAseq studies have also failed to detect stem cells in human ovary [4], testes [5], prostate [6] as well as in the uterus [7]. However, this negative data probably has a technical basis and we pointed out this in case of ovarian and prostate stem cells [6, 8, 9]. The stem cells are of smaller size compared to the mature cells and hence exist as a distinct population. Uterus is one of the most plastic organs, undergoes regular break-down and remodeling but the true identity of uterine stem cells remains unknown [10] despite more than three decades of research. It was proposed that mesenchymal stem cells (MSCs) are the uterine stem cells [11], but this was challenged by our group [12]. MSCs are not stem cells, rather they are 'paracrine providers' [13].

Our research over a decade has shown that two populations of stem cells exist in testes, ovaries, and uterus and include most primitive very small embryonic-like stem cells (VSELs) and tissue specific progenitors that include spermatogonial stem cells (SSCs) in testes, ovarian stem cells (OSCs) in the ovaries, and endometrial stem cells (EnSCs) in the uterus. We have earlier published a review on various aspects of VSELs in the ovaries and testes [12]. Presence of VSELs along with SSCs in the testes was discussed [14] and reported in testes [15], ovaries [16], and uterus [17–20]. VSELs exist in reproductive tissues and also in increased numbers in cancers of ovaries [21] and testes [22]. This chapter is focused on the VSELs biology in reproductive tissues, protocols for their isolation, how they function, and their role in initiating diseases including cancers.

Isolation of Two Populations of Stem Cells from Testes, Ovaries, and Uterus

VSELs can be studied by flow cytometry as small sized (2–6 μm) cells with the surface phenotype as LIN-CD45-SCA1+ in mice and as LIN-CD45-CD133+ is humans. However, a controversy surfaced in 2013 when few groups could not detect these stem cells following the published protocols [23]. Shortcomings during the validation studies were pointed out [24], but the damage was done and affected the field adversely. VSELs still struggle to get acknowledged widely [25]. More than 50 independent groups have now acknowledged their existence in various adult tissues [26]. Working for more than a decade on the VSELs in reproductive tissues with the first paper being on human testicular VSELs [27], our group has recently reported a robust protocol to enrich these stem cells from any solid tissue including reproductive tissues that can easily be replicated by anyone and it does not require use of sophisticated instruments. Single cells suspension obtained after enzymatic digestion of any solid tissue is first spun at 200–400 g for 15 min when majority of somatic mature cells pellet down. However, the stem cells remain suspended in the supernatant and can be enriched by further centrifuging the supernatant at 1000 g. This high speed does not affect the stem cells which have minimal cytoplasm compared to mature somatic cells with abundant cytoplasm holding several organelles necessary to maintain their active metabolic state.

Using this method, we recently reported enrichment of the two populations of stem cells from multiple adult tissues [28], adult pancreas [29], testes [15], and uterus [17]. The enriched stem cells can be further used for detailed characterization. Flow cytometry studies on the enriched cell population showed almost 10-folds enrichment of stem cells upon centrifuging at 1000 g. Stem cells are located in the ovary surface epithelium and can be collected by gently scraping of the ovarian surface into a Petri dish. Mice ovaries are small in size and stem cells can be collected by gentle trypsinization of the ovary surface [30]. Single cell RNAseq studies have failed to detect stem cells in ovary, testes, uterus as well as in the prostate basically because cells are invariably collected by centrifugation at 200-400 g, stem cells get discarded unknowingly, and never get subjected to scRNAseq [8].

Characterization of VSELs in Reproductive Tissues

Stem cells enriched as described above can be viewed after H&E staining and appear as small, distinctly spherical stem cells with high nucleo-cytoplasmic ratio and dark blue stained nuclei. These stem cells express pluripotent markers including nuclear OCT-4, SOX-2, NANOG, SSEA-1 (SSEA-4 in humans), and also markers specific for primordial germ cells (STELLA, FRAGILIS) along with SCA-1 and c-KIT. Two populations of stem cells can easily be distinguished based on their size and OCT-4 expression, smaller VSELs express nuclear OCT-4A whereas slightly bigger progenitors express cytoplasmic OCT-4B. Being a transcription factor, nuclear OCT-4 is crucial to maintain pluripotent state of VSELs and once they initiate differentiation, nuclear OCT-4A is no longer required, shifts to the cytoplasm, and eventually gets degraded as cells become further differentiated. Cytoplasmic OCT-4, SCA-1, and c-KIT are expressed by progenitors residing in different tissues. However, the 'progenitors' become tissue committed and express specific markers including MVH in the ovaries, GFRA in the testes, and Axin-2 in the uterus. DNA/RNA/proteins can also be isolated from the enriched stem cells population for OMICS studies. We have regularly detected pluripotent (Oct-4A, Sox2, Nanog) and primordial germ cells specific (Stella, Fragilis) transcripts by RT-PCR suggesting the presence of pluripotent stem cells with a developmental link to PGCs as suggested in the literature [26]. Progenitors express Oct-4B, Sca-1, c-Kit, and tissue specific transcripts (Mvh, Gfra, Axin-2). These enriched stem/progenitor cells can also be studied by flow cytometry as 2–6 μm, viable cells with the surface phenotype as LIN-CD45-SCA-1+. Ultrastructure of bone marrow VSELs is reported [31] and gene expression profiling of small sized SSEA-4 positive VSELs are reported in the human ovary [32]. Further advanced techniques including scRNAseq should be employed for analysis of whole transcriptome/epigenome to further understand these stem/progenitor cells. Presence of VSELs in reproductive tissues needs to be acknowledged by the scientific community. Ovarian VSELs have been reported by Virant-Klun's and our group, and their presence were recently confirmed by Tilly's group in mouse ovary [33] and by Silvestris et al. [34] in human ovaries.

Functional Potential of VSELs in Reproductive Tissues

VSELs and OSCs in Ovaries

Tilly's group recently confirmed the presence of LIN-CD45-SCA-1+ VSELs in adult mice ovaries and further showed that the VSELs are negative for MVH by flow cytometry [33]. Tilly's group and others working in the field have worked on MVH positive female germ-line stem cells (FGSCs) also termed OSCs which expand in cultures, can differentiate into oocytes, and upon transplantation, live births have been reported. As a result, they questioned the biological relevance of ovarian VSELs [33]. Similarly, Silvestris et al. [34] also commented that small-sized stem cells were observed in the cultures but did not change over time in vitro. One needs to appreciate that in vitro and in vivo situations are distinct from each other. FGSCs/OSCs may be able to replicate over months in vitro, but once they expand and initiate differentiation in vivo, they most likely enter a point of no return. As described above, ovarian progenitors 'OSCs' have a fixed life-span and are continuously replenished in vivo by the immortal VSELs by undergoing asymmetrical cell divisions. Thus, VSELs are the sole and infinite source of OSCs that further differentiate into oocytes. It will be incorrect to discuss and compare the potential of VSELs and OSCs to differentiate into oocytes as VSELs only role is to differentiate into OSCs which further differentiate into oocytes. Moreover, VSELs express nuclear MVH which is expressed on cell surface in the OSCs. Nuclear expression of MVH is the reason why Tilly's group [33] failed to detect MVH positive VSELs. MVH expression in VSELs also strengthens the postulate that VSELs are developmentally linked to PGCs as MVH (DDX-4) is also expressed by the migratory primordial germ cells in both humans and mice [35, 36]. DDX-4 expression on FGSCs/OSCs has been debated [37], but it has been used to enrich the OSCs where it is expressed on the cell surface unlike the oocytes where it is cytoplasmic in nature [38, 39]. In Fig. 12.1 we can see Hematoxylin eosin (HE) stained VSELs which were retrieved by scraping of ovarian surface epithelium (OSE).

Primordial germ cells are natural precursors for gametes [40]. As discussed above, VSELs are developmentally equivalent to PGCs and thus spontaneously differentiate into oocyte-like structures, and there is no need to add any additional growth factors/cytokines. Virant-Klun's [41] and our [42] groups have reported that the VSELs lodged in the OSE can differentiate in vitro into oocyte-like structures. This

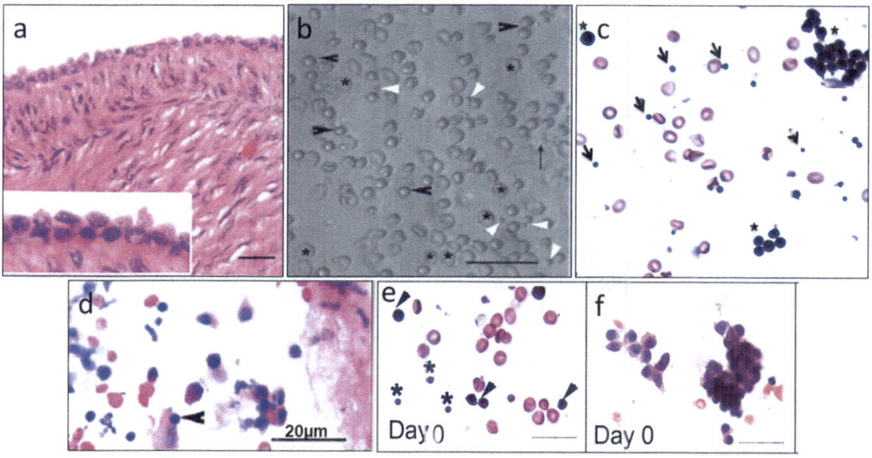

Fig. 12.1 Stem cells in adult (perimenopausal) human ovary. **a** Hematoxylin and Eosin (H&E) stained section of adult ovary showed no follicles but the ovary surface epithelium was prominent. **b** The ovary surface is gently scraped in a dish and the stem cells were clearly observed including VSELs (white arrowhead), OSCs (black arrowhead), and red blood cells (asterix). **c** The smears were stained with H&E and stem cells were clearly visualized. VSELs (black arrow), OSCs (asterix), were clearly visualized. **d–f** Additional fields showing stem cells in ovary surface epithelial cells smears. Stem cells are distinctly spherical with diameters up to 5 μm. Scale Bars: 20 μm

is in contrast to using embryonic/induced pluripotent stem cells/OSCs to differentiate into oocytes which requires a cocktail of growth factors and was discussed [43]. The oocytes obtained in vitro are capable of undergoing cortical reaction in the presence of sperm [44]. DDX-4 positive OSCs have been maintained and expanded in vitro for long term, differentiated into oocytes and upon transplantation, GFP positive OSCs differentiate into oocytes and eggs that were fertilized to produce viable offspring [45]. The major bottleneck while attempting to differentiate ES/iPS cells in to gametes is their maturation into PGC-like cells and thus VSELs (equivalent to PGCs) are novel stem cell candidates to differentiate into oocytes [46].

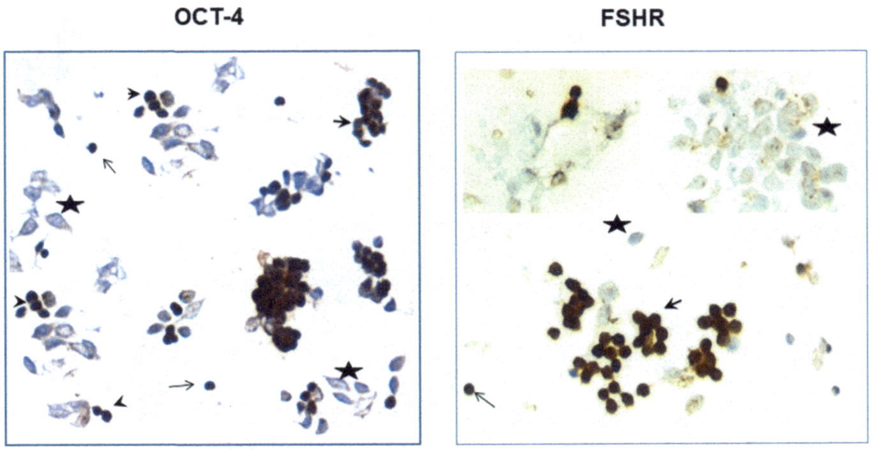

Fig. 12.2 Stem cells in ovarian surface epithelium. Immuno-expression of OCT-4 and FSHR on the stem cells located in the ovarian surface epithelium cell smears. It is clearly evident that the stem cells are positive for OCT-4 and FSHR and epithelial cells are distinctly negative. These OCT-4 and FSHR expressing stem cells most likely initiate ovarian cancers since cancer cells are reported to express OCT-4 and FSHR

In Fig. 12.2 we can see the OCT-4 and follicle-stimulating hormone receptor (FSHR) expression in stem cells located in the ovarian surface epithelium (OSE) cell smears.

Being quiescent in nature, initially Ratajczak's group showed that bone marrow VSELs survived total body irradiation in mice bone marrow[47]. Later we have reported that VSELs survive in adult ovaries after chemotherapy [48]. Similarly, Virant-Klun's group has reported presence of stem cells in the ovary with no apparent follicles [41]. These stem cells that survive in non-functional ovaries can be manipulated to restore ovarian function upon transplanting niche cells, 'paracrine providers' mesenchymal stromal cells [16]. Rather than rejuvenating aged eggs by transplanting mitochondria collected from the stem cells as advocated by Tilly's group [49], it will be ideal to regenerate non-functional ovaries by manipulating endogenous, tissue resident VSELs [50, 51]. This is clearly evident in published literature as several studies have reported improved ovarian function by transplanting 'paracrine providers' or MSCs (from various sources) in animal models and humans [52–54]. A baby has also been born by transplanting autologous MSCs in non-functional ovaries [55]. A clinical trial is also registered to treat women with POF by transplanting MSCs (NCT02603744). MSCs only help to rejuvenate and help endogenous stem cells to make new follicles and this further evident by the fact that rather than MSCs, even the extracellular vesicles obtained from MSCs help improve ovarian function [56, 57].

Presence of stem cells in ovaries is questioned by few who fail to understand why menopause occurs when stem cells exist? Menopause occurs with advanced age despite the presence of stem cells as the niche gets compromised with age. We need to invariably think of stem cells in the context of their niche which controls stem cells function (proliferation/differentiation). This has been discussed in details in the literature [58, 59]. Studies are available showing that transplanting stem cells from aged ovaries in young ovaries differentiate normally to generate oocytes [60]. Similarly transplanting mesenchymal stromal cells in non-functional aged ovaries can be of benefit.

VSELs and SSCs in Testes

We recently reviewed testicular stem cells and their role during spermatogenesis, infertility and cancer. Various models explaining spermatogonial stem cell biology have been proposed including the A_s, fragmentation and hierarchical model. 'A_s model' for spermatogonial multiplication and stem cell renewal in mice suggests that A_s are SSCs (reserve, rarely divide) that give rise to A_{pr} and A_{al} spermatogonial cells. In human testes, A_{dark}, and A_{pale} SSCs exist wherein A_{pale} cells divide regularly and are historically defined as the 'active' stem cell pool. There is lack of clarity on the two new models including the 'fragmentation' and the 'hierarchical' model since the stem cells identified in these models are located far away from the basal lamina (natural niche for testicular stem cells) of the seminiferous tubules [61]. We propose that VSELs are the most primitive stem cells in the testis, they undergo ACD to self-renew and give rise to the SSCs which undergo SCD and further clonal expansion prior to initiating spermatogenesis [62].

In Fig. 12.3 we can see the presence of two populations of stem cells in testicular smears obtained from day 15 (D15) testes: VSELs and SSCs. These two populations of testicular stem cells were studied by flow cytometry as viable cells (7AAD negative) based on size and SCA-1 expression (Fig. 12.4).

In the field of testicular stem cells, scientific community does not wish to think beyond spermatogonial stem cells (SSCs). A recent grand-theme review by Schlatt's

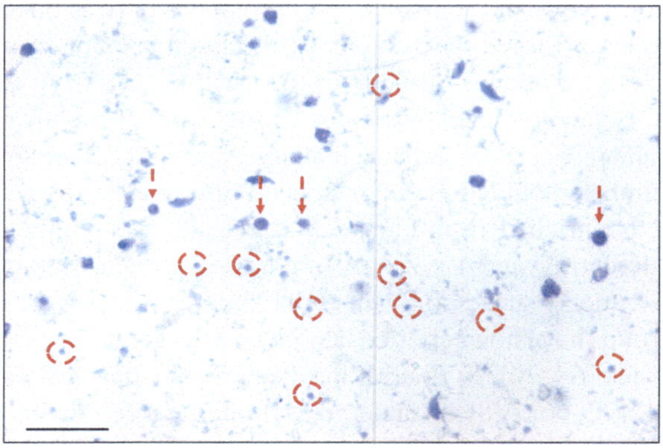

Fig. 12.3 Testicular stem cells. Note the presence of two populations of stem cells in testicular smears obtained from D15 testes. Smaller VSELs are in red circles whereas the SSCs are marked by red arrows. Details of methods to prepare these smears were published recently [22]. Scale Bar: 20 μm

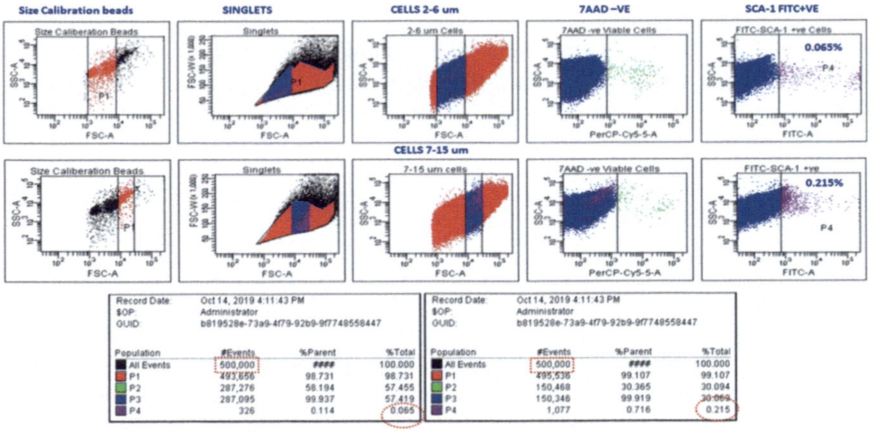

Fig. 12.4 Two populations of testicular stem cells studied by flow cytometry as viable cells (7AAD negative) based on size and SCA-1 expression. VSELs were in the size range of 2–6 μm whereas SSCs were >7 μm. As evident, VSELs (0.065%) were few in numbers compared to the SSCs (0.215%) upon evaluating 5 lakh events. These results were published recently [22]

group focused on the SSCs in adult mammalian testes [63]. We wrote a commentary [14] but the response was very dismal [64]. They concluded that PGCs express pluripotent markers and exist only in the fetal gonads whereas SSCs are the most primitive stem cells in adult testes. We agree with this viewpoint but have data to show that PGCs survive as a distinct population in adult testes as VSELs that are more primitive and give rise to SSCs throughout life by undergoing asymmetrical cell divisions similar to as in the ovaries as discussed above. It is evident that utter disbelief exists in the field regarding our findings published a decade ago describing the presence of VSELs in addition to SSCs in adult human testes [27]. VSELs are pluripotent, relatively quiescent, small size, and express nuclear OCT-4A, whereas SSCs divide rapidly, express cytoplasmic OCT-4B, and form chains (clonal expansion). Now since only one (our) group is reporting on testicular VSELs, considerable doubt exists on the veracity of the findings. But seeing is believing, and we have discussed above our methods to enrich VSELs from adult testes.

Being quiescent, testicular VSELs survive oncotherapy in both mice [65, 66] and human [67] testes. Anand et al. [65] reported that not only VSELs survived, but their numbers doubled in chemoablated mouse testes by flow cytometry. Further, Sertoli cells enriched from chemoablated testes were functionally affected as judged by microarray analysis. As a result, transplanting mesenchymal stromal cells 'paracrine providers' successfully restored spermatogenesis in chemoablated testes. Several groups have now reported beneficial effects of transplanting MSCs in non-functional testes [68–70]. Further, it has been shown that VSELs & SSCs express FSHR and ER [22, 66]. VSELs undergo asymmetrical cell divisions to self-renew and give rise to SSCs that in turn undergo symmetrical divisions and clonal expansion. VSELs numbers in chemoablated mouse testes were increased upon treatment with FSH.

These results have relevance in the context of recent reports suggesting to treat male infertility by treating with FSH [71–73]. A large number of male partners visiting infertility clinics are labelled as suffering from 'idiopathic infertility' and only a subgroup with hypogonadotropic hypogonadisms, who fail to produce FSH, luteinizing hormone (LH), and gonadotropin-releasing hormone (GnRH) get treated with FSH. But it is now being advocated that even the men with 'idiopathic infertility' should be treated with FSH [72, 73] just as FSH is widely used in the female partners to activate the ovaries.

Testicular VSELs enriched from busulphan treated mice testes were put in culture on a Sertoli cells bed, spontaneously differentiated into sperm [74]. Initially it was observed that the VSELs underwent asymmetrical cell divisions to give rise to the SSCs that further differentiated into sperm. It was intriguing to note that no additional growth factors were required in vitro for differentiation of sperm On the contrary, it has proved extremely difficult to differentiate hES/iPS cells into gametes and the major bottleneck remains to differentiate them into PGCLCs. In order to further prove the potential of VSELs to differentiate into male germ cells VSELs enriched from the bone marrow were placed on a Sertoli cells bed for 2 weeks. Male germ cells were detected in the culture expressing GFRA, DAZL, and STRA8 [75].

Our group also studied how the mouse testicular stem cells in adult life get affected by neonatal exposure to endocrine disrupting chemicals including estradiol and diethylstilbestrol [22]. Huge body of evidences from experimental, epidemiological, and animal- studies indicates that exposure to these chemicals during critical time of growth and development lead to disease onset in adult life. To date, little is known about underlying mechanisms by which endocrine disrupting chemicals cause deteriorating effects which are passed down to subsequent generations. As a result of endocrine disruption, VSELs increase in numbers and their differentiation into c-KIT positive spermatogonial cells gets blocked, meiosis specific transcripts are also downregulated with no signs of apoptosis, p53 levels are lowered, and NP95 (a chromatin modulator) expression gets massively down regulated. These results suggest that excessive self-renewal and compromised differentiation of VSELs is responsible for disrupted spermatogenesis, reduced sperm count, infertility, and cancer initiation, being reported in young men in recent times [76]. Extensive self-renewal of VSELs in the testes could be forerunner to testicular cancer. Simultaneously blocked differentiation of the stem cells resulted in reduced sperm count and infertility.

VSELs and EnSCs in Uterus

Uterus is one of the most plastic organs wherein endometrium grows, differentiates to accept the growing embryo otherwise breaks down on regular basis. Stem cells were reported to exist in uterus almost three decades ago [77] but till now specific markers to detect endometrial stem cells remain elusive [10]. Taylor's group has shown that bone marrow could be a possible source of uterine stem cells [78]. Gargett's group has greatly contributed toward understanding of stem cells in the uterus and menstrual blood [11, 79, 80], their possible role to initiate endometriosis, and their regenerative potential. Their focus has remained on mesenchymal stem/stromal cells and have developed serum-free cultures to expand endometrial MSCs in vitro for clinical translation. They also compiled various publications available in literature suggesting ability of menstrual blood to differentiate into multiple lineages (neuronal: neuronal and glial-like cells and endoderm: hepatocytes) suggesting their pluripotent potential and plasticity [79].

MSCs exist as pericytes in multiple adult tissues [81, 82]. We had discussed earlier that MSCs could not be the true endometrial stem cells which are expected to undergo asymmetrical cell divisions to self-renew and give rise to progenitors that further expand in numbers before initiating differentiation. MSCs are not stem cells rather have regenerative property and act as 'paracrine providers' [83, 84]. MSCs from any source like adipose tissue, cord blood, bone marrow, etc. can be transplanted to achieve regeneration, and there is no need to isolate them from menstrual blood for endometrial regeneration.

MSCs are a source of growth factors/cytokines crucial for endogenous stem cells to bring about regeneration. Rather than MSCs, even the extracellular vesicles secreted by the MSCs or exosomes derived from them have also provided similar beneficial effects in multiple adult tissues [57, 85, 86] including endometrium [87]. Esfandyari et al. [88] discussed MSCs as a bio-organ to treat female infertility including PCOS, POF, endometriosis, Asherman syndrome, and preeclampsia. MSCs exert beneficial effects in both males and females by providing a paracrine support to the tissue-resident stem/progenitor cells. It needs to be understood that the MSCs (or extracellular vesicles derived from them) will help to rejuvenate, whereas tissue-resident stem cells will ensure regeneration. Thus, it becomes essential to differentiate between a stem cell and a niche providing stromal cell (MSC) and was discussed recently [51].

Our group has reported presence of two populations of stem cells including VSELs and slightly bigger progenitors EnSCs (endometrial stem cells) in adult mouse uterus [19]. These stem cells express OCT-4 and FSHR (Fig. 12.5). We recently published a robust protocol to isolate these stem cells from adult uterus [17]. VSELs undergo asymmetrical cell divisions to self-renew and give rise to EnSCs which expand rapidly, clonally expand with incomplete cytokinesis, and further differentiate into both epithelial and myometrial cells. Asymmetrical and symmetrical divisions were clearly demonstrated by studying NUMB expression in the dividing doublets of OCT-4 positive uterine stem cells [18]. Uterine stem cells express Erα, Erβ, PR, and also FSHR [17], and as a result can respond to these hormones directly. Studies have been conducted in bilaterally ovariectomized mice how the uterine stem cells are affected by physiological levels of estradiol (E) and progesterone (P) [19] and also by pharmacological levels of hormones (E, P, and FSH) on myometrium [20] and endometrium [18]. Variation of VSELs under normal physiological conditions across estrus cycle has also been reported [17].

Uterine stem cells can be enriched by centrifuging the (Fig. 12.6) and detected by flow-cytometry (Fig. 12.7).

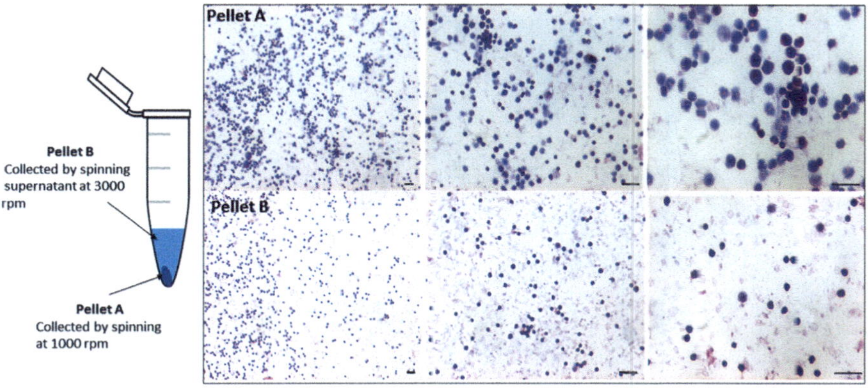

Fig. 12.5 Uterine stem cells in adult mice. They can be enriched by centrifuging the somatic cells at 250 g and later the stem cells were enriched by centrifuging at 1000 g. Similar to the testes and ovaries, even uterine VSELs are distinctly spherical, darkly stained, and small-sized cells. This work was recently published [17]. Scale Bars: 20 μm

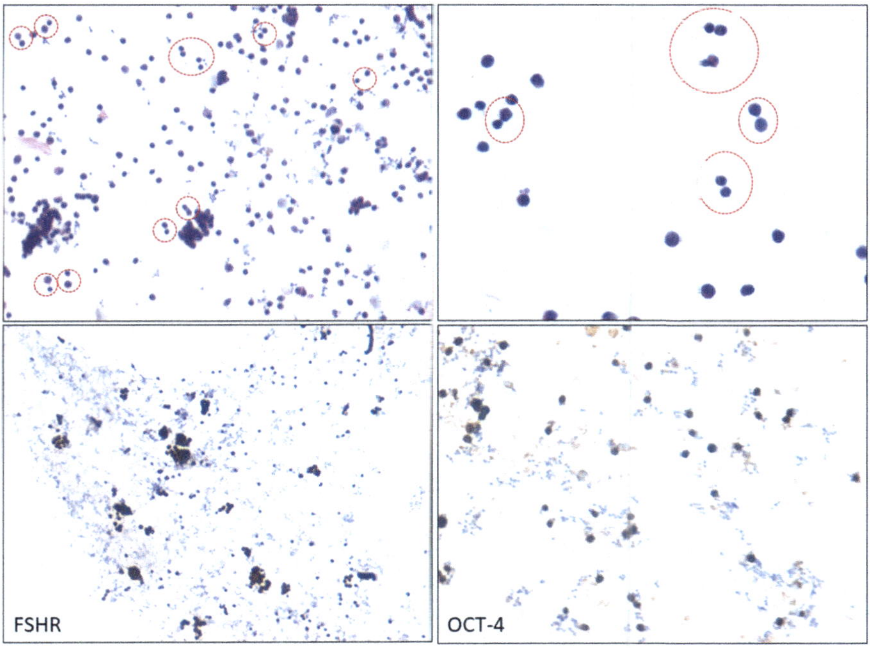

Fig. 12.6 Uterine stem cell smears showing dividing stem cells and few clumps of cells showing clonal expansion. The stem cells express FSHR and OCT-4

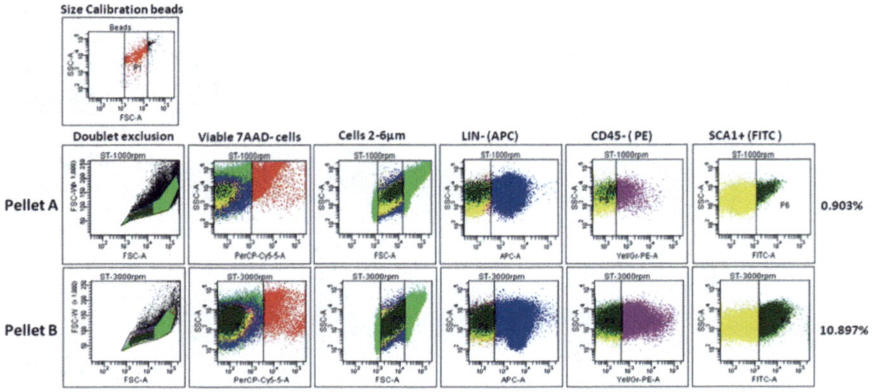

Pellet A

Population	#Events	%Parent	%Total
All Events	300,000	####	100.000
P1	289,665	96.555	96.555
P2	275,323	95.049	91.774
P3	143,853	52.249	47.951
P4	16,828	11.698	5.609
P5	13,915	82.690	4.638
P6	2,709	19.468	0.903

Pellet B

Population	#Events	%Parent	%Total
All Events	300,000	####	100.000
P1	298,212	99.404	99.404
P2	291,862	97.871	97.287
P3	175,443	60.112	58.481
P4	113,232	64.541	37.744
P5	79,402	70.123	26.467
P6	32,692	41.173	10.897

Fig. 12.7 Detection of VSELs in adult uterus by flow cytometry. VSELs (2–6 μm) are studied as viable cells (7 AAD: 7-Aminoactinomycin D negative) after doublets exposure as cells with surface phenotype of LIN-CD45-SCA-1+. Three lakh events were studied in both the pellets collected by centrifuging at 300 g and 1000 g. Almost 10 folds enrichment of VSELs by centrifuging at 1000 g. This work was recently published [17]

Gunjal et al. [19] reported two populations of stem cells including VSELs and EnSCs in mouse uterus based on their size and OCT-4 expression. The stem cells are distinctly spherical in shape, stain dark with Hematoxylin, and show high nucleo-cytoplasmic ratio with minimal cytoplasm. VSELs were studied by flow cytometry as LIN-CD45-SCA-1+cells and stem cell specific transcripts Oct-4, Oct-4A, Sox-2, Nanog, and Sca-1 by RT-PCR. VSELs survived bilateral ovariectomy in the atrophied mice uterus and showed distinct variation after treatment with E, P, E+P, and after withdrawal of E+P for 48 h compared to no treatment to bilaterally ovariectomized group. Increased numbers (>2 folds) of Oct-4A and Sca-1 positive VSELs were detected upon withdrawal of E+P, whereas EnSCs were maximal (>10 folds increase) in E+P treated group.

In order to study the effect of E, P, and FSH on endometrial and myometrial stem cells, we treated bilaterally ovariectomized mice with E (2 μg/day), P (1 mg/Kg day), or FSH (5 IU per day) for seven days each. OCT-4A positive VSELs were detected in the myometrium and showed dark Hematoxylin stained nuclei. Increased expression of PCNA and cytoplasmic OCT-4 was observed upon various treatment and numbers of VSELs were much more increased in numbers after P and FSH treatment compared to E. Stem cells expressing OCT-4, SOX-2, and NANOG have been reported in the side population in human myometrium [89], and their numbers are increased in cases of leiomyoma. Thus, we had concluded that VSELs probably initiate leiomyoma [20, 90].

Similar mouse model was used to study the effect of these hormones on the endometrial stem cells. Presence of VSELs in ovariectomized uterus was confirmed by histological studies, immuno-localization, and RT-PCR. Estradiol treatment led to increased hypertrophy of epithelial cells. Treatment with P and FSH resulted in marked hyperplasia and overcrowding of epithelial cells. Stem cells were enriched by centrifuging at 1000 g were subjected to RT-PCT, and transcripts specific for Oct-4A, ERα, PR, Fshr1, and Fshr3 were detected. Presence of alternately spliced Fshr transcripts was further confirmed by Western blotting. Stem cells were greatly increased in numbers after treatment with P and FSH and were found to express OCT-4, PCNA, and FSHR. We also successfully showed the VSELs divide by asymmetrical cell divisions to self-renew and give rise to the EnSCs which undergo symmetrical cell divisions and clonal expansion prior to further differentiation into epithelial cells. ACD and SCD were further confirmed by the selective expression of NUMB (differentiation marker) in the EnSCs, whereas OCT-4 expressing VSELs remained negative for NUMB [18].

A distinct variation in numbers of VSELs was observed under physiological conditions as well. Maximal numbers of VSELs were detected during estrus and metestrus stages of estrus cycle based on flow cytometry and qRT-PCR studies [17]. Expression of FSHR on the uterine stem cells was intriguing and supports extra-gonadal action of FSH. Another interesting observation was a direct action of E, P, and FSH on the endometrial stem cells. Current understanding is that estrogen acts indirectly on the epithelial cells leading to their proliferation. Estrogen acts on the stromal cells which secrete growth factors including IGF-1, FGF that stimulate epithelial cells to undergo proliferation [91–93]. Our findings suggest that the hormones exert a direct action on the stem cells located in the basal region of epithelial cells lining the lumen and the glands in addition to the myometrium whereas the stromal cells provide a niche to the stem cells and are a source of growth factors/cytokines crucial for stem cells proliferation/differentiation.

VSELs Role in Initiating Reproductive Health Related Diseases Including Cancers

Incidence of infertility and various diseases associated with reproductive health including PCOS, POF, endometriosis, fibroids, and reproductive cancers have increased in recent times. Global efforts to delineate genetic basis for various pathologies and various OMICS studies have not yielded any convincing data of clinical relevance. Besides germline/inherited genetic changes, studies are now being focused on somatic/acquired mutations in order to understand underlying etiology leading to endometriosis [94]. A large fraction of infertile men still remains idiopathic, and the underlying reason for decreased sperm count in recent times remains a puzzle. Also, cancers generally considered a disease of aged people is affecting young men and women in recent times. Endocrine disruption during fetal and perinatal period are held responsible for the increased incidence of these diseases, and several studies have reported increased levels of endocrine disrupting chemicals in circulation of patients with infertility, PCOS, endometriosis, etc.

It is likely that various diseases arise due to a dysfunction of tissue-resident stem cells by endocrine disruption, which will never get revealed when DNA/RNA/proteins are isolated from intact tissues since stem cells comprise less than 1% of the total cells. Stem cells express estrogen receptors and thus become direct targets for endocrine disruptors. Studies have been undertaken by our group in mice which suggest that it is the normal body tissue-resident stem cells that get affected by perinatal exposure to endocrine disruption and their altered proliferation/differentiation lead to various pathologies in adult life including cancer (discussed ahead).

Stem cells function in a subtle manner to maintain life-long tissue homeostasis. This is ensured by occasional asymmetrical divisions of stem cells whereby small VSELs divide to self-renew and also to give rise to slightly bigger progenitors. The progenitors in turn undergo rapid symmetrical divisions and clonal expansion before initiating differentiation. This stem cells activity is regulated by the niche or microenvironment which provides the crucial paracrine support for stem cells to undergo proliferation/differentiation. One needs to think of the stem cells and their niche together as the 'seed and soil' concept. The niche is provided to the stem cells by Sertoli cells in the testes, ovary surface epithelial cells to the ovarian stem cells, and MSCs to the uterine stem cells. The niche gets compromised and results in menopause with age. Also the niche is affected resulting in uncontrolled expansion of stem cells leading to cancer initiation [95].

OCT-4 is a major regulator of pluripotent state and besides embryo development, and OCT-4 also has an important role during tumorigenesis. Evidence is piling up regarding OCT-4 role in cancer cells and also in tumor initiating cells [96, 97]. Nuclear OCT-4 along with other embryonic markers are expressed by VSELs in adult tissues and cytoplasmic OCT-4 by the immediate descendants or tissue specific progenitors and eventually gets degraded as cells differentiate further. Since embryonic markers including OCT-4 are upregulated in various pathologies including cancer, it can be inferred that disrupted stem cells (VSELs) biology results in various pathologies. Extensive global efforts have been undertaken, and several genes have been reported to be involved in various pathologies, but we have focused below on the possible implication of OCT-4 in various pathologies with an intention to correlate various pathologies to disrupted function of VSELs.

OCT-4 and Testicular Cancer

Kaushik et al. [22] found that mice pups when exposed neonatally to diethylstilbestrol (DES, 2 μg/day on days 1–5 to mouse pups), led to almost 7 folds increase in VSELs numbers along with 5 folds reduction in c-KIT positive spermatogonia by flow cytometry. This was associated with 9 folds upregulation of *Oct-4A*, 14 folds of *Oct*-4, and more than 40 folds increase of *Sox2* and *Nanog*. Reduction in *c-KIT* (5 folds) and meiosis specific transcripts was also noted. Results suggest excessive self-renewal of VSELs initiates testicular cancer, and blocked differentiation leads to reduction in sperm count and infertility. We had earlier discussed that testicular cancer possibly initiates from VSELs [15]. This is further supported by earlier reports suggesting OCT-4 is a sensitive and specific maker for intra-tubular germ cell neoplasia [98–100].

OCT-4 and Ovarian Cancer

Ovarian cancer is the leading cause of death and affects a large fraction of women worldwide. Similar increased numbers of VSELs were reported in human borderline cases of ovarian cancers [21]. About 90% of ovarian cancers arise in the ovary surface epithelium and are malignant in nature. Ovary surface epithelium also houses ovarian stem cells (VSELs and OSCs) as described above. Similar small sized stem cells were observed in large numbers in cases of ovarian serous carcinoma slides [101–103]. Zuber et al. [104] have discussed expression of embryonic markers by ovarian cancer stem cells. Stem cells in adult human ovaries are of great interest to reproductive medicine for improved understanding of the mechanisms leading to ovarian infertility due to PCOS or POF and cancer, yet they represent a difficult scientific subject, because of the disbelief regarding their existence. VSELs from normal ovarian tissue remain quiescent and differentiate into oocyte-like structures in vitro, whereas those from ovarian cancer divide rapidly and expand in numbers. Understanding how the VSELs isolated from the normal and cancerous tissue differ from each other can unravel how cancer initiates and may provide better insight for treatment. Despite extensive research, underlying etiology and whether PCOS and POF are stem cell disease remains to be understood.

OCT-4 and Endometriosis

Endometriosis is a chronic, benign and an inflammatory disease which causes growth of endometrium outside the uterus, in ectopic locations primarily the pelvic peritoneum, ovaries, and rectovaginal septum. It affects almost 6–10% of women of reproductive age and is associated with dysmenorrhea, dyspareunia, chronic pelvic pain, irregular uterine bleeding, and/or infertility [80]. Different theories are postulated to be involved in the pathogenesis of endometriosis including retrograde menstruation, coelomic metaplasia, immune dysfunction, hormone imbalance, oxidative stress, and inflammation and stem/progenitor cells [105]. Human endometrium side-population (SP) cells can generate endometrium-like tissue in NOD-SCID mice beneath the kidney capsule [106] supporting a stem cell origin for the disease. Ectopic endometrium expresses significantly higher levels of pluripotency markers Oct4, SSEA-1, Sox2, Nanog, Musashi, c-KIT [107, 108].

OCT-4 and Uterine Leiomyomas

Leiomyomas are also known as uterine fibroids or myomas are the most common benign gynecologic tumors affecting the myometrium in women of reproductive age. Both clinical and mice studies suggest that leiomyoma has a stem cell basis [109–111]. Side population cells have been detected in leiomyoma which when transplanted in mice, give rise to fibroids. Ono et al. [112] reported higher expression of OCT-4A (but not cytoplasmic OCT4B) in myometrial SP cells isolated from human myometrial tissue from 18 patients undergoing hysterectomy and concluded that OCT-4A positive cells could be involved in uterine biology as well as in pathology.

OCT-4 and Endometrial Cancer

Endometrial cancer is the most common gynecologic malignant tumor. OCT4 is overexpressed in endometrial carcinoma [113, 114]. Sun et al. [115] identified that the CD133+ and CXCR4+ endometrial cancer cells exhibited stronger expression of embryonic markers including c-Myc, Sox-2, Nanog, Oct4A, ABCG2, BMI-1, CK-18, and Nestin.

It becomes evident that various pathologies are associated with disrupted OCT-4 expression suggestive of a possible involvement of VSELs and tissue committed progenitors in disease initiation. This needs to be investigated and hopefully scRNAseq will help unravel these well-kept secrets of Mother Nature in times to come.

Brief summary of the chapter highlights is presented in Fig. 12.8. Since this chapter was submitted, we published two related articles which readers may with to refer [116, 117].

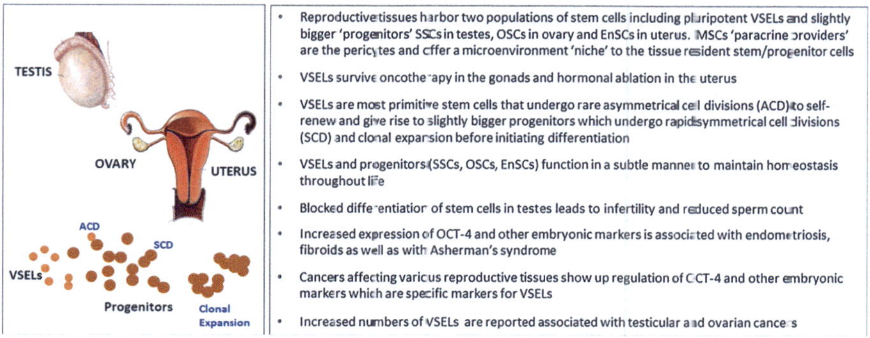

Fig. 12.8 Brief summary of the chapter highlights. All the reproductive tissues harbor two populations of stem cells including VSELs and tissue specific progenitors. VSELs (very small embryonic-like stem cells), OCT-4 (octamer-binding transcription factor 4), SSCs (spermatogonial stem cells), OSCs (ovarian stem cells), EnSCs (endometrial stem cells), ACD (asymmetric cell divisions), and SCD (symmetrical cell divisions)

Acknowledgements We are thankful to our colleagues who have contributed to various aspects of biology summarized here, but we may not have quoted their work as we included the most recent articles. NIRRH Accession number OTH/995/11-2020.

References

1. Li L, Clevers H (2010) Coexistence of quiescent and active adult stem cells in mammals. Science 327:542–545. https://doi.org/10.1126/science.1180794
2. De Rosa L, De Luca M (2012) Cell biology: dormant and restless skin stem cells. Nature 489:215–217. https://doi.org/10.1038/489215a
3. Post Y, Clevers H (2019) Defining adult stem cell function at its simplest: the ability to replace lost cells through mitosis. Cell Stem Cell 25:174–183. https://doi.org/10.1016/j.stem.2019.07.002
4. Wagner DE, Klein AM (2020) Lineage tracing meets single-cell omics: opportunities and challenges. Nat Rev Genet 21:410–427. https://doi.org/10.1038/s41576-020-0223-2
5. Suzuki S, Diaz VD, Hermann BP (2019) What has single-cell RNA-seq taught us about mammalian spermatogenesis? Biol Reprod 101:617–634. https://doi.org/10.1093/biolre/ioz088
6. Karthaus WR, Hofree M, Choi D, Linton EL, Turkekul M, Bejnood A, Carver B, Gopalan A, Abida W, Laudone V, Biton M, Chaudhary O, Xu T, Masilionis I, Manova K, Mazutis L, Pe'er D, Regev A, Sawyers CL (2020) Regenerative potential of prostate luminal cells revealed by single-cell analysis. Science 368:497–505. https://doi.org/10.1126/science.aay0267
7. Saatcioglu HD, Kano M, Horn H, Zhang L, Samore W, Nagykery N, Meinsohn M-C, Hyun M, Suliman R, Poulo J, Hsu J, Sacha C, Wang D, Gao G, Lage K, Oliva E, Morris Sabatini ME, Donahoe PK, Pépin D (2019) Single-cell sequencing of neonatal uterus reveals an Misr2+ endometrial progenitor indispensable for fertility. eLife 8:e46349. https://doi.org/10.7554/eLife.46349
8. Bhartiya D, Sharma D (2020) Ovary does harbor stem cells—size of the cells matter! J Ovarian Res 13:39. https://doi.org/10.1186/s13048-020-00647-2

9. Bhartiya D, Kausik A, Singh P, Sharma D (2020) Will single-cell RNAseq decipher stem cells biology in normal and cancerous tissues? Hum Reprod Update dmaa058. https://doi.org/10.1093/humupd/dmaa058
10. Santamaria X, Mas A, Cervelló I, Taylor H, Simon C (2018) Uterine stem cells: from basic research to advanced cell therapies. Hum Reprod Update 24:673–693. https://doi.org/10.1093/humupd/dmy028
11. Gargett CE, Schwab KE, Deane JA (2016) Endometrial stem/progenitor cells: the first 10 years. Hum Reprod Update 22:137–163. https://doi.org/10.1093/humupd/dmv051
12. Bhartiya D, Shaikh A, Anand S, Patel H, Kapoor S, Sriraman K, Parte S, Unni S (2016) Endogenous, very small embryonic-like stem cells: critical review, therapeutic potential and a look ahead. Hum Reprod Update 23:41–76. https://doi.org/10.1093/humupd/dmw030
13. De Luca M, Aiuti A, Cossu G, Parmar M, Pellegrini G, Robey PG (2019) Advances in stem cell research and therapeutic development. Nat Cell Biol 21:801–811. https://doi.org/10.1038/s41556-019-0344-z
14. Bhartiya D, Anand S, Kaushik A (2020) Pluripotent very small embryonic-like stem cells co-exist along with spermatogonial stem cells in adult mammalian testis. Hum Reprod Update 26:136–137. https://doi.org/10.1093/humupd/dmz030
15. Kaushik A, Bhartiya D (2020) Additional evidence to establish existence of two stem cell populations including VSELs and SSCs in adult mouse testes. Stem Cell Rev Rep 16:992–1004. https://doi.org/10.1007/s12015-020-09993-6
16. Bhartiya D, Patel H (2018) Ovarian stem cells-resolving controversies. J Assist Reprod Genet 35:393–398. https://doi.org/10.1007/s10815-017-1080-6
17. Singh P, Bhartiya D (2021) Pluripotent stem (VSELs) and progenitor (EnSCs) cells exist in adult mouse uterus and show cyclic changes across estrus cycle. Reprod Sci 28:278–290. https://doi.org/10.1007/s43032-020-00250-2
18. James K, Bhartiya D, Ganguly R, Kaushik A, Gala K, Singh P, Metkari SM (2018) Gonadotropin and steroid hormones regulate pluripotent very small embryonic-like stem cells in adult mouse uterine endometrium. J Ovarian Res 11:83. https://doi.org/10.1186/s13048-018-0454-4
19. Gunjal P, Bhartiya D, Metkari S, Manjramkar D, Patel H (2015) Very small embryonic-like stem cells are the elusive mouse endometrial stem cells–a pilot study. J Ovarian Res 8:9. https://doi.org/10.1186/s13048-015-0138-2
20. Bhartiya D, James K (2017) Very small embryonic-like stem cells (VSELs) in adult mouse uterine perimetrium and myometrium. J Ovarian Res 10:29. https://doi.org/10.1186/s13048-017-0324-5
21. Kenda Suster N, Virant-Klun I (2019) Presence and role of stem cells in ovarian cancer. World J Stem Cells 11:383–397. https://doi.org/10.4252/wjsc.v11.i7.383
22. Kaushik A, Anand S, Bhartiya D (2020) Altered biology of Testicular VSELs and SSCs by neonatal endocrine disruption results in defective spermatogenesis, reduced fertility and tumor initiation in adult mice. Stem Cell Rev Rep 16:893–908. https://doi.org/10.1007/s12015-020-09996-3
23. Abbott A (2013) Doubt cast over tiny stem cells. Nature 499:390. https://doi.org/10.1038/499390a
24. Ratajczak MZ, Zuba-Surma E, Wojakowski W, Suszynska M, Mierzejewska K, Liu R, Ratajczak J, Shin DM, Kucia M (2014) Very small embryonic-like stem cells (VSELs) represent a real challenge in stem cell biology: recent pros and cons in the midst of a lively debate. Leukemia 28:473–484. https://doi.org/10.1038/leu.2013.255
25. Bhartiya D (2017) Pluripotent stem cells in adult tissues: struggling to be acknowledged over two decades. Stem Cell Rev Rep 13:713–724. https://doi.org/10.1007/s12015-017-9756-y
26. Ratajczak MZ, Ratajczak J, Kucia M (2019) Very small embryonic-like stem cells (VSELs)—an update and future directions. Circ Res 124:208–210. https://doi.org/10.1161/CIRCRESAHA.118.314287
27. Bhartiya D, Kasiviswanathan S, Unni SK, Pethe P, Dhabalia JV, Patwardhan S, Tongaonkar HB (2010) Newer insights into premeiotic development of germ cells in adult human testis

using Oct-4 as a stem cell marker. J Histochem Cytochem 58:1093–1106 https://doi.org/10.1369/jhc.2010.956870
28. Bhartiya D, Ali Mohammad S, Guha A, Singh P, Sharma D, Kaushik A (2019) Evolving definition of adult stem/progenitor cells. Stem Cell Rev Rep 15:456–458 https://doi.org/10.1007/s12015-019-09879-2
29. Mohammad SA, Metkari S, Bhartiya D (2020) Mouse pancreas stem/progenitor cells get augmented by streptozotocin and regenerate diabetic pancreas after partial pancreatectomy. Stem Cell Rev Rep 16:144–158. https://doi.org/10.1007/s12015-019-09919-x
30. Parte S, Patel H, Sriraman K, Bhartiya D (2015) Isolation and characterization of stem cells in the adult mammalian ovary. Methods Mol Biol 1235:203–229. https://doi.org/10.1007/978-1-4939-1785-3_16
31. Kucia M, Halasa M, Wysoczynski M, Baskiewicz-Masiuk M, Moldenhawer S, Zuba-Surma E, Czajka R, Wojakowski W, Machalinski B, Ratajczak MZ (2007) Morphological and molecular characterization of novel population of CXCR4+ SSEA-4+ Oct-4+ very small embryonic-like cells purified from human cord blood: preliminary report. Leukemia 21:297–303. https://doi.org/10.1038/sj.leu.2404470
32. Virant-Klun I, Stimpfel M, Skutella T (2012) Stem cells in adult human ovaries: from female fertility to ovarian cancer. Curr Pharm Des 18:283–292. https://doi.org/10.2174/133161212799040394
33. Martin JJ, Woods DC, Tilly JL (2019) Implications and current limitations of oogenesis from female germline or oogonial stem cells in adult mammalian ovaries. Cells 8:93. https://doi.org/10.3390/cells8020093
34. Silvestris E, Cafforio P, D'Oronzo S, Felici C, Silvestris F, Loverro G (2018) In vitro differentiation of human oocyte-like cells from oogonial stem cells: single-cell isolation and molecular characterization. Hum Reprod 33:464–473 https://doi.org/10.1093/humrep/dex377
35. Castrillon DH, Quade BJ, Wang TY, Quigley C, Crum CP (2000) The human VASA gene is specifically expressed in the germ cell lineage. Proc Natl Acad Sci USA 97:9585–9590. https://doi.org/10.1073/pnas.160274797
36. Toyooka Y, Tsunekawa N, Takahashi Y, Matsui Y, Satoh M, Noce T (2000) Expression and intracellular localization of mouse Vasa-homologue protein during germ cell development. Mech Dev 93:139–149. https://doi.org/10.1016/s0925-4773(00)00283-5
37. Zarate-Garcia L, Lane SIR, Merriman JA, Jones KT (2016) FACS-sorted putative oogonial stem cells from the ovary are neither DDX4-positive nor germ cells. Sci Rep 6:27991. https://doi.org/10.1038/srep27991
38. Woods DC, Tilly JL (2013) Isolation, characterization and propagation of mitotically active germ cells from adult mouse and human ovaries. Nat Protoc 8:966–988. https://doi.org/10.1038/nprot.2013.047
39. Woods DC, Tilly JL (2013) An evolutionary perspective on adult female germline stem cell function from flies to humans. Semin Reprod Med 31:24–32. https://doi.org/10.1055/s-0032-1331794
40. Magnúsdóttir E, Surani MA (2014) How to make a primordial germ cell. Development 141:245–252. https://doi.org/10.1242/dev.098269
41. Virant-Klun I, Zech N, Rozman P, Vogler A, Cvjeticanin B, Klemenc P, Malicev E, Meden-Vrtovec H (2008) Putative stem cells with an embryonic character isolated from the ovarian surface epithelium of women with no naturally present follicles and oocytes. Differ Res Biol Divers 76:843–856. https://doi.org/10.1111/j.1432-0436.2008.00268.x
42. Parte S, Bhartiya D, Telang J, Daithankar V, Salvi V, Zaveri K, Hinduja I (2011) Detection, characterization, and spontaneous differentiation in vitro of very small embryonic-like putative stem cells in adult mammalian ovary. Stem Cells Dev 20:1451–1464. https://doi.org/10.1089/scd.2010.0461
43. Bhartiya D, Patel H, Parte S (2018) Improved understanding of very small embryonic-like stem cells in adult mammalian ovary. Hum Reprod 33:978–979. https://doi.org/10.1093/humrep/dey039

44. Virant-Klun I (2018) Functional testing of primitive oocyte-like cells developed in ovarian surface epithelium cell culture from small VSEL-like stem cells: can they be fertilized one day? Stem Cell Rev Rep 14:715–721. https://doi.org/10.1007/s12015-018-9832-y
45. Zou K, Yuan Z, Yang Z, Luo H, Sun K, Zhou L, Xiang J, Shi L, Yu Q, Zhang Y, Hou R, Wu J (2009) Production of offspring from a germline stem cell line derived from neonatal ovaries. Nat Cell Biol 11:631–636. https://doi.org/10.1038/ncb1869
46. Bhartiya D, Anand S, Patel H, Parte S (2017) Making gametes from alternate sources of stem cells: past, present and future. Reprod Biol Endocrinol 15:89. https://doi.org/10.1186/s12958-017-0308-8
47. Ratajczak J, Wysoczynski M, Zuba-Surma E, Wan W, Kucia M, Yoder MC, Ratajczak MZ (2011) Adult murine bone marrow-derived very small embryonic-like stem cells differentiate into the hematopoietic lineage after coculture over OP9 stromal cells. Exp Hematol 39:225–237. https://doi.org/10.1016/j.exphem.2010.10.007
48. Sriraman K, Bhartiya D, Anand S, Bhutda S (2015) Mouse ovarian very small embryonic-like stem cells resist chemotherapy and retain ability to initiate oocyte-specific differentiation. Reprod Sci 22:884–903. https://doi.org/10.1177/1933719115576727
49. Truman AM, Tilly JL, Woods DC (2017) Ovarian regeneration: The potential for stem cell contribution in the postnatal ovary to sustained endocrine function. Mol Cell Endocrinol 445:74–84. https://doi.org/10.1016/j.mce.2016.10.012
50. Bhartiya D (2017) Letter to the editor: rejuvenate eggs or regenerate ovary? Mol Cell Endocrinol 446:111–113. https://doi.org/10.1016/j.mce.2017.03.008
51. Bhartiya D (2019) Fertility preservation for female cancer patients by manipulating ovarian stem cells that survive oncotherapy. Onco Fertil J 2:53. https://doi.org/10.4103/tofj.tofj_12_19
52. Na J, Kim GJ (2020) Recent trends in stem cell therapy for premature ovarian insufficiency and its therapeutic potential: a review. J Ovarian Res 13:74. https://doi.org/10.1186/s13048-020-00671-2
53. Wang Z, Wei Q, Wang H, Han L, Dai H, Qian X, Yu H, Yin M, Shi F, Qi N (2020) Mesenchymal stem cell therapy using human umbilical cord in a rat model of autoimmune-induced premature ovarian failure. Stem Cells Int 2020:3249495. https://doi.org/10.1155/2020/3249495
54. Malard PF, Peixer MAS, Grazia JG, Brunel H dos SS, Feres LF, Villarroel CL, Siqueira LGB, Dode MAN, Pogue R, Viana JHM, Carvalho JL (2020) Intraovarian injection of mesenchymal stem cells improves oocyte yield and in vitro embryo production in a bovine model of fertility loss. Sci Rep 10:8018. https://doi.org/10.1038/s41598-020-64810-x
55. Edessy M, Hosni H, Shady Y, Waf Y, Bakry S, Kamel M (2016) Autologous stem cells therapy, the first baby of idiopathic premature ovarian failure. Acta Medial Int 3:19–23. https://doi.org/10.5530/ami.2016.1.7
56. Bahrehbar K, Rezazadeh Valojerdi M, Esfandiari F, Fathi R, Hassani S-N, Baharvand H (2020) Human embryonic stem cell-derived mesenchymal stem cells improved premature ovarian failure. World J Stem Cells 12:857–878. https://doi.org/10.4252/wjsc.v12.i8.857
57. Zhang Q, Sun J, Huang Y, Bu S, Guo Y, Gu T, Li B, Wang C, Lai D (2019) Human amniotic epithelial cell-derived exosomes restore ovarian function by transferring microRNAs against apoptosis. Mol Ther Nucleic Acids 16:407–418. https://doi.org/10.1016/j.omtn.2019.03.008
58. Ferraro F, Celso CL, Scadden D (2010) Adult stem cels and their niches. Adv Exp Med Biol 695:155–168. https://doi.org/10.1007/978-1-4419-7037-4_11
59. Spradling A, Drummond-Barbosa D, Kai T (2001) Stem cells find their niche. Nature 414:98–104. https://doi.org/10.1038/35102160
60. Niikura Y, Niikura T, Tilly JL (2009) Aged mouse ovaries possess rare premeiotic germ cells that can generate oocytes following transplantation into a young host environment. Aging 1:971–978. https://doi.org/10.18632/aging.100105
61. de Rooij DG (2017) The nature and dynamics of spermatogonial stem cells. Development 144:3022–3030. https://doi.org/10.1242/dev.146571
62. Bhartiya D, Anand S, Patel H, Kaushik A, Pramodh S (2019) Testicular stem cells, spermatogenesis and infertility. In: Singh R (ed) Molecular mechanisms in spermatogenesis and infertility, 1st edn. CRC Press, Boca Raton, pp 17–29

63. Sharma S, Wistuba J, Pock T, Schlatt S, Neuhaus N (2019) Spermatogonial stem cells: updates from specification to clinical relevance. Hum Reprod Update 25:275–297. https://doi.org/10.1093/humupd/dmz006
64. Sharma S, Wistuba J, Neuhaus N, Schlatt S (2020) Reply: Pluripotent very small embryonic-like stem cells co-exist along with spermatogonial stem cells in adult mammalian testis. Hum Reprod Update 26:138. https://doi.org/10.1093/humupd/dmz031
65. Anand S, Bhartiya D, Sriraman K, Mallick A (2016) Underlying mechanisms that restore spermatogenesis on transplanting healthy niche cells in busulphan treated mouse testis. Stem Cell Rev Rep 12:682–697. https://doi.org/10.1007/s12015-016-9685-1
66. Patel H, Bhartiya D (2016) Testicular stem cells express follicle-stimulating hormone receptors and are directly modulated by FSH. Reprod Sci 23:1493–1508. https://doi.org/10.1177/1933719116643593
67. Kurkure P, Prasad M, Dhamankar V, Bakshi G (2015) Very small embryonic-like stem cells (VSELs) detected in azoospermic testicular biopsies of adult survivors of childhood cancer. Reprod Biol Endocrinol 13:122. https://doi.org/10.1186/s12958-015-0121-1
68. Eliyasi Dashtaki M, Hemadi M, Saki G, Mohammadiasl J, Khodadadi A (2020) Spermatogenesis recovery potentials after transplantation of adipose tissue-derived mesenchymal stem cells cultured with growth factors in experimental azoospermic mouse models. Cell J 21:401–409. https://doi.org/10.22074/cellj.2020.6055
69. Sagaradze GD, Basalova NA, Efimenko AYu, Tkachuk VA (2020) Mesenchymal stromal cells as critical contributors to tissue regeneration. Front Cell Dev Biol 8:576176. https://doi.org/10.3389/fcell.2020.576176
70. Bhartiya D (2018) Stem cells survive oncotherapy & can regenerate non-functional gonads: a paradigm shift for oncofertility. Indian J Med Res 148:S38–S49. https://doi.org/10.4103/ijmr.IJMR_2065_17
71. Behre HM (2019) Clinical Use of FSH in Male Infertility. Front Endocrinol 10:322. https://doi.org/10.3389/fendo.2019.00322
72. Simoni M, Brigante G, Rochira V, Santi D, Casarini L (2020) Prospects for FSH treatment of male infertility. J Clin Endocrinol Metab 105:2105–2118. https://doi.org/10.1210/clinem/dgaa243
73. Simoni M, Santi D (2020) FSH treatment of male idiopathic infertility time for a paradigm change. Andrology 8:535–544. https://doi.org/10.1111/andr.12746
74. Anand S, Patel H, Bhartiya D (2015) Chemoablated mouse seminiferous tubular cells enriched for very small embryonic-like stem cells undergo spontaneous spermatogenesis in vitro. Reprod Biol Endocrinol 13:33. https://doi.org/10.1186/s12958-015-0031-2
75. Shaikh A, Anand S, Kapoor S, Ganguly R, Bhartiya D (2017) Mouse bone marrow VSELs exhibit differentiation into three embryonic germ lineages and germ & hematopoietic cells in culture. Stem Cell Rev Rep 13:202–216. https://doi.org/10.1007/s12015-016-9714-0
76. Bhartiya D, Kaushik A (2021) Testicular stem cell dysfunction due to environmental insults could be responsible for deteriorating reproductive health of men. Reprod Sci 28:649–658. https://doi.org/10.1007/s43032-020-00411-3
77. Padykula HA (1991) Regeneration in the primate uterus: the role of stem cells. Ann N Y Acad Sci 622:47–56. https://doi.org/10.1111/j.1749-6632.1991.tb37849.x
78. Taylor HS (2004) Endometrial cells derived from donor stem cells in bone marrow transplant recipients. J Am Med Assoc 292:81–85. https://doi.org/10.1001/jama.292.1.81
79. Bozorgmehr M, Gurung S, Darzi S, Nikoo S, Kazemnejad S, Zarnani A-H, Gargett CE (2020) Endometrial and menstrual blood mesenchymal stem/stromal cells: biological properties and clinical application. Front Cell Dev Biol 8:497. https://doi.org/10.3389/fcell.2020.00497
80. Cousins FL, Dorien FO, Gargett CE (2018) Endometrial stem/progenitor cells and their role in the pathogenesis of endometriosis. Best Pract Res Clin Obstet Gynaecol 50:27–38. https://doi.org/10.1016/j.bpobgyn.2018.01.01
81. Crisan M, Yap S, Casteilla L, Chen C-W, Corselli M, Park TS, Andriolo G, Sun B, Zheng B, Zhang L, Norotte C, Teng P-N, Traas J, Schugar R, Deasy BM, Badylak S, Buhring H-J, Giacobino J-P, Lazzari L, Huard J, Péault B (2008) A perivascular origin for mesenchymal

stem cells in multiple human organs. Cell Stem Cell 3:301–313. https://doi.org/10.1016/j.stem.2008.07.003
82. Caplan AI (2008) All MSCs are pericytes? Cell Stem Cell 3:229–230. https://doi.org/10.1016/j.stem.2008.08.008
83. Bhartiya D (2013) Are mesenchymal cells indeed pluripotent stem cells or just stromal cells? OCT-4 and VSELs biology has led to better understanding. Stem Cells Int 2013:547501. https://doi.org/10.1155/2013/547501
84. Bhartiya D (2016) An update on endometrial stem cells and progenitors. Hum Reprod Update 22:529–530. https://doi.org/10.1093/humupd/dmw010
85. Abbaszadeh H, Ghorbani F, Derakhshani M, Movassaghpour A, Yousefi M (2020) Human umbilical cord mesenchymal stem cell-derived extracellular vesicles: a novel therapeutic paradigm. J Cell Physiol 235:706–717. https://doi.org/10.1002/jcp.29004
86. Zhao R, Chen X, Song H, Bie Q, Zhang B (2020) Dual role of MSC-derived exosomes in tumor development. Stem Cells Int 2020:8844730. https://doi.org/10.1155/2020/8844730
87. Marinaro F, Sánchez-Margallo FM, Álvarez V, López E, Tarazona R, Brun MV, Blázquez R, Casado JG (2019) Meshes in a mess: mesenchymal stem cell-based therapies for soft tissue reinforcement. Acta Biomater 85:60–74. https://doi.org/10.1016/j.actbio.2018.11.042
88. Esfandyari S, Chugh RM, Park H-S, Hobeika E, Ulin M, Al-Hendy A (2020) Mesenchymal stem cells as a bio organ for treatment of female infertility. Cells 9:2253. https://doi.org/10.3390/cells9102253
89. Yin P, Ono M, Moravek MB, Coon JS, Navarro A, Monsivais D, Dyson MT, Druschitz SA, Malpani SS, Serna VA, Qiang W, Chakravarti D, Kim JJ, Bulun SE (2015) Human uterine leiomyoma stem/progenitor cells expressing CD34 and CD49b initiate tumors in vivo. J Clin Endocrinol Metab 100:E601-606. https://doi.org/10.1210/jc.2014-2134
90. Bhartiya D (2017) Shifting gears from embryonic to very small embryonic-like stem cells for regenerative medicine. Indian J Med Res 146:15–21. https://doi.org/10.4103/ijmr.IJMR_1485_16
91. Marquardt RM, Kim TH, Shin J-H, Jeong J-W (2019) Progesterone and estrogen signaling in the endometrium: what goes wrong in endometriosis? Int J Mol Sci 20:3822. https://doi.org/10.3390/ijms20153822
92. Pierro E, Minici F, Alesiani O, Miceli F, Proto C, Screpanti I, Mancuso S, Lanzone A (2001) Stromal-epithelial interactions modulate estrogen responsiveness in normal human endometrium. Biol Reprod 64:831–838. https://doi.org/10.1095/biolreprod64.3.831
93. Cooke PS, Buchanan DL, Young P, Setiawan T, Brody J, Korach KS, Taylor J, Lubahn DB, Cunha GR (1997) Stromal estrogen receptors mediate mitogenic effects of estradiol on uterine epithelium. Proc Natl Acad Sci U S A 94:6535–6540. https://doi.org/10.1073/pnas.94.12.6535
94. Montgomery GW, Mortlock S, Giudice LC (2020) Should genetics now be considered the pre-eminent etiologic factor in endometriosis? J Minim Invasive Gynecol 27:280–286. https://doi.org/10.1016/j.jmig.2019.10.020
95. Pasquier J, Rafii A (2012) Role of the microenvironment in ovarian cancer stem cell maintenance. BioMed Res Int 2013:e630782. https://doi.org/10.1155/2013/630782
96. Trosko JE (2006) From adult stem cells to cancer stem cells: Oct-4 gene, cell-cell communication, and hormones during tumor promotion. Ann N Y Acad Sci 1089:36–58. https://doi.org/10.1196/annals.1386.018
97. Tai M-H, Chang C-C, Kiupel M, Webster JD, Olson LK, Trosko JE (2005) Oct4 expression in adult human stem cells: evidence in support of the stem cell theory of carcinogenesis. Carcinogenesis 26:495–502. https://doi.org/10.1093/carcin/bgh321
98. Looijenga LHJ, Van der Kwast TH, Grignon D, Egevad L, Kristiansen G, Kao C-S, Idrees MT (2020) Report from the International Society of Urological Pathology (ISUP) consultation conference on molecular pathology of urogenital cancers: IV: current and future utilization of molecular-genetic tests for testicular germ cell tumors. Am J Surg Pathol 44:e66–e79. https://doi.org/10.1097/PAS.0000000000001465
99. Oosterhuis JW, Looijenga LHJ (2019) Human germ cell tumours from a developmental perspective. Nat Rev Cancer 19:522–537. https://doi.org/10.1038/s41568-019-0178-9

100. Jones TD, Ulbright TM, Eble JN, Cheng L (2004) OCT4: a sensitive and specific biomarker for intratubular germ cell neoplasia of the testis. Clin Cancer Res Off J Am Assoc Cancer Res 10:8544–8547. https://doi.org/10.1158/1078-0432.CCR-04-0688
101. Virant-Klun I, Leicht S, Hughes C, Krijgsveld J (2016) Identification of maturation-specific proteins by single-cell proteomics of human oocytes. Mol Cell Proteomics MCP 15:2616–2627. https://doi.org/10.1074/mcp.M115.056887
102. Virant-Klun I, Stimpfel M (2016) Novel population of small tumour-initiating stem cells in the ovaries of women with borderline ovarian cancer. Sci Rep 6:34730. https://doi.org/10.1038/srep34730
103. Kenda Šuster N, Frković Grazio S, Virant-Klun I, Verdenik I, Smrkolj Š (2017) Cancer stem cell-related marker NANOG expression in ovarian serous tumors: a clinicopathological study of 159 cases. Int J Gynecol Cancer Off J Int Gynecol Cancer Soc 27:2006–2013. https://doi.org/10.1097/IGC.0000000000001105
104. Zuber E, Schweitzer D, Allen D, Parte S, Kakar SS (2020) Stem cells in ovarian cancer and potential therapies. Proc Stem Cell Res Oncog 8:e1001. 14343/PSCRO:2020.8e1001
105. Laganà AS, Garzon S, Götte M, Viganò P, Franchi M, Ghezzi F, Martin DC (2019) The pathogenesis of endometriosis: molecular and cell biology insights. Int J Mol Sci 20:5615. https://doi.org/10.3390/ijms20225615
106. Masuda H, Matsuzaki Y, Hiratsu E, Ono M, Nagashima T, Kajitani T, Arase T, Oda H, Uchida H, Asada H, Ito M, Yoshimura Y, Maruyama T, Okano H (2010) Stem cell-like properties of the endometrial side population: implication in endometrial regeneration. PloS One 5:e10387. https://doi.org/10.1371/journal.pone.0010387
107. Othman ER, Meligy FY, Sayed AA-R, El-Mokhtar MA, Refaiy AM (2018) Stem cell markers describe a transition from somatic to pluripotent cell states in a rat model of endometriosis. Reprod Sci 25:873–881. https://doi.org/10.1177/1933719117697124
108. Song Y, Xiao L, Fu J, Huang W, Wang Q, Zhang X, Yang S (2014) Increased expression of the pluripotency markers sex-determining region Y-box 2 and Nanog homeobox in ovarian endometriosis. Reprod Biol Endocrinol 12:42. https://doi.org/10.1186/1477-7827-12-42
109. Prusinski Fernung LE, Yang Q, Sakamuro D, Kumari A, Mas A, Al-Hendy A (2018) Endocrine disruptor exposure during development increases incidence of uterine fibroids by altering DNA repair in myometrial stem cells. Biol Reprod 99:735–748. https://doi.org/10.1093/biolre/ioy097
110. Brakta S, Mas A, Al-Hendy A (2018) The ontogeny of myometrial stem cells in OCT4-GFP transgenic mouse model. Stem Cell Res Ther 9:333. https://doi.org/10.1186/s13287-018-1079-7
111. Mas A, Stone L, O'Connor PM, Yang Q, Kleven D, Simon C, Walker CL, Al-Hendy A (2017) Developmental exposure to endocrine disruptors expands murine myometrial stem cell compartment as a prerequisite to leiomyoma tumorigenesis. Stem Cells 35:666–678. https://doi.org/10.1002/stem.2519
112. Ono M, Kajitani T, Uchida H, Arase T, Oda H, Nishikawa-Uchida S, Masuda H, Nagashima T, Yoshimura Y, Maruyama T (2010) OCT4 expression in human uterine myometrial stem/progenitor cells. Hum Reprod 25:2059–2067. https://doi.org/10.1093/humrep/deq163
113. Zhang R, Jiao J, Chu H, Yang H, Wang L, Gao N (2018) Expression of microRNA-145, OCT4, and SOX2 in double primary endometrioid endometrial and ovarian carcinomas. Histol Histopathol 33:859–870. https://doi.org/10.14670/HH-11-986
114. Pityński K, Banas T, Pietrus M, Milian-Ciesielska K, Ludwin A, Okon K (2015) SOX-2, but not Oct4, is highly expressed in early-stage endometrial adenocarcinoma and is related to tumour grading. Int J Clin Exp Pathol 8:8189–8198
115. Sun Y, Yoshida T, Okabe M, Zhou K, Wang F, Soko C, Saito S, Nikaido T (2017) Isolation of stem-like cancer cells in primary endometrial cancer using cell surface markers CD133 and CXCR4. Transl Oncol 10:976–987. https://doi.org/10.1016/j.tranon.2017.07.007
116. Bhartiya D, Kaushik A (2021) Testicular stem cell dysfunction due to environmental insults could be responsible for deteriorating reproductive health of men. Reprod Sci 28(3):649–658

117. Bhartiya D, Patel H, Kaushik A, Singh P, Sharma D (2021) Endogenous, tissue-resident stem/progenitor cells in gonads and bone marrow express FSHR and respond to FSH via FSHR-3. J Ovarian Res 14(1):145. https://doi.org/10.1186/s13048-021-00883-0

Chapter 13
Amniotic Membrane: A Unique Combination of Stem-Like Cells, Extracellular Matrix with Indispensable Potential for Regenerative Medicine

Taja Ramuta Železnik, Larisa Tratnjek, and Mateja Kreft Erdani

Abstract

Introduction: Human amniotic membrane (hAM) has a 100-year long history of use in the clinic, especially in ophthalmology and dermatology. In this chapter, we reveal the unique properties of hAM, such as the promotion of epithelization, decrease of scarring and fibrosis, pro- and anti-angiogenic properties, low immunogenicity, immunomodulatory activity, and even its anticancer and antimicrobial properties, all of which are beneficial for use in regenerative medicine.MethodsWe summarize new knowledge about the potential use of the amniotic membrane in regenerative medicine.

Results: We expose hAM as a unique combination of stem-like cells and the extracellular matrix that is easy to obtain and ethically acceptable to use, further demonstrating its indispensable potential for regenerative medicine. Moreover, we present hAM- derived preparations, which are being developed and tested for use as therapeutics or drug delivery tools in various fields of tissue engineering and regenerative medicine. These include hAM-derived cells and tissues, hAM homogenates, hAM extracts, and hAM-containing hydrogels as well as hAM secretomes in the form of conditioned media or extracellular vesicles.

Conclusion: Although recent in vitro and in vivo studies with these hAM-derived preparations show promising results, in the future special attention should be dedicated to the functional characterization, standardization, and storage of these preparations in order to guarantee the high quality of hAM-derived products and speed up their translation from the laboratory bench to the bedside.

Keywords Amniotic membrane · Cancer · Extracellular matrix · Infection · Regenerative medicine · Stem cells · Tissue engineering

T. Ramuta Železnik · L. Tratnjek · M. Kreft Erdani (✉)
Institute of Cell Biology, Faculty of Medicine, University of Ljubljana, Vrazov trg 2, SI-1000 Ljubljana, Slovenia
e-mail: mateja.erdani@mf.uni-lj.si

I. Virant-Klun (ed.), *Stem Cells in Reproductive Tissues and Organs*, Stem Cell Biology and Regenerative Medicine 70, https://doi.org/10.1007/978-3-030-90111-0_13

Abbreviations

bFGF	Basic fibroblast growth factor
Cdk4	Cyclin-dependent kinase 4
DPPA3	Developmental pluripotency-associated protein 3
ECM	Extracellular matrix
EGF	Epidermal growth factor
hAEC	Human amniotic membrane epithelial cells
hAM	Human amniotic membrane
hAMSC	Human amniotic membrane mesenchymal stromal cells
HGF	Hepatocyte growth factor
HGFR	Hepatocyte growth factor receptor
HLA	Human leukocyte antigen
hPAM	Human placental amniotic membrane
hRAM	Human reflected amniotic membrane
hUC-AM	Human umbilical cord amniotic membrane
IL-1	Interleukin 1
KGF	Keratinocyte growth factor
KLF-4	Kruppel-like factor-4
MHC	Major histocompatibility complex
MIG	Monokine induced by gamma interferon
MIP1α	Macrophage inflammatory protein 1α
OCT-4	Octamer-binding protein 4
PBMC	Peripheral blood mononuclear cells
PDGF	Platelet-derived growth factor
PEDF	Pigment epithelium-derived factor
PROM1	Prominin 1
SOX-2	Sex-determining region Y (SRY)-related HMG-box gene 2
SSEA-3	Stage-specific embryonic antigens 3
TDGF-1	Teratocarcinoma-derived growth factor 1
TGFα	Transforming growth factor α
TGFβ	Transforming growth factor β
Th1	T helper 1 cell
TIMP	Tissue inhibitor of metalloproteinases
TRA1-60	Tumor rejection antigens
VEGF	Vascular endothelial growth factor

Introduction

The shortage of tissues and organs for organ transplantation is a major public health challenge with serious constraints due to the problems with accessing enough tissue (only 146.840 organs are transplanted annually, Global Observatory on Donation and

Transplantation, 2018) and the risk of rejection by the patient's immune system, so there is a great need for new therapies capable of regenerating tissues, which would consequently reduce reliance on transplants. Regenerative medicine is a rapidly evolving field that employs an interdisciplinary approach to replace the damaged tissue or to stimulate regeneration of the original tissue. It combines biological, engineering, and medical knowledge for the development and clinical use of cells, tissues, or organs. Four main therapeutic approaches are used in regenerative medicine, and they are based on the use of (a) biological carriers (scaffolds), (b) cells, (c) combinations of scaffolds and cells, i.e. tissue engineering, and (d) secretome [1–4].

The production of novel biomaterials or scaffolds that would improve or promote tissue regeneration and function is the focus of many research studies [4, 5]. One of the promising candidates for use in tissue engineering and regenerative medicine is the human amniotic membrane (hAM), the innermost layer of placenta, which has many properties that make it suitable for use in regenerative medicine, such as the promotion of epithelization, inhibition of fibrosis, and immunomodulatory properties [2, 6].

The Structure of Human Amniotic Membrane

Human amniotic membrane (hAM) is composed of human amniotic membrane epithelial cells (hAEC), basal lamina, and hAM stroma (Fig. 13.1). The latter is further divided into compact layer, layer of human amniotic membrane mesenchymal stromal cells (hAMSC), and spongy layer [6, 7]. Despite the uniform structure, there are differences in hAM based on its anatomical region in the placenta. Namely, hAM is anatomically divided into human placental amniotic membrane (hPAM) that adheres to the chorionic plate, human reflected amniotic membrane (hRAM) that adheres to the chorion laeve, and human umbilical cord amniotic membrane (hUC-AM) that adheres to the umbilical cord [8–12]. Han et al. [8] demonstrated that there are stark differences in gene expression between the hPAM and hRAM at term and the same research group also reported differences in microRNAome and miR-143 regulation of prostaglandin-endoperoxidase synthase 2 in hAM regions at term [9]. Similarly, Centurione et al. [10] demonstrated the heterogeneity in expression of pluripotent markers octamer-binding protein 4 (OCT-4) and sex-determining region Y (SRY)-related HMG-box gene 2 (SOX- 2) in hAEC isolated from different hAM regions. Next, Banerjee et al. [11, 12] demonstrated that there are differences in morphology of hAEC, mitochondrial activity, content of reactive oxygen species, ATP, and lactate concentrations between hPAM and hRAM. These findings indicate that special attention should be dedicated to the hAM collection when the regenerative effect of hAM is evaluated. Consequently, the hAM region with the best regenerative effect could be determined.

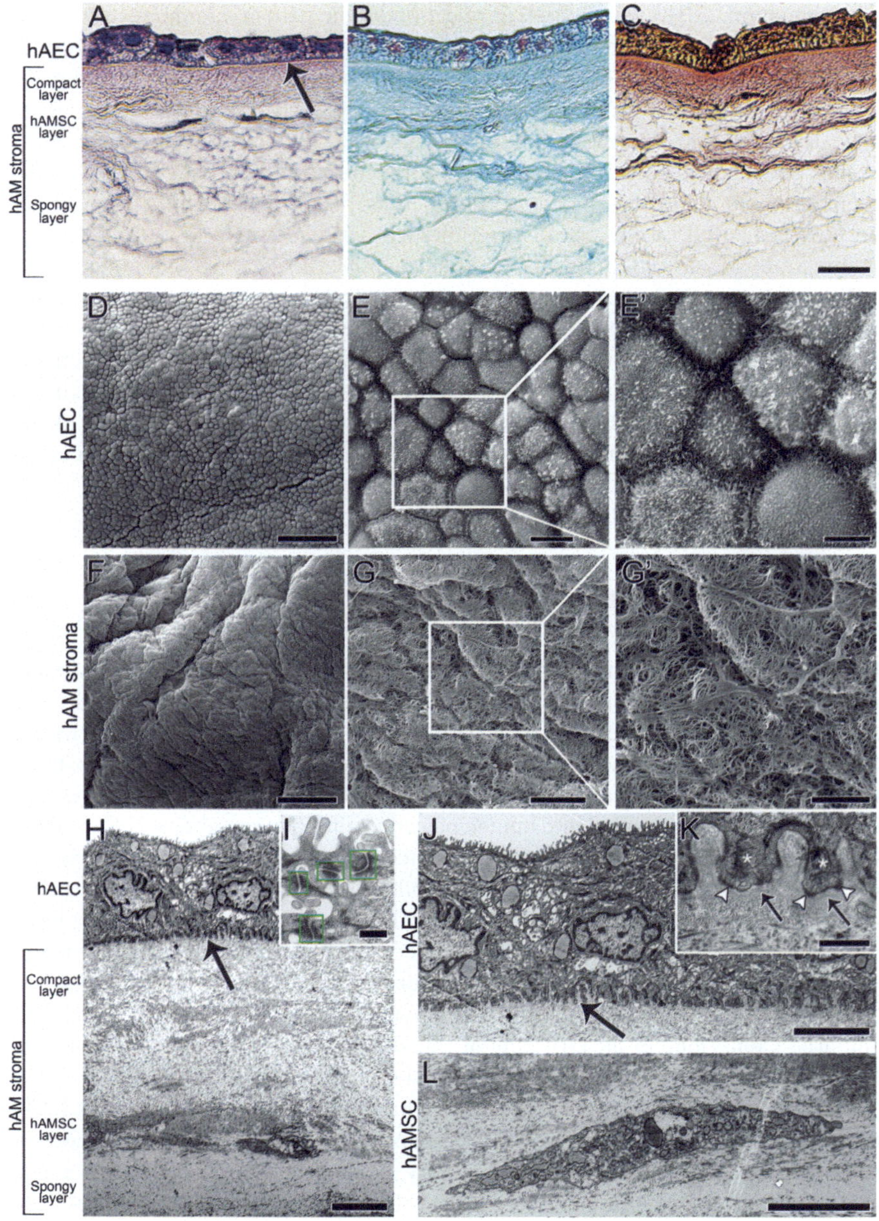
A
B
C
hAEC
Compact layer
hAMSC layer
Spongy layer
hAM stroma
D
E
E'
hAEC
F
G
G'
hAM stroma
H
I
J
K
L
hAEC
Compact layer
hAM stroma
hAMSC layer
Spongy layer
hAEC
hAMSC

◀**Fig. 13.1** The structure of the hAM from term vaginal delivery, examined using standard histology techniques (**A-C**), scanning electron microscopy (**D-G'**), and transmission electron microscopy (**H–L**). The hAM consists of a monolayer of hAEC, basal lamina (black arrows), and hAM stroma, further composed of the compact layer, hAMSC layer and spongy layer. The hematoxylin–eosin staining (**A**) of the hAM shows hAEC and hAMSC with stained nuclei (blue), thick amniotic basal lamina (arrow), and the underlying hAM stroma. Alcian Blue (**B**) and Weigert-Van Gieson (**C**) staining show the localization of acidic proteoglycans (blue, **B**) and collagen fibers (light red, **C**) in the hAM stroma. Scanning electron microscopy shows the apical surface of the hAEC enlarged with microvilli (**D-E'**). Collagen fibrils are densely packed in the compact layer of the hAM stroma (F-G'). Transmission electron microscopy shows a cross-section of the hAEC and the hAM stroma, separated with thick basal lamina (black arrows, **H**, **J**, **K**). The hAEC exhibit numerous microvilli at the apical surface and irregular nuclei, while the cytoplasm contains numerous vesicles (**H**, **J**). Adjacent hAEC are connected via numerous desmosomes (green squares, **I**). The surface of hAEC basal plasma membrane is enlarged by foot processes called pedicels (asterisks, **K**) and attached to the basal lamina (black arrows **H**, **J**, **K**) with hemidesmosomes (arrowheads, **J**). Cells comprising the hAMSC layer are mostly spindle-shaped (**L**). White framed areas in images **E** and **G** are shown in the magnified view in images **E'** and **G'**, respectively. Scale bars, 20 μm (**A-C**), 100 μm (**D**, **F**), 10 μm (**E**, **G**), 5 μm (**E'**, **G'**), 6 μm (**H**, **J**, **L**), 400 nm (**I**), and 600 nm (**K**)

Human Amniotic Membrane Epithelial Cells

hAEC form a monolayer of cuboidal cells, which are in contact with amniotic fluid (Fig. 1A–E, J, K) [6]. The cells express the following markers: CD9, CD10, CD13, CD24, CD29, CD44, CD49e, CD54, CD73, CD79, CD90, CD105, CD140b, CD166, CD324 and also POU5F1, CFC1, DPPA3, PROM1, and PAX6 [6, 10, 13–18]. Since hAM is formed from the epiblast prior to gastrulation, hAEC retain some of the pluripotent properties of early epiblast cells [13]. Several research groups demonstrated that hAEC expresses embryonic stem cell surface markers, namely, tumor rejection antigens 1–60 (TRA1-60) [13, 19–21], TRA1-81 [13, 20, 22, 23], stage-specific embryonic antigens 3 (SSEA-3) [13, 19–21], SSEA4 [13, 16, 19–25], teratocarcinoma-derived growth factor 1 (TDGF-1) [26], and GCTM-2 [16]. Moreover, researchers also found that hAEC express the following transcription factors, characteristic for stem cells: OCT-4 [13, 16, 19, 21–25, 27–30], NANOG [13, 16, 19, 21, 23, 25, 28–30], SOX-2 [13, 16, 19, 21, 25, 28, 29], and SOX-3 [30].

While most hAEC express stem cell surface markers in the early second trimester, only a subpopulation of hAEC expresses these markers in term placentas [13, 31, 32]. Furthermore, at term, only 1–3% hAEC express NANOG, 9% express SSEA-3, 44% express SSEA-4, 10% express TRA markers [13, 31], and only 4% of hAEC co-express SSEA-4, TRA1-60, and TRA1-81 [20]. The percentage of stem cell marker-positive hAEC further decreases from passage 4 in the currently used culture conditions [6, 31, 33].

hAEC are capable of differentiation into all three germ layers: ectoderm, mesoderm, and endoderm [13, 16]. It was demonstrated that they are capable of osteogenic [14, 16], cardiac [13, 34, 35], hepatic [13, 35–39], chondrogenic [16, 40], skeletal myogenic [14], pancreatic [13, 14], and adipogenic [14, 16] differentiation in vitro [6, 13]. Moreover, researchers have also been able to differentiate hAEC into neurons and glial cells [13, 16, 41–44], surfactant producing alveolar epithelial cells [45], and insulin secreting pancreatic β-islet like-cells [46]. Interestingly, cultured hAEC can obtain a mesenchymal stem cell-like phenotype by being capable of osteogenic, adipogenic, and chondrogenic differentiation and by expressing cell surface markers CD73 [14, 15, 18, 47–52], CD90 [14, 15, 18, 47, 50–53], and CD105 [14, 15, 18, 47, 49–53]. On the other hand, hAEC do not express the hematopoietic cell surface markers CD45, CD34, CD14, CD79, and HLA-DR [54].

Human Amniotic Membrane Mesenchymal Stromal Cells

There are two cell populations in the hAMSC layer, namely amniotic membrane mesenchymal stromal cells (hAMSC) and fibroblasts (Fig. 1A–C, H, L). Both populations express cell surface markers CD29, CD44, CD79, CD90, and CD105, but they can be distinguished based on the expression of CD34, since the hAMSC are CD34-positive and the fibroblasts are CD34-negative [55, 56]. Furthermore, the hAMSC also express the CD73, CD90, CD105, CD117, CD133, CD146, CD201, and Globo H markers [6, 15, 17, 46, 57–60], while there are some inconsistencies in the literature regarding the expression of CD271 [48, 57, 61]. Similar to hAEC, a subpopulation of hAMSC expresses some markers of pluripotent cells, namely OCT-3, OCT-4, SSEA-1, SSEA-3, SSEA-4, SOX-2, NANOG, Kruppel-like factor-4 (KLF-4), and REX-1 [6, 15, 17, 46, 57–60, 62, 63]. However, the expression of pluripotency markers, e.g. SOX-2, Nanog, and KLF-4, decreases during culture [64].

Moreover, it has been shown that hAMSC express only low levels of TRA1-60 and TRA1-80 or even none at all [48, 65]. Reports on the expression of SSEA-3 and SSEA-4 in hAMSC show conflicting results, some of them showing that only a small percentage of hAMSC express these markers, while the others show that as many as 43% of hAMSC express these markers [48, 64–66].

hAMSC have the capability of differentiating into all three germ layers. Studies have shown that they have adipogenic [14, 57, 58], chondrogenic [14, 52, 57, 58, 67, 68], osteogenic [14, 15, 41, 46, 52, 57–59], skeletal myogenic [16, 58], angiogenic [58], neurogenic [14, 41], pancreatic [46], and cardiomyogenic [59, 69] differentiation potential.

hAM Extracellular Matrix

The integrity and mechanical strength of hAM is provided by its extracellular matrix (ECM), which is very abundant in the connective tissue of hAM stroma (Fig. 1A–C, F–H) and much less in the epithelial hAEC layer [70, 71]. In the compact layer of hAM stroma collagen types I, III, V, and VI and fibronectin predominate, while the layer of hAMSC contains collagen types I, III, and VI, nidogen, laminin-5, and fibronectin in addition to cells. The spongy layer consist of loosely arranged components of ECM, namely collagen types I, III, and IV, proteoglycans, and hyaluronic acid [6, 55, 72–78].

An important component of the hAM is also a specialized form of ECM, namely the basal lamina (Fig. 1A, H, J, K), which lies between the epithelial and connective tissue of hAM and consists of collagen types III, IV, and V, laminin-5, fibronectin, and nidogen, which are tightly interconnected [55, 72].

What Makes hAM Suitable for Use in Clinical Practice?

hAM is a versatile tissue and its potential for use in regenerative medicine has been recognized more than a century ago. hAM is a biological barrier, which supports the development of the fetus by preparing an anatomically, physiologically, and immunologically privileged space [79]. Together with the ECM, the hAM-derived cells provide a molecular cocktail that promotes tissue regeneration with no or only slight induction of the immune response in the host (Table 13.1) [80, 81].

hAM Promotes Epithelization

Tissue regeneration is a result of the interaction of different cell types, depending on the target tissue (e.g. epithelial cells, immune cells, fibroblasts, platelets), and the components of the ECM. It is regulated by biochemical mediators, namely cytokines and growth factors [82, 83]. hAM expresses a plethora of molecules that contribute to epithelization, namely epidermal growth factor (EGF), transforming growth factor α (TGF-α), TGF-β, hepatocyte growth factor (HGF), HGF receptor (HGFR), keratinocyte growth factor (KGF), KGF receptor (KGFR), basic fibroblast growth factor (bFGF), and platelet-derived growth factor (PDGF) [84, 85]. Koizumi et al. [84] and Gicquel et al. [86] demonstrated that these molecules are present in the intact and de-epithelized hAM, however, in the latter in much lower concentrations, which suggests they originate from hAEC. Furthermore, Jin et al. [85] showed that conditioned medium from hAEC also significantly promotes cell migration, indicating that most of the above-mentioned growth factors are involved in paracrine signaling.

ECM of hAM is significantly involved in wound healing since it provides the structural support for the repairing epithelial tissue [82, 87]. Therefore, hAM can serve as a scaffold that promotes proliferation and differentiation of epithelial cells [88–91].

Table 13.1 Overview of molecules expressed or secreted by hAM-derived cells and components of the ECM of hAM that contribute to properties of hAM that are desirable in regenerative medicine

Effect of hAM on other cells	hAM-derived cells	Extracellular matrix
Promotion of epithelization	EGF	Collagen types I, III, IV, V, VI
	TGF-α	Fibronectin
	TGF-β	Elastin
	HGF	Nidogen
	HGFR	Hyaluronic acid
	KGF	
	KGFR	
	bFGF	
	PDGF	
Decreased scarring and fibrosis	TIMP1-4	Hyaluronic acid
	Suppression of IL-1, IL-6, IL-8, TGF-β in the target cells	
Pro-angiogenic effect	VEGF	
	Angiogenin	
	Angiopoietin-2	
	IL-2	
	IL-8	
	interferon-γ	
	bFGF	
	EGF	
	HGF	
	PDGF	
Anti-angiogenic effect	Thrombospondin-1	
	Heparin sulfate proteoglycan	
	PEDF	
	Endostatin	
	TIMP1-4	
Antimicrobial effect	α-defensins (HNP1-3)	
	β-defensins (HBD-1-3)	
	SLPI	
	Elafin	
	H2A, H2B histones	

hAM Decreases Scarring and Fibrosis

Ineffective and/or inappropriate tissue repair following wound healing response can result in scarring and fibrosis that causes tissue dysfunction [92, 93]. Researchers are investigating the use of various stem cells, including hAM-derived stem-like cells, to promote wound healing while minimizing fibrosis [92].

Fibrosis is characterized by excess deposition of ECM. Some proteinases, e.g. metalloproteinases-9, -12, -13, can have pro-fibrotic functions [94]. hAM inhibits fibrosis by secretion of tissue inhibitor of metalloproteinases (TIMP)-1, -2, -3, 4 [95–98]. Both hAEC and hAMSC are supposed to secrete all 4 TIMPs [96]. Furthermore, the anti-fibrotic effect is supported also by the anti-inflammatory activity of hAM since hAM suppresses expression of interleukin-1 (IL-1), IL-6, IL-8, and also TGF-β in the target cells [99–103].

Pro- and Anti-Angiogenic Activity of hAM

hAEC secrete some of the pro-angiogenic factors, namely vascular endothelial growth factor (VEGF), angiogenin, angiopoietin-2, IL-6, IL-8, interferon-γ, bFGF, EGF, HGF, and PDGF [104–107]. Interestingly, hAEC also secrete anti-angiogenic factors, e.g. thrombospondin-1, heparin sulfate proteoglycan, pigment epithelium-derived factor (PEDF), and endostatin. Furthermore, the secretion of TIMP-1–4 by hAM-derived cells also contributes to the anti-angiogenic effect of hAM [96, 106]. Importantly, Wolbank et al. [104] demonstrated that the angiogenic factor profile released from hAM depends on the preparation, namely whether hAM is non-preserved (intact/fresh), glycerol-preserved or cryo-preserved. Moreover, they have also shown that the loss of cell viability during preparation of hAM is associated with the loss of cellular factors secreted from the hAM.

Furthermore, Niknejad et al. [105] used a dorsal skinfold chamber model on male rats, in which a layer of dorsal skin was removed and the hAM was implanted with the epithelial or mesenchymal side facing up. They demonstrated that the stroma of hAM promoted the number of vessel sprouts and their length, while the contact with hAEC resulted in an anti-angiogenic effect. Moreover, they also performed an in vitro aorta ring assay and showed that angiogenesis was detected only when the hAEC were removed from the hAM. Next, Danieli et al. [108] investigated the effect of hAMSC on the regeneration of infarcted rat hearts, and they demonstrated that the injection of the conditioned medium of hAMSC into infarcted rat hearts limited the infarct size, reduced cardiomyocyte apoptosis, and ventricular remodeling and also promoted capillary formation at the infarct border zone. A recent study by Nasiry et al. [109] also demonstrated a pro-angiogenic activity of hAM, as they showed that a bioengineered 3D hAM scaffold promotes healing in ischemic wounds in diabetic type 1 rats by improving the length density of blood vessels.

Low Immunogenicity of hAM

hAM does not trigger an allogenic or xenogenic immune reaction in recipients (2). hAM cells express the major histocompatibility complex (MHC) class Ia antigens, namely HLA-A, -B, -C, and class Ib antigens, HLA-G, and HLA-E [80, 81, 110–112]. Furthermore, these cells are considered to be poorly immunogenic due to low or limited expression of MHC class II antigens, namely HLA-DR and co- stimulatory molecules (CD80, CD86, CD275) [81, 113]. However, Magatti et al. [81] pointed out that hAM-derived stem-like cells may not be immune-privileged and that the immune tolerance may not be the result of their lack of immunogenicity but their immune-suppressive properties.

Immunomodulatory Activity of hAM

hAM-derived cells downregulate inflammation by reducing migration of neutrophils and macrophages, inhibition of NK cell cytotoxicity and reduction of IFN-γ expression. Furthermore, they also inhibit generation and maturation of dendritic cells and/or inflammatory M1 macrophages and shift the differentiation towards an anti-inflammatory M2 phenotype. Moreover, the treatment with hAM-derived cells results in reduced secretion of pro-inflammatory cytokines (IL-1α, IL-1β, IL-12, IL-8, TNF-α, MIP1α, MIP1β, MIG, Rantes, IP-10) and an increased secretion of the anti-inflammatory cytokine IL-10. Researchers also showed that co-culture with hAM-derived cells or their conditioned medium lead to decreased expression of co-stimulatory proteins (CD80, CD86, CD40) [6, 81, 110, 114–121].

hAM-derived cells are also capable of suppressing proliferation of activated T helper cells (CD4 +) and T cytotoxic cells (CD8 +) in a dose-dependent manner in the in vitro conditions, namely when in cell-cell contact, a transwell system or treated with conditioned medium. Moreover, hAM-derived cells reduce the number of T cells, namely T helper 1 (Th1), Th9, and Th17 and related cytokines, while they promote the induction of regulatory T cells in mix lymphocyte cultures. Consequently, hAM-derived cells favor the emergence of regulatory T cells (Table 13.2) [15, 53, 110, 111, 114, 122–127]. Furthermore, the conditioned medium of hAEC induced apoptosis in murine B cells and inhibited lipopolysaccharide-induced proliferation of B cells [2, 81, 115].

Interestingly, hAM-derived cells are capable of immunostimulation as well. When low concentrations of hAEC and hAMSC were cultured with unstimulated allogeneic peripheral blood mononuclear cells (PBMC), the hAM-derived cells stimulated PBMC's proliferation. Similarly, low concentrations of hAM-derived cells also strongly induced proliferation of T cells (Table 13.2) [15, 81, 122, 128].

To summarize, these studies demonstrate the ambivalent nature of hAM, which is able to inhibit and stimulate the immune system. Further studies are needed to deduce which factors in the microenvironment determine whether the application of hAM

Table 13.2 Immunomodulatory activity of hAM Based on the microenvironment, hAM is able to suppress or stimulate the immune system

Immunomodulatory effect of hAM	
Suppression of inflammation	Suppression of IL-1α, IL-1β, IL-12, IL-8, TNF-α, MIP1α, MIP1β, MIG, Rantes, IP-10 in the target cells
	Decreased expression of co-stimulatory proteins (CD80, CD86, CD40) in the target cells
	Promotion of IL-10 expression in the target cells
	Suppression of proliferation of activated T helper cells (CD4 +) and T cytotoxic cells (CD8 +)
	Reduction in the number of Th1, Th9 and Th17 cells and related cytokines
	Promotion of induction of the regulatory T cells
Immunostimulation	Stimulation of PBMC and T cell proliferation

leads to its pro- or anti-inflammatory activity. Moreover, special attention should be dedicated to the characterization of all immune cells involved in these studies.

Anticancer Activity of hAM

hAEC and hAMSC are capable of inducing cell cycle arrest in cancer cells. Namely, Bu et al. [129] showed that hAEC induced cell cycle arrest in G0/G1 phase in epithelial ovarian cancer cells and also inhibited growth of these cells in a tumor-bearing nude mouse model. Similarly, Magatti et al. [130] demonstrated that hAMSC decrease proliferation of cancer cells of hematopoietic and non-hematopoietic origin, namely through induction of cell cycle arrest in G0/G1 phase. Moreover, hAMSC down-regulate the expression of genes encoding proteins associated with cell cycle progression (cyclin D2, cyclin E1, cyclin H, CDK4, CDK6, CDK2) and up-regulate the expression of negative regulators of the cell cycle (p15, p21). Recently we demonstrated that hAM-derived cells and hAM scaffolds decreased the proliferation of muscle-invasive bladder cancer cells. Moreover, hAM scaffolds decreased the invasive potential of bladder cancer cells and the expression of epithelial-mesenchymal transition markers. Importantly, individual bladder cancer cells grown on the hAM scaffolds even expressed epithelial markers E-cadherin and occludin [131]. The development of new blood vessels for the tumor cells and the manipulation of the immune system that would otherwise prevent tumor invasion is crucial for tumor growth and metastasis. hAM prevents tumor metastasis also through its anti-angiogenic and immunoregulatory activity [79].

Not only the cells but also the hAM-conditioned medium is capable of inducing inhibition of DNA synthesis, reducing the viability and number of hepatocarcinoma cells and inducing cell cycle arrest in the G2/M phase by reduced expression of cyclin D1 and Ki-67 and promoted expression of p53 and p21 [132]. Moreover, hAM-derived cells and their conditioned medium affect metabolic activity [133] and induce apoptosis in several cancer cell lines and mouse xenograft models [79, 134–138].

On the other hand, Kim et al. [139] treated the breast cancer cell lines MCF-7 and MDA-MB-231 with the hAMSC conditioned medium and demonstrated that the treatment increased proliferation and migration of cancer cells. Next, Meng et al. [140] treated the pulmonary adenocarcinoma cell line SPC-A-1 and gastric carcinoma cell line BGC-823 with hAMSC conditioned medium and while the treatment resulted in decreased cancer cell motility, the proliferation of cancer cells was enhanced. Together these results once again demonstrate the ambivalent nature of hAM-derived preparations and point to the need for further research to elucidate the exact mechanisms behind the hAM's pro- and anti-cancer activity using well-established and relevant biomimetic in vitro models.

Antimicrobial Activity of hAM

hAM secretes natural antibacterial peptides that are necessary for the protection of pregnancy against bacterial, fungal and viral infections [141, 142]. The antimicrobial activity of hAM and its derivatives (e.g. hAM-derived cells, homogenate, extract) has been demonstrated on several Gram-positive and Gram-negative bacteria and also on fungi [141, 143–152]. hAEC secrete human α- and β-defensins, which are small antimicrobial peptides that play an important role in innate immunity [142, 153–156] and hAMSC secrete only the human β-defensin-3 (HBD-3) [142, 157]. β-defensins are constitutively expressed (e.g. HBD-1) or induced in the presence of proinflammatory cytokines or bacterial products (e.g. HBD-2) [142, 158, 159]. Furthermore, hAEC also secrete whey acidic peptide (WAP) motif-containing proteins, namely secretory leukocyte protease inhibitor (SLPI) and elafin [142, 160]. Additional antimicrobial activity of hAM is provided by the H2A and H2B histones, which are constitutively secreted from hAEC and are capable of neutralizing endotoxins [161].

Novel Approaches Using hAM as a Therapeutic Agent

The first therapeutic use of hAM was reported in 1910 for skin transplantation [162, 163]. Today hAM is most commonly used in ophthalmology, namely for treatment of ocular surface wounds, corneal ulcers, pterygium, glaucoma, neoplasia, strabismus, limbal stem cell deficiency, and oculoplastics [162, 164]. In addition, hAM is also frequently used for wound healing, especially in patients with acute and chronic wounds, burns, chronic lower extremity diabetic ulcers, fistulas, and venous leg ulcers [165–167]. Interestingly, hAM has been used also in peritoneal, intra-oral

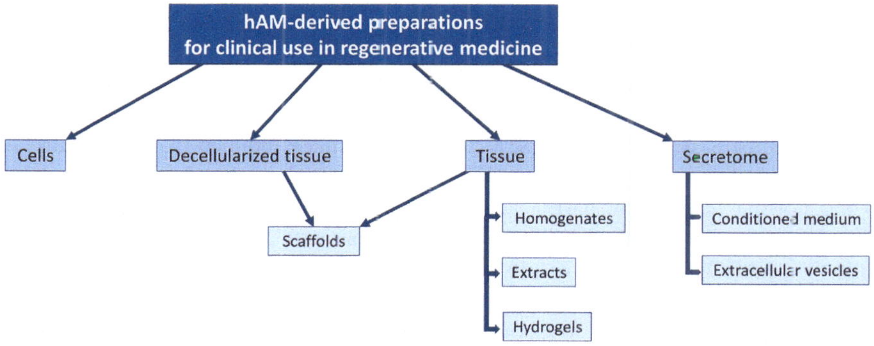

Fig. 13.2 hAM-derived preparations used in regenerative medicine

and genital reconstruction, urology, dentistry, ligament and tendon healing, cartilage restoration, osteoarthritis and vascular reconstruction [2, 7, 168–170].

Since the use of hAM in ophthalmology, wound healing and orthopedics had been extensively covered in several review articles [162, 165, 169, 170] we will hereafter focus only on novel hAM-derived preparations that show potential for clinical use in regenerative medicine and are presented also in Fig. 13.2.

hAM-Derived Cells

hAEC and hAMSC offer a good alternative to embryonic stem cells or bone marrow stem cells, which are more difficult to acquire and might cause ethical unease [171]. Due to their stem cell-like differentiation capability, immunomodulatory activity and low immunogenicity the hAM-derived stem-like cells show great potential for use in regenerative medicine [33]. There are several studies that demonstrated the beneficial effects hAEC and hAMSC in liver [172, 173], lung [174], heart [175, 176], and kidney [177] regeneration. However, in the following chapter we will focus on the two research areas where new studies with hAEC and hAMSC are emerging, namely regenerative medicine of the reproductive tract and the nervous system.

The Potential of hAM-Derived Cells in Infertility Treatment

Infertility affects 8–12% of couples of reproductive age. Stem cells from different sources have been proposed as potential candidates for infertility treatment [178, 179], and several studies have also examined the regenerative potential of hAM-derived cells in the reproductive system. Indeed, stem cells promote the inhibition of premature ovarian aging function. Due to their stem-like cell properties, Ding et al.

[23] used hAM-derived cells to improve ovarian function damaged by chemotherapy (cyclophosphamide) in a mice model.

Namely, the hAEC and hAMSC were injected intravenously into the tail vein of mice and their results show that the capacity of the hAM-derived cells to restore the follicle numbers, depends on the severity of the damage induced by a chemotherapeutic agent and that in general, hAMSC more effectively restored follicle numbers to the normal levels than hAEC [23]. Furthermore, hAMSC proved to be more effective in improving the proliferation rate of patients' human ovarian granular cells than hAEC [23]. These results are further supported by Liu et al. [180], who demonstrated that hAMSC administered by intraperitoneal injection, improved the follicular microenvironment to recover ovarian function in premature ovarian failure in mice. Namely, following the hAMSC transplantation, the oestrus cycle was recovered and, fertility rate and population of follicles increased. Moreover, the newborn mice showed normal growth and development and were also fertile.

hAEC were also used to improve fertility in the mouse model of intrauterine adhesion, which is a consequence of the damaged endometrium trauma usually accompanied by fibrosis that results in the complete or partial obstruction of the uterine cavity or cervical canal. Following mechanical injury to the uterus, hAEC were transplanted by intraperitoneal injection, which resulted in thicker endometrium, increased number of endometrial glands, reduced fibrosis, and increased generation of microvessels. Moreover, the treatment leads to improved angiogenesis and increased stromal cell proliferation and also increased pregnancy rate and fetus number in treated mice, which demonstrated the potential of hAEC to repair the uterus after injury [181].

The Potential of hAM-Derived Cells for Treatment of Neurological Disorders

Stem cells also show great potential for the treatment of several neurological disorders [182]. Studies have shown that stem cells have the potential to repair or replace damaged or degenerative neurons and consequently improve neurological functions [171]. The administration of hAEC in in vivo models of stroke, namely acute ischemia (mice, rats, and marmoset monkeys) and intracerebral hemorrhage (rats and rabbits) lead to reduced cerebral apoptosis and inflammation, induced neural differentiation, systemic immunosuppression, reduced microglial activation, and increased neural cell survival and regeneration [27, 183–185]. In in vivo models of spinal cord injury (in rats, monkeys), the application of hAEC resulted in the promotion of axonal growth and regeneration, survival, and neural differentiation and remyelination of nerve fibers and inhibited atrophy of the axotomized red nucleus and reduced microglial activation, glial scar formation, and immunological reaction at the lesion site [42, 186–190]. hAEC transplantation had beneficial effect also in rat models of Parkinson's disease [35, 43, 191, 192]. In in vivo models of cerebral palsy (in preterm fetal sheep and perinatal mouse), the hAEC application lead to reduced microglial activation, apoptosis and astrogliosis, vascular leakage and increased microglial activity, oligodendrocytes, and MBP-positive cells [193–196]. Moreover, the immunomodulatory activity of hAEC was beneficial in mouse models of multiple sclerosis, since hAEC suppressed the proliferation of splenocytes and T cells, increased the proportion of Th2 cells, and the number of Treg and naïve CD4 + T cells, increased the production of IL-2 and IL-5 and decreased the T cell response and inflammatory factors [118, 171, 197, 198].

Kim et al. [199] examined the effect of tail vein-injected hAMSC in an Alzheimer's disease mouse model and concluded that hAMSC-treated mice showed evidence of improved spatial learning, which correlated with the fewer toxic extracellular deposits of amyloid-beta (Aβ plaques) observed in the brain. Furthermore, they also recorded a higher number of phagocytic microglial cells associated with Aβ plaques, a larger amount of Aβ-degrading enzymes, a lower level of proinflammatory cytokines (IL- 1, TNFα), and a higher level of anti-inflammatory cytokines (IL-10, TGF-β). Importantly, these effects lasted up to 12 weeks after hAMSC application. Similarly, Kim et al. [200] demonstrated that intracerebral injection of hAEC into the Tg2576 transgenic mice model of Alzheimer's disease alleviated cognitive impairment. hAMSC have been used also in an in vivo model of multiple sclerosis (experimental autoimmune encephalomyelitis mouse model), resulting in a significant improvement in disease severity and inhibition of proinflammatory cytokines, decrease in the number of CD4 + and CD8 + T cells and promotion of production of neuron-repair factors in the central nervous system [201].

To conclude, the in vitro and in vivo studies show great capacity of hAM-derived stem-like cells for use in regenerative medicine. However, in order to ensure an optimal clinical outcome, it is necessary to standardize the procedures for characterization, preparation and application of hAM-derived stem-like cells, paying particular attention to the range of cell passages considered suitable for clinical use, adequate expression of cell markers, and functional tests.

hAM As a Scaffold

Biomaterials suitable for use in tissue engineering and regenerative medicine must mimic the microenvironment of the natural ECM and therefore, must possess appropriate biochemical and biophysical properties, such as molecular compatibility, high porosity, and suitable mechanical strength [202, 203]. The use of hAM as a scaffold has been investigated in several studies, and among other beneficial properties, another advantage of using hAM as a scaffold is its biological origin, as biological grafts are less prone to infections than synthetic grafts [204, 205].

hAM can be used as a scaffold in several ways. Our research group seeded the normal porcine urothelial (NPU) cells on the hAEC of the hAM, on the basal lamina of deepithelized hAM and on the stroma of hAM. After 3 weeks in culture, we demonstrated that all hAM scaffolds enable the development of the urothelium. However, the NPU cells grew the fastest and reached the highest level of differentiation, comparable to that of the native urothelium, when grown on the stroma of hAM [206]. Recently, we also demonstrated that hAM allows integration of urinary bladder fibroblasts into its stroma and enables the formation of the differentiated urothelial tissue equivalents in vitro, as well as promotes regeneration of the urothelium when used as a wound dressing [207]. Moreover, another study by our research group demonstrated that hAM scaffolds decrease proliferation, invasive potential, and expression of epithelial-mesenchymal markers in muscle-invasive bladder cancer cells [131], which demonstrates the potential of using hAM scaffolds as an implantable device in bladder cancer management and regenerative urology.

Similarly, Iranpour et al. [208] seeded the adipose-derived stromal (ASC) cells and a human immortalized keratinocyte cell line (HaCaT) on the basal lamina of deepithelized hAM and on the stroma of hAM and cultured the constructs for 3 weeks. They demonstrated that the growth and viability of ASC and HaCaT were the highest on the denuded hAM scaffold.

hAM has been used as a scaffold for chondrocytes and used in in vitro cartilage repair experiments. Chondrocytes isolated from human articular cartilage were seeded on the hAEC of the hAM and on the hAM stroma. While they did not grow on the epithelial side of hAM, the hAM stroma proved to be a suitable scaffold, which allowed the maintenance of the chondrogenic phenotype. Moreover, the in vitro human osteoarthritis cartilage repair trial indicated excellent integration of the new tissue, which indicates that hAM supports chondrocyte proliferation [47].

hAM has been used also as a scaffold for stem cells. Researchers compared the suitability of deepithelized hAM or a porcine small intestine submucosa as a scaffold for seeding the mesenchymal stem cells isolated from the limbal stroma. Both scaffolds maintained cell viability, actin cytoskeleton, nuclei morphology, and mesenchymal phenotype and did not cause cell death. However, their results show that the denuded hAM is preferable for long-term culture as it is more successful in maintaining the mesenchymal stem cell phenotype [209]. Dorazehi et al. [210] seeded the bone marrow-derived mesenchymal stem cells on the deepithelized hAM and treated them with the mural embryonic cerebrospinal fluid. They showed that the combination of deepithelized hAM and mural embryonic cerebrospinal fluid improves the cultivation and differentiation of bone marrow-derived mesenchymal stem cells. The suitability of hAM as a scaffold for stem cell-derived tissue was demonstrated also by Parveen et al. [211], who demonstrated that the hAM supports the differentiation and improves cardiomyogenesis of human induced stem cell-derived cardiomyocytes.

To summarize, there might not be an optimal hAM scaffold for all cell types, and the selection of the most suitable scaffold may depend on the target tissue.

hAM Extract, hAM Homogenate, and hAM-Derived Extracellular Vesicles

hAM is a thin membrane that is prone to folding and tearing, which is a limitation when trying to incorporate hAM into routine clinical applications [212].

Therefore, researchers are developing new hAM preparations, e.g. hAM extracts, hAM homogenates, and hAM-derived extracellular vesicles (EVs), which are more compelling to clinicians.

Choi et al. [213] prepared a hAM extract-loaded double-layered wound dressing, comprised of polyvinyl alcohol (PVA) solution and sodium alginate. Compared to a single-layered commercially available wound dressing product, the hAM extract-loaded wound dressing enhanced the wound healing effect, when used to treat full-thickness wounds on dorsum of rats.

hAM extract can be used also as a rich source of growth factors in cell culture. Vojdani et al. [214] investigated the effect of hAM extract on the umbilical cord mesenchymal stem cells, and they reported that the addition of hAM extract increased the proliferation rate and decreased duplication time without change in cell morphology. The researchers concluded that the hAM extract could be used to support preparation of umbilical cord mesenchymal stem cells for cell therapy approaches.

Bacterial infections of the damaged tissue are a serious complication in regenerative medicine [215, 216]. Our research group prepared a hAM homogenate that has broad-spectrum activity against Gram-positive and Gram-negative uropathogenic bacteria, including some multiple-resistant clinical strains. Moreover, we have shown that the manner of preparation and storage crucially affect its antimicrobial activity, which demonstrates the importance of choosing the appropriate preservation procedure to ensure the optimal antimicrobial efficacy of hAM homogenate [143, 144].

EVs are lipid membrane vesicles, which contain various RNA species, cytosolic proteins, and transmembrane proteins. EVs can mediate cell–cell communication and are involved in various processes, e.g. cell differentiation, proliferation, stress response, and immune signaling and as such they show great potential for use in regenerative medicine [217]. Gao et al. [218] isolated exosomes, EVs that originate in multivesicular bodies, from hAMSC, and injected them into wounds of rats. Their results show improved wound healing in comparison to the control treatment, which they attribute to microRNA miR-135a that was derived from the hAMSC EVs.

hAM extracts, hAM homogenates, and hAM-derived EVs show great potential for use in tissue engineering and regenerative medicine, especially when combined with novel biomaterials. Furthermore, novel easier-to-use hAM-derived preparations will also facilitate quicker translation of hAM-derived products into the clinic.

hAM-Containing Hydrogel

Hydrogels can be engineered into 3D scaffolds that mimic the native ECM and support cell growth, which makes them one of the pertinent biomaterials for tissue engineering [219, 220]. Rahman et al. [90] developed a hydrogel containing hAM powder and *Aloe vera* powder that could be used as a burn wound healing product. They showed that the hydrogel was non-cytotoxic to brine shrimps (*Artemia salina*) and promoted cell attachment and proliferation of HaCaT cell line, derived from human skin, and HFF1 cell line, derived from human fibroblasts. Moreover, the preparation also promoted migration of HaCaT cells in an in vitro scratch assay. Next, they used the hydrogel in a rat model with induced burn and they demonstrated that the preparation accelerated wound closure through re-epithelization and wound contraction and also promoted angiogenesis. The treatment resulted in a thicker regenerated epidermis, an increased number of blood vessels, and more proliferating keratinocytes in the epidermis. In a similar study researchers prepared a hAM and collagen-based hydrogel, which they used for cutaneous burn wound healing in rats. They showed that the hydrogel was non-cytotoxic, compatible with human blood cells, and caused no skin irritation. Furthermore, the hydrogel promoted rapid wound healing through re-epithelization and wound contraction [91].

Chen et al. [221] tested the ability of gelatin methacryloyl hydrogel eye pads loaded with hAM extract to prevent symblepharon in rabbits. Namely, after ocular alkali burn, the first group of rabbits was treated with the hAM extract-loaded hydrogel, the second group was treated with hAM transplantation, and the third group received no treatment. After one week, the epithelial healing rate was higher in rabbits in the treated groups, and after 3 and 4 weeks no symblepharon was found in rabbits treated with hAM extract-loaded hydrogel, while they detected some symblepharon in the other two groups. Therefore, the researchers concluded that treatment with hAM extract-loaded hydrogel after chemical injury prevented symblepharon in rabbits.

Murphy et al. [222] developed a solubilized hAM, which they combined with the hyaluronic acid hydrogel. They used the composite for wound treatment in a full-thickness murine wound model and demonstrated that the preparation promoted re-epithelization and prevented wound contraction, which resulted in accelerated wound closure. Furthermore, the treated wounds had thicker regenerated skin, higher count of blood vessels, and more proliferating keratinocytes within the epidermis [212]. This research group performed another study, in which they used 1) solubilized hAM in combination with hyaluronic acid hydrogel and 2) lyophilized hAM powder for treatment of a full-thickness porcine skin wound model. They demonstrated that both preparations promoted rapid wound healing due to increased re-epithelization and minimal contraction, which resulted in the formation of a mature epidermis and dermis comparable to the healthy skin [222].

Novel approaches using hAM as a Drug Delivery Tool

The aim of drug delivery systems is to improve the pharmacological activities of therapeutics and to decrease their side effects and overcome the issues of limited solubility, low bioavailability and lack of selectivity [223, 224].

Li et al. [225] investigated the antifibrotic effect of the freeze-dried bilayered fibrin-binding hAM as a drug delivery system on glaucoma surgery in rabbit model. They loaded the hAM with 5-fluorouracil and applied it to rabbits that received ocular trabeculectomy. Their results showed that the application of 5-fluorouracil-loaded hAM was beneficial, since it resulted in wound healing without scar formation. Similarly, Hu et al. [226] evaluated the effect of hAM loaded with 5-fluorouracil poly (lactic-co-glycolic acid) (PLGA) nanoparticles in the experimental trabeculectomy in rabbits. They also concluded that the hAM preparation may function as an effective anti-scarring implant and could lead to improved long-term surgical outcomes.

Francisco et al. [227] loaded the decellularized hAM scaffold with 15-deoxy-Δ12,14-prostaglandin J2 (15d-J2) and compared its efficacy in improving ventricular function in a rat model of post-infarct ventricular dysfunction. They concluded that the use of decellularized hAM + 15dJ2 resulted in an increase in the ejection fraction and prevented ventricular dilation in rats.

Bonomi et al. [228] investigated whether the hAMSC could be used for drug delivery in cancer therapy. Since an important property of mesenchymal stromal cells is their ability to home toward sites of injury, hAMSC have great potential to serve as a vehicle for drug delivery. The researchers demonstrated that paclitaxel-primed hAMSC are highly resistant to the cytotoxic effects of the drug, are able to uptake the drug, and could be used as an extended-release delivery system. Importantly, a sufficient amount of paclitaxel was released from the hAMSC to inhibit tumor cell proliferation.

Several researchers investigated whether hAM could be used as a tool for the delivery of antibiotics [143, 229–233]. Kim et al. [231] evaluated ocular penetration and drug levels in tears after topical instillation of a fluoroquinolone antibiotic ofloxacin in rabbits after hAM transplantation. Their results show that hAM transplantation enhances ofloxacin penetration in rabbits with epithelial defects in corneas. Mencucci et al. [233] impregnated hAM patches with an aminoglycoside antibiotic netilmicin and showed that antibiotic uptake was dose-dependent and occurred rapidly, while the antimicrobial effect of netilmicin-loaded hAM patches was detectable for at least 3 days after treatment. Next, Resch et al. [230] have loaded hAM patches with 3% ophthalmic solution of ofloxacin and demonstrated that they can serve as a drug delivery tool since they allow slow release for up to 7 h in vitro. Similarly, Yelchuri et al. [232] have shown that hAM patches are a suitable vehicle for drug delivery of another fluoroquinolone antibiotic, moxifloxacin. Moreover, their results show that hAM patches impregnated with the antibiotic allow the sustained release of moxifloxacin up to 7 weeks in in vitro conditions. hAM can also serve as a vehicle for delivery of a beta-lactam antibiotic cefazolin, since the 3-h treatment of hAM patches with cefazolin resulted in high drug entrapment and allowed a 5-day release of the drug [234]. Furthermore, a study performed by our research group has also shown that hAM patches allow the uptake and prolonged retention of the aminoglycoside antibiotic gentamicin [143].

A drug delivery tool must demonstrate sufficient biocompatibility and biodegradability, good drug loading capacity, low toxicity, and good stability in physiological conditions [235]. hAM has the potential to be a suitable delivery tool, especially since it also possesses properties that contribute to tissue regeneration.

Benefits and Potential Pitfalls of using hAM in the Clinic

Currently, hAM is most commonly used in the clinic as a scaffold in the form of patches, but recent studies show that many new hAM-derived preparations are being developed that may reach the market in the coming years. Novel hAM-derived preparations not only focus on preserving the beneficial biological properties of hAM, but also strive to ease the administration of these products, which may accelerate their translation into the clinic.

Use of hAM and hAM-derived preparations has many benefits. Apart from the many favorable properties that make it applicable in tissue engineering and regenerative medicine, hAM is also a cost- effective solution. Moreover, the use of hAM is ethically acceptable, and this is a major advantage that can contribute to a faster and smoother translation of hAM and hAM-derived preparations into the clinic. Namely, while the use of other stem cells may cause ethical unease due to the sources from which stem cells are isolated, the hAM is usually discarded after birth and therefore raises less concerns. Importantly, unlike embryonic stem cells, hAM-derived cells do not form teratomas when transplanted in vivo, which is of the utmost importance when planning to use hAM for therapeutic purposes [7, 236].

Different hAM-derived preparations may highlight various beneficial properties of hAM, e.g. while the hAM scaffolds and hAM hydrogels would be the most appropriate for wound healing, the hAM-derived stem-like cells would be more appropriate for regeneration of nerves, heart or reproductive organs. Therefore, special attention must be dedicated to the selection of the most suitable hAM-derived preparation to achieve an optimal clinical outcome.

However, we must also address the potential pitfalls of the hAM use. Namely, since hAM is a biological material, there is a certain variability between donors. This can be minimized by standardization, namely by including donors of similar ages, gestation age and by excluding donors with any prior conditions or illnesses (e.g. diabetes) and fetal congenital anomalies. Moreover, we must also standardize hAM preparation and storage procedures to ensure the high quality of hAM-derived preparations.

In summary, hAM is a unique combination of stem-like cells and ECM and the development of novel hAM-derived preparations that are easy to use in the clinic and the standardization of preparation and storage procedures will ensure that hAM and hAM-derived preparations reach their full potential and become truly indispensable in regenerative medicine.

References

1. Dzobo K, Thomford NE, Senthebane DA, Shipanga H, Rowe A, Dandara C, Pillay M, Motaung KSCM (2018) Advances in regenerative medicine and tissue engineering: innovation and transformation of medicine. Stem Cells Int 2018:2495848. https://doi.org/10.1155/2018/2495848

2. Wassmer C-H, Berishvili E (2020) Immunomodulatory properties of amniotic membrane derivatives and their potential in regenerative medicine. Curr Diab Rep 20:31. https://doi.org/10.1007/s11892-020-01316-w
3. Giwa S, Lewis JK, Alvarez L, Langer R, Roth AE, Church GM, Markmann JF, Sachs DH, Chandraker A, Wertheim JA, Rothblatt M, Boyden ES, Eidbo E, Lee WPA, Pomahac B, Brandacher G, Weinstock DM, Elliott G, Nelson D, Acker JP, Uygun K, Schmalz B, Weegman BP, Tocchio A, Fahy GM, Storey KB, Rubinsky B, Bischof J, Elliott JAW, Woodruff TK, Morris GJ, Demirci U, Brockbank KGM, Woods EJ, Ben RN, Baust JG, Gao D, Fuller B, Rabin Y, Kravitz DC, Taylor MJ, Toner M (2017) The promise of organ and tissue preservation to transform medicine. Nat Biotechnol 35:530–542. https://doi.org/10.1038/nbt.3889
4. Mao AS, Mooney DJ (2015) Regenerative medicine: current therapies and future directions. Proc Natl Acad Sci U S A 112:14452–14459. https://doi.org/10.1073/pnas.1508520112
5. Atala A (2012) Regenerative medicine strategies. J Pediatr Surg 47:17–28. https://doi.org/10.1016/j.jpedsurg.2011.10.013
6. Parolini O, Alviano F, Bagnara GP, Bilic G, Bühring H-J, Evangelista M, Hennerbichler S, Liu B, Magatti M, Mao N, Miki T, Marongiu F, Nakajima H, Nikaido T, Portmann-Lanz CB, Sankar V, Soncini M, Stadler G, Surbek D, Takahashi TA, Redl H, Sakuragawa N, Wolbank S, Zeisberger S, Zisch A, Strom SC (2008) Concise review: isolation and characterization of cells from human term placenta: outcome of the first international Workshop on Placenta derived stem cells. Stem Cells 26:300–311. https://doi.org/10.1634/stemcells.2007-0594
7. Ramuta TŽ, Kreft ME (2018) Human amniotic membrane and amniotic membrane-derived cells: how far are we from their use in regenerative and reconstructive urology? Cell Transplant 27:77–92. https://doi.org/10.1177/0963689717725528
8. Han YM, Romero R, Kim J-S, Tarca AL, Kim SK, Draghici S, Kusanovic JP, Gotsch F, Mittal P, Hassan SS, Kim CJ (2008) Region-specific gene expression profiling: novel evidence for biological heterogeneity of the human amnion. Biol Reprod 79:954–961. https://doi.org/10.1095/biolreprod.108.069260
9. Kim SY, Romero R, Tarca AL, Bhatti G, Lee J, Chaiworapongsa T, Hassan SS, Kim CJ (2011) miR-143 regulation of prostaglandin-endoperoxidase synthase 2 in the amnion: implications for human parturition at term. PloS One 6:e24131. https://doi.org/10.1371/journal.pone.0024131
10. Centurione L, Passaretta F, Centurione MA, Munari SD, Vertua E, Silini A, Liberati M, Parolini O, Di Pietro R (2018) Mapping of the human placenta: experimental evidence of amniotic epithelial cell heterogeneity. Cell Transplant 27:12–22, https://doi.org/10.1177/0963689717725078
11. Banerjee A, Weidinger A, Hofer M, Steinborn R, Lindenmair A, Hennerbichler-Lugscheider S, Eibl J, Redl H, Kozlov AV, Wolbank S (2015) Different metabolic activity in placental and reflected regions of the human amniotic membrane. Placenta 36:1329–1332. https://doi.org/10.1016/j.placenta.2015.08.015
12. Banerjee A, Lindenmair A, Hennerbichler S, Steindorf P, Steinborn R, Kozlov AV, Redl H, Wolbank S, Weidinger A (2018) Cellular and site-specific mitochondrial characterization of vital human amniotic membrane. Cell Transplant 27:3–11. https://doi.org/10.1177/0963689717735332
13. Miki T, Lehmann T, Cai H, Stolz DB, Strom SC (2005) Stem cell characteristics of amniotic epithelial cells. Stem Cells 23:1549–1559. https://doi.org/10.1634/stemcells.2004-0357
14. Portmann-Lanz CB, Schoeberlein A, Huber A, Sager R, Malek A, Holzgreve W, Surbek DV (2006) Placental mesenchymal stem cells as potential autologous graft for pre- and perinatal neuroregeneration. Am J Obstet Gynecol 194:664–673. https://doi.org/10.1016/j.ajog.2006.01.101
15. Wolbank S, Peterbauer A, Fahrner M, Hennerbichler S, van Griensven M, Stadler G, Redl H, Gabriel C (2007) Dose-dependent immunomodulatory effect of human stem cells from amniotic membrane: a comparison with human mesenchymal stem cells from adipose tissue. Tissue Eng 13:1173–1183. https://doi.org/10.1089/ten.2006.0313

16. Ilancheran S, Michalska A, Peh G, Wallace EM, Pera M, Manuelpillai U (2007) Stem cells derived from human fetal membranes display multilineage differentiation potential. Biol Reprod 77:577–588. https://doi.org/10.1095/biolreprod.106.055244
17. Bilic G, Zeisberger SM, Mallik AS, Zimmermann R, Zisch AH (2008) Comparative characterization of cultured human term amnion epithelial and mesenchymal stromal cells for application in cell therapy. Cell Transplant 17:955–968. https://doi.org/10.3727/096368908786576507
18. Stadler G, Hennerbichler S, Lindenmair A, Peterbauer A, Hofer K, van Griensven M, Gabriel C, Redl H, Wolbank S (2008) Phenotypic shift of human amniotic epithelial cells in culture is associated with reduced osteogenic differentiation in vitro. Cytotherapy 10:743–752. https://doi.org/10.1080/14653240802345804
19. Evron A, Goldman S, Shalev E (2011) Human amniotic epithelial cells cultured in substitute serum medium maintain their stem cell characteristics for up to four passages. Int J Stem Cells 4:123–132. https://doi.org/10.15283/ijsc.2011.4.2.123
20. Bryzek A, Czekaj P, Plewka D, Komarska H, Tomsia M, Lesiak M, Sieroń AL, Sikora J, Kopaczka K (2013) Expression and co-expression of surface markers of pluripotency on human amniotic cells cultured in different growth media. Ginekol Pol 84:1012–1024. https://doi.org/10.17772/gp/1673
21. García-Castro IL, García-López G, Ávila-González D, Flores-Herrera H, Molina-Hernández A, Portillo W, Ramón-Gallegos E, Díaz NF (2015) Markers of pluripotency in human amniotic epithelial cells and their differentiation to progenitor of cortical neurons. PLoS One 10:e0146082. https://doi.org/10.1371/journal.pone.0146082
22. Resca E, Zavatti M, Maraldi T, Bertoni L, Beretti F, Guida M, La Sala GB, Guillot PV, David AL, Sebire NJ, De Pol A, De Coppi P (2015) Enrichment in c-Kit improved differentiation potential of amniotic membrane progenitor/stem cells. Placenta 36:18–26. https://doi.org/10.1016/j.placenta.2014.11.002
23. Ding C, Li H, Wang Y, Wang F, Wu H, Chen R, Lv J, Wang W, Huang B (2017) Different therapeutic effects of cells derived from human amniotic membrane on premature ovarian aging depend on distinct cellular biological characteristics. Stem Cell Res Ther 8:173. https://doi.org/10.1186/s13287-017-0613-3
24. Jiang L-W, Chen H, Lu H (2016) Using human epithelial amnion cells in human de-epidermized dermis for skin regeneration. J Dermatol Sci 81:26–34. https://doi.org/10.1016/j.jdermsci.2015.10.018
25. Maymó JL, Riedel R, Pérez-Pérez A, Magatti M, Maskin B, Dueñas JL, Parolini O, Sánchez-Margalet V, Varone CL (2018) Proliferation and survival of human amniotic epithelial cells during their hepatic differentiation. PloS One 13:e0191489. https://doi.org/10.1371/journal.pone.0191489
26. Simat SF, Chua KH, Abdul Rahman H, Tan AE, Tan GC (2008) The stemness gene expression of cultured human amniotic epithelial cells in serial passages. Med J Malaysia 63 Suppl A:53–54
27. Liu T, Wu J, Huang Q, Hou Y, Jiang Z, Zang S, Guo L (2008) Human amniotic epithelial cells ameliorate behavioral dysfunction and reduce infarct size in the rat middle cerebral artery occlusion model. Shock 29:603–611. https://doi.org/10.1097/SHK.0b013e318157e845
28. Zhou K, Koike C, Yoshida T, Okabe M, Fathy M, Kyo S, Kiyono T, Saito S, Nikaido T (2013) Establishment and characterization of immortalized human amniotic epithelial cells. Cell Reprogramm 15:55–67. https://doi.org/10.1089/cell.2012.0021
29. Kim M-S, Yu JH, Lee M-Y, Kim AL, Jo MH, Kim M, Cho S-R, Kim Y-H (2016) Differential expression of extracellular matrix and adhesion molecules in fetal-origin amniotic epithelial cells of preeclamptic pregnancy. PLoS One 11:e0156038. https://doi.org/10.1371/journal.pone.0156038
30. Zou G, Liu T, Guo L, Huang Y, Feng Y, Duan T (2018) MicroRNA-32 silences WWP2 expression to maintain the pluripotency of human amniotic epithelial stem cells and β islet-like cell differentiation. Int J Mol Med 41:1983–1991. https://doi.org/10.3892/ijmm.2018.3436

31. Miki T, Mitamura K, Ross MA, Stolz DB, Strom SC (2007) Identification of stem cell marker-positive cells by immunofluorescence in term human amnion. J Reprod Immunol 75:91–96. https://doi.org/10.1016/j.jri.2007.03.017
32. Izumi M, Pazin BJ, Minervini CF, Gerlach J, Ross MA, Stolz DB, Turner ME, Thompson RL, Miki T (2009) Quantitative comparison of stem cell marker-positive cells in fetal and term human amnion. J Reprod Immunol 81:39–43. https://doi.org/10.1016/j.jri.2009.02.007
33. Miki T (2018) Stem cell characteristics and the therapeutic potential of amniotic epithelial cells. Am J Reprod Immunol 80:e13003. https://doi.org/10.1111/aji.13003
34. Cargnoni A, Di Marcello M, Campagnol M, Nassuato C, Albertini A, Parolini O (2009) Amniotic membrane patching promotes ischemic rat heart repair. Cell Transplant 18:1147–1159. https://doi.org/10.3727/096368909X12483162196764
35. Kakishita K, Nakao N, Sakuragawa N, Itakura T (2003) Implantation of human amniotic epithelial cells prevents the degeneration of nigral dopamine neurons in rats with 6-hydroxydopamine lesions. Brain Res 980:48–56. https://doi.org/10.1016/s0006-8993(03)02875-0
36. Sakuragawa N, Enosawa S, Ishii T, Thangavel R, Tashiro T, Okuyama T, Suzuki S (2000) Human amniotic epithelial cells are promising transgene carriers for allogeneic cell transplantation into liver. J Hum Genet 45:171–176. https://doi.org/10.1007/s100380050205
37. Takashima S, Ise H, Zhao P, Akaike T, Nikaido T (2004) Human amniotic epithelial cells possess hepatocyte-like characteristics and functions. Cell Struct Funct 29:73–84. https://doi.org/10.1247/csf.29.73
38. Strom SC, Bruzzone P, Cai H, Ellis E, Lehmann T, Mitamura K, Miki T (2006) Hepatocyte transplantation: clinical experience and potential for future use. Cell Transplant 15(Suppl 1):S105-110. https://doi.org/10.3727/000000006783982395
39. Miki T, Marongiu F, Ellis ECS, Dorko K, Mitamura K, Ranade A, Gramignoli R, Davila J, Strom SC (2009) Production of hepatocyte-like cells from human amnion. Methods Mol Biol 481:155–168. https://doi.org/10.1007/978-1-59745-201-4_13
40. Zhou J, Yu G, Cao C, Pang J, Chen X (2011) Bone morphogenetic protein-7 promotes chondrogenesis in human amniotic epithelial cells. Int Orthop 35:941–948. https://doi.org/10.1007/s00264-010-1116-3
41. Sakuragawa N, Kakinuma K, Kikuchi A, Okano H, Uchida S, Kamo I, Kobayashi M, Yokoyama Y (2004) Human amnion mesenchyme cells express phenotypes of neuroglial progenitor cells. J Neurosci Res 78:208–214. https://doi.org/10.1002/jnr.20257
42. Sankar V, Muthusamy R (2003) Role of human amniotic epithelial cell transplantation in spinal cord injury repair research. Neuroscience 118:11–17. https://doi.org/10.1016/s0306-4522(02)00929-6
43. Kakishita K, Elwan MA, Nakao N, Itakura T, Sakuragawa N (2000) Human amniotic epithelial cells produce dopamine and survive after implantation into the striatum of a rat model of Parkinson's disease: a potential source of donor for transplantation therapy. Exp Neurol 165:27–34. https://doi.org/10.1006/exnr.2000.7449
44. Elwan MA, Sakuragawa N (1997) Evidence for synthesis and release of catecholamines by human amniotic epithelial cells. Neuroreport 8:3435–3438. https://doi.org/10.1097/00001756-199711100-00004
45. Carbone A, Paracchini V, Castellani S, Di Gioia S, Seia M, Colombo C, Conese M (2014) Human amnion-derived cells: prospects for the treatment of lung diseases. Curr Stem Cell Res Ther 9:297–305. https://doi.org/10.2174/1574888x0904140429142451
46. Wei JP, Zhang TS, Kawa S, Aizawa T, Ota M, Akaike T, Kato K, Konishi I, Nikaido T (2003) Human amnion-isolated cells normalize blood glucose in streptozotocin-induced diabetic mice. Cell Transplant 12:545–552. https://doi.org/10.3727/000000003108747000
47. Díaz-Prado S, Rendal-Vázquez ME, Muiños-López E, Hermida-Gómez T, Rodríguez-Cabarcos M, Fuentes-Boquete I, de Toro FJ, Blanco FJ (2010) Potential use of the human amniotic membrane as a scaffold in human articular cartilage repair. Cell Tissue Bank 11:183–195. https://doi.org/10.1007/s10561-009-9144-1

48. Koike C, Zhou K, Takeda Y, Fathy M, Okabe M, Yoshida T, Nakamura Y, Kato Y, Nikaido T (2014) Characterization of amniotic stem cells. Cell Reprogramm 16:298–305. https://doi.org/10.1089/cell.2013.0090
49. Tabatabaei M, Mosaffa N, Nikoo S, Bozorgmehr M, Ghods R, Kazemnejad S, Rezania S, Keshavarzi B, Arefi S, Ramezani-Tehrani F, Mirzadegan E, Zarnani A-H (2014) Isolation and partial characterization of human amniotic epithelial cells: the effect of trypsin. Avicenna J Med Biotechnol 6:10–20
50. Yu S-C, Xu Y-Y, Li Y, Xu B, Sun Q, Li F, Zhang X-G (2015) Construction of tissue engineered skin with human amniotic mesenchymal stem cells and human amniotic epithelial cells. Eur Rev Med Pharmacol Sci 19:4627–4635
51. Roy R, Kukucka M, Messroghli D, Kunkel D, Brodarac A, Klose K, Geißler S, Becher PM, Kang SK, Choi Y-H, Stamm C (2015) Epithelial-to-mesenchymal transition enhances the cardioprotective capacity of human amniotic epithelial cells. Cell Transplant 24:985–1002. https://doi.org/10.3727/096368913X675151
52. Topoluk N, Hawkins R, Tokish J, Mercuri J (2017) Amniotic mesenchymal stromal cells exhibit preferential osteogenic and chondrogenic differentiation and enhanced matrix production compared with adipose mesenchymal stromal cells. Am J Sports Med 45:2637–2646. https://doi.org/10.1177/0363546517706138
53. Pratama G, Vaghjiani V, Tee JY, Liu YH, Chan J, Tan C, Murthi P, Gargett C, Manuelpillai U (2011) Changes in culture expanded human amniotic epithelial cells: implications for potential therapeutic applications. PLoS One 6:e26136. https://doi.org/10.1371/journal.pone.0026136
54. Dominici M, Le Blanc K, Mueller I, Slaper-Cortenbach I, Marini F, Krause D, Deans R, Keating A, Prockop D, Horwitz E (2006) Minimal criteria for defining multipotent mesenchymal stromal cells. The International Society for Cellular Therapy position statement. Cytotherapy 8:315–317. https://doi.org/10.1080/14653240600855905
55. Niknejad H, Peirovi H, Jorjani M, Ahmadiani A, Ghanavi J, Seifalian AM (2008) Properties of the amniotic membrane for potential use in tissue engineering. Eur Cell Mater 15:88–99. https://doi.org/10.22203/ecm.v015a07
56. Lee P-H, Tu C-T, Hsiao C-C, Tsai M-S, Ho C-M, Cheng N-C, Hung T-M, Shih DT-B (2016) Antifibrotic activity of human placental amnion membrane-derived CD34+ mesenchymal stem/progenitor cell transplantation in mice with thioacetamide-induced liver injury. Stem Cells Transl Med 5:1473–1484. https://doi.org/10.5966/sctm.2015-0343
57. Soncini M, Vertua E, Gibelli L, Zorzi F, Denegri M, Albertini A, Wengler GS, Parolini O (2007) Isolation and characterization of mesenchymal cells from human fetal membranes. J Tissue Eng Regen Med 1:296–305. https://doi.org/10.1002/term.40
58. Alviano F, Fossati V, Marchionni C, Arpinati M, Bonsi L, Franchina M, Lanzoni G, Cantoni S, Cavallini C, Bianchi F, Tazzari PL, Pasquinelli G, Foroni L, Ventura C, Grossi A, Bagnara GP (2007) Term amniotic membrane is a high throughput source for multipotent mesenchymal stem cells with the ability to differentiate into endothelial cells in vitro. BMC Dev Biol 7:11. https://doi.org/10.1186/1471-213X-7-11
59. Zhao P, Ise H, Hongo M, Ota M, Konishi I, Nikaido T (2005) Human amniotic mesenchymal cells have some characteristics of cardiomyocytes. Transplantation 79:528–535. https://doi.org/10.1097/01.tp.0000149503.92433.39
60. Kim J, Kang HM, Kim H, Kim MR, Kwon HC, Gye MC, Kang SG, Yang HS, You J (2007) Ex vivo characteristics of human amniotic membrane-derived stem cells. Cloning Stem Cells 9:581–594. https://doi.org/10.1089/clo.2007.0027
61. Iaffaldano L, Nardelli C, Raia M, Mariotti E, Ferrigno M, Quaglia F, Labruna G, Capobianco V, Capone A, Maruotti GM, Pastore L, Di Noto R, Martinelli P, Sacchetti L, Del Vecchio L (2013) High aminopeptidase N/CD13 levels characterize human amniotic mesenchymal stem cells and drive their increased adipogenic potential in obese women. Stem Cells Dev 22:2287–2297. https://doi.org/10.1089/scd.2012.0499
62. Battula VL, Treml S, Abele H, Bühring H-J (2008) Prospective isolation and characterization of mesenchymal stem cells from human placenta using a frizzled-9-specific monoclonal antibody. Differ Res Biol Divers 76:326–336. https://doi.org/10.1111/j.1432-0436.2007.00225.x

63. Nogami M, Tsuno H, Koike C, Okabe M, Yoshida T, Seki S, Matsui Y, Kimura T Nikaido T (2012) Isolation and characterization of human amniotic mesenchymal stem cells and their chondrogenic differentiation. Transplantation 93:1221–1228. https://doi.org/10.1097/TP.0b013e3182529b76
64. Ghamari S-H, Abbasi-Kangevari M, Tayebi T, Bahrami S, Niknejad H (2020) The bottlenecks in translating placenta-derived amniotic epithelial and mesenchymal stromal cells into the clinic: current discrepancies in marker reports. Front Bioeng Biotechnol 8:180. https://doi.org/10.3389/fbioe.2020.00180
65. Roubelakis M, Trohatou O, Anagnou N (2012) Amniotic fluid and amniotic membrane stem cells: marker discovery. Stem Cells Int 2012:107836. https://doi.org/10.1155/2012/107836
66. Magatti M, Pianta S, Silini A, Parolini O (2016) Isolation, culture, and phenotypic characterization of mesenchymal stromal cells from the amniotic membrane of the human term placenta. Methods Mol Biol 1416:233–244. https://doi.org/10.1007/978-1-4939-3584-0_13
67. Wei JP, Nawata M, Wakitani S, Kametani K, Ota M, Toda A, Konishi I, Ebara S, Nikaido T (2009) Human amniotic mesenchymal cells differentiate into chondrocytes. Clon Stem Cells 11:19–26. https://doi.org/10.1089/clo.2008.0027
68. Naseer N, Bashir S, Latief N, Latif F, Khan SN, Riazuddin S (2018) Human amniotic membrane as differentiating matrix for in vitro chondrogenesis. Regen Med 13:821–832. https://doi.org/10.2217/rme-2018-0017
69. Ventura C, Cantoni S, Bianchi F, Lionetti V, Cavallini C, Scarlata I, Foroni L, Maioli M, Bonsi L, Alviano F, Fossati V, Bagnara GP, Pasquinelli G, Recchia FA, Perbellini A (2007) Hyaluronan mixed esters of butyric and retinoic Acid drive cardiac and endothelial fate in term placenta human mesenchymal stem cells and enhance cardiac repair in infarcted rat hearts. J Biol Chem 282:14243–14252. https://doi.org/10.1074/jbc.M609350200
70. Arrizabalaga JH, Nollert MU (2018) Human amniotic membrane: a versatile scaffold for tissue engineering. ACS Biomater Sci Eng 4:2226–2236. https://doi.org/10.1021/acsbiomaterials.8b00015
71. Cirman T, Beltram M, Schollmayer P, Rožman P, Kreft ME (2014) Amniotic membrane properties and current practice of amniotic membrane use in ophthalmology in Slovenia. Cell Tissue Bank 15:177–192. https://doi.org/10.1007/s10561-013-9417-6
72. Parry S, Strauss JF (1998) Premature rupture of the fetal membranes. N Engl J Med 338:663–670. https://doi.org/10.1056/NEJM199803053381006
73. Lei J, Priddy LB, Lim JJ, Massee M, Koob TJ (2017) Identification of extracellular matrix components and biological factors in micronized dehydrated human amnion/chorion membrane. Adv Wound Care 6:43–53. https://doi.org/10.1089/wound.2016.0699
74. Meinert M, Eriksen GV, Petersen AC, Helmig RB, Laurent C, Uldbjerg N, Malmström A (2001) Proteoglycans and hyaluronan in human fetal membranes. Am J Obstet Gynecol 184:679–685. https://doi.org/10.1067/mob.2001.110294
75. Takashima S, Yasuo M, Sanzen N, Sekiguchi K, Okabe M, Yoshida T, Toda A, Nikaido T (2008) Characterization of laminin isoforms in human amnion. Tissue Cell 40:75–81. https://doi.org/10.1016/j.tice.2007.09.001
76. Malak TM, Ockleford CD, Bell SC, Dalgleish R, Bright N, Macvicar J (1993) Confocal immunofluorescence localization of collagen types I, III, IV, V and VI and their ultrastructural organization in term human fetal membranes. Placenta 14:385–406. https://doi.org/10.1016/s0143-4004(05)80460-6
77. Alitalo K, Kurkinen M, Vaheri A, Krieg T, Timpl R (1980) Extracellular matrix components synthesized by human amniotic epithelial cells in culture. Cell 19:1053–1062. https://doi.org/10.1016/0092-8674(80)90096-3
78. Hieber AD, Corcino D, Motosue J, Sandberg LB, Roos PJ, Yu SY, Csiszar K, Kagan HM, Boyd CD, Bryant-Greenwood GD (1997) Detection of elastin in the human fetal membranes: proposed molecular basis for elasticity. Placenta 18:301–312. https://doi.org/10.1016/s0143-4004(97)80065-3
79. Niknejad H, Khayat-Khoei M, Peirovi H, Abolghasemi H (2014) Human amniotic epithelial cells induce apoptosis of cancer cells: a new anti-tumor therapeutic strategy. Cytotherapy 16:33–40. https://doi.org/10.1016/j.jcyt.2013.07.005

80. Kubo M, Sonoda Y, Muramatsu R, Usui M (2001) Immunogenicity of human amniotic membrane in experimental xenotransplantation. Invest Ophthalmol Vis Sci 42:1539–1546
81. Magatti M, Vertua E, Cargnoni A, Silini A, Parolini O (2018) The immunomodulatory properties of amniotic cells: the two sides of the coin. Cell Transplant 27:31–44. https://doi.org/10.1177/0963689717742819
82. Olczyk P, Mencner Ł, Komosinska-Vassev K (2014) The role of the extracellular matrix components in cutaneous wound healing. BioMed Res Int 2014:747584. https://doi.org/10.1155/2014/747584
83. Broughton G, Janis JE, Attinger CE (2006) The basic science of wound healing. Plast Reconstr Surg 117:12S-34S. https://doi.org/10.1097/01.prs.0000225430.42531.c2
84. Koizumi NJ, Inatomi TJ, Sotozono CJ, Fullwood NJ, Quantock AJ, Kinoshita S (2000) Growth factor mRNA and protein in preserved human amniotic membrane. Curr Eye Res 20:173–177
85. Jin E, Kim T-H, Han S, Kim S-W (2016) Amniotic epithelial cells promote wound healing in mice through high epithelialization and engraftment. J Tissue Eng Regen Med 10:613–622. https://doi.org/10.1002/term.2069
86. Gicquel J-J, Dua HS, Brodie A, Mohammed I, Suleman H, Lazutina E, James DK, Hopkinson A (2009) Epidermal growth factor variations in amniotic membrane used for ex vivo tissue constructs. Tissue Eng Part A 15:1919–1927. https://doi.org/10.1089/ten.tea.2008.0432
87. Hodde JP, Johnson CE (2007) Extracellular matrix as a strategy for treating chronic wounds. Am J Clin Dermatol 8:61–66. https://doi.org/10.2165/00128071-200708020-00001
88. Insausti CL, Alcaraz A, García-Vizcaíno EM, Mrowiec A, López-Martínez MC, Blanquer M, Piñero A, Majado MJ, Moraleda JM, Castellanos G, Nicolás FJ (2010) Amniotic membrane induces epithelialization in massive posttraumatic wounds. Wound Repair Regen Off Publ Wound Heal Soc Eur Tissue Repair Soc 18:368–377. https://doi.org/10.1111/j.1524-475X.2010.00604.x
89. Riau AK, Beuerman RW, Lim LS, Mehta JS (2010) Preservation, sterilization and de-epithelialization of human amniotic membrane for use in ocular surface reconstruction. Biomaterials 31:216–225. https://doi.org/10.1016/j.biomaterials.2009.09.034
90. Rahman MS, Islam R, Rana MM, Spitzhorn L-S, Rahman MS, Adjaye J, Asaduzzaman SM (2019) Characterization of burn wound healing gel prepared from human amniotic membrane and Aloe vera extract. BMC Complement Altern Med 19:115. https://doi.org/10.1186/s12906-019-2525-5
91. Rana MM, Rahman MS, Ullah MA, Siddika A, Hossain ML, Akhter MS, Hasan MZ, Asaduzzaman SM (2020) Amnion and collagen-based blended hydrogel improves burn healing efficacy on a rat skin wound model in the presence of wound dressing biomembrane. Biomed Mater Eng 31:1–17. https://doi.org/10.3233/BME-201076
92. Jones RE, Foster DS, Hu MS, Longaker MT (2019) Wound healing and fibrosis: current stem cell therapies. Transfusion (Paris) 59:884–892. https://doi.org/10.1111/trf.14836
93. White ES, Mantovani AR (2013) Inflammation, wound repair, and fibrosis: reassessing the spectrum of tissue injury and resolution. J Pathol 229:141–144. https://doi.org/10.1002/path.4126
94. Giannandrea M, Parks WC (2014) Diverse functions of matrix metalloproteinases during fibrosis. Dis Model Mech 7:193–203. https://doi.org/10.1242/dmm.012062
95. Koh JW, Shin YJ, Oh JY, Kim MK, Ko JH, Hwang JM, Wee WR, Lee JH (2007) The expression of TIMPs in cryo-preserved and freeze-dried amniotic membrane. Curr Eye Res 32:611–616. https://doi.org/10.1080/02713680701459441
96. Hao Y, Ma DH, Hwang DG, Kim WS, Zhang F (2000) Identification of antiangiogenic and antiinflammatory proteins in human amniotic membrane. Cornea 19:348–352. https://doi.org/10.1097/00003226-200005000-00018
97. Rowe TF, King LA, MacDonald PC, Casey ML (1997) Tissue inhibitor of metalloproteinase-1 and tissue inhibitor of metalloproteinase-2 expression in human amnion mesenchymal and epithelial cells. Am J Obstet Gynecol 176:915–921. https://doi.org/10.1016/s0002-9378(97)70621-5

98. Riley SC, Leask R, Denison FC, Wisely K, Calder AA, Howe DC (1999) Secretion of tissue inhibitors of matrix metalloproteinases by human fetal membranes, decidua and placenta at parturition. J Endocrinol 162:351–359. https://doi.org/10.1677/joe.0.1620351
99. SantAnna LB, Hage R, Cardoso MAG, Arisawa EAL, Cruz MM, Parolini O, Cargnoni A, SantAnna N (2016) Antifibrotic effects of human amniotic membrane transplantation in established biliary fibrosis induced in rats. Cell Transplant 25:2245–2257. https://doi.org/10.3727/096368916X692645
100. Tseng SC, Li DQ, Ma X (1999) Suppression of transforming growth factor-beta isoforms, TGF-beta receptor type II, and myofibroblast differentiation in cultured human corneal and limbal fibroblasts by amniotic membrane matrix. J Cell Physiol 179:325–335. https://doi.org/10.1002/(SICI)1097-4652(199906)179:3%3c325::AID-JCP10%3e3.0.CO;2-X
101. Solomon A, Wajngarten M, Alviano F, Anteby I, Elchalal U, Pe'er J, Levi-Schaffer F (2005) Suppression of inflammatory and fibrotic responses in allergic inflammation by the amniotic membrane stromal matrix. Clin Exp Allergy J Br Soc Allergy Clin Immunol 35:941–948. https://doi.org/10.1111/j.1365-2222.2005.02285.x
102. Tseng SCG, Espana EM, Kawakita T, Di Pascuale MA, Li W, He H, Liu T-S, Cho T-H, Gao Y-Y, Yeh L-K, Liu C-Y (2004) How does amniotic membrane work? Ocul Surf 2:177–187. https://doi.org/10.1016/s1542-0124(12)70059-9
103. Hortensius RA, Ebens JH, Harley BAC (2016) Immunomodulatory effects of amniotic membrane matrix incorporated into collagen scaffolds. J Biomed Mater Res A 104:1332–1342. https://doi.org/10.1002/jbm.a.35663
104. Wolbank S, Hildner F, Redl H, van Griensven M, Gabriel C, Hennerbichler S (2009) Impact of human amniotic membrane preparation on release of angiogenic factors. J Tissue Eng Regen Med 3:651–654. https://doi.org/10.1002/term.207
105. Burgos H (1986) Angiogenic factor from human term placenta. Purification and partial characterization. Eur J Clin Invest 16:486–493. https://doi.org/10.1111/j.1365-2362.1986.tb02166.x
106. Niknejad H, Paeini-Vayghan G, Tehrani FA, Khayat-Khoei M, Peirovi H (2013) Side dependent effects of the human amnion on angiogenesis. Placenta 34:340–345. https://doi.org/10.1016/j.placenta.2013.02.001
107. Koob TJ, Lim JJ, Massee M, Zabek N, Denozière G (2014) Properties of dehydrated human amnion/chorion composite grafts: Implications for wound repair and soft tissue regeneration. J Biomed Mater Res B Appl Biomater 102:1353–1362. https://doi.org/10.1002/jbm.b.33141
108. Danieli P, Malpasso G, Ciuffreda MC, Cervio E, Calvillo L, Copes F, Pisano F, Mura M, Kleijn L, de Boer RA, Viarengo G, Rosti V, Spinillo A, Roccio M, Gnecchi M (2015) Conditioned medium from human amniotic mesenchymal stromal cells limits infarct size and enhances angiogenesis. Stem Cells Transl Med 4:448–458. https://doi.org/10.5966/sctm.2014-0253
109. Nasiry D, Khalatbary AR, Abdollahifar M-A, Amini A, Bayat M, Noori A, Piryaei A (2020) Engraftment of bioengineered three-dimensional scaffold from human amniotic membrane-derived extracellular matrix accelerates ischemic diabetic wound healing. Arch Dermatol Res. https://doi.org/10.1007/s00403-020-02137-3
110. Magatti M, Caruso M, De Munari S, Vertua E, De D, Manuelpillai U, Parolini O (2015) Human amniotic membrane-derived mesenchymal and epithelial cells exert different effects on monocyte-derived dendritic cell differentiation and function. Cell Transplant 24:1733–1752. https://doi.org/10.3727/096368914X684033
111. Banas RA, Trumpower C, Bentlejewski C, Marshall V, Sing G, Zeevi A (2008) Immunogenicity and immunomodulatory effects of amnion-derived multipotent progenitor cells. Hum Immunol 69:321–328. https://doi.org/10.1016/j.humimm.2008.04.007
112. Houlihan JM, Biro PA, Harper HM, Jenkinson HJ, Holmes CH (1995) The human amnion is a site of MHC class Ib expression: evidence for the expression of HLA-E and HLA-G. J Immunol 154:5665–5674
113. Insausti CL, Blanquer M, García-Hernández AM, Castellanos G, Moraleda JM (2014) Amniotic membrane-derived stem cells: immunomodulatory properties and potential clinical application. Stem Cells Cloning Adv Appl 7:53–63. https://doi.org/10.2147/SCCAA.S58696

114. Magatti M, Vertua E, De Munari S, Caro M, Caruso M, Silini A, Delgado M, Parolini O (2017) Human amnion favours tissue repair by inducing the M1-to-M2 switch and enhancing M2 macrophage features. J Tissue Eng Regen Med 11:2895–2911. https://doi.org/10.1002/term.2193
115. Li H, Niederkorn JY, Neelam S, Mayhew E, Word RA, McCulley JP, Alizadeh H (2005) Immunosuppressive factors secreted by human amniotic epithelial cells. Invest Ophthalmol Vis Sci 46:900–907. https://doi.org/10.1167/iovs.04-0495
116. Navas A, Magaña-Guerrero FS, Domínguez-López A, Chávez-García C, Partido G, Graue-Hernández EO, Sánchez-García FJ, Garfias Y (2018) Anti-inflammatory and anti-fibrotic effects of human amniotic membrane mesenchymal stem cells and their potential in corneal repair. Stem Cells Transl Med 7:906–917. https://doi.org/10.1002/sctm.18-0042
117. Li J, Koike-Soko C, Sugimoto J, Yoshida T, Okabe M, Nikaido T (2015) Human amnion-derived stem cells have immunosuppressive properties on NK cells and monocytes. Cell Transplant 24:2065–2076. https://doi.org/10.3727/096368914X685230
118. Liu YH, Vaghjiani V, Tee JY, To K, Cui P, Oh DY, Manuelpillai U, Toh B-H, Chan J (2012) Amniotic epithelial cells from the human placenta potently suppress a mouse model of multiple sclerosis. PloS One 7:e35758. https://doi.org/10.1371/journal.pone.0035758
119. Moodley Y, Ilancheran S, Samuel C, Vaghjiani V, Atienza D, Williams ED, Jenkin G, Wallace E, Trounson A, Manuelpillai U (2010) Human amnion epithelial cell transplantation abrogates lung fibrosis and augments repair. Am J Respir Crit Care Med 182:643–651. https://doi.org/10.1164/rccm.201001-0014OC
120. Alhomrani M, Correia J, Zavou M, Leaw B, Kuk N, Xu R, Saad MI, Hodge A, Greening DW, Lim R, Sievert W (2017) The human amnion epithelial cell secretome decreases hepatic fibrosis in mice with chronic liver fibrosis. Front Pharmacol 8:748. https://doi.org/10.3389/fphar.2017.00748
121. Tan JL, Chan ST, Wallace EM, Lim R (2014) Human amnion epithelial cells mediate lung repair by directly modulating macrophage recruitment and polarization. Cell Transplant 23:319–328. https://doi.org/10.3727/096368912X661409
122. Magatti M, De Munari S, Vertua E, Gibelli L, Wengler GS, Parolini O (2008) Human amnion mesenchyme harbors cells with allogeneic T-cell suppression and stimulation capabilities. Stem Cells 26:182–192. https://doi.org/10.1634/stemcells.2007-0491
123. Roelen DL, van der Mast BJ, in't Anker PS, Kleijburg C, Eikmans M, van Beelen E, de Groot-Swings GMJS, Fibbe WE, Kanhai HHH, Scherjon SA, Claas FHJ (2009) Differential immunomodulatory effects of fetal versus maternal multipotent stromal cells. Hum Immunol 70:16–23. https://doi.org/10.1016/j.humimm.2008.10.016
124. Pianta S, Bonassi Signoroni P, Muradore I, Rodrigues MF, Rossi D, Silini A, Parolini O (2015) Amniotic membrane mesenchymal cells-derived factors skew T cell polarization toward treg and downregulate Th1 and Th17 cells subsets. Stem Cell Rev 11:394–407. https://doi.org/10.1007/s12015-014-9558-4
125. Parolini O, Souza-Moreira L, O'Valle F, Magatti M, Hernandez-Cortes P, Gonzalez-Rey E, Delgado M (2014) Therapeutic effect of human amniotic membrane-derived cells on experimental arthritis and other inflammatory disorders. Arthritis Rheumatol 66:327–339. https://doi.org/10.1002/art.38206
126. Pianta S, Magatti M, Vertua E, Bonassi Signoroni P, Muradore I, Nuzzo AM, Rolfo A, Silini A, Quaglia F, Todros T, Parolini O (2016) Amniotic mesenchymal cells from pre-eclamptic placentae maintain immunomodulatory features as healthy controls. J Cell Mol Med 20:157–169. https://doi.org/10.1111/jcmm.12715
127. Kang JW, Koo HC, Hwang SY, Kang SK, Ra JC, Lee MH, Park YH (2012) Immunomodulatory effects of human amniotic membrane-derived mesenchymal stem cells. J Vet Sci 13:23–31. https://doi.org/10.4142/jvs.2012.13.1.23
128. Karlsson H, Erkers T, Nava S, Ruhm S, Westgren M, Ringdén O (2012) Stromal cells from term fetal membrane are highly suppressive in allogeneic settings in vitro. Clin Exp Immunol 167:543–555. https://doi.org/10.1111/j.1365-2249.2011.04540.x

129. Bu S, Zhang Q, Wang Q, Lai D (2017) Human amniotic epithelial cells inhibit growth of epithelial ovarian cancer cells via TGF-β1-mediated cell cycle arrest. Int J Oncol 51:1405–1414. https://doi.org/10.3892/ijo.2017.4123
130. Magatti M, De Munari S, Vertua E, Parolini O (2012) Amniotic membrane-derived cells inhibit proliferation of cancer cell lines by inducing cell cycle arrest. J Cell Mol Med 16:2208–2218. https://doi.org/10.1111/j.1582-4934.2012.01531.x
131. Ramuta TŽ, Jerman UD, Tratnjek L, Janev A, Magatti M, Vertua E, Bonassi Signoroni P, Silini AR, Parolini O, Kreft ME (2020) The cells and extracellular matrix of human amniotic membrane hinder the growth and invasive potential of bladder urothelial cancer cells. Front Bioeng Biotechnol 8:554530. https://doi.org/10.3389/fbioe.2020.554530
132. Riedel R, Pérez-Pérez A, Carmona-Fernández A, Jaime M, Casale R, Dueñas JL, Guadix P, Sánchez-Margalet V, Varone CL, Maymó JL (2019) Human amniotic membrane conditioned medium inhibits proliferation and modulates related microRNAs expression in hepatocarcinoma cells. Sci Rep 9:14193. https://doi.org/10.1038/s41598-019-50648-5
133. Mamede AC, Laranjo M, Carvalho MJ, Abrantes AM, Pires AS, Brito AF, Moura P, Maia CJ, Botelho MF (2014) Effect of amniotic membrane proteins in human cancer cell lines: an exploratory study. J Membr Biol 247:357–360. https://doi.org/10.1007/s00232-014-9642-3
134. Mamede AC, Guerra S, Laranjo M, Carvalho MJ, Oliveira RC, Gonçalves AC, Alves R, Prado Castro L, Sarmento-Ribeiro AB, Moura P, Abrantes AM, Maia CJ, Botelho MF (2015) Selective cytotoxicity and cell death induced by human amniotic membrane in hepatocellular carcinoma. Med Oncol 32:257. https://doi.org/10.1007/s12032-015-0702-z
135. Mamede AC, Guerra S, Laranjo M, Santos K, Carvalho MJ, Carvalheiro T, Moura P, Paiva A, Abrantes AM, Maia CJ, Botelho MF (2016) Oxidative stress, DNA, Cell Cycle/Cell cycle associated proteins and multidrug resistance proteins: targets of human amniotic membrane in hepatocellular carcinoma. Pathol Oncol Res 22:689–697. https://doi.org/10.1007/s12253-016-0053-x
136. Niknejad H, Yazdanpanah G, Ahmadiani A (2016) Induction of apoptosis, stimulation of cell-cycle arrest and inhibition of angiogenesis make human amnion-derived cells promising sources for cell therapy of cancer. Cell Tissue Res 363:599–608. https://doi.org/10.1007/s00441-016-2364-3
137. Kang N-H, Yi B-R, Lim SY, Hwang K-A, Baek YS, Kang K-S, Choi K-C (2012) Human amniotic membrane-derived epithelial stem cells display anticancer activity in BALB/c female nude mice bearing disseminated breast cancer xenografts. Int J Oncol 40:2022–2028. https://doi.org/10.3892/ijo.2012.1372
138. Niknejad H, Yazdanpanah G, Mirmasoumi M, Abolghasemi H, Peirovi H, Ahmadiani A (2013) Inhibition of HSP90 could be possible mechanism for anti-cancer property of amniotic membrane. Med Hypotheses 81:862–865. https://doi.org/10.1016/j.mehy.2013.08.018
139. Kim S-H, Bang SH, Kang SY, Park KD, Eom JH, Oh IU, Yoo SH, Kim C-W, Baek SY (2015) Human amniotic membrane-derived stromal cells (hAMSC) interact depending on breast cancer cell type through secreted molecules. Tissue Cell 47:10–16. https://doi.org/10.1016/j.tice.2014.10.003
140. Meng M-Y, Li L, Wang W-J, Liu F-F, Song J, Yang S-L, Tan J, Gao H, Zhao Y-Y, Tang W-W, Han R, Zhu K, Liao L-W, Hou Z-L (2019) Assessment of tumor promoting effects of amniotic and umbilical cord mesenchymal stem cells in vitro and in vivo. J Cancer Res Clin Oncol 145:1133–1146. https://doi.org/10.1007/s00432-019-02859-6
141. Tehrani FA, Modaresifar K, Azizian S, Niknejad H (2017) Induction of antimicrobial peptides secretion by IL-1β enhances human amniotic membrane for regenerative medicine. Sci Rep 7:17022. https://doi.org/10.1038/s41598-017-17210-7
142. King AE, Paltoo A, Kelly RW, Sallenave J-M, Bocking AD, Challis JRG (2007) Expression of natural antimicrobials by human placenta and fetal membranes. Placenta 28:161–169. https://doi.org/10.1016/j.placenta.2006.01.006
143. Ramuta TŽ, Starčič Erjavec M, Kreft ME (2020) Amniotic Membrane Preparation Crucially Affects Its Broad-Spectrum Activity Against Uropathogenic Bacteria. Front Microbiol 11:469. https://doi.org/10.3389/fmicb.2020.00469

144. Šket T, Ramuta TŽ, Starčič Erjavec M, Kreft ME (2019) Different effects of amniotic membrane homogenate on the growth of uropathogenic Escherichia coli, Staphylococcus aureus and Serratia marcescens. Infect Drug Resist 12:3365–3375. https://doi.org/10.2147/IDR.S215006
145. Talmi YP, Sigler L, Inge E, Finkelstein Y, Zohar Y (1991) Antibacterial properties of human amniotic membranes. Placenta 12:285–288. https://doi.org/10.1016/0143-4004(91)90010-d
146. Tehrani FA, Ahmadiani A, Niknejad H (2013) The effects of preservation procedures on antibacterial property of amniotic membrane. Cryobiology 67:293–298. https://doi.org/10.1016/j.cryobiol.2013.08.010
147. Yadav MK, Go YY, Kim SH, Chae S-W, Song J-J (2017) Antimicrobial and antibiofilm effects of human amniotic/chorionic membrane extract on Streptococcus pneumoniae. Front Microbiol 8:1948. https://doi.org/10.3389/fmicb.2017.01948
148. Mao Y, Hoffman T, Johnson A, Arnold Y, Danilkovitch A, Kohn J (2016) Human cryopreserved viable amniotic membrane inhibits the growth of bacteria associated with chronic wounds. J Diabet Complications 8:23–30
149. Mao Y, Hoffman T, Singh-Varma A, Duan-Arnold Y, Moorman M, Danilkovitch A, Kohn J (2017) Antimicrobial peptides secreted from human cryopreserved viable amniotic membrane contribute to its antibacterial activity. Sci Rep 7:13722. https://doi.org/10.1038/s41598-017-13310-6
150. Mao Y, Singh-Varma A, Hoffman T, Dhall S, Danilkovitch A, Kohn J (2018) The effect of cryopreserved human placental tissues on biofilm formation of wound-associated pathogens. J Funct Biomater 9:3. https://doi.org/10.3390/jfb9010003
151. Kjaergaard N, Hein M, Hyttel L, Helmig RB, Schønheyder HC, Uldbjerg N, Madsen H (2001) Antibacterial properties of human amnion and chorion in vitro. Eur J Obstet Gynecol Reprod Biol 94:224–229. https://doi.org/10.1016/s0301-2115(00)00345-6
152. Wang X, Xie J, Tan L, Huo J, Xie H (2012) Epithelium of human fresh amniotic membrane has antimicrobial effects in vitro. Afr J Microbiol Res 6:4533–4537. https://doi.org/10.5897/AJMR12.127
153. Klotman ME, Chang TL (2006) Defensins in innate antiviral immunity. Nat Rev Immunol 6:447–456. https://doi.org/10.1038/nri1860
154. Svinarich DM, Gomez R, Romero R (1997) Detection of human defensins in the placenta. Am J Reprod Immunol 38:252–255. https://doi.org/10.1111/j.1600-0897.1997.tb00511.x
155. Buhimschi IA, Jabr M, Buhimschi CS, Petkova AP, Weiner CP, Saed GM (2004) The novel antimicrobial peptide beta3-defensin is produced by the amnion: a possible role of the fetal membranes in innate immunity of the amniotic cavity. Am J Obstet Gynecol 191:1678–1687. https://doi.org/10.1016/j.ajog.2004.03.081
156. Denison FC, Kelly RW, Calder AA, Riley SC (1999) Secretory leukocyte protease inhibitor concentration increases in amniotic fluid with the onset of labour in women: characterization of sites of release within the uterus. J Endocrinol 161:299–306. https://doi.org/10.1677/joe.0.1610299
157. Zaga-Clavellina V, Ruiz M, Flores-Espinosa P, Vega-Sanchez R, Flores-Pliego A, Estrada-Gutierrez G, Sosa-Gonzalez I, Morales-Méndez I, Osorio-Caballero M (2012) Tissue-specific human beta-defensins (HBD)-1, HBD-2 and HBD-3 secretion profile from human amniochorionic membranes stimulated with Candida albicans in a two-compartment tissue culture system. Reprod Biol Endocrinol 10:70. https://doi.org/10.1186/1477-7827-10-70
158. Harder J, Meyer-Hoffert U, Teran LM, Schwichtenberg L, Bartels J, Maune S, Schröder JM (2000) Mucoid Pseudomonas aeruginosa, TNF-alpha, and IL-1beta, but not IL-6, induce human beta-defensin-2 in respiratory epithelia. Am J Respir Cell Mol Biol 22:714–721. https://doi.org/10.1165/ajrcmb.22.6.4023
159. Krisanaprakornkit S, Weinberg A, Perez CN, Dale BA (1998) Expression of the peptide antibiotic human beta-defensin 1 in cultured gingival epithelial cells and gingival tissue. Infect Immun 66:4222–4228
160. Zhang Q, Shimoya K, Moriyama A, Yamanaka K, Nakajima A, Nobunaga T, Koyama M, Azuma C, Murata Y (2001) Production of secretory leukocyte protease inhibitor by human

amniotic membranes and regulation of its concentration in amniotic fluid. Mol Hum Reprod 7:573–579. https://doi.org/10.1093/molehr/7.6.573
161. Kim HS, Cho JH, Park HW, Yoon H, Kim MS, Kim SC (2002) Endotoxin-neutralizing antimicrobial proteins of the human placenta. J Immunol 168:2356–2364. https://doi.org/10.4049/jimmunol.168.5.2356
162. Malhotra C, Jain AK (2014) Human amniotic membrane transplantation: different modalities of its use in ophthalmology. World J Transplant 4:111–121. https://doi.org/10.5500/wjt.v4.i2.111
163. Silini AR, Cargnoni A, Magatti M, Pianta S, Parolini O (2015) The long path of human placenta, and its derivatives, in regenerative medicine. Front Bioeng Biotechnol 3:162. https://doi.org/10.3389/fbioe.2015.00162
164. Walkden A (2020) Amniotic membrane transplantation in ophthalmology: an updated perspective. Clin Ophthalmol 14:2057–2072. https://doi.org/10.2147/OPTH.S208008
165. Kogan S, Sood A, Granick MS (2018) Amniotic membrane adjuncts and clinical applications in wound healing: a review of the literature. Wounds Compend Clin Res Pract 30:168–173
166. ElHeneidy H, Omran E, Halwagy A, Al-Inany H, Al-Ansary M, Gad A (2016) Amniotic membrane can be a valid source for wound healing. Int J Womens Health 8:225–231. https://doi.org/10.2147/IJWH.S96636
167. Su Y-N, Zhao D-Y, Li Y-H, Yu T-Q, Sun H, Wu X-Y, Zhou X-Q, Li J (2020) Human amniotic membrane allograft, a novel treatment for chronic diabetic foot ulcers: a systematic review and meta-analysis of randomised controlled trials. Int Wound J 17:753–764. https://doi.org/10.1111/iwj.13318
168. Fairbairn NG, Randolph MA, Redmond RW (2014) The clinical applications of human amnion in plastic surgery. J Plast Reconstr Aesthetic Surg JPRAS 67:662–675. https://doi.org/10.1016/j.bjps.2014.01.031
169. Riboh JC, Saltzman BM, Yanke AB, Cole BJ (2016) Human amniotic membrane-derived products in sports medicine: basic science, early results, and potential clinical applications. Am J Sports Med 44:2425–2434. https://doi.org/10.1177/0363546515612750
170. Huddleston HP, Cohn MR, Haunschild ED, Wong SE, Farr J, Yanke AB (2020) Amniotic product treatments: clinical and basic science evidence. Curr Rev Musculoskelet Med 13:148–154. https://doi.org/10.1007/s12178-020-09614-2
171. Xu H, Zhang J, Tsang KS, Yang H, Gao W-Q (2019) Therapeutic potential of human amniotic epithelial cells on injuries and disorders in the central nervous system. In: Stem Cells International. https://www.hindawi.com/journals/sci/2019/5432301/. Accessed 11 Feb 2021
172. Miki T (2016) A rational strategy for the use of amniotic epithelial stem cell therapy for liver diseases. Stem Cells Transl Med 5:405–409. https://doi.org/10.5966/sctm.2015-0304
173. Pietrosi G, Fernández-Iglesias A, Pampalone M, Ortega-Ribera M, Lozano JJ, García-Calderó H, Abad-Jordà L, Conaldi PG, Parolini O, Vizzini G, Luca A, Bosch J, Gracia-Sancho J (2020) Human amniotic stem cells improve hepatic microvascular dysfunction and portal hypertension in cirrhotic rats. Liver Int Off J Int Assoc Study Liver 40:2500–2514. https://doi.org/10.1111/liv.14610
174. Hodges R, Lim R, Jenkin G, Wallace E (2012) Amnion epithelial cells as a candidate therapy for acute and chronic lung injury. Stem Cells Int 2012:709763. https://doi.org/10.1155/2012/709763
175. Razavi Tousi SMT, Faghihi M, Nobakht M, Molazem M, Kalantari E, Darbandi Azar A, Aboutaleb N (2016) Improvement of heart failure by human amniotic mesenchymal stromal cell transplantation in Rats. J Tehran Heart Cent 11:123–138
176. Bollini S, Silini AR, Banerjee A, Wolbank S, Balbi C, Parolini O (2018) Cardiac restoration stemming from the placenta tree: insights from fetal and perinatal cell biology. Front Physiol 9:385. https://doi.org/10.3389/fphys.2018.00385
177. Cetinkaya B, Unek G, Kipmen-Korgun D, Koksoy S, Korgun ET (2019) Effects of human placental amnion derived mesenchymal stem cells on proliferation and apoptosis mechanisms in chronic kidney disease in the rat. Int J Stem Cells 12:151–161. https://doi.org/10.15283/ijsc18067

178. Volarevic V, Bojic S, Nurkovic J, Volarevic A, Ljujic B, Arsenijevic N, Lako M, Stojkovic M (2014) Stem cells as new agents for the treatment of infertility: current and future perspectives and challenges. BioMed Res Int 2014:507234. https://doi.org/10.1155/2014/507234
179. Wang J, Liu C, Fujino M, Tong G, Zhang Q, Li X-K, Yan H (2019) Stem cells as a resource for treatment of infertility-related diseases. Curr Mol Med 19:539–546. https://doi.org/10.2174/1566524019666190709172636
180. Liu R, Zhang X, Fan Z, Wang Y, Yao G, Wan X, Liu Z, Yang B, Yu L (2019) Human amniotic mesenchymal stem cells improve the follicular microenvironment to recover ovarian function in premature ovarian failure mice. Stem Cell Res Ther 10:299. https://doi.org/10.1186/s13287-019-1315-9
181. Li B, Zhang Q, Sun J, Lai D (2019) Human amniotic epithelial cells improve fertility in an intrauterine adhesion mouse model. Stem Cell Res Ther 10:257. https://doi.org/10.1186/s13287-019-1368-9
182. Song C-G, Zhang Y-Z, Wu H-N, Cao X-L, Guo C-J, Li Y-Q, Zheng M-H, Han H (2018) Stem cells: a promising candidate to treat neurological disorders. Neural Regen Res 13:1294–1304. https://doi.org/10.4103/1673-5374.235085
183. Teo L, Bourne JA (2014) A reproducible and translatable model of focal ischemia in the visual cortex of infant and adult marmoset monkeys. Brain Pathol 24:459–474. https://doi.org/10.1111/bpa.12129
184. Dong W, Chen H, Yang X, Guo L, Hui G (2010) Treatment of intracerebral haemorrhage in rats with intraventricular transplantation of human amniotic epithelial cells. Cell Biol Int 34:573–577. https://doi.org/10.1042/CBI20090248
185. Zhou H, Mu Z, Chen X, Shi Z, Zha Z, Liu Y, Xu Z (2015) HAEC in the treatment of brain hemorrhage: a preliminary observation in rabbits. Int J Clin Exp Pathol 8:6772–6778
186. Wu Z, Hui G, Lu Y, Wu X, Guo L (2006) Transplantation of human amniotic epithelial cells improves hindlimb function in rats with spinal cord injury. Chin Med J (Engl) 119:2101–2107
187. Roh D-H, Seo M-S, Choi H-S, Park S-B, Han H-J, Beitz AJ, Kang K-S, Lee J-H (2013) Transplantation of human umbilical cord blood or amniotic epithelial stem cells alleviates mechanical allodynia after spinal cord injury in rats. Cell Transplant 22:1577–1590. https://doi.org/10.3727/096368912X659907
188. Meng X, Li C, Dong Z, Liu J, Li W, Liu Y, Xue H, Chen D (2008) Co-transplantation of bFGF-expressing amniotic epithelial cells and neural stem cells promotes functional recovery in spinal cord-injured rats. Cell Biol Int 32:1546–1558. https://doi.org/10.1016/j.cellbi.2008.09.001
189. Xue H, Zhang X-Y, Liu J-M, Song Y, Li Y-F, Chen D (2013) Development of a chemically extracted acellular muscle scaffold seeded with amniotic epithelial cells to promote spinal cord repair. J Biomed Mater Res A 101:145–156. https://doi.org/10.1002/jbm.a.34311
190. Wang T-G, Xu J, Zhu A-H, Lu H, Miao Z-N, Zhao P, Hui G-Z, Wu W-J (2016) Human amniotic epithelial cells combined with silk fibroin scaffold in the repair of spinal cord injury. Neural Regen Res 11:1670–1677. https://doi.org/10.4103/1673-5374.193249
191. Yang X, Xue S, Dong W, Kong Y (2009) Therapeutic effect of human amniotic epithelial cell transplantation into the lateral ventricle of hemiparkinsonian rats. Chin Med J (Engl) 122:2449–2454
192. Yang X, Song L, Wu N, Liu Z, Xue S, Hui G (2010) An experimental study on intracerebroventricular transplantation of human amniotic epithelial cells in a rat model of Parkinson's disease. Neurol Res 32:1054–1059. https://doi.org/10.1179/016164110X12681290831207
193. Leaw B, Zhu D, Tan J, Muljadi R, Saad MI, Mockler JC, Wallace EM, Lim R, Tolcos M (2017) Human amnion epithelial cells rescue cell death via immunomodulation of microglia in a mouse model of perinatal brain injury. Stem Cell Res Ther 8:46. https://doi.org/10.1186/s13287-017-0496-3
194. Yawno T, Schuilwerve J, Moss TJM, Vosdoganes P, Westover AJ, Afandi E, Jenkin G, Wallace EM, Miller SL (2013) Human amnion epithelial cells reduce fetal brain injury in response to intrauterine inflammation. Dev Neurosci 35:272–282. https://doi.org/10.1159/000346683

195. Yawno T, Sabaretnam T, Li J, McDonald C, Lim R, Jenkin G, Wallace EM, Miller SL (2017) Human amnion epithelial cells protect against white matter brain injury after repeated endotoxin exposure in the preterm ovine fetus. Cell Transplant 26:541–553. https://doi.org/10.3727/096368916X693572
196. Barton SK, Melville JM, Tolcos M, Polglase GR, McDougall ARA, Azhan A, Crossley KJ, Jenkin G, Moss TJM (2015) Human amnion epithelial cells modulate ventilation-induced white matter pathology in preterm lambs. Dev Neurosci 37:338–348. https://doi.org/10.1159/000371415
197. Liu Y-H, Chan J, Vaghjiani V, Murthi P, Manuelpillai U, Toh B-H (2014) Human amniotic epithelial cells suppress relapse of corticosteroid-remitted experimental autoimmune disease. Cytotherapy 16:535–544. https://doi.org/10.1016/j.jcyt.2013.10.007
198. McDonald CA, Payne NL, Sun G, Moussa L, Siatskas C, Lim R, Wallace EM, Jenkin G, Bernard CCA (2015) Immunosuppressive potential of human amnion epithelial cells in the treatment of experimental autoimmune encephalomyelitis. J Neuroinflammation 12:112. https://doi.org/10.1186/s12974-015-0322-8
199. Kim K-S, Kim HS, Park J-M, Kim HW, Park M-K, Lee H-S, Lim DS, Lee TH, Chopp M, Moon J (2013) Long-term immunomodulatory effect of amniotic stem cells in an Alzheimer's disease model. Neurobiol Aging 34:2408–2420. https://doi.org/10.1016/j.neurobiolaging.2013.03.029
200. Kim KY, Suh Y-H, Chang K-A (2020) Therapeutic effects of human amniotic epithelial stem cells in a transgenic mouse model of alzheimer's disease. Int J Mol Sci 21:2658. https://doi.org/10.3390/ijms21072658
201. Shu J, He X, Li H, Liu X, Qiu X, Zhou T, Wang P, Huang X (2018) The beneficial effect of human amnion mesenchymal cells in inhibition of inflammation and induction of neuronal repair in EAE mice. J Immunol Res 2018:5083797. https://doi.org/10.1155/2018/5083797
202. Xu Y, Chen C, Hellwarth PB, Bao X (2019) Biomaterials for stem cell engineering and biomanufacturing. Bioact Mater 4:366–379. https://doi.org/10.1016/j.bioactmat.2019.11.002
203. Rahmati M, Pennisi CP, Budd E, Mobasheri A, Mozafari M (2018) Biomaterials for regenerative medicine: historical perspectives and current trends. Adv Exp Med Biol 1119:1–19. https://doi.org/10.1007/5584_2018_278
204. Brennan EP, Reing J, Chew D, Myers-Irvin JM, Young EJ, Badylak SF (2006) Antibacterial activity within degradation products of biological scaffolds composed of extracellular matrix. Tissue Eng 12:2949–2955. https://doi.org/10.1089/ten.2006.12.2949
205. FitzGerald JF, Kumar AS (2014) Biologic versus synthetic mesh reinforcement: what are the Pros and Cons? Clin Colon Rectal Surg 27:140–148. https://doi.org/10.1055/s-0034-1394155
206. Jerman UD, Veranič P, Kreft ME (2014) Amniotic membrane scaffolds enable the development of tissue-engineered urothelium with molecular and ultrastructural properties comparable to that of native urothelium. Tissue Eng Part C Methods 20:317–327. https://doi.org/10.1089/ten.TEC.2013.0298
207. Jerman UD, Veranič P, Cirman T, Kreft ME (2020) Human amniotic membrane enriched with urinary bladder fibroblasts promote the re-epithelization of urothelial injury. Cell Transplant 29:963689720946668. https://doi.org/10.1177/0963689720946668
208. Iranpour S, Mahdavi-Shahri N, Miri R, Hasanzadeh H, Bidkhori HR Naderi-Meshkin H, Zahabi E, Matin MM (2018) Supportive properties of basement membrane layer of human amniotic membrane enable development of tissue engineering applications. Cell Tissue Bank 19:357–371. https://doi.org/10.1007/s10561-017-9680-z
209. Sous Naasani LI, Rodrigues C, Azevedo JG, Damo Souza AF, Buchner S, Wink MR (2018) Comparison of human denuded amniotic membrane and porcine small intestine submucosa as scaffolds for limbal mesenchymal stem cells. Stem Cell Rev Rep 14 744–754. https://doi.org/10.1007/s12015-018-9819-8
210. Dorazehi F, Nabiuni M, Jalali H (2018) Potential use of amniotic membrane—derived scaffold for cerebrospinal fluid applications Int J Mol Cell Med 7:91–101. https://doi.org/10.22088/IJMCM.BUMS.7.2.91

211. Parveen S, Singh SP, Panicker MM, Gupta PK (2019) Amniotic membrane as novel scaffold for human iPSC-derived cardiomyogenesis. In Vitro Cell Dev Biol Anim 55:272–284. https://doi.org/10.1007/s11626-019-00321-y
212. Murphy SV, Skardal A, Song L, Sutton K, Haug R, Mack DL, Jackson J, Soker S, Atala A (2017) Solubilized amnion membrane hyaluronic acid hydrogel accelerates full-thickness wound healing. Stem Cells Transl Med 6:2020–2032. https://doi.org/10.1002/sctm.17-0053
213. Choi WY, Jeon HG, Chung Y, Lim JJ, Shin DH, Kim JM, Ki BS, Song S-H, Choi S-J, Park K-H, Shim SH, Moon J, Jung SJ, Kang HM, Park S, Chung HM, Ko JJ, Cha KY, Yoon TK, Kim H, Lee DR (2013) Isolation and characterization of novel, highly proliferative human CD34/CD73-double-positive testis-derived stem cells for cell therapy. Stem Cells Dev 22:2158–2173. https://doi.org/10.1089/scd.2012.0385
214. Vojdani Z, Babaei A, Vasaghi A, Habibagahi M, Talaei-Khozani T (2016) The effect of amniotic membrane extract on umbilical cord blood mesenchymal stem cell expansion: is there any need to save the amniotic membrane besides the umbilical cord blood? Iran J Basic Med Sci 19:89–96
215. Johnson CT, García AJ (2015) Scaffold-based anti-infection strategies in bone repair. Ann Biomed Eng 43:515–528. https://doi.org/10.1007/s10439-014-1205-3
216. Ahmed W, Zhai Z, Gao C (2019) Adaptive antibacterial biomaterial surfaces and their applications. Mater Today Bio 2:100017. https://doi.org/10.1016/j.mtbio.2019.100017
217. De Jong OG, Van Balkom BWM, Schiffelers RM, Bouten CVC, Verhaar MC (2014) Extracellular vesicles: potential roles in regenerative medicine. Front Immunol 5:608. https://doi.org/10.3389/fimmu.2014.00608
218. Gao S, Chen T, Hao Y, Zhang F, Tang X, Wang D, Wei Z, Qi J (2020) Exosomal miR-135a derived from human amnion mesenchymal stem cells promotes cutaneous wound healing in rats and fibroblast migration by directly inhibiting LATS2 expression. Stem Cell Res Ther 11:56. https://doi.org/10.1186/s13287-020-1570-9
219. Mantha S, Pillai S, Khayambashi P, Upadhyay A, Zhang Y, Tao O, Pham HM, Tran SD (2019) Smart hydrogels in tissue engineering and regenerative medicine. Mater Basel 12:3323. https://doi.org/10.3390/ma12203323
220. Guan X, Avci-Adali M, Alarçin E, Cheng H, Kashaf SS, Li Y, Chawla A, Jang HL, Khademhosseini A (2017) Development of hydrogels for regenerative engineering. Biotechnol J 12:10. https://doi.org/10.1002/biot.201600394
221. Chen J, Wang M-W, Xu J-J, Wu X-Y, Yao J (2020) Gelatin methacryloyl hydrogel eye pad loaded with amniotic extract prevents symblepharon in rabbit eyes. Eur Rev Med Pharmacol Sci 24:10134–10142. https://doi.org/10.26355/eurrev_202010_23233
222. Murphy SV, Skardal A, Nelson RA, Sunnon K, Reid T, Clouse C, Kock ND, Jackson J, Soker S, Atala A (2020) Amnion membrane hydrogel and amnion membrane powder accelerate wound healing in a full thickness porcine skin wound model. Stem Cells Transl Med 9:80–92. https://doi.org/10.1002/sctm.19-0101
223. Li C, Wang J, Wang Y, Gao H, Wei G, Huang Y, Yu H, Gan Y, Wang Y, Mei L, Chen H, Hu H, Zhang Z, Jin Y (2019) Recent progress in drug delivery. Acta Pharm Sin B 9:1145–1162. https://doi.org/10.1016/j.apsb.2019.08.003
224. Tiwari G, Tiwari R, Sriwastawa B, Bhati L, Pandey S, Pandey P, Bannerjee SK (2012) Drug delivery systems: an updated review. Int J Pharm Investig 2:2–11. https://doi.org/10.4103/2230-973X.96920
225. Li W, Chen W-J, Liu W, Liang L, Zhang M-C (2012) Homemade lyophilized cross linking amniotic sustained-release drug membrane with anti-scarring role after filtering surgery in rabbit eyes. Int J Ophthalmol 5:555–561. https://doi.org/10.3980/j.issn.2222-3959.2012.05.03
226. Hu F, Zeng X-Y, Xie Z-L, Liu L-L, Huang L (2015) Clinical outcomes of amniotic membrane loaded with 5-FU PLGA nanoparticles in experimental trabeculectomy. Int J Ophthalmol 8:29–34. https://doi.org/10.3980/j.issn.2222-3959.2015.01.05
227. Francisco JC, Uemura L, Simeoni RB, da Cunha RC, Mogharbel BF, Simeoni PRB, Naves G, Napimoga MH, Noronha L, Carvalho KAT, Moreira LFP, Guarita-Souza LC (2020) Acellular

human amniotic membrane scaffold with 15d-PGJ2 nanoparticles in postinfarct rat model. Tissue Eng Part A 26:1128–1137. https://doi.org/10.1089/ten.TEA.2019.0340
228. Bonomi A, Silini A, Vertua E, Signoroni PB, Coccè V, Cavicchini L, Sisto F, Alessandri G, Pessina A, Parolini O (2015) Human amniotic mesenchymal stromal cells (hAMSCs) as potential vehicles for drug delivery in cancer therapy: an in vitro study. Stem Cell Res Ther 6:155. https://doi.org/10.1186/s13287-015-0140-z
229. Resch MD, Resch BE, Csizmazia E, Imre L, Németh J, Révész P, Csányi E (2010) Permeability of human amniotic membrane to ofloxacin in vitro. Invest Ophthalmol Vis Sci 51:1024–1027. https://doi.org/10.1167/iovs.09-4254
230. Resch MD, Resch BE, Csizmazia E, Imre L, Németh J, Szabó-Révész P, Csányi E (2011) Drug reservoir function of human amniotic membrane. J Ocul Pharmacol Ther Off J Assoc Ocul Pharmacol Ther 27:323–326. https://doi.org/10.1089/jop.2011.0007
231. Kim HS, Sah WJ, Kim YJ, Kim JC, Hahn TW (2001) Amniotic membrane, tear film, corneal, and aqueous levels of ofloxacin in rabbit eyes after amniotic membrane transplantation. Cornea 20:628–634. https://doi.org/10.1097/00003226-200108000-00014
232. Yelchuri ML, Madhavi B, Gohil N, Sajeev HS, Venkatesh Prajna N, Srinivasan S (2017) In vitro evaluation of the drug reservoir function of human amniotic membrane using moxifloxacin as a model drug. Cornea 36:594–599. https://doi.org/10.1097/ICO.0000000000001168
233. Mencucci R, Menchini U, Dei R (2006) Antimicrobial activity of antibiotic-treated amniotic membrane: an in vitro study. Cornea 25:428–431. https://doi.org/10.1097/01.ico.0000214207.06952.23
234. Sara SH, Prajna NV, Senthilkumari S (2019) Human amniotic membrane as a drug carrier—an in-vitro study using fortified cefazolin ophthalmic solution. Indian J Ophthalmol 67:472–475. https://doi.org/10.4103/ijo.IJO_1336_18
235. Liu D, Yang F, Xiong F, Gu N (2016) The smart drug delivery system and its clinical potential. Theranostics 6:1306–1323. https://doi.org/10.7150/thno.14858
236. Rennie K, Gruslin A, Hengstschläger M, Pei D, Cai J, Nikaido T, Bani-Yaghoub M (2012) Applications of amniotic membrane and fluid in stem cell biology and regenerative medicine. Stem Cells Int 2012:721538. https://doi.org/10.1155/2012/721538

Chapter 14
Human Umbilical Cord Blood Mesenchymal Stem Cell Transplantation in Kidney Injury Animal Models: A Critical Review

Martina Perše and Željka Večerić-Haler

Abstract

Introduction: Kidney disease is a global health problem associated with high mortality. The available therapy is mainly supportive. In recent years, regenerative medicine has shown great promise for kidney repair. Mesenchymal stem cells derived from the human umbilical cord (hUC-MSCs) have become the preferred stem cell type due to the technical and ethical reasons. To get insight into the effects of hUC-MSCs on kidney injury, we have reviewed the literature on the use of hUC-MSCs in various animal models of kidney injury.
Methods: An electronic search was carried out in the PubMed and Scopus database using the keywords "kidney injury", "stem cells", and "cord blood". Of the 106 potentially relevant publications, 35 were included.
Results: Results reported that hUC-MSCs prolonged survival and improved functional and morphological parameters in the kidney of various animal models of kidney injury. However, the outcome of MSCs treatment was affected by numerous factors such as characteristics of animal models, stem cell manufacturing process, MSCs therapy timing, route or dose, and many others. Importantly, there is a lack of reports evaluating multiple treatment effects of MSCs as well as possible adverse effects of these therapies.
Conclusions: The variability of animal models for kidney injury, the inconsistency in quality control and reporting on the applied MSC composition and the lack of reports of possible side effects result in many studies that are not reproducible, cannot be critically evaluated or extrapolated to humans. In order to ensure quality and reproducibility and to avoid the unnecessary use of animals, a planned system is essential to ensure that stem cell products used in preclinical research meet the minimum standards and undergo the same quality control scheme required for clinical and commercial use.

M. Perše (✉)
Medical Faculty, Medical Experimental Center, University of Ljubljana, Ljubljana, Slovenia
e-mail: martina.perse@mf.uni-lj.si

Ž. Večerić-Haler
Department of Nephrology, University Medical Center Ljubljana, Ljubljana, Slovenia

I. Virant-Klun (ed.), *Stem Cells in Reproductive Tissues and Organs*, Stem Cell Biology and Regenerative Medicine 70, https://doi.org/10.1007/978-3-030-90111-0_14

Keywords Animal models · Kidney injury · Umbilical cord blood · Stem cells

Abbreviations

Ad-MSC	Adipose mesenchymal stem cells (Ad-MSC)
AKI	Acute kidney injury
pAKT	Serine-threonine kinase Akt
AQP2	Aquaporin 2
ASK1	Apoptosis signal-regulating kinase 1
BAX	Bcl-2-associated X protein
BCL-2	B-cell lymphoma 2
BCL-XL	B-cell lymphoma-extra large
BM-MSC	Bone marrow-derived mesenchymal stem cells
BMP7	Bone morphogenic protein 7
BUN	Blood urea nitrogen test
BW	Body weight
ECFC	Endothelial colony-forming cells
CAT	Catalase
CD	Cluster of differentiation
CK	Cytokeratin
CKD	Chronic kidney disease
CM	Culture medium
CM-Dil	Cell tracker fluorescent dye
Cr	Creatinine
DIO	3,3′-Dioctadecyloxacarbocyanine perchlorate ($DiOC_{18}(3)$)
DNA	Deoxyribonucleic acid
EGF	Epidermal growth factor
EMT	Epithelial-mesenchymal transition
EV	Extracellular vesicle
EX	Exosome
FAP	Fibroblast activation protein alpha
FEN	Flap structure-specific endonuclease
FGF	Fibroblast growth factor
HGF	Hepatocyte growth factor
GFP	Green fluorescent protein
GFR	Glomerular filtration rate
GSH	Glutathione
GVHD	Graft-versus-host disease
HO-1	Heme oxygenase-1
GST	Glutathione-S-transferase
h	Human
HIF-1α	Hypoxia inducible factor-1α
HLA	Human leukocyte antigen

HLA-A	Major histocompatibility complex, Class I, A
HLA-DR	Major histocompatibility complex, Class II, DR
HLA-II	HLA, Class II antigen
HMGB1	High mobility group protein B1
Hsp 47	Heat shock protein 47
hUC	Human umbilical cord
IA	Intraarterially
IGF1	Insulin like growth factor 1
IL	Interleukin
IFN-γ	Interferon gamma
IR	Ischemia reperfusion
ISCT	International Society for Cellular Therapy
IV	Intravenouslly
LC3B	Microtubule associated protein 1 light chain 3 beta
mALB	Microalbumin
β2 mG	Macroglobulin
miRNA	Micro RNA
MCP1	Monocyte chemoattractant factor 1
MDA	Malondialdehyde
MSC	Mesenchymal stem cell
NF-κB	Nuclear factor-kappa B
NK	Natural killer cells
NOD/SCID	Nonobese diabetes/Severe combined immunodeficient (mice)
MAPK	Mitogen-activated protein kinase
MnSOD	Manganese superoxide dismutase
p38 MAPK	P38 Mitogen-activated protein kinase
PCNA	Proliferating cell nuclear antigen
PCR	Polymerase chain reaction
PKH26	Red fluorescent cell linker kit for general cell membrane labeling
PUMA	P53-Upregulated modulator of apoptosis (BBC3)
RAGE	Advanced glycation end product receptor
RANTES	C-C motif chemokine ligand 5 (CCL5)
RNA	Ribonucleic acid
ROS	Reactive oxygen species
RSC	Subcapsular
SC	Subcutaneously
SIRT3	Sirtuin 3
α-SMA	α-Smooth muscle actin
TNFα	Tumor necrosis factor alpha
βTRCP	PIkappaBalpha-E3 receptor subunit (BTRC)
TREG	Regulatory T cells
TUNEL	Transferase-mediated dUTP nick-end labeling assay
TXNIP	Thioredoxin-interacting protein
UC	Umbilical cord
UC-MSC	Umbilical cord blood mesenchymal stem cell

STZ	Streptozotocin
S100A4	S100 calcium-binding protein A4
TGF-β	Transforming growth factor beta
VEGF	Vascular endothelial growth factor
WJ-EPC	Wharton's jelly endothelial progenitor cells
WT1	Wilms' Tumor gene 1
YAP	Yes1 associated transcriptional regulator

Introduction

Acute kidney injury (AKI) is the most common nephrological problem worldwide, with high morbidity and mortality. AKI develops suddenly and leads to a range of renal impairment from minor changes in renal function to end-stage renal disease [1]. In contrast, chronic kidney disease (CKD) develops gradually, over months to years. Patients who have successfully recovered from AKI are at higher risk of developing CKD, especially older people with cardiovascular disease, diabetes, or high blood pressure. Treatment is supportive and in case of end stage renal disease neccessitates the start of renal replacement therapy, i.e. haemodialysis or kidney transplantation. Despite recent advances in renal replacement therapy, the five-year mortality rate for patients with AKI is still exceeding 50%. There is therefore an urgent need for new therapeutic interventions and strategies to improve the survival chances of patients with kidney failure [2].

Since the first description of bone marrow-derived mesenchymal stem cells (BM-MSCs) in the 1970s [3], MSCs research has made numerous advances. Nowadays, MSCs can be isolated from various tissues such as adipose (Ad-MSC), dental, peripheral blood, salivary gland, skin, synovial fluid, placenta, umbilical cord blood (UC-MSC), amniotic fluid, etc. [4]. Currently, BM-MSCs are still among the most frequently used MSCs, despite their ethical concerns, technical and health constrains. On the other hand, UC-MSCs are a safe source of large quantities of stem cells that can be easily obtained by non-invasive methods, have no ethical concerns and have similar characteristics to BM-MSCs [5].

Nevertheless, over the years, stem cell-based therapy has gained great interest in the treatment of renal disease [6]. The therapeutic potential of MSCs has been investigated in numerous animal models. Many animal studies have reported beneficial and protective effects of MSCs in the treatment of kidney injury [6]. The reported beneficial treatment effects of MSCs in animal models have contributed to the approval of numerous clinical trials with inconsistent results [7–10]. Indeed, some recently published articles have raised some important questions that need to be considered when interpreting the results of preclinical studies [8, 11].

This chapter briefly summarizes the animal studies in which the effects of hUC-MSCs on kidney injury were investigated with the aim of critically reassessing the quality of the animal studies and identifying considerations that need to be taken into account when interpreting such results. For the reasons mentioned above, we have focused our investigation on hUC-MSCs.

hUC-MSCs and Kidney Injury in Animal Models

We conducted a PubMed and Scopus search using keywords "kidney injury", "stem cells", "umbilical blood", and identified 106 potentially relevant publications (between July and September, 2020). All titles and abstracts of publications were reviewed. Among them, 35 publications investigated the effects of hUC-MSCs or hUC-MSCs derived conditioned media (CM) or exosomes (Ex) on animal models of kidney injury. Additional references were selected from relevant articles

Species used in publications were mice and rats. Effects of hUC-MSCs treatment were investigated in various rodent models of kidney injury such as cisplatin nephrotoxicity [12–20], ischemia/reperfusion kidney injury [21–28], ureteral obstruction model [29–31], sepsis induced kidney injury [32, 33], lupus nephritis [18, 34, 35], diabetic nephropathy [36–41], or other induced nephropathy [42–44].

hUC-MSCs Treatment Ameliorate Kidney Injury in Animal Models

Of 35 studies only few studies found no renoprotective effect of hUC-MSCs [12, 15, 20, 44]. No study reported negative effects. In most cases hUC-MSCs were found to improve function (estimated by blood urea nitrogen (BUN), serum creatinine (Cr), glomerular filtration rate (GFR)) and morphological changes in the kidney. A few studies investigated the effect on survival and found that hUC-MSCs prolonged survival [15, 19, 32].

hUC-MSCs reduced capillary alterations, the infiltration of neutrophils in the renal interstitium [19], while did not affect the number of lymphocytes ($CD3^+$) [22], the proportion of total intrarenal lymphocytes, total T cells, total B cells, $CD4^+$, and $CD8^+$ T cells, activated $CD4^+$ and $CD8^+$ T cells, and activated B cells [13, 26]. Few studies reported an inhibitory effect of hUC-MSCs on the number of macrophages/monocytes ($CD68^+$) [22, 31, 37], while the other found no effect on the percentage of the intrarenal macrophages [13]. Nevertheless, hUC-MSCs significantly decreased the percentage of natural killer (NK) cells in the injured kidney [26] and increased the percentage of regulatory $CD4^+CD25^+$ T cells [13, 26], that play beneficial role in the pathogenesis of kidney injury [45].

It was found that treatment of hUC-MSCs attenuates the number of apoptotic cells (detected by TUNEL or caspase-3) and increases tubular cell proliferation (highlighted by staining and quantification of PCNA or Ki-67 (an index of renal regeneration) (see Tables 14.1, 14.2 and 14.3).

Table 14.1 Examples of hUC-MSCs treatment in ischemia reperfusion (IR) animal models

Strain, IR induction	MSCs dose, volume, route	End	Parameters investigated and results in comparison to untreated IR-induced animals	References
Wistar h0: Unilateral L 60 min	H?: 1 × 10^6 in 0.2 ml saline Iv	D30	Blood: ↓BUN, ↓Cr (normal levels) (reversible) Urine:↓uric acid, ↓urine albumin (normal levels) Kidney: morphology normal, ↓MDA, ↑GSH, ↑GST, ↑CAT	[21]
Wistar–Kyoto h0: Bilateral 45 min	P3-P5 h6: 1 × 10^6 in 2 ml saline ip	D2 D7 D49	D2: peak (BUN, Cr), D7: recovery, D49: long term effects Blood: BUN, Cr; Urine: Cr clearance, ↓FENa Kidney: ↓histology score, ↑AQP2, ↓*$CD68^+$, $CD3^+$, ↓TGF-β1, ≈β-gal, ↑Klotho protein, ↑HO-1, ↑MnSOD, ↑miR-29a, ↑miR-34a, ≈p21, ↓p16, ↑PCNA Blood: BUN, Cr; Urine: Cr clearance, ↓FENa Kidney: ↓histology score, ↑AQP2, ↓*$CD68^+$, ≈TGF-β1, ≈β-gal, ≈HO-1, ≈MnSOD, ≈p21, ↓p16, ≈PCNA Blood: BUN, Cr; Urine: Cr clearance, ↓FENa, ↑osmolality Kidney: ↓histology score, ↓*$CD68^+$, $CD3^+$, ↑β-gal, ↑Klotho protein, ↑HO-1, ↑MnSOD, ↑miR-29a, ↑miR-34a, ≈TGF-β1, ≈p21, ≈p16, ↑PCNA	[22]
Sprague–Dawley h0:Bilateral 60 min	P3-6, MV h0:30 μg in 0.5 ml vehicle iv	D2 Wk2	D1: ↓TUNEL, ↑PCNA, ↑Vimentin, ↑HGF, ≈TGF-β1, ≈IGF-1, ≈EGF Wk2: Blood: ↓BUN, ↓Cr (reversible model – accelerate kidney recovery) Kidney: ↓collagen deposit (Masson), ↓αSMA	[23]
NOD.CB17-$Prkdc^{scid}$/J D0: bilateral 30 min	hECFCs P3-P4 10^6 in 100μL PBS	D1 D3	D1: Blood: ↓BUN, ↓Cr; (reversible model – accelerate kidney recovery) Kidney: ↓histology score, ↓αSMA, ↑megalin, ≈macrophages (F4/80), ≈PCNA, ↓TUNEL, ↓caspase-3, ≈superoxid production D3: Blood: ≈BUN, ↓Cr Kidney: ↓histology score, ↓αSMA, ↑megalin, ↓macrophages (F4/80), ↓PCNA, ↑TUNEL, ≈caspase-3, ≈superoxid production	[24]
C57BL/6 D0: unilateral L nephrectomy R 40 min	hWJ-EPC 5 × 10^5 rsc	h12 D7	Blood: ↓BUN, ↓Cr (results shown for time point 12 h only) Kidney: ↓histology score, ↓TUNEL, ↓ROS, ↓MIP-2, ↓KC, ↓S100A4 (fibroblast marker), ↓PUMA, ↑Bcl-2	[25]
C57BL/6 D0: Bilateral 27 min	24 h prior and at h0: 1 × 10^6 volume nr ip	D1 D2	Blood: D1, D2, ↓BUN, ↓Cr (**but 4 × higher than baseline levels**) Kidney: ≈histology score; ≈$CD45^+$ (leucocytes infiltration) IL-2, IL-6, IL-10, IL-17, MCP-1, RANTES, TNFα, ↓IFNγ, ↑VEGF; ≈T and B cells, CD4 + CD8 + T cell, ↓NK, ↑Treg	[26]

(continued)

Table 14.1 (continued)

Strain, IR induction	MSCs dose, volume, route	End	Parameters investigated and results in comparison to untreated IR-induced animals	References
Sprague–Dawley h0:Bilateral 60 min	P4: h16: 1×10^6 in 0.5 ml saline; ia	h72 D3	Blood at h0, h16, h24, h48, h72: from h24 onwards ↓BUN, Cr (decline also in control at h72 – reversible) Kidney: ↓hystology score, ↑PCNA, ↓caspase-3, ↓IL-1β, ≈IL-6, ↓TNFα	[28]

Legend h—human; D—day; UC—umbilical cord blood; CM—culture media; sc—subcutenously; rsc—subcapsular; ia—intraarterially; iv—intravenouslly; TUNEL—transferase-mediated dUTP nick-end labeling assay; PCNA—proliferating cell nuclear antigen; EMT—epithelial-mesenchymal transition; BW—body weight; WJ-EPC—Wharton's jelly endothelial progenitor cells; ECFC—endothelial colony-forming cells

Table 14.2 Examples of hUC-MSCs treatment in nephrotoxin induced animal models

Strain, nephrotoxin induction	MSCs dose, route, volume	End	Parameters investigated and results in comparison to untreated cisplatin animals	References
NOD-SCID C57BL6 × 129 D0: 16 mg/kg sc cisplatin	D1: 5×10^5 P6 Iv	D4	Blood: ↓BUN, Kidney: ↓histology score, ↑mitochondrial morphology	[12]
SIRT3$^{-/-}$ D0: 16 mg/kg sc cisplatin	D1: 5×10^5 P6 Iv	D4	SIRT3$^{-/-}$ severe renal dysfunction and no effect was found Blood: BUN, Kidney: histology score	[12]
BALB/cOlaHsd D0: 17 mg/kg, ip cisplatin	D1: 5×10^5 in 0.2 ml iv	D4 D14	MSCs without ATG pretreatment had no effect MSCs with ATG pretreatment improved survival and renal functional and structural parameters; Blood: ≈BUN, ↓Cr; Kidney: ↓histology score, ≈casp-3; ↑HO-1, GPx, ↓SOD-1, SAA3	[15]
D0: 18 mg/kg, ip cisplatin	hUCB–CM D1: CM 0.5 ml (1x) D0, D1, D3 (3x) 2x/day for 5 days	D4	Blood: ≈BUN, Cr, histology score, ↓BW 3 different protocols of hUC CM were used (D0, D1, D3 or repeated 2 × per day for 5 days) —No effect Findings of Bi et al. [46] were not reproducible in their lab	[20]
C57BL/6 D0: 22 mg/kg, sc cisplatin (20% mortality on D6)	D1: 1×10^6 P6 (hUC) iv versus ip	D3	Blood:↓BUN; Kidney: histology score, ↓MCP-1, IL-6, TNFα, ↑IL-10, VEGF, ≈IL-2, ↓caspse-3; ↑Treg in kidney and spleen Kidney: ≈total T and B cells, ≈CD4$^+$T cells, ≈CD8$^+$T cells, ≈macrophages (D3, D6)	[13]
C57BL/6 D0: 20 mg/kg, sc cisplatin	D1 versus D3: 1×10^6 P6	D3 versus D6	**Early** Blood: ↓BUN, Cr; Kidney: ↓histology score ↓MCP-1, ≈IL-6, ≈TNFα, ≈IL-10, ↑VEGF, ≈IL-2, ↓caspse-3;↑Ki-67 **Late**: Blood: ↓BUN, Cr; Kidney: ≈histology score MCP-1, IL-6, TNFα, IL-10, VEGF, ≈IL-2, ≈caspse-3; ≈Ki-67; Treg in kidney and spleen	[13]
Sprague–Dawley D0: 5 mg/kg, ip cispatin	hUC **exosomes** 0.5 h prior cisplatin 0.2 mg, rsc	D3	Blood:↓BUN, Cr, TNFα, IL-1β, IL6; Kidney: ↓histology score,TUNEL, ↑PCNA, Bax, LC3B, BCL-2 (autophagy)	[14]

(continued)

Table 14.2 (continued)

Strain, nephrotoxin induction	MSCs dose, route, volume	End	Parameters investigated and results in comparison to untreated cisplatin animals	References
Rat D0: 6 mg/kg, ip cisplatin	D1: 2×10^6 in 0.5 ml saline Iv	D5 D10 6 wk 8 wk	(D0, D2, D3, D5) Blood: ↓ BUN, Cr; Kidney: ↑PCNA, Bcl-2, ↓TUNEL, ↓caspase-3, Bax, IL-1β, TNFα, MDA, ↓histology score, ↓MDA, ↑cytochrome C (in mitochondria) D10: both groups returned to baseline level (BUN, Cr) 6wk and 8wk: structural restoration better after MSCs, ↑Bax/Bcl-2 ratio, ↓TGFβ1, ↓collagen deposit (Masson), mRNA↓vimentin, ↓N-cadherin, ↓Slug ↑E-cadherin	[16]
Sprauge Dawley D0: 6 mg/kg, ip cisplatin	D1: 200 μg Ex 40–100 nm rsc	D5	D3-D5: ↓Cr, ↓BUN (progressive increase 3x-peak) D5: kidney: ↓histology score, ↑Bcl-2, ↓Bax, ↓p38MSPK, ↓caspase-3, ↓TUNEL, ↑GSH, ↓MDA, ↓8-OH6G	[17]
NOD-SCID, D0: 12.7 mg/kg, sc, cisplatin	D1: 5×10^5 P6 Iv	D4	Blood: ↓BUN; Kidney: ↓histology score, ↓TUNEL, ↓peroxynitrite (oxidative stress), pAkt, ↓PCNA, EM: perituburar microvascular capillary changes; Survival D9-14; hUC (86%) versus control (0%)	[19]
Wistar rat Adenine ig for 4 wks reversible	hUC versus ICAhUC D0: 8×10^6 in 1 ml saline, P4 iv	D3 D7 D14	selfrecovery at the end—reversible model Blood: ↓BUN, ↓Cr, ↑IL-10, ↓IL-6, ↓TNFα Kidney: histology, ↑IL-10, ↓IL-6, ↓TNFα, ↑bFGF, ↑BMP-7 icahUC better than MSC alone—almost all time points	[43]
Wistar D0: 0.05 ml/kg CCl_4systemic	hUC versus hUC + G-CFS h6 or h24: 2×10^6 ip	D2	Blood: ↓BUN, ≈Cr, ≈Na, ≈potassium Kidney: ≈histology score, ↓GST Systemic	[44]

Legend h—human; D—day; UC—umbilical cord blood, CM—culture media; sc—subcutenously; rsc—subcapsular; ia—intraarterially; iv—intravenouslly; TUNEL—transferase-mediated dUTP nick-end labeling assay; PCNA—proliferating cell nuclear antigen; EMT—epithelial-mesenchymal transition, BW—body weight, mALD microalbumin; β2 mG macroglobulin; HIF-1α—hypoxia inducible factor-1α; p—peak

Table 14.3 Examples of hUC-MSCs treatment in diabetic nephropathy and ureteral obstruction animal models

Strain, STZ dose and route	MSCs dose, route, volume	End after MSCs	Parameters investigated and results in comparison to untreated animal model	References
Sprague Dawley D0: 60 mg/kg ip STZ	Wk6: hUC 2×10^6 in 0.5 ml PBS iv	2wk	(↓)BW; Urine: ↓(↑) protein, ↓(↑) protein/Cr ratio Blood: ↓(↑) BUN, ↓Cr, ↓Cr clearance; ↓IL-1β, ↓IL-6, ↓TNF-α Kidney: ↓histology score; ↓IL-1β, ↓I L-6, ↓TNF-α, ↓TGF-β, ↓F4/80, ↓$CD3^+$, ↓collagen IV, ↓αSMA	[36]
Sprague Dawley D0: 60 mg/kg ip STZ	Wk6: P5 hUC 2×10^6 in 0.5 ml PBS iv 1x/wk for 2wk	2wk	≈(↓)BW; Urine: ↓(↑) protein, ↓(↑) protein/Cr ratio; Blood: ≈(↑)glucose, ↓(↑) BUN, ↓(↑)Cr Kidney: ↓(↑)histology score (PAS), ↓(↑) Masson, ↓(↑)TUNEL, ↓(↑)HMGB1, ≈(↑)Bax, ↑(↑)Bcl-xl, ↓(↑)TXNIP, ↑(≈)TRX1,↓(≈) p-ASK1/ASK1, ↓(↑)p-P38,	[37]
Sprague Dawley D0: 50 mg/kg iv STZ glucose?	D2: hUC 5×10^5 iv	4wk	≈(↓)BW, ≈(↑)kidney weight, Blood: ≈(↑)glucose, ≈Cr, ↓(≈)urinary protein Kidney: ≈αSMA, ↓(≈)TGF-β, ↑(≈)E-cadherin, ↑(≈)BMP-7, ↓(≈)Hsp47, collagen (Masson)	[40]
Sprague Dawley D0: 50 mg/kg iv STZ glucose?	Wk4: hUC 1×10^6 iv	4wk	≈BW, ↓(≈)kidney weight; Blood: ≈(↑)glucose, ↓(≈)Cr, ↓(≈)urinary protein Kidney: protein: ≈fibronectin, ↓(≈)αSMA, ↑(↑)E-cadherin mRNA: ↓(≈)fibronectin, ↓(≈)αSMA, ≈E-cadherin	[39]
NOD adaptive feeding D0: diabetes onset	hUC versus hUC + resveratr D3: 1×10^6 in 0.3 ml PBS Iv	8wk	↑(↓)Body weight (but lower than control) Blood:↓(near) glucose, ↓(↑)BUN, ↓(↑)Cr; Urine: ↓(↑)albumin excretion rate Kidney: ↑(↓)podocyte number, ↑(↓)nephrin, ↑(↓)WT1 (podocyte proteins), ↓(↑)RAGE, ↓(↑)MCP-1, ↓(↑)NF-κB; Some animals died	[41]
Strain, Model induction	**MSCs dose, route, volume**	**End**	**Parameters investigated and results in comparison to untreated animal model**	References
Sprague Dawley D0: left ureter ligation	D6, D9, D12: hUC-Ex 200 μg;	D14	Blood: ↓(↑)Cr, ↓(↑)BUN Kidney: ↓Collagen I, ↓FAP, ↓α-SMA, ↓TGF-β1, ↑(↓)CK1δ, ↑(↓)βTRCP, ↓(≈)YAP	[29]

(continued)

Table 14.3 (continued)

Strain, STZ dose and route	MSCs dose, route, volume	End after MSCs	Parameters investigated and results in comparison to untreated animal model	References
Sprague Dawley D0: left ureter ligation	D0:0.5 mL CM from 5×10^6 MSCs, ia	D14	Kidney: ↑(↓) histology score, ↑(↓)Masson, ↑(↑)PCNA, ↑(↓)TUNEL ↑(↓)GSH, ↑(↓)MDA, ↑(↓)ROS, ↑(↓)Collagen I, ↑(↓)α-SMA, ↑(↓)TGF-β1, ↑(↓)TNFα, ↑(↓)E-cadherin	[30]
Sprague Dawley D0: left ureter ligation	D0: 0.1 mL CM, ia	D14	Blood: ↓(↑)BUN, ↓(↑)Cr, ↓(↑)TNF-α, ↓(↑) IL-6, ↓(↑)IL-1β, Kidney: ↓(↑)histology score, ↓(↑)TNF-α, ↓(↑)IL-6, ↓(↑)IL-1β, ↓(↑)MCP-1, ↓(↑)collagen I, ↓(↑)αSMA, ↓(↑)TLR4, ↓(↑)NFκB, ↓(↑)CD3$^+$, ↓(↑)CD68$^+$, ↑(↓)E-cadherin	[31]
Sepsis CLP C57BL/6 H0: CLP	H3: hUC-Ex 120 μg in 100μL PBS	h24	Blood: ↓(↑)Cr, BUN Kidney: ↓histology score, ↓TUNEL, ↓caspase-3, ↓(↑)IL-1β, ≈(↑)TNFα, ↑miR146, ↓IRAK1 ↑Survival (h72): 28% versus 45% (hUC-Ex)	[32]

Legend h—human; D—day; UC—umbilical cord blood; CM—culture media; sc—subcutenously; rsc—subcapsular; ia—intraarterially, iv—intravenouslly; TUNEL—transferase mediated dUTP nick-end labeling assay; PCNA—proliferating cell nuclear antigen; EMT—epithelial-mesenchymal transition; BW—body weight; RAGE—advanced glycation end product receptor; MCP-1—monocyte chemoattractant factor-1; NF-κB—nuclear factor-kappa B; NOD mice—nonobese diabetes; WT1—Wilms' Tumor gene-1; STZ—Streptozotocin; α-SMA—α-smooth muscle actin; Hsp 47—heat shock protein 47; BMP-7—bone morphogenic protein-7 hUC-MSCs treatment influenced the expression of various genes and proteins implicated in apoptotic signaling pathways such as high mobility group protein B1 (HMGB1), thioredoxin-interacting protein (TXNIP), apoptosis signal-regulating kinase 1 (ASK1), P38 MAPK, Bcl-xl [37], Bax, Bcl-2, cytochrome C [16], serine-threonine kinase Akt (pAkt) [19], mitochondiral mass and function, mitochondrial biogenesis, Sirtuin 3 (SIRT3) activity, antioxidant defence, and energy supply (ATP production) [12]. hUC-MSCsreduced oxidative stress markers such as malondialdehyde (MDA, a marker of lipid peroxidation) [16, 21] or peroxynitrite, a marker of protein oxidation [19], and enhance antioxidant enzymes such as glutathione (GSH), catalase (CAT), glutathione-S-transferase (GST) [21], heme oxygenase-1 (HO-1), and manganese superoxide dismutase (MnSOD) [22]. hUC-MSCs affect also Klotho protein expression and expression of some senescence-related miRs in kidney tissue (mainly miR-29a and miR-34a) [22].

It has been shown that hUC-MSCs modulate inflammatory states through the complex network of numerous pro-inflammatory and anti-inflammatory factors that play a role in the pathogenesis of kidney injury. However, the results of the cytokines and other inflammatory factors in the studies were not consistent (see Tables 14.1, 14.2 and 14.3).

hUC-MSC was found to have protective effects also on collagen accumulation [16], pro-fibrotic factor (TGF-β) [22, 36], expression of several epithelial-mesenchymal transition associated genes, such as Slug, Vimentin, and E-cadherin was restored by hUC-MSCs [16] (see Table 14.3).

Strategies to Improve Renoprotective Effects of hUC-MSCs

To improve renoprotective properties researchers investigated various strategies. For instance hUC-MSCs were injected in combination with vitamin E [47] or resveratrol [41] which improved MSCs renoprotective effects. In combination with resveratrol hUC-MSCs were found to protect podocyte function and increase podocyte-related proteins (nephrin, WT1), reduce inflammatory factors (MCP-1, RAGE, NF-κB) [41]. Pretreatment of hUC-MSCs with icariin [43] resulted in improved effects as well, while pretreatment of hUC-MSCs with Poly-inosinic acid:cytidylic acid (I:C) showed no beneficial effect [34]. Renoprotective effects of hUC-MSCs were improved when immunosupression therapy was used [15].

hUC-MSCs Renoprotective Effects Are Better than Those of MSCs from Other Sources

Renoprotective effectiveness of hUC-MSCs was compared to other sources of hMSCs. It was reported that hUC-MSCs treatment improved kidney function and survival to a higher degree than treatment with hBM-MSCs (BUN: 58 ± 7 mg/dl (hUC) versus 76 ± 8 mg/dl (hBM); survival D9-14; 70% (hUC) versus 30% (hBM)) [19] or hAdMSC [22].

Klotho protein expression and expression of some senescence-related miRs in kidney tissue (mainly miR-29a and miR-34a) were protected by hUC-MSCs but not by hAd-MSC treatment [22]. When renoprotective effects of hUC-MSCs treatment were compared to effects of syngeneic MSCs (mouse BM) no significant difference was observed [13].

The Timing of MSCs Treatment Influences the Outcome of Treatment

In almost all cases of experimental AKI, MSCs treatment was given as a single dose and in the induction phase of kidney injury, i.e. at a time when no clinical symptoms and signs of kidney dysfunction were observed. Therefore, MSCs have mostly been evaluated as a preemptive treatment strategy applied at the earliest stage of development of AKI [13]. However, only one study evaluated the effect of MCSs treatment when kidney dysfunction was established and found that the treatment was not effective, suggesting that the timing of MSC treatment is a very important factor in the outcome of treatment [13]. As there is a lack of studies investigating long-term effects of MSCs, it is not known whether the beneficial effects observed in short-term studies (which assess the effect 1–7 days after single administration of MSCs) represent a therapeutic effect or only a supportive effect with temporarily improved parameters.

Delivery Method, Cell Dose, and the Fate of Injected Cells

Almost all studies reported on dose and route of administration and found similar effects of hUC-MSCs, regardless of dose or administration route [26, 38]. The most commonly used delivery method was the systemic route, such as intravenous and intra-arterial, and less frequently localized, such as intraperitoneal or intrarenal (subcapsular) delivery. Number of systemically injected cells were usually in a range of previously reported for mice (from 5×10^5 to 1×10^6) and rats (1×10^6 to 2×10^6) (examples can be seen in the Tables 14.1, 14.2 and 14.3). Importantly, the volume of the injected suspension was lacking in some studies, although it is known that in patients with kidney injury hydration (i.e. volume expansion) is the primary approach to prevent and/or improve kidney injury and may therefore significantly affect the results and severity of experimental nephrotoxicity.

However, no study investigated possible complications related to the route of administration or cell dose. It should be clarified that different routes of administration of stem cell products provide an enormously different number of cells that reach the kidney tissue. Intra-arterial injection with first-pass kidney exposure indicates that the MSCs product (stem cells and medium) reaches the renal arteries in a relatively original and undiluted form, e.g. injection into the left heart, suprarenal aorta, or renal arteries. In contrast, intra-venous or intra-arterial injection with second pass kidney exposure indicates that MSCs product reaches the renal vessels after metabolic and morphological changes following dilution (or entrappment) either in the pulmonary or peripheral circulation, e.g. injection into the right heart, pulmonary artery, carotid artery, subclavian, infrarenal artery, or intravenously.

Based on this, it is expected that the method of administration and the cell dose are associated not only with the treatment effect but also with complications due to the large size of the hMSCs (range between 16 and 53 μm in suspension). For example, intravenously injected MSCs become trapped in the lungs and can cause embolism [48, 49], while intra-arterial injections can also lead to blockages in the vascular system due to entrapment in the small size peripheral arterial vasculature [50].

Furthermore, it has recently been recognized that we still do not have a systematic and reliable test to determine the amount of MSCs that have reached the target tissue, nor to establish their effectiveness. Various methods have been used to detect and confirm the presence of injected cells in kidneys and/or other organs, such as radioactivity imaging, human DNA polymerase chain reaction (PCR), in situ hybridization analysis for biodistribution of human MSCs, labeling dyes such as PKH26, GFP and DIO, etc. However, it is important to note that there is no single assay or method that can help identify MSCs in mixed populations of the organism. All these methods only confirm the presence of the dye or a specific DNA sequence but cannot confirm that the cell caring the material is injected MSC or whether the cell is vital. In most cases, the cells have been labeled with lipid (PKH26) or RNA dyes (CM -Dil), which are known to be less reliable methods for tracking cells (for more information see [11, 51]).

Factors to Be Considered When Interpreting the Results of Animal Studies

From the so far reported results, we can conclude that a single administration of hUC-MSCs to a xenogeneic organism does not cause adverse effects, at least not in the short term. However, the effects on long-term or following multiple treatments are not known.

Although most studies have shown renoprotective effects of hUC-MSCs, it is important to note that there are many factors to consider before a conclusion can be drawn about the efficacy of hUC-MSC therapy.

hUC-MSC Quality, Culturing, and Minimal Criteria

It is well known that MSCs are a heterogeneous cell population and may vary from laboratory to laboratory. To avoid heterogeneity between laboratories, minimum criteria for phenotypic and functional characterization of MSCs have been proposed. hMSCs must be plastic-adherent, have the ability to differentiate into osteoblasts, adipocytes, chondrocytes, lack expression of CD11b, CD14, CD19, CD79a, CD34, CD45, HLA-DR surface antigens, but in more than 95% of the MSC population CD105, CD90, and CD73 surface antigens must be expressed [52] (Table 14.4).

An increasing number of studies report that also the duration of cultivation (number of passages), cultivation conditions (serum, growth factors, oxygen status, availability of cytokines), and cryopreservation (media, reagents, freezing temperature) can influence the differentiation potential, morphological, and phenotypic characteristics and thus the outcome of in vivo studies (i.e. potency, homing, immunological response, coagulation complications) [4].

Importantly, complications in human trials like thromboembolism and subsequent health complications or even death [53, 54] led to the investigation of the mechanisms responsible for coagulation after MSCs injection [55]. It was discovered that in vitro cultured MSCs may express procoagulant factors such as tissue factor, collagen1A and fibronectin1, which can trigger a coagulation cascade in the circulation after administration of MSCs. Complications were associated with the increased number of passages of MSCs regardless of species or source of MSCs. Coagulation complications may occur after 5 passages of cultivation of MSCs. A higher number of passages increases the risk of coagulation complications (P5-P12) [56, 57] and reduces the potency of MSCs and the efficiency of grafting [4].

Additional mechanisms of adverse effects, such as transfusion-like reactions, which have already been observed in humans after the transplantation of stem cell products [58], can be caused by contaminating (so-called *passenger*) lymphocytes [59]. Assays to confirm the identity of stem cell transplants assure us that the cellular component is indeed predominantly based on MSCs. However, to a limited extent and in allowance with the ISCT criteria [52], contamination with donor leukocytes is permitted (up to 2% of the MSC product). Given that two-thirds of clinical trials use non-cryopreserved MSCs, such products could therefore contain significant amounts of live lymphocytes. Interestingly, relatively small amounts of living

Table 14.4 Minimal criteria for defining human MSCs for preclinical studies [52]

1. Adherence to plastic in standard culturd conditions		
2. Phenotype	Positive (95%) CD105 CD90 CD73	Negative (2%) CD45 CD34 CD14 or CD11b CD79α or CD19 HLA-DR
3. In vitro differentiation: oseteoblasts, adipocytes, chondroblasts (demonstrated by staining of in vitro cell culture)		

lymphocytes have been reported responsible for the occurrence of transfusion-related Graft-versus-Host Disease (GVHD) in humans and could also be associated with the late-stage GVHD-like response in mice described by our group after hUC-MSCs application [11]. It is assumed that the pathological process is triggered by the transfusion of viable T cells present in stem cells or blood products. In the case of an immunocompetent recipient, these lymphocytes normally remain in the circulation for several days until the recipient's immune system finally removes them. However, if the recipient's immune system is immunocompromised, the lymphocytes could engraft and proliferate. It is not known whether and how such mechanisms could occur after application of hUC-MSCs in the case of thrombotic microangiopathy, which was already described after transplantation of non-UC MSCs [60].

The present review of the studies revealed that the researchers did not provide accurate and consistent information on the method of MSCs production, growing conditions, number of passages, composition of the MSCs-culture medium, safety tests, and description of the batch release criteria. Furthermore, in the individual rare cases of studies where data were reported, it is clear that the cellular product generally did not meet the minimum release criteria set by the ISCT [52].

To ensure quality and reproducibility in stem cell research, a planned system is essential to ensure that stem cell products meet the standards required for research and translational application. There are several regulatory issues related to the safety, efficacy, and quality of MSCs products that need to be considered when preparing such a therapy for clinical application. This standards should also be followed in the experimental environment to allow successful interpretation of the results and their translation to humans. Initially, safety testing is essential, including tests for potential microbial, fungal, endotoxin, mycoplasma and viral contamination, karyotype tests and in vitro functional tests, which should serve as guidance for clinical efficacy. Without adequate quality control of a stem cell product, it cannot be guaranteed that the stem cells are what they are expected to be. If we take into account the heterogeneity already mentioned and the inherent limitations of animal models, the additional deviations of the cell product content can completely prevent critical conclusions and the successful transfer of knowledge from preclinical to clinical use.

Characteristics and Challenges of Animal Models and Interpretation of the Results

Research of kidney disease involves experimental models ranging from the simplest to the most complex. In the case of acute kidney injury, for example, three basic types of animal models are used: ischemia/reperfusion, toxic, and sepsis models. This classification is somehow artificial because human and animal models usually contain a mixed type of acute kidney injury involving components of all three mechanisms. However, it is important to note that each animal model has its own pathogenesis and underlying molecular mechanisms that offer a unique perspective on the disease and can be studied and evaluated as a potential therapeutic target. In addition, each model has its own characteristics, challenges, and factors that influence the extent and severity of the disease and consequently the outcome of the study.

Thus, depending on the protocol of the respective animal model, the renal damage may be reversible or irreversible, and may initially be induced locally in the kidney or systemically. For example, in the ischemia/reperfusion model, which is a technically demanding model, the injury is primarily induced in the kidney (by clamping of renal artery for a considerable period of time). However, the extent and reversibility of the injury depends on numerous factors (e.g., surgeon skills, duration of surgery, species, strain, sex, microbiological state, body temperature during surgery, hydration state, duration of ischemia) [61]. On the other hand, sepsis models affect the entire organism and are usually irreversible. However, even these models can be very different as they can be induced by a variety of protocols such as intraperitoneal administration of fecal solution, systemic LPS, or bacterial administration or protocols involving surgical treatment such as caecal ligation and puncture, and thus each model has its own characteristics and challenges [62]. Finally, toxic models like the cisplatin model are dose-dependent. Cisplatin-associated damage can therefore range from primary kidney localized damage, which can potentially be reversible, to irreversible systemic damage with multiorgan failure and death [45].

It is important to stress that in the context of experimental nephrotoxicity, renal function is usually monitored by serum creatinine (Cr) and blood urea nitrogen (BUN). These biomarkers are insensitive and non-specific, especially BUN given its widely variable non-glomerular filtration related determinants (primarily urea generation and tubular reabsorption). However, mainly due to the practicability and simplicity of their determination, BUN and Cr are the most commonly used markers for the assessment of renal function also in the clinical setting. Put it simply, when renal function deteriorates, the values of BUN and Cr increase many times over. When renal function is restored, the values return to the baseline levels.

After careful review of all identified studies, we found heterogeneity in the protocols for each animal model, which may have significantly influenced the molecular mechanisms and also the interpretation of the effects of MSCs therapy. For example, in most studies, treatment with MSCs resulted in a significant attenuation of BUN and/or Cr values. Does this mean that renal function was actually restored? If the answer is yes, the answer, at least from a clinician's perspective, is often oversimplified and requires a critical overview. At this point, it is important to recall that in animals and humans different types and severities of injury can coexist, especially in the situation of multi-organ damage. For example, the ischemia/reperfusion model was in most cases a kidney localized reversible model that peaked at BUN and Cr levels 2–3 days after surgery, from which point a decrease was observed and the animal was able to self-recover [22]. On the other hand, if a lethal dose of cisplatin is used to induce nephrotoxicity, a kidney injury is exceptionally only toxic. In this case, other prerenal mechanisms due to reduced mean arterial blood pressure (as a consequence of concomitant inappetence, starvation and dehydration) interact with the toxicity, aggravating the tissue damage and leading to irreversibility of the situation with the onset of severe acute tubular- or even cortical necrosis. Thus, the severity of the disease is most likely one of the most important factors that can influence the outcome of treatment.

This is clearly shown in a study using two cispatin protocols, one of which was reversible and the other was lethal [13]. When nephrotoxic, non-lethal dose of cisplatin was used (which reduced the severity of the disease as measured by BUN, Cr), the administration of MSCs resulted in a decrease in BUN and Cr levels, which returned to baseline, suggesting that renal function has indeed recovered. Similarly, when lethal cisplatin dose was used (22 mg/kg; s/c) administration of hUC-MSCs also significantly attenuated BUN levels in comparison to the untreated cisplatin group (i.e. 50–70 mg/dL vs. 70–90 mg/dL). It is however important to note that in the second case the values of BUN were still very far from normal values (BUN values above 30 mg/dl were considered abnormal), indicating that renal function had not yet recovered at the time of the study (day 3). In addition, such restorations of glomerular filtration rate observed in the experimental environment, which are considered "significant improvements", would not lead to dialysis independence when extrapolated to humans and would be fatal without renal replacement therapy.

Therefore, the heterogeneity of the protocols can contribute to a better understanding of the underlying mechanisms, but only if all details are given. Otherwise, the role of factors involved in the pathogenesis of a particular disease may be misinterpreted.

Some studies lacked not only reporting on the animal model protocol, but also on the condition and characterization of the MSCs culture. It is therefore very important that the reporting includes all the necessary details of the animal model, the experimental design, and cell manufacturing process.

Immune Microenvironment May Affect Injected MSCs

A review of the literature showed that most of the so far studies were performed on immunocompetent animals without the use of immunosuppression. Although it is assumed that human MSCs in the naïve state do not express antigens involved in the immune recognition process (i.e. the MHC class II and the costimulatory molecules CD80 (B7-1), CD86 (B7-2), CD40, or CD40L) [63], animal studies investigating the immune response in immunocompetent recipients have clearly shown that cellular and humoral responses against xenogeneic MSCs can develop (for more information see [11]). Immune responses to allogeneic MSCs have also been observed (excellently explained elsewhere [64–67]).

The severity and toxicity of the microenvironment of the damaged tissue or organism can be very important for the effectiveness of MSCs and the outcome of treatment. We and others have shown that MSCs can both enhance or inhibit an immune response in accordance with the recipient's immunological and physiological microenvironment. MSCs are known to migrate towards the injured kidney [68], and there is evidence that MSCs can be polarized toward a proinflammatory phenotype capable of releasing proinflammatory cytokines [69]. In this respect, the effect of MSCs could be more strongly influenced by the patients' uremic milieu (very typical of advanced renal failure) and the effect of immunosuppressive drugs such as calcineurin inhibitors or steroids, that are frequently prescribed to patients with renal disease and could influence MSC-mediated immune responses.

In addition, the harmful microenvironment for MSCs can affect not only MSC survival, but also the protein content of the MSCs and the functional properties of the secreted extracellular vesicles (EVs). EVs secreted by MSCs (exosomes and microvesicles) have a content that includes cytokines, growth factors, signal lipids, mRNA, miR, i.e. molecules involved in cell signaling or cell-to-cell communication over short and long distances in the body.

In vitro studies have clearly shown that the microenvironment of MSCs influences the protein composition of EVs. Namely, when hUC-MSCs were cultured in the presence of IFN-γ for 24 h, MSCs produced and secreted EVs containing the HLA-A (MHCI) molecule and both α and β units of the proteasome complex required for antigen presentation and activation of T cells (recognition of foreign and rejection). When hUCB-MSCs were stimulated with IFNγ for 48 h, the EVs also contained HLA-II proteins [70]. Thus, when MSCs are exposed to a harmful microenvironment such as an environment with mass of cytotoxic factors leading to an inflammatory cytokine storm (as in the case of a systemic inflammatory response associated with multi-organ failure), the differentiation potentials and functional properties of MSCs may change, affecting the protein composition of MSCs and producing EVs with complement factors and HLA-II proteins.

Conclusion

In summary, the therapeutic potential of hUC-MSCs has been investigated in numerous animal models of kidney injury, including irreversible failure of kidney function. While research to date has focused on a number of positive effects of MSCs treatment, there are very few studies reporting possible side effects or adverse outcomes of such therapies. With few exceptions, the positive effects of MSCs in kidney research have been explained by improvements in surrogate biomarkers of renal damage, such as serum BUN or Cr, which are often influenced by many other parameters of the diseased animal's metabolism that go beyond glomerular filtration. The inherent specificities of animal models and the inconsistency in quality control of cellular products and reporting on the composition of MSCs branches add to the confusion, preventing critical evaluation of results obtained at the preclinical level and their successful transfer to the clinical ground. Compliance with at least existing standards and the introduction of stricter regulation in the field of preclinical stem cell research must aim at scientific progress that is not confusing, difficult to interpret or even meaningless leading to unjustified use of animals.

References

1. Ronco C, Bellomo R, Kellum JA (2019) Acute kidney injury. Lancet 394:1949–1964. https://doi.org/10.1016/S0140-6736(19)32563-2
2. Webster AC, Nagler EV, Morton RL, Masson P (2017) Chronic kidney disease. Lancet 389:1238–1252. https://doi.org/10.1016/S0140-6736(16)32064-5
3. Friedenstein AJ (1976) Precursor cells of mechanocytes. Int Rev Cytol 47:327–359. https://doi.org/10.1016/s0074-7696(08)60092-3
4. Ullah I, Subbarao RB, Rho GJ (2015) Human mesenchymal stem cells—current trends and future prospective. Biosci Rep 35:e00191. https://doi.org/10.1042/BSR20150025
5. Vieira Paladino F, de Moraes RJ, da Silva A, Goldberg AC (2019) The immunomodulatory potential of Wharton's jelly mesenchymal stem/stromal cells. Stem Cells Int 2019:3548917
6. Wang Y, He J, Pei X, Zhao W (2013) Systematic review and meta-analysis of mesenchymal stem/stromal cells therapy for impaired renal function in small animal models. Nephrology (Carlton) 18:201–208. https://doi.org/10.1111/nep.12018
7. Lukomska B, Stanaszek L, Zuba-Surma E, Legosz P, Sarzynska S, Drela K (2019) Challenges and controversies in human mesenchymal stem cell therapy. Stem Cells Int 2019:9628536. https://doi.org/10.1155/2019/9628536
8. Marcheque J, Bussolati B, Csete M, Perin L (2019) Concise reviews: stem cells and kidney regeneration: an update. Stem Cells Transl Med 8:82–92. https://doi.org/10.1002/sctm.18-0115
9. Squillaro T, Peluso G, Galderisi U (2016) Clinical trials with mesenchymal stem cells: an update. Cell Transplant 25:829–848. https://doi.org/10.3727/096368915X689622
10. Bochon B, Kozubska M, Surygała G, Witkowska A, Kuźniewicz R, Grzeszczak W et al (2019) Mesenchymal stem cells-potential applications in kidney diseases. Int J Mol Sci 20:2462. https://doi.org/10.3390/ijms20102462
11. Večerić-Haler Ž, Cerar A, Perše M (2017) (Mesenchymal) Stem cell-based therapy in cisplatin-induced acute kidney injury animal model: risk of immunogenicity and tumorigenicity. Stem Cells Int 2017:7304643. https://doi.org/10.1155/2017/7304643
12. Perico L, Morigi M, Rota C, Breno M, Mele C, Noris M et al (2017) Human mesenchymal stromal cells transplanted into mice stimulate renal tubular cells and enhance mitochondrial function. Nat Commun 8:983. https://doi.org/10.1038/s41467-017-00937-2
13. Park JH, Jang HR, Kim DH, Kwon GY, Lee JE, Huh W et al (2017) Early, but not late treatment with human umbilical cord blood-derived mesenchymal stem cells attenuates cisplatin nephrotoxicity through immunomodulation. Am J Physiol Renal Physiol 313:F984–F996. https://doi.org/10.1152/ajprenal.00097.2016
14. Wang B, Jia H, Zhang B, Wang J, Ji C, Zhu X et al (2017) Pre-incubation with hucMSC-exosomes prevents cisplatin-induced nephrotoxicity by activating autophagy. Stem Cell Res Ther 8:75. https://doi.org/10.1186/s13287-016-0463-4
15. Večerić-Haler Ž, Erman A, Cerar A, Motaln H, Kološa K, Lah Turnšek T et al (2016) Improved protective effect of umbilical cord stem cell transplantation on cisplatin-induced kidney injury in mice pretreated with antithymocyte globulin. Stem Cells Int 2016:3585362. https://doi.org/10.1155/2016/3585362
16. Peng X, Xu H, Zhou Y, Wang B, Yan Y, Zhang X et al (2013) Human umbilical cord mesenchymal stem cells attenuate cisplatin-induced acute and chronic renal injury. Exp Biol Med (Maywood) 238:960–970. https://doi.org/10.1177/1535370213497176
17. Zhou Y, Xu H, Xu W, Wang B, Wu H, Tao Y et al (2013) Exosomes released by human umbilical cord mesenchymal stem cells protect against cisplatin-induced renal oxidative stress and apoptosis in vivo and in vitro. Stem Cell Res Ther 4:34. https://doi.org/10.1186/scrt194
18. Chang JW, Hung SP, Wu HH, Wu WM, Yang AH, Tsai HL et al (2011) Therapeutic effects of umbilical cord blood-derived mesenchymal stem cell transplantation in experimental lupus nephritis. Cell Transplant 20:245–257. https://doi.org/10.3727/096368910X520056
19. Morigi M, Rota C, Montemurro T, Montelatici E, Lo Cicero V, Imberti B et al (2010) Life-sparing effect of human cord blood-mesenchymal stem cells in experimental acute kidney injury. Stem Cells 28:513–522. https://doi.org/10.1002/stem.293

20. Gheisari Y, Ahmadbeigi N, Naderi M, Nassiri SM, Nadri S, Soleimani M (2011) Stem cell-conditioned medium does not protect against kidney failure. Cell Biol Int 35:209–213. https://doi.org/10.1042/CBI20100183
21. Fahmy SR, Soliman AM, El Ansary M, Elhamid SA, Mohsen H (2017) Therapeutic efficacy of human umbilical cord mesenchymal stem cells transplantation against renal ischemia/reperfusion injury in rats. Tissue Cell 49:369–375. https://doi.org/10.1016/j.tice.2017.04.006
22. Rodrigues CE, Capcha JM, de Bragança AC, Sanches TR, Gouveia PQ, de Oliveira PA et al (2017) Human umbilical cord-derived mesenchymal stromal cells protect against premature renal senescence resulting from oxidative stress in rats with acute kidney injury. Stem Cell Res Ther 8:19. https://doi.org/10.1186/s13287-017-0475-8
23. Ju GQ, Cheng J, Zhong L, Wu S, Zou XY, Zhang GY et al (2015) Microvesicles derived from human umbilical cord mesenchymal stem cells facilitate tubular epithelial cell dedifferentiation and growth via hepatocyte growth factor induction. PLoS ONE 10:e0121534. https://doi.org/10.1371/journal.pone.0121534
24. Burger D, Viñas JL, Akbari S, Dehak H, Knoll W, Gutsol A et al (2015) Human endothelial colony-forming cells protect against acute kidney injury: role of exosomes. Am J Pathol 185:2309–2323. https://doi.org/10.1016/j.ajpath.2015.04.010
25. Liang CJ, Shen WC, Chang FB, Wu VC, Wang SH, Young GH et al (2015) Endothelial progenitor cells derived from Wharton's Jelly of human umbilical cord attenuate ischemic acute kidney injury by increasing vascularization and decreasing apoptosis, inflammation, and fibrosis. Cell Transplant 24:1363–1377. https://doi.org/10.3727/096368914X681720
26. Jang HR, Park JH, Kwon GY, Lee JE, Huh W, Jin HJ et al (2014) Effect of preemptive treatment with human umbilical cord blood-derived mesenchymal stem cells on the development of renal ischemia-reperfusion injury in mice. Am J Physiol Renal Physiol 307:F1149–F1161. https://doi.org/10.1152/ajprenal.00555.2013
27. Chen Y, Qian H, Zhu W, Zhang X, Yan Y, Ye S et al (2011) Hepatocyte growth factor modification promotes the amelioration effects of human umbilical cord mesenchymal stem cells on rat acute kidney injury. Stem Cells Dev 20:103–113. https://doi.org/10.1089/scd.2009.0495
28. Cao H, Qian H, Xu W, Zhu W, Zhang X, Chen Y et al (2010) Mesenchymal stem cells derived from human umbilical cord ameliorate ischemia/reperfusion-induced acute renal failure in rats. Biotechnol Lett 32:725–732. https://doi.org/10.1007/s10529-010-0207-y
29. Ji C, Zhang J, Zhu Y, Shi H, Yin S, Sun F et al (2020) Exosomes derived from hucMSC attenuate renal fibrosis through CK1δ/β-TRCP-mediated YAP degradation. Cell Death Dis 11:327. https://doi.org/10.1038/s41419-020-2510-4
30. Liu B, Ding F, Hu D, Zhou Y, Long C, Shen L et al (2018) Human umbilical cord mesenchymal stem cell conditioned medium attenuates renal fibrosis by reducing inflammation and epithelial-to-mesenchymal transition via the TLR4/NF-κB signaling pathway in vivo and in vitro. Stem Cell Res Ther 9:7. https://doi.org/10.1186/s13287-017-0760-6
31. Liu B, Ding FX, Liu Y, Xiong G, Lin T, He DW et al (2018) Human umbilical cord-derived mesenchymal stem cells conditioned medium attenuate interstitial fibrosis and stimulate the repair of tubular epithelial cells in an irreversible model of unilateral ureteral obstruction. Nephrology (Carlton) 23:728–736. https://doi.org/10.1111/nep.13099
32. Zhang R, Zhu Y, Li Y, Liu W, Yin L, Yin S et al (2020) Human umbilical cord mesenchymal stem cell exosomes alleviate sepsis-associated acute kidney injury via regulating microRNA-146b expression. Biotechnol Lett 42:669–679. https://doi.org/10.1007/s10529-020-02831-2
33. Lee FY, Chen KH, Wallace CG, Sung PH, Sheu JJ, Chung SY et al (2017) Xenogeneic human umbilical cord-derived mesenchymal stem cells reduce mortality in rats with acute respiratory distress syndrome complicated by sepsis. Oncotarget 8:45626–45642. https://doi.org/10.18632/oncotarget.17320
34. Huang S, Wang D, Gu F, Zhang Z, Deng W, Chen W et al (2016) No significant effects of Poly(I:C) on human umbilical cord-derived mesenchymal stem cells in the treatment of B6.MRL-Fas(lpr) mice. Curr Res Transl Med 64:55–60. https://doi.org/10.1016/j.retram.2016.03.002

35. Gu Z, Akiyama K, Ma X, Zhang H, Feng X, Yao G et al (2010) Transplantation of umbilical cord mesenchymal stem cells alleviates lupus nephritis in MRL/lpr mice. Lupus 19:1502–1514. https://doi.org/10.1177/0961203310373782
36. Xiang E, Han B, Zhang Q, Rao W, Wang Z, Chang C et al (2020) Human umbilical cord-derived mesenchymal stem cells prevent the progression of early diabetic nephropathy through inhibiting inflammation and fibrosis. Stem Cell Res Ther 11:336. https://doi.org/10.1186/s13287-020-01852-y
37. Chen L, Xiang E, Li C, Han B, Zhang Q, Rao W et al (2020) Umbilical cord-derived mesenchymal stem cells ameliorate nephrocyte injury and proteinuria in a diabetic nephropathy rat model. J Diabetes Res 2020:8035853. https://doi.org/10.1155/2020/8035853
38. Maldonado M, Huang T, Yang L, Xu L, Ma L (2017) Human umbilical cord Wharton jelly cells promote extra-pancreatic insulin formation and repair of renal damage in STZ-induced diabetic mice. Cell Commun Signal 15:43. https://doi.org/10.1186/s12964-017-0199-5
39. Park JH, Park J, Hwang SH, Han H, Ha H (2012) Delayed treatment with human umbilical cord blood-derived stem cells attenuates diabetic renal injury. Transplant Proc 44:1123–1126. https://doi.org/10.1016/j.transproceed.2012.03.044
40. Park JH, Hwang I, Hwang SH, Han H, Ha H (2012) Human umbilical cord blood-derived mesenchymal stem cells prevent diabetic renal injury through paracrine action. Diabetes Res Clin Pract 98:465–473. https://doi.org/10.1016/j.diabres.2012.09.034
41. Xian Y, Lin Y, Cao C, Li L, Wang J, Niu J et al (2019) Protective effect of umbilical cord mesenchymal stem cells combined with resveratrol against renal podocyte damage in NOD mice. Diabetes Res Clin Pract 156:107755. https://doi.org/10.1016/j.diabres.2019.05.034
42. Fang TC, Pang CY, Chiu SC, Ding DC, Tsai RK (2012) Renoprotective effect of human umbilical cord-derived mesenchymal stem cells in immunodeficient mice suffering from acute kidney injury. PLoS ONE 7:e46504. https://doi.org/10.1371/journal.pone.0046504
43. Li W, Wang L, Chu X, Cui H, Bian Y (2017) Icariin combined with human umbilical cord mesenchymal stem cells significantly improve the impaired kidney function in chronic renal failure. Mol Cell Biochem 428(1–2):203–212. https://doi.org/10.1007/s11010-016-2930-8
44. Koc Y, Sokmen M, Unsal A, Cigerli S, Ozagari A, Basturk T et al (2012) Effects of human umbilical cord stem cells and granulocyte colony-stimulating factor (G-CSF) on carbon tetrachloride-induced nephrotoxicity. Nephrourol Mon 4:545–550. https://doi.org/10.5812/numonthly.2979
45. Perše M, Večerić-Haler Ž (2018) Cisplatin-induced rodent model of kidney injury: characteristics and challenges. Biomed Res Int 2018:1462802. https://doi.org/10.1155/2018/1462802
46. Bi B, Schmitt R, Israilova M, Nishio H, Cantley LG (2007) Stromal cells protect against acute tubular injury via an endocrine effect. J Am Soc Nephrol 18:2486–2496. https://doi.org/10.1681/ASN.2007020140
47. Guo Q, Wang J (2018) Effect of combination of vitamin E and umbilical cord-derived mesenchymal stem cells on inflammation in mice with acute kidney injury. Immunopharmacol Immunotoxicol 40:168–172. https://doi.org/10.1080/08923973.2018.1424898
48. Fischer UM, Harting MT, Jimenez F, Monzon-Posadas WO, Xue H, Savitz SI et al (2009) Pulmonary passage is a major obstacle for intravenous stem cell delivery: the pulmonary first-pass effect. Stem Cells Dev 18:683–692. https://doi.org/10.1089/scd.2008.0253
49. Harting MT, Jimenez F, Xue H, Fischer UM, Baumgartner J, Dash PK et al (2009) Intravenous mesenchymal stem cell therapy for traumatic brain injury. J Neurosurg 110:1189–1197. https://doi.org/10.3171/2008.9.JNS08158
50. Furlani D, Ugurlucan M, Ong L, Bieback K, Pittermann E, Westien I et al (2009) Is the intravascular administration of mesenchymal stem cells safe? Mesenchymal stem cells and intravital microscopy. Microvasc Res 77:370–376. https://doi.org/10.1016/j.mvr.2009.02.001
51. Morales-Kastresana A, Telford B, Musich TA, McKinnon K, Clayborne C, Braig Z et al (2017) Labeling extracellular vesicles for nanoscale flow cytometry. Sci Rep 7:1878. https://doi.org/10.1038/s41598-017-01731-2

52. Dominici M, Le Blanc K, Mueller I, Slaper-Cortenbach I, Marini F, Krause D et al (2006) Minimal criteria for defining multipotent mesenchymal stromal cells. The International Society for Cellular Therapy position statement. Cytotherapy 8:315–317. https://doi.org/10.1080/14653240600855905
53. Jung JW, Kwon M, Choi JC, Shin JW, Park IW, Choi BW et al (2013) Familial occurrence of pulmonary embolism after intravenous, adipose tissue-derived stem cell therapy. Yonsei Med J 54:1293–1296. https://doi.org/10.3349/ymj.2013.54.5.1293
54. Cyranoski D (2010) Korean deaths spark inquiry. Nature 468:485. https://doi.org/10.1038/468485a
55. Tatsumi K, Ohashi K, Matsubara Y, Kohori A, Ohno T, Kakidachi H et al (2013) Tissue factor triggers procoagulation in transplanted mesenchymal stem cells leading to thromboembolism. Biochem Biophys Res Commun 431:203–209. https://doi.org/10.1016/j.bbrc.2012.12.134
56. Liao L, Shi B, Chang H, Su X, Zhang L, Bi C et al (2017) Heparin improves BMSC cell therapy: anticoagulant treatment by heparin improves the safety and therapeutic effect of bone marrow-derived mesenchymal stem cell cytotherapy. Theranostics 7:106–116. https://doi.org/10.7150/thno.16911
57. Moll G, Rasmusson-Duprez I, von Bahr L, Connolly-Andersen AM, Elgue G, Funke L et al (2012) Are therapeutic human mesenchymal stromal cells compatible with human blood? Stem Cells 30:1565–1574. https://doi.org/10.1002/stem.1111
58. Kopolovic I, Ostro J, Tsubota H, Lin Y, Cserti-Gazdewich CM, Messner HA et al (2015) A systematic review of transfusion-associated graft-versus-host disease. Blood 126:406–414. https://doi.org/10.1182/blood-2015-01-620872
59. Moosavi MM, Duncan A, Stowell SR, Roback JD, Sullivan HC (2020) Passenger lymphocyte syndrome; a review of the diagnosis, treatment, and proposed detection protocol. Transfus Med Rev 34:178–187. https://doi.org/10.1016/j.tmrv.2020.06.004
60. Koch M, Lehnhardt A, Hu X, Brunswig-Spickenheier B, Stolk M, Bröcker V et al (2013) Isogeneic MSC application in a rat model of acute renal allograft rejection modulates immune response but does not prolong allograft survival. Transpl Immunol 29:43–50. https://doi.org/10.1016/j.trim.2013.08.004
61. Wei Q, Dong Z (2012) Mouse model of ischemic acute kidney injury: technical notes and tricks. Am J Physiol Renal Physiol 303:F1487–F1494. https://doi.org/10.1152/ajprenal.00352.2012
62. Korneev KV (2019) Mouse models of sepsis and septic shock. Mol Biol (Mosk) 53:799–814. https://doi.org/10.1134/S0026898419050100
63. Le Blanc K, Tammik C, Rosendahl K, Zetterberg E, Ringdén O (2003) HLA expression and immunologic properties of differentiated and undifferentiated mesenchymal stem cells. Exp Hematol 31:890–896. https://doi.org/10.1016/s0301-472x(03)00110-3
64. Caplan H, Olson SD, Kumar A, George M, Prabhakara KS, Wenzel P et al (2019) Mesenchymal stromal cell therapeutic delivery: translational challenges to clinical application. Front Immunol 10:1645. https://doi.org/10.3389/fimmu.2019.01645
65. Griffin MD, Ryan AE, Alagesan S, Lohan P, Treacy O, Ritter T (2013) Anti-donor immune responses elicited by allogeneic mesenchymal stem cells: what have we learned so far? Immunol Cell Biol 91:40–51. https://doi.org/10.1038/icb.2012.67
66. Lohan P, Treacy O, Griffin MD, Ritter T, Ryan AE (2017) Anti-donor immune responses elicited by allogeneic mesenchymal stem cells and their extracellular vesicles: are we still learning? Front Immunol 8:1626. https://doi.org/10.3389/fimmu.2017.01626
67. Lohan P, Coleman CM, Murphy JM, Griffin MD, Ritter T, Ryan AE (2014) Changes in immunological profile of allogeneic mesenchymal stem cells after differentiation: should we be concerned? Stem Cell Res Ther 5:99. https://doi.org/10.1186/scrt488
68. Wise AF, Ricardo SD (2012) Mesenchymal stem cells in kidney inflammation and repair. Nephrology (Carlton) 17:1–10. https://doi.org/10.1111/j.1440-1797.2011.01501.x
69. Raicevic G, Rouas R, Najar M, Stordeur P, Boufker HI, Bron D et al (2010) Inflammation modifies the pattern and the function of Toll-like receptors expressed by human mesenchymal stromal cells. Hum Immunol 71:235–244. https://doi.org/10.1016/j.humimm.2009.12.005

70. Kilpinen L, Impola U, Sankkila L, Ritamo I, Aatonen M, Kilpinen S et al (2013) Extracellular membrane vesicles from umbilical cord blood-derived MSC protect against ischemic acute kidney injury, a feature that is lost after inflammatory conditioning. J Extracell Vesicles 2013:2. https://doi.org/10.3402/jev.v2i0.21927

Chapter 15
Stem Cells in Human Breast Milk and Neonate

Jure Bedenk

Abstract

Introduction: Human breast milk is the essential feed for all newborn babies because it includes many nutrients and cells that contribute to a healthy development. However, on a finer scale, breast milk is much more than that—it is believed to be one of the human fluids where stem cells can be found. In this chapter, current knowledge about stem cells, breast milk, and breast milk stem cells (BmSCs) will be discussed, as well as the potential use of BmSCs in the near future.
Methods: This chapter was written with the help of a thorough literature review that was found on PubMed and Web of Science using the key words *breast milk* and *stem cells*, as well as websites dedicated to the topic of breast milk and its composition.
Results: All the knowledge we found regarding the stem cells in human breast milk is astonishing; however, we can say that this research area is still lacking, especially on the side of determining the exact type and quantity of BmSCs present in the breast milk and their usability for therapeutic medicine and bioengineering.
Conclusions: Currently, research on BmSCs is still in the beginning. In the years since their discovery, not much research has been made, although there are signs of their potential in medicine, especially for pre-term born children, for which breast milk itself is regarded as stem cell therapy.

Keywords Stem cells · Breast milk · Regenerative medicine · Infants

Abbreviations

BmSCs	Breast milk stem cells
CD33	Cluster of differentiation 33
CD34	Cluster of differentiation 34

J. Bedenk (✉)
Department of Obstetrics and Gynecology, University Medical Center Ljubljana, Ljubljana, Slovenia
e-mail: jure.bedenk@kclj.si

I. Virant-Klun (ed.), *Stem Cells in Reproductive Tissues and Organs*, Stem Cell Biology and Regenerative Medicine 70, https://doi.org/10.1007/978-3-030-90111-0_15

CD44	Cluster of differentiation 44
CD45	Cluster of differentiation 45
CD73	Cluster of differentiation 73
CD90	Cluster of differentiation 90
CD117	Cluster of differentiation 117
CD133	Cluster of differentiation 133
CD146	Cluster of differentiation 146
CD271	Cluster of differentiation 271
CK14	Cytokeratin 14
EPCAM	Epithelial cell adhesion molecule
ESCs	Embryonic stem cell
hBmSCs	Human breast milk stem cells
HGF	Hepatocyte growth factor
HLA-DR	Human leukocyte antigen—DR isotype
iPSCs	Induced pluripotent stem cells
OCT-4	Octamer-binding transcription factor 4
PCR	Polymerase chain reaction
qPCR	Quantitative real-time polymerase chain reaction
SCa1	Stem cells antigen-1
SCID	Severely compromised immuno-deficient
SMA	Smooth muscle actin
SOX2	SRY-box transcription factor 2
VEGF	Vascular endothelial growth factor

Stem Cells

Multicellular organisms have a special type of cells that are able to differentiate into various cell types. These cells are called stem cells. They are also immortal, meaning that they can proliferate into the same stem cells over and over, thus able to be cultivated for a long time before being differentiated into the wanted cell type. Stem cells are present in the embryos and in adult organisms; however, these cells are not the same and have different properties. Finding of these cells was quite a coincidence. Researchers at the University in Toronto, Canada, were conducting experiments on irradiated mice, where they injected bone marrow cells into them [1]. After a certain period, they observed small lumps in their spleen, which were dubbed as spleen colonies. They postulated that these colonies arose from the marrow cells or even one single marrow cell from the cells they injected. In later studies, the same party confirmed that these colonies were in fact formed from a single cell, thus finding the first stem cells [2]. Later, the ability of these cells to self-renew was proven [3] and stem cells were cemented in the biological history.

The main two properties of stem cells are their ability to go through numerous cycles of cell proliferation without differentiating and their ability to differentiate into different and specialized cell types. This "infinite" proliferation is enabled by the use of the enzyme telomerase, which is able to restore telomeres at the end of DNA strands in cells, thus prolonging the ability of these cells to divide [4, 5]. The ability to differentiate into various cell types is called potency. There are different types of potency, dependent on their potential to divide into different cell types. Currently, we know of totipotent (able to divide into embryonic and extraembryonic cells), pluripotent (able to divide into cells of all three germ layers), multipotent (able to differentiate into different cell types of the related family), oligopotent (able to differentiate into a few different types of cells), and unipotent (able to differentiate only in one cell type) stem cells.

Stem cells can be divided into three groups, dependent on their origin—embryonic stem cells (ESCs), adult stem cells, and induced pluripotent stem cells (iPSCs).

Embryonic Stem Cells

Embryonic stem cells come from the inner cell mass of a blastocyst before it is implanted in the uterus. These cells are pluripotent and can differentiate into cells of all three germ lines—ectoderm, endoderm, and mesoderm [6]. Most research is done on mouse or human embryonic stem cells and requires a very optimal culture medium or else they quickly differentiate into other cell types. ESCs express certain markers, which are unique to them, for example, Nanog, OCT-4, and SOX2 [7]. Because they have a pluripotent potential and can thus be used to differentiate into all possible cells of the human body, ESCs have the highest potential for the use in regenerative medicine. However, the use of ESCs has a high obstacle—it is unethical at this time to use cells or tissues from an unborn child and many countries have limitations or even moratoria on the use of ESCs in research. Additionally, they have a high potential to cause tumors when applied into mice.

Adult Stem Cells

Adult stem cells are cells that can be obtained from the adult individual and are used by the body to maintain and repair the tissue of their origin. Most known and used are the bone marrow stem cells, adipose tissue stem cells, and blood cell (hematopoieic) stem cells. Usually the cells are obtained from the same individual that uses them in therapy subsequently. This poses the least risk when transferring the cells back as there is a low chance of autoimmune response. These stem cells are rarely pluripotent (umbilical cord blood), mostly they are multipotent (mesenchymal, adipose-derived, endothelial, etc.) [8]. Also, in difference to ESCs, adult stem cells have a limit in their use, as there are more and more genetic problems within these cells with ongoing ageing. Adult stem cells are used for threating leukemia and certain blood and bone cancers with the use of bone marrow transplantation, though usually a close relative donor is required. The use of these cells in research and in therapy is not controversial, though it is harder to obtain them, as it commonly needs an invasive technique of extraction. Because of this, more and more research is done on searching for stem cells present in body fluids that are easy to obtain and extract.

Induced Pluripotent Stem Cells

Induced pluripotent stem cells were discovered in Japan in 2006 and are actually somatic cells, which are reprogrammed with the help of certain transcription factors into cells having a pluripotent differentiation ability [9]. They share certain similarities with ESCs like pluripotency and differentiation potential; however, their chromatin is more methylated and the gene expression is different. Because of this, there lingers a question if in fact reprogramming of somatic cells is complete. Despite this, the iPSCs are successfully used in experiments in the field of medicine and even shows some therapeutic advantages, the biggest of those is the ability to reprogram a cell from the same individual that needs treatment, thus circumventing autoimmunity [10]. However, too little is still known about the iPSCs for them to be usable in current therapeutic medicine.

Stem cells are thought to be the holy grail of regenerative medicine and may well be the future of medicine itself. Their ability to differentiate in many different cell types, dependent on their potency, is remarkable. Stem cell therapy is still in its infancy, though bone marrow transplant, a form of stem cell therapy, has been used for years in leukemia patients. In time, many additional different cell therapies could arise, especially for neurodegenerative diseases that are at this time mainly incurable. However, every boon comes with its limitations. Stem cells are very hard to obtain from adults (bone marrow, adipose tissue, tendons, hair follicles, etc.) and even those cells that are obtained are usually only multipotent, which limits their ability of use. On the other hand, there are many totipotent and pluripotent stem cells available in the embryo, but the use of these cells is severely unethical. As stated before, scientists are constantly trying to find a way to either reverse the ability of multipotent stem cells to become pluri- or even totipotent (like iPSCs) or more commonly search for the wanted stem cells in different parts of human body and fluids. One of those fluids, where stem cells were recently found, is breast milk. These cells are called human breast milk stem cells (hBmSCs, sometimes also referred as BSCs or hBSCs in literature).

Breast Milk and Its Composition

Mammals differ from other classes in the animal kingdom in a very specific way—females produce milk that is the primary food of newborns. Milk is produced via the mammary glands that are present inside the breasts. In essence, there are two types of cells found in the glands—alveolar and myoepithelial cells. Due to hormonal changes, alveolar cells are expanded during pregnancy and change into milk-secreting cells during the lactation. In this period, the alveolar cells take up the majority of cells present in the breast [11]. The main product of these cells is milk. Milk has a varying composition among mammalian species and is appropriate only for the species the milk was synthesized for [12]. Even more, it has been proven that not only is breast milk different in composition between different women but its components vary with each mother–child dyad and even with the age of the infant [13]. Human breast milk is very important for newborns because of its unique biochemical and cellular composition. This composition allows infants to be optimally nurtured, protected, and developed [12, 14]. Despite this, certain women still decide not to breastfeed because of various unjust reasons, one of those being the availability of breast milk substitutes, which are regarded as prestigious compared to breastfeeding. Because of this, children can have a lower chance of survival, good health, and development [15]. The World Health Organization (WHO), United Nations (UN), and hospitals are trying to facilitate breastfeeding as a natural and recommended practice for all infants. The UN convention even acknowledged breastfeeding as a legal right of every child [16].

Despite breast milk being studied quite extensively, there is still little known about the type and quantity of cells inside it [17]. Among the first found were leukocytes or immune cells, which are predominant in the colostrum, but in a very low quantity in mature human milk [18]. The majority of cells in the breast milk of a healthy woman are the luminal and myoepithelial cells [19]. However, the most intriguing cells that have been proven to exist in breast milk are the stem cells, which were discovered in 2007 by Cregan et al. [20].

Biochemical Composition of Milk

There are three stages of breast milk—colostrum, which is present in the first 3–5 days after birth, transitional milk, present until 2–3 week after birth, and mature milk after this period [12]. Each of those has a different composition. Colostrum's composition provides enhanced immunological protection and a nutritional and developmental support to the newborn. Additionally, it contains cell proliferation-inducing factors that are supposedly important for the development of the gastrointestinal tract in the newborn and to stimulate hematopoiesis and immune maturation [12, 21]. There are three sources from which the nutritional components of human milk derive; a part of those are from the synthesis by milk-secreting cells, others are obtained from the diet, and the last are present in the maternal stores. The biochemicals in milk can be divided into macronutrients and micronutrients.

Macronutrients in breast milk are proteins, fats, and sugars. Proteins in human milk can be divided to whey and casein complexes, each of those consists a broad array of peptides and proteins [22]. Proteins that are most abundant are casein, α-lactalbumin, lactoferrin, secretory immunoglobulin IgA, lysozyme, and serum albumin. Fat is the most variable macronutrient in human milk; however, high concentrations of palmitic and oleic acids are usually present. It is even varied between the foremilk and hindmilk of a single feed. Lactose is the basic disaccharide in human milk. Its concentration is also the least variable of the macronutrients in milk. Other common carbohydrates in human milk are oligosaccharides [21].

In comparison to macronutrients, micronutrients in human milk vary on the diet of the weaning mother and the body stores she has. Usually it contains the vitamins A, B1, B2, B6, B12, and D, as well as iodine. Vitamin K and D are present in the breast milk in very low numbers, so it is vital for mothers to supplement it [21].

Bioactive Composition of Milk

The bioactive composition of breast milk can be divided into growth factors, immunological factors, and cellular components [23]. The growth and immunological factors are elements that have an effect on the biological processes that concurrently affect body functions and with it health of an individual. Usually, these components are released by mammary cells in the breast epithelium, cells inside breast milk or even from maternal serum [21]. It is important to note that breast milk does not only have a value as a nutritional fluid, but a fluid with many medicinal factors that allow normal survival and health to the infant. Because of this, it is critical that every infant receives mothers' milk when possible.

Breast milk contains a large variety of cells, which includes epithelial cells, leukocytes, stem cells, and even bacteria. Li et al. [24] have found that there is an average of 5, 3 $\times$ 10^5 viable cells/mL in breast milk; however, there exists a difference between cell counts among the colostrum (13 $\times$ 10^5 cells/mL), transitional (4, 7 $\times$ 10^5 cells/mL), and mature milk (3 $\times$ 10^5 cells/mL). Among them, the most researched are leukocytes because of their ability to grant the infant immunity and to infiltrate the tissues of them. Despite this, leukocytes are not the most abundant cells in breast milk—they are in fact in minority compared to other cell types. Epithelial cells are the most abundant cell type in breast milk; however, their function and abilities were still not discovered to its fullest [25]. Epithelial cells are divided into luminal and myoepithelial cells. First express epithelial cell adhesion molecule (EPCAM), while second express smooth muscle actin (SMA) and cytokeratin 14 (CK14) [25]. Also, stem cell like cells were identified in breast milk by Cregan et al. [20], expressing the marker Nestin. Their presence was further confirmed by other investigators [24, 26–33]. They are commonly referred to as breast milk stem cells and are able to differentiate into cells of all three germinal layers [27, 28, 31, 34, 35]. It is believed that BmSCs can integrate into the tissues of neonate and cause widespread benefits to it. Also, because BmSCs are not tumorigenic, they are a perfect candidate for stem cell therapy [13]. Because breast milk is not sterile, it also contains bacteria. On average, an infant consumes 10^7–10^8 bacteria in 800 mL of breast milk every day. This consumption is very important as it is believed that an early colonization of the gut by bacteria found in breast milk may be crucial for the prevention of disease and even health later in life [23]. It is suggested that bacteria are transported to the breast milk from the gastrointestinal microbiota of mothers via the entero-mammary pathway [36].

Breast Milk Stem Cells

Stem cells in breast milk were first predicted to exist by McGregor and Rogo [37]. One year later, Cregan et al. [20] identified hBmSCs that were present in breast milk of mothers with pre- and term infants. These cells were nestin positive, nestin being a well-known marker for neural, bone marrow, pancreatic, and epithelial stem cells [38]. These cells also expressed markers that are present on ESCs [30]. They supposedly have a pluripotent nature, which enables them to differentiate to all cells of all three germ layers, just like the human embryonic stem cells [39]. This is important because after a child is born, only adult stem cells remain, which can only differentiate into cells and tissues from which the stem cells originate. BmSCs thus represent a viable alternative to ESCs, which can be used in regenerative medicine [26]. It is predicted that BmSCs also have a function in breast regeneration, especially in the time of preparing it for lactation, as well as help the infant gaining tolerance to maternal cell antigens with establishing a microchimeric state [27]. Because of the same origin of mammary gland and nervous system cells, BmSCs have a high capability to develop into neural cells and are able to differentiate into all three neural lineages, which may be used for cell replacement therapies of brain diseases [30]. BmSCs also express mesenchymal stem cell markers [30, 33, 40]. It is predicted that progenitors of stem cells are present in niches in the breast in a dormant state, waiting for a signal for them to start asymmetric division to the morphogenesis of the alveolar or ductal cells during pregnancy and lactation [34]. Moreover, it was discovered that there is a subpopulation of cells that can be cultured from breast milk that express stem cell markers and have multipotent properties. They could also be, in culture, differentiated into two types of epithelial cells—CK18$^+$ luminal cells or CK14$^+$ myoepithelial cells [41, 42]. BmSCs truly appear to be multipotent, as they were differentiated in vitro into osteoblast-like cells, chondrocytes, adipocytes, cardiomyocytes, pancreatic beta-like cells, hepatocyte-like cells, glia-like cells, neuron-like cells, lactocytes, and myoepithelial cells [29, 33, 39]. Though BmSCs are comparable to ESCs, there are currently still problems, other than research, in their usage potential. Additionally, the amount of stem cells present in the breast milk, their phenotype and the expression of pluripotency markers greatly vary between breast milk donors [34]. Also, the origin of BmSCs is still not discovered fully. It is though suggested that BmSCs have the same origin as hematopoieic stem cells because of other blood-derived cells present in breast milk [29, 43]. It is proposed that hBmSCs are responsible for the remodeling of the breast for supporting a milk-secretory organ and even the proliferation, development, and epigenetic regulation of tissues in infants.

Distribution of Stem Cells from Breast Milk to the Organs and Tissues of the Infant

It is hypothesized that BmSCs follow the same route of distribution through the systemic circulation in the infant as other cells in the breast milk through the gastrointestinal tract. From there, they enter the bloodstream, where they circulate, inhabit different tissues and organs, where they are ready to assist tissue repair and grow [26]. This is especially important for preterm infants, where most of the tissues and organs are underdeveloped. Later, it was successfully determined that stem cells are present in the milk of mothers with preterm infants. Additionally, there is a differential expression of stem cell specific markers between preterm and term breast milk, as well as the percentage of cells expressing the markers between the two different breast milk [26]. This means that the cell contents of breast milk is adjusted according to the age of the infant, which is, as told before, very important for preterm babies. Because of the ability of BmSCs to inhabit tissues of the neonate, it is believed that a special state of microchimerism is established in these tissues—this means that mother's stem cells migrate into different tissues and organs in the infant and reside there, having influence on the infants' organs as stated before [23]. In some instances, the breastfeeding is also thought of as a stem cell therapy [38]. In mice, it was proven that BmSCs from genetically modified mice breast milk were able to migrate into the brain, thymus, pancreas, liver, spleen, and kidneys of normal mice [44]. This further eludes to the assumption that BmSCs have a regenerative potential in the tissues of the infant and even an effect on the regeneration of immune cells of the infant [27].

Methods Used to Determine Stem Cells in Breast Milk

Currently, most used methods to prove the presence of BmSCs in breast milk are quantitative real-time polymerase chain reaction (qPRC), immunofluorescence and flow cytometry. qPCR is used for determining the gene expression in cells, immunofluorescence for the localization of specific markers on these cells, while flow cytometry uses antibodies for identification of specific cell populations [26].

qPCR is a very versatile, accurate, and common method applied in molecular biology used to detect several analytes that include DNA, RNA, and protein [45]. It uses polymerase chain reaction (PCR) for the amplification of targeted molecules in real time. The method can be used with non-specific fluorescent dyes that intercalate into double stranded DNA or use sequence specific DNA probes labeled with fluorescent reporters. With this method important genes for multi- and pluripotency can be analyzed, especially with the modern use of single cell qPCR, which allows the quantification of targets in single cells.

Immunofluorescence permits the visualization of virtually all components present in any given tissue or cell. To achieve this, a combination of specific antibodies tagged with fluorophores is used [46]. This method is very specific because the antibodies adhere only to the specific target, which then emits fluorescence under the fluorescent microscope. It can also be used for basic quantification because the strength of fluorescence can be measured. In the case of BmSCs, this method is utilized to target specific markers on cells, which can later be identified as stem cells.

Flow cytometry is a method that rapidly analyzes single cells and particles as they travel through a single or multiple lasers while in a buffered solution. Each of single cells or particles that travel through the lasers is analyzed for visible light scatter and fluorescence parameters. The scatter is measured in two different directions, which allow the identification of relative size of the cell and the internal complexity or granularity of the cell. The fluorescence measurement is prepared in the same way as other fluorescence methods, with the help of fluorescent proteins, dyes, or antibodies. It is most useful when simultaneous characterization of mixed population of cells from fluids or even solid tissues is needed [47]. This is why this method is preferred in the search for BmSCs in breast milk. Breast milk has a lot of cellular and chemical components, so it is very hard to purify cells out of it with a sufficient quality, especially to detect cells with pluripotency markers. Recently a novel method was developed, which allows a more reliable approach of identification and quantification of cells in the breast milk as well as a more accurate identification of certain subpopulations of putative stem cells present in breast milk [32].

Characterization of Stem Cells in Breast Milk

Breast milk does not contain only one type of stem cells, so it was important to characterize them. The first stem cells were differentiated into adipogenic, chondrogenic, and osteogenic lineages, which led the researchers to believe that the stem cells present are multipotent mesenchymal [33]. This means that they have the potential to differentiate and, due to this, be used in regenerative medicine. Additionally, some scientists found that breast milk stem cells also express embryonic cell markers and that they are able to differentiate into all three neural lineages (oligodendrocytes, astrocytes, and neurons) [30]. In Australia, a team was able to differentiate the BmSCs into all three germ layers, which would mean that BmSCs are pluripotent [39]. However, they were not able to induce teratoma formations in severely compromised immuno-deficient (SCID) mice in vivo, which is a standard for testing pluripotency. This may mean that there exists a fundamental difference between pluripotent cells in embryo and adult. Also, it may mean that the teratoma assay is not an appropriate method for assessing pluripotency of non-tumorigenic cells. Currently, not much progress has been done of the characterization of stem cells in breast milk—they are regarded as their own entity; however, they may only be a new type of existing adult stem cell groups.

Markers of Stem Cells in Breast Milk

There are several markers that are used to identify stem cells in breast milk. They are directed at hematopoietic, mesenchymal, neuro-epithelial, and even multipotent genes. Mostly used are pluripotency markers OCT-4, Nanog, SOX2, and TRA60-1 [28]; however, other markers are also routinely tested—markers for hematopoieic stem cells (CD33, CD34, CD45, CD117), mesenchymal stem cells (CD44, CD73, CD90, SCa1), neural stem cells (Nestin, CD133), and embryonic stem cells (SOX2, OCT-4, Nanog, HLA-DR) —and are present in colostrum and in mature breast milk, although in varying percentages. This leads to the assumption that the difference of stem cell marker expression is dependent on certain characteristics of mothers, such as birth of child preterm or at term, BMI of mothers, nutritional needs of the infant and even their change in bra size [25, 26]. Mostly all markers were more expressed in the colostrum, meaning that stem cells are most abundant in the colostrum [28]. Other researches show similar results [48], although Indumathi et al. [43] showed that only scarce amounts of mesenchymal and hematopoieic stem cells are present in breast milk. This leads to the assumption that stem cells of all types decline in number with each day lactation is prolonged [24]. Despite this, Li et al. [24] found that mesenchymal and pluripotent stem cells in breast milk are not affected by the gestational age. Why is there such a variance between the studies conducted on the total cell counts and expression of stem cell markers? It is still not known, however it may be caused by different stages of lactation in mothers when their milk sample was taken, the storage time and even the methods used [38]. Most embryonic stem cell markers were also confirmed with immunocytochemical and immunohystochemical staining [34].

Cultivation of Breast Milk Stem Cells

Stem cells can be cultured in flasks with the ability to attach to the flask and form cell colonies. It is important to note though, that stem cells, which were able to be cultured, came from the colostrum. Stem cells from mature milk do not adhere to the flask, neither do they form colonies in any of the tested mediums [28]. The reason behind this may be that after seven days of lactation, there is such a sudden drop of stem cell number that they are unable to form a colony. After cultivating the cells, flow cytometry was preformed and it was again proven that stem cells are present, except for the hematopoietic stem cells, which cannot form colonies. Most of the markers were expressed and even more so after cultivation. This means that BmSCs from colostrum can be cultured and in time be a useful resource for cell therapy in curing diseases, such as diabetes, leukemia, nerve damage, and neurodegenerative diseases.

Possible Use of Breast Milk Stem Cells in Medicine

Use of stem cells in regenerative medicine is highly regarded. Even though the research on breast milk stem cells is still in its infancy, we can see the opportunity of their use from research on other available stem cells. This includes cell replacement therapies, cell renewal, and many ongoing research trials that search for prevention and treatment of diseases. Another possibility of their use is bioengineering [38]. There is still a lack of protocols on techniques to differentiate BmSCs into various cell lineages; however, certain methods for differentiation of progenitor cells were successfully used. Cells were differentiated and expressed certain mesenchymal stem cell markers (CD44, CD90, CD271, and CD146), as well as embryonic stem cell markers (OCT-4, SOX2, TRA60-1, and Nanog), although in a smaller subpopulation [35]. Some researchers even found that BmSCs express markers of undifferentiated pluripotent human stem cells [29]. It is believed that BmSCs can be used as a novel source of stem cells that can be transplanted for use in therapies, especially as they show high plasticity, have an unusually low potential to form tumors, as well as have a very low chance to form teratomas [39]. This is also a main benefit in comparison to iPSCs, which are unstable and form teratomas when injected into mice [49]. Another bonus point is the relative ease of harvesting them from breast milk because the breast milk sampling is non-invasive. It is therefore a promising source of cells for various autologous transplantation. Stroke therapy is currently one of the most attractive fields of BmSCs application, as BmSCs were suggested as a promising source of stem cells that are available for the treatment of stroke associated pathology [50]. Also, the in vitro differentiation of BmSCs into neuronal and glial cells is possible [39]. Because the livers are important for the regulation of blood glucose level, metabolism of proteins and lipids and the detoxification of urea, it was tested if BmSCs are able to differentiate into hepatocytes. Sani et al. [51] were able to accomplish this differentiation. Because of relatively easy accessibility of BmSCs, these cells can also be used for comparison with breast cancer cells to compare the balance of gene expression or in the role of proliferation-responsive cell populations in tumorigenesis [34]. Hassiotou et al. [52] also compared BmSCs with breast cancer cells and found out that in both the expression of pluripotency genes was upregulated in comparison to non-lactating breast. Interestingly, the expression was highest in the pregnant women. Also, it is thought that the difference between women to initiate milk production is down to the development of their breasts, for which BmSCs may be accountable. In women with underdeveloped breasts, a more targeted monitoring could be applied to increase their chance of better lactation [44]. The use of BmSCs can especially show helpful in treatment of preterm neonates, where most of the organs are not developed thoroughly. This could also mitigate the almost obligatory use of breastmilk in preterm infants, which is in certain countries hard to obtain because of unavailable banks for breast milk or the inefficient storage of it. With this, high amounts of preterm infant deaths could be avoided.

Currently, there are a number of limitations that are in the way of clinical application of treating strokes. Mainly, they are the lack of knowledge in the behavior of the stem cells in vivo, especially their restorative effects, efficiency of therapy, and the safety of application. For further use of BmSCs, research has to be done on the breast milk that derives from induced lactation of women without infants. Thus, studies on animals regarding BmSCs use is still in high demand before any therapeutic use is possible [34]. These studies will importantly show their ability of use in treating diseases of different organs in infants and in adults. BmSCs were also isolated from bovine milk [53], where they predict their use in veterinary purposes for restoring milk production, manipulate lactation yields and regenerative medicine.

Research is important but curiously lacking in the field of BmSCs. Since their discovery in 2007 not much new has been discovered. As of this writing, only 21 original articles were found on Web of Science, using key words *breast milk* and *stem cells* (Table 15.1). In this table, short findings are given for a brief overview of the current situation on BmSCs research. There is still more to discover about the presence of BmSCs in breast milk; however, currently each recent discovery prompts more new questions than answers. Most importantly, new research has to be made on the functionality of BmSCs, for example, how are the BmSCs involved in the development of an infant, how do they target certain organs, how does storage effect stem cell viability, and so on [38]. Due to ESCs being very controversial, especially in medicinal use, BmSCs are a spark of potential that may yet change the way we look at stem cell therapies and especially the difficulty to obtain stem cells. Promising results are also found in the area of iPSCs, where first epithelial breast milk cells were transformed into functioning iPSCs that were able to differentiate into all three germ layers [54].

Conclusion

In conclusion, BmSCs are a promising new type of stem cells that can be used in a therapeutic way for adults and infants, especially preterm ones. Despite its low quantity of research, it is gaining momentum and exceptional results may await us in the near future. BmSCs will have an important say in this.

Table 15.1 Original articles on breast milk stem cells—authors, the year of publication, methods used, and findings

Researchers and year of publication	Methods used	Findings
Cregan et al. [20]	Immunofluorescent RT-PCR	Discovered Nestine positive putative stem cells in human breast milk
Patki et al. [33]	Immunocytochemistry and flow cytometry	Confirmed mesenchymal nature of BmSCs
Fan et al. [48]	RT-PCR and flow cytometry	Demonstrated the presence of stem cells in breast milk, however they were unable to culture them
Hassiotou et al. [39]	Flow cytometry, qPCR, western blotting, in vitro differentiation, immunohystochemistry and teratoma formation assay	Stem cells from breastmilk express same transcription factors as ESCs ex vivo and are able to differentiate into cells of mammary, mesodermal and endodermal lineage
Indumathi et al. [43]	Flow cytometry	Breast milk possesses small amounts of mesenchymal and hematopoieic stem cells
Hosseini et al. [30]	Immunocytochemistry, differentiation and sphere-forming assay	Demonstrated that BmSCs express mesenchymal and embryonic stem cell makers and differentiated BmSCs into neural cell lineages
Sani et al. [35]	Flow cytometry, immunocytochemistry, PCR, osteogenia and adipogenic differentiation	BmSCs express mesenchymal stem cell markers, as well as certain embryonic stem cell markers, cells were differentiated into adipocytes and osteoblasts
Twigger et al. [25]	qPCR, immunohystochemistry, flow cytometry	Stem cell gene expression is varied because of gestational age, BMI of mothers and their bra size
Kaingade et al. [40]	Flow cytometry	Identified VEGF and HGF expression in BmSCs, the BmSCs were cultured successfully
Pichiri et al. [55]	Immunohystochemistry	Evidence of stem cells in breast milk
Sharp et al. [56]	RT-PCR	Certain stem cell markers were expressed higher in colostrum and involution milk than in mature milk
Briere et al. [26]	qPCR, flow cytometry	Higher count of stem cell populations in breast milk was recorded for preterm infants than for term infants

(continued)

Table 15.1 (continued)

Researchers and year of publication	Methods used	Findings
Kaingade et al. [31]	Flow cytometry	In comparison to Sharp et al. [56] , stem cell markers were expressed more in mature milk than in colostrum
Sani et al. [51]	Immunocytochemistry, RT-PCR	Differentiation of BmSCs to hepatocyte like cells
Aydın et al. [57]	Flow cytometry, qPCR and immunohystochemistry	Proven in mice that BmSCs from breast milk fed to pups reach their brain and differentiate into neuron and glial cells
Nosrati Tirkani et al. [58]	Cell culture	More colonies were obtained from breast milk of mothers with preterm delivered infants than in term infants
Li et al. [24]	Flow cytometry and qPCR	Confirmed the presence of stem cells in breast milk and discovered that pre-term birth, maternal parity, BMI and mode of delivery influence the numbers of certain cells in breast milk
Keller et al. [32]	Immunocytochemistry and flow cytometry	Improved the method for the detection of viable cells in human breast milk, especially for putative stem cells
Khamis et al. [59]	qPCR and immunohystochemistry	Proven in mice that BmSCs, if applied early, can prevent diabetic testicular dysfunction and in turn prevent sterility
Goudarzi et al. [28]	Flow cytometry	Found that the lactation stage can influence the population of stem cells and with this their cultivation. They propose the use of colostrum for future therapeutic use due to a high reservoir of heterogeneous stem cell population
Borhani-Haghighi et al. [60]	Immunosorbent assay	Proven in rats that BmSCc can be administered to the site of spinal cord injury and have therapeutic effects

References

1. McCulloch EA, Till JE (1960) The radiation sensitivity of normal mouse bone marrow cells, determined by quantitative marrow transplantation into irradiated mice. Radiat Res 13:115–125. https://doi.org/10.2307/3570877
2. Becker AJ, McCulloch EA, Till JE (1963) Cytological demonstration of the clonal nature of spleen colonies derived from transplanted mouse marrow cells. Nature 197:452–454. https://doi.org/10.1038/197452a0
3. Siminovitch L, Mcculloch E, Till J (1963) The distribution of colony-forming cells among spleen colonies. J Cell Comp Physiol. https://doi.org/10.1002/JCP.1030620313
4. Shay JW, Bacchetti S (1997) A survey of telomerase activity in human cancer. Eur J Cancer 33:787–791. https://doi.org/10.1016/S0959-8049(97)00062-2
5. Wright WE, Piatyszek MA, Rainey WE, Byrd W, Shay JW (1996) Telomerase activity in human germline and embryonic tissues and cells. Dev Genet 18:173–179. https://doi.org/10.1002/(SICI)1520-6408(1996)18:2%3c173::AID-DVG10%3e3.0.CO;2-3
6. Thomson JA, Itskovitz-Eldor J, Shapiro SS, Waknitz MA, Swiergiel JJ, Marshall VS, Jones JM (1998) Embryonic stem cell lines derived from human blastocysts. Science 282:1145–1147. https://doi.org/10.1126/science.282.5391.1145
7. Zhao W, Ji X, Zhang F, Li L, Ma L (2012) Embryonic stem cell markers. Molecules 17:6196–6246. https://doi.org/10.3390/molecules17066196
8. Gardner R (2002) Stem cells: potency, plasticity and public perception. J Anat 200:277–282. https://doi.org/10.1046/j.1469-7580.2002.00029.x
9. Takahashi K, Yamanaka S (2006) Induction of pluripotent stem cells from mouse embryonic and adult fibroblast cultures by defined factors. Cell 126:663–676. https://doi.org/10.1016/j.cell.2006.07.024
10. Robinton DA, Daley GQ (2012) The promise of induced pluripotent stem cells in research and therapy. Nature 481:295–305. https://doi.org/10.1038/nature10761
11. Inman JL, Robertson C, Mott JD, Bissell MJ (2015) Mammary gland development: cell fate specification, stem cells and the microenvironment. Development 142:1028–1042. https://doi.org/10.1242/dev.087643
12. Hassiotou F, Geddes D (2013) Anatomy of the human mammary gland: current status of knowledge. Clin Anat 26:29–48. https://doi.org/10.1002/ca.22165
13. Hassiotou F, Geddes DT, Hartmann PE (2013) Cells in human milk: state of the science. J Hum Lact 29:171–182. https://doi.org/10.1177/0890334413477242
14. Section on Breastfeeding (2012) Breastfeeding and the use of human milk. Pediatrics 129:e827–e841. https://doi.org/10.1542/peds.2011-3552
15. Rollins NC, Bhandari N, Hajeebhoy N, Horton S, Lutter CK, Martines JC, Piwoz EG, Richter LM, Victora CG (2016) Why invest, and what it will take to improve breastfeeding practices? Lancet 387:491–504. https://doi.org/10.1016/S0140-6736(15)01044-2
16. Office of the United Nations High Commissioner for Human Rights (1989) Convention on the Rights of the Child. https://www.ohchr.org/EN/ProfessionalInterest/Pages/CRC.aspx. Accessed 1 Oct 2020
17. Allah SHA, Shalaby SM, El-Shal AS, Nabtety SME, Khamis T, Rhman SAAE, Ghareb MA, Kelani HM (2016) Breast milk MSCs: an explanation of tissue growth and maturation of offspring. IUBMB Life 68:935–942. https://doi.org/10.1002/iub.1573
18. Hassiotou F, Hepworth AR, Metzger P, Tat Lai C, Trengove N, Hartmann PE, Filgueira L (2013) Maternal and infant infections stimulate a rapid leukocyte response in breastmilk. Clin Transl Immunol 2:e3. https://doi.org/10.1038/cti.2013.1
19. Qian J, Chen T, Lu W, Wu S, Zhu J (2010) Breast milk macro- and micronutrient composition in lactating mothers from suburban and urban Shanghai. J Paediatr Child Health 46:115–120. https://doi.org/10.1111/j.1440-1754.2009.01648.x
20. Cregan MD, Fan Y, Appelbee A, Brown ML, Klopcic B, Koppen J, Mitoulas LR, Piper KME, Choolani MA, Chong Y-S, Hartmann PE (2007) Identification of nestin-positive putative

mammary stem cells in human breastmilk. Cell Tissue Res 329:129–136. https://doi.org/10.1007/s00441-007-0390-x
21. Ballard O, Morrow AL (2013) Human milk composition: nutrients and bioactive factors. Pediatr Clin North Am 60:49–74. https://doi.org/10.1016/j.pcl.2012.10.002
22. Liao Y, Alvarado R, Phinney B, Lönnerdal B (2011) Proteomic characterization of human milk whey proteins during a twelve-month lactation period. J Proteome Res 10:1746–1754. https://doi.org/10.1021/pr101028k
23. Witkowska-Zimny M, Kaminska-El-Hassan E (2017) Cells of human breast milk. Cell Mol Biol Lett 22. https://doi.org/10.1186/s11658-017-0042-4
24. Li S, Zhang L, Zhou Q, Jiang S, Yang Y, Cao Y (2019) Characterization of stem cells and immune cells in preterm and term mother's milk. J Hum Lact 35:528–534. https://doi.org/10.1177/0890334419838986
25. Twigger A-J, Hepworth AR, Tat Lai C, Chetwynd E, Stuebe AM, Blancafort P, Hartmann PE, Geddes DT, Kakulas F (2015) Gene expression in breastmilk cells is associated with maternal and infant characteristics. Sci Rep 5. https://doi.org/10.1038/srep12933
26. Briere C-E, Jensen T, McGrath JM, Young EE, Finck C (2017) Stem-like cell characteristics from breast milk of mothers with preterm infants as compared to mothers with term infants. Breastfeed Med 12:174–179. https://doi.org/10.1089/bfm.2017.0002
27. Cacho NT, Lawrence RM (2017) Innate immunity and breast milk. Front Immunol 8. https://doi.org/10.3389/fimmu.2017.00584
28. Goudarzi N, Shabani R, Ebrahimi M, Baghestani A, Dehdashtian E, Vahabzadeh G, Soleimani M, Moradi F, Katebi M (2020) Comparative phenotypic characterization of human colostrum and breast milk-derived stem cells. Hum Cell 33:308–317. https://doi.org/10.1007/s13577-019-00320-x
29. Hassiotou F, Hartmann PE (2014) At the dawn of a new discovery: the potential of breast milk stem cells1234. Adv Nutr 5:770–778. https://doi.org/10.3945/an.114.006924
30. Hosseini SM, Talaei-khozani T, Sani M, Owrangi B (2014) Differentiation of human breast-milk stem cells to neural stem cells and neurons. Neurol Res Int 2014. https://doi.org/10.1155/2014/807896
31. Kaingade P, Somasundaram I, Sharma A, Patel D, Marappagounder D (2017) Cellular components, including stem-like cells, of preterm mother's mature milk as compared with those in her colostrum: a pilot study. Breastfeed Med 12:446–449. https://doi.org/10.1089/bfm.2017.0063
32. Keller T, Wengenroth L, Smorra D, Probst K, Kurian L, Kribs A, Brachvogel B (2019) Novel DRAQ5™/SYTOX® blue based flow cytometric strategy to identify and characterize stem cells in human breast milk. Cytometry B Clin Cytom 96:480–489. https://doi.org/10.1002/cyto.b.21748
33. Patki S, Kadam S, Chandra V, Bhonde R (2010) Human breast milk is a rich source of multipotent mesenchymal stem cells. Hum Cell 23:35–40. https://doi.org/10.1111/j.1749-0774.2010.00083.x
34. Ninkina N, Kukharsky MS, Hewitt MV, Lysikova EA, Skuratovska LN, Deykin AV, Buchman VL (2019) Stem cells in human breast milk. Hum Cell 32:223–230. https://doi.org/10.1007/s13577-019-00251-7
35. Sani M, Hosseini SM, Salmannejad M, Aleahmad F, Ebrahimi S, Jahanshahi S, Talaei-Khozani T (2015) Origins of the breast milk-derived cells; an endeavor to find the cell sources. Cell Biol Int 39:611–618. https://doi.org/10.1002/cbin.10432
36. Rescigno M, Urbano M, Valzasina B, Francolini M, Rotta G, Bonasio R, Granucci F, Kraehenbuhl J-P, Ricciardi-Castagnoli P (2001) Dendritic cells express tight junction proteins and penetrate gut epithelial monolayers to sample bacteria. Nat Immunol 2:361–367. https://doi.org/10.1038/86373
37. McGregor JA, Rogo LJ (2016) Breast milk: an unappreciated source of stem cells. J Hum Lact 22:270–271. https://doi.org/10.1177/0890334406290222
38. Briere C-E, McGrath JM, Jensen T, Matson A, Finck C (2016) Breast milk stem cells: current science and implications for preterm infants. Adv Neonatal Care 16:410–419. https://doi.org/10.1097/ANC.0000000000000338

39. Hassiotou F, Beltran A, Chetwynd E, Stuebe AM, Twigger A-J, Metzger P, Trengove N, Lai CT, Filgueira L, Blancafort P, Hartmann PE (2012) Breastmilk is a novel source of stem cells with multilineage differentiation potential. Stem Cells Dayt Ohio 30:2164–2174. https://doi.org/10.1002/stem.1188
40. Kaingade PM, Somasundaram I, Nikam AB, Sarang SA, Patel JS (2016) Assessment of growth factors secreted by human breastmilk mesenchymal stem cells. Breastfeed Med 11:26–31. https://doi.org/10.1089/bfm.2015.0124
41. Thomas E, Zeps N, Cregan M, Hartmann P, Martin T (2011) 14–3-3σ (sigma) regulates proliferation and differentiation of multipotent p63-positive cells isolated from human breastmilk. Cell Cycle 10:278–284. https://doi.org/10.4161/cc.10.2.14470
42. Thomas E, Lee-Pullen T, Rigby P, Hartmann P, Xu J, Zeps N (2012) Receptor activator of NF-κB ligand promotes proliferation of a putative mammary stem cell unique to the lactating epithelium. Stem Cells Dayt Ohio 30:1255–1264. https://doi.org/10.1002/stem.1092
43. Indumathi S, Dhanasekaran M, Rajkumar JS, Sudarsanam D (2013) Exploring the stem cell and non-stem cell constituents of human breast milk. Cytotechnology 65:385–393. https://doi.org/10.1007/s10616-012-9492-8
44. Hassiotou F, Heath B, Ocal O, Filgueira L, Geddes D, Hartmann P, Wilkie T (2014) Breastmilk stem cell transfer from mother to neonatal organs (216.4). FASEB J 28:216.4. https://doi.org/10.1096/fasebj.28.1_supplement.216.4
45. Ståhlberg A, Kubista M (2018) Technical aspects and recommendations for single-cell qPCR. Mol Aspects Med 59:28–35. https://doi.org/10.1016/j.mam.2017.07.004
46. Im K, Mareninov S, Diaz MFP, Yong WH (2019) An introduction to performing immunofluorescence staining. Methods Mol Biol Clifton NJ 1897:299–311. https://doi.org/10.1007/978-1-4939-8935-5_26
47. McKinnon KM (2018) Flow cytometry: an overview. Curr Protoc Immunol 120:5.1.1–5.1.11. https://doi.org/10.1002/cpim.40
48. Fan Y, Chong YS, Choolani MA, Cregan MD, Chan JKY (2010) Unravelling the mystery of stem/progenitor cells in human breast milk. PLoS ONE 5. https://doi.org/10.1371/journal.pone.0014421
49. Carpenter MK, Frey-Vasconcells J, Rao MS (2009) Developing safe therapies from human pluripotent stem cells. Nat Biotechnol 27:606–613. https://doi.org/10.1038/nbt0709-606
50. Stonesifer C, Corey S, Ghanekar S, Diamandis Z, Acosta SA, Borlongan CV (2017) Stem cell therapy for abrogating stroke-induced neuroinflammation and relevant secondary cell death mechanisms. Prog Neurobiol 158:94–131. https://doi.org/10.1016/j.pneurobio.2017.07.004
51. Sani M, Ebrahimi S, Aleahmad F, Salmannejad M, Hosseini SM, Mazarei G, Talaei-Khozani T (2017) Differentiation potential of breast milk-derived mesenchymal stem cells into hepatocyte-like cells. Tissue Eng Regen Med 14:587–593. https://doi.org/10.1007/s13770-017-0066-x
52. Hassiotou F, Hepworth AR, Beltran AS, Mathews MM, Stuebe AM, Hartmann PE, Filgueira L, Blancafort P (2013) Expression of the pluripotency transcription factor OCT4 in the normal and aberrant mammary gland. Front Oncol 3. https://doi.org/10.3389/fonc.2013.00079
53. Pipino C, Mandatori D, Buccella F, Lanuti P, Preziuso A, Castellani F, Grotta L, Di Tomo P, Marchetti S, Di Pietro N, Cichelli A, Pandolfi A, Martino G (2018) Identification and characterization of a stem cell-like population in bovine milk: a potential new source for regenerative medicine in veterinary. Stem Cells Dev 27:1587–1597. https://doi.org/10.1089/scd.2018.0114
54. Tang C, Lu C, Ji X, Ma L, Zhou Q, Xiong M, Zhou W (2019) Generation of two induced pluripotent stem cell (iPSC) lines from human breast milk using episomal reprogramming system. Stem Cell Res 39:101511. https://doi.org/10.1016/j.scr.2019.101511
55. Pichiri G, Lanzano D, Piras M, Dessì A, Reali A, Puddu M, Noto A, Fanos V, Coni C, Faa G, Coni P (2016) Human breast milk stem cells: a new challenge for perinatologists. J Pediatr Neonatal Individ Med JPNIM 5:e050120–e050120. https://doi.org/10.7363/050120
56. Sharp JA, Lefèvre C, Watt A, Nicholas KR (2016) Analysis of human breast milk cells: gene expression profiles during pregnancy, lactation, involution, and mastitic infection. Funct Integr Genomics 16:297–321. https://doi.org/10.1007/s10142-016-0485-0

57. Aydın MŞ, Yiğit EN, Vatandaşlar E, Erdoğan E, Öztürk G (2018) Transfer and integration of breast milk stem cells to the brain of suckling pups. Sci Rep 8:14289. https://doi.org/10.1038/s41598-018-32715-5
58. Nosrati Tirkani A, Arjmand MH, Jalili-Nik M, Hamidi Alamdari D, Shah Farhat A (2019) Comparison of colony-forming efficiency between breast milk stem/progenitor cells of mothers with preterm and full-term delivery. Iran J Neonatol IJN 10:38–44. https://doi.org/10.22038/ijn.2019.32358.1452
59. Khamis T, Abdelalim AF, Abdallah SH, Saeed AA, Edress NM, Arisha AH (2020) Early intervention with breast milk mesenchymal stem cells attenuates the development of diabetic-induced testicular dysfunction via hypothalamic Kisspeptin/Kiss1r-GnRH/GnIH system in male rats. Biochim Biophys Acta BBA—Mol Basis Dis 1866:165577. https://doi.org/10.1016/j.bbadis.2019.165577
60. Borhani-Haghighi M, Navid S, Mohamadi Y (2020) The therapeutic potential of conditioned medium from human breast milk stem cells in treating spinal cord injury. Asian Spine J 14:131–138. https://doi.org/10.31616/asj.2019.0026

Index

I. Virant-Klun (ed.), *Stem Cells in Reproductive Tissues and Organs*, Stem Cell Biology and Regenerative Medicine 70, https://doi.org/10.1007/978-3-030-90111-0

The manufacturer's authorised representative in the EU is Springer Nature Customer Service Centre GmbH, Europaplatz 3, 69115 Heidelberg, Germany. If you have any concerns regarding our products, please contact ProductSafety@springernature.com

Printed and bound by CPI Group (UK) Ltd, Croydon, CR0 4YY
30/09/2026
02225498-0015